Connectivity of Communication Networks

Guoqiang Mao

Connectivity of Communication Networks

Springer

Guoqiang Mao
School of Computing and Communications
The University of Technology Sydney
Sydney, NSW, Australia

ISBN 978-3-319-52988-2 ISBN 978-3-319-52989-9 (eBook)
DOI 10.1007/978-3-319-52989-9

Library of Congress Control Number: 2017933967

Printed on acid-free paper

This Springer imprint is published by Springer Nature
The registered company is Springer International Publishing AG
The registered company address is: Gewerbestrasse 11, 6330 Cham, Switzerland

Preface

Connectivity is one of the most fundamental properties, if not the most fundamental property, of communication networks. A communication network is said to be *connected* if there is a path from any node to any other node in the network. A communication network is essentially an organized collection of communication links between pairs of devices, ultimately serving to allow information exchange between human beings, between human and machine, and between machine and machine. It is the way the communication devices are deployed and collection of communication links is organized that gives rise to the connectivity properties of a communication network.

This book introduces a number of recent developments dealing with connectivity of communication networks, ranging from connectivity of large static networks and connectivity of highly dynamic networks to connectivity of small- to medium-sized networks. A fundamental problem being studied in this book is, the conditions under which a network with nodes randomly deployed and connections between nodes established following a prescribed connection model becomes connected. Any communication network is used to transport information from some nodes in the network to some other nodes. Therefore, connectivity studies play an important role in the design, deployment, and management of a network. Furthermore, in wireless networks, a direct connection can only occur between two nearby nodes. Therefore, connectivity of a network conveys topological information of the network, which can be used to infer topology-related information such as location of nodes and boundary of the network. This book also introduces some applications of connectivity studies in network optimization, in network localization, and in estimating distances between nodes.

This book is organized into 14 chapters. Chapter 1 provides an overview of the fundamental concepts, models, tools, and methodologies used for connectivity studies. The rest of the chapters can be divided into four parts: Part (Chaps. 2–7), connectivity of large static networks; Part (Chaps. 8–10), connectivity of highly dynamic networks; Part (Chaps. 11–13), connectivity of small- to medium-sized networks; and Part (Chap. 14), applications of connectivity studies.

Part deals with connectivity of large wireless networks with stationary nodes. Considering a large network with nodes distributed in a region following either uniform or Poisson distribution and a pair of nodes directly connected following some prescribed connection model *independent* of other pairs of nodes, we first study the sufficient and necessary condition for the network to be connected. This study reveals that a connected network is a very demanding requirement in the sense that as the number of nodes in the network increases, every node has to increase its transmission power or equivalently transmission range, in order to keep the network connected. This observation motivates us to further study whether less stringent requirement suffices if we can tolerate a small percentage of disconnected nodes. This study leads to an interesting finding that indeed if, instead of requiring all nodes to be connected via a multi-hop path, only a large fraction of nodes need to be connected where this fraction can be any positive number smaller than one, a significantly less amount of transmission power or transmission range is required. Furthermore, we study the conditions required for a stationary network to *percolate*. In an infinite network, the network is said to *percolate* if there exists a component of infinite size in the network, where a *component* is a maximal set of nodes in the network such that there is a path between every pair of nodes in the set. In Part , we also study the well-known phase transition phenomenon in network connectivity. Informally, a *phase transition* is defined as a phenomenon where a small change in the local parameters of a system results in an abrupt change in the macroscopic behavior of the system. Our particular interest in this book is studying the changes in the transmission range or transmission power required to transform an almost disconnected network to an almost connected network. Finally, in a wireless network, the presence of interference among nodes challenges the assumption used in the early studies on the independence of connections. This motivates us to further study the network connectivity in the presence of interference where we use the widely deployed carrier sense multiple access wireless networks as the subject of our study.

Part focuses on the study of connectivity of large highly dynamic networks. In dynamic networks with mobile nodes, it is possible that two nodes may never be part of the same connected component but they are still able to communicate with each other within a finite time interval. This occurs because a mobile node can carry information over a physical distance as it moves and transmits the information to another node when new connection opportunities arise. Therefore, information propagates in dynamic networks via a combination of wireless communications and node mobility. Consequently, connectivity in a dynamic network should be more broadly defined as a node *is connected to* another node if information from the first node can propagate to the second node within a prescribed amount of time. In this part of the book, we study information propagation process in one-dimensional and two-dimensional networks with vehicular networks and mobile ad hoc networks being used as two major examples of dynamic networks in the study.

Part deals with connectivity of small- to medium-sized networks. The study of large networks is mainly based on asymptotic analysis, and many conclusions obtained are applicable to networks with a sufficiently large number of nodes only.

In many real communication networks that we may encounter, the number of nodes in the network is not necessarily large that warrants the use of asymptotic analysis. Therefore, the third part of this book is dedicated to studying connectivity of small- to medium-sized networks. Our study in this part focuses on the analysis of three related probabilistic measures:

1. $\Pr(k)$, the probability that a randomly selected node is k-hops apart from another randomly selected node, i.e., the length of the shortest path from the first node to the second node, measured by the number of hops, is k
2. $\Pr(k|\boldsymbol{x})$, the probability that a node at a displacement $\boldsymbol{x}$ apart from another node is connected to that node in exactly k-hops
3. $\Pr(\boldsymbol{x}|k)$, the spatial distribution of the nodes k-hops apart from another designated node.

These three probabilities are grossly referred to as *the probabilities of k-hop connection* or *hop count statistics*. The analysis on the probabilities of k-hop connection plays a foundational role in our understanding of connectivity of small- to medium-sized networks. We study the probabilities of k-hop connection in both one-dimensional and two-dimensional networks. Furthermore, as we go deeper in connectivity studies, it becomes clear that existing tools used for connectivity studies fall short of answering an important category of problems: how to measure the *quality of connectivity* of a wireless network which has a realistic number of nodes, *not necessarily* large enough to warrant the use of asymptotic analysis, and has unreliable connections, reflecting the inherent unreliable characteristics of wireless communications. This motivates us to propose a new measure of network connectivity which, compared with existing well-known connectivity measures derived from the algebraic graph theory concentrating on describing the connectivity between nodes with directed connections, focuses more on the characterization of the quality of *end-to-end* connections in the network.

Part introduces some applications of connectivity studies. Among the broad range of applications of connectivity studies, we give three examples on the analysis of key performance measures of vehicular networks and its subsequent use in network design and optimization, on the use of connectivity information to estimate the distance between a pair of neighboring nodes, and on connectivity-based wireless localization algorithms.

The target audience for this book includes professionals who are designers and/or planners of communication networks, researchers (academics and graduate students), and those who would like to learn about the field. The format and flow of information have been organized such that it can be used as a textbook for graduate courses and research-oriented courses that deal with the design and analysis of communication networks.

Sydney, NSW, Australia — Guoqiang Mao

Contents

Chapter 1
Introduction

Abstract This chapter provides an overview of the fundamental concepts, models, tools, and methodologies used for connectivity studies. We start with an introduction to the major connection models used to describe the establishment of communication links between devices or nodes. These include the Erdős–Rényi connection model, the unit disk connection model, the log-normal connection model, the random connection model, and the SINR connection model. The relationship of these models and their impact on connectivity studies are discussed. Next, we cover the network models that have been widely used to capture the spatial distribution of nodes. Particularly, the dense network model, the extended network model, and the infinite network model are discussed. Finally, we briefly introduce the main tools and methodologies used for connectivity studies, including Continuum percolation theory, branching process, and algebraic graph theory, and present the main results, established using these tools, on network connectivity.

This chapter provides an overview of the fundamental concepts, models, tools, and methodologies used for connectivity studies. We shall start with an introduction to the major connection models used to describe the establishment of communication links between devices or nodes. Next, we shall cover the network models that have been widely used to capture the spatial distribution of nodes. Finally, we briefly introduce the main tools and methodologies used for connectivity studies and present the main results, established using these tools, on network connectivity.

1.1 Connection Models

A network can be conveniently represented by a graphic model by using one vertex in the graph to uniquely represent a node in the network and using one edge in the graph to uniquely represent a link in the network, and the converse. The obtained graph is called the *underlying graph* of the network. Properties of a network can be investigated by studying its underlying graph. The two terms "network" and its "underlying graph" are used almost interchangeably in this book. When we need

G. Mao, *Connectivity of Communication Networks*,
DOI 10.1007/978-3-319-52989-9_1

to refer to the location of network nodes, we prefer the term "network"; when the properties of the network can be studied using its underlying graph and without consideration of the nodes' location, we prefer the term "graph."

There are five models that are widely used to describe the establishment of an edge between a pair of vertices. In these models, the edges may be directional or undirectional.

1.1.1 Erdős–Rényi Connection Model

There are two closely related variants of the Erdős–Rényi connection model that are widely used for generating random graphs: the $G(n, m)$ model and the $G(n, p)$ model.

In the $G(n, m)$ model, a graph is chosen uniformly at random from the collection of all graphs that have n vertices and m edges. Consider the $G(4, 3)$ model for example. There are a maximum of six possible edges between vertices and the total number of graphs with four vertices and three edges are 20. Each of these 20 graphs is included with probability $1/20$.

In the $G(n, p)$ model, a graph is constructed by connecting vertices randomly. Each edge is included in the graph with probability p, independent of every other edge. It follows that the expected number of edges in $G(n, p)$ is $\binom{n}{2}p$ and the number of edges in $G(n, p)$ is a binomially distributed random integer with parameters $\binom{n}{2}$ and p.

The random graph whose edges are established following the Erdős–Rényi connection model is known as the Erdős–Rényi graph. The study of the Erdős–Rényi graphs was pioneered by two Hungarian mathematicians Paul Erdős and Alfréd Rényi. They established many properties of the Erdős–Rényi graph. Particularly, when the model parameters are chosen properly, many properties of the Erdős–Rényi graph arise with very high probability. The Erdős–Rényi graphs were however abstract mathematical objects. Particularly, the establishment of edges does not depend on the geometric positions of vertices—a phenomenon that is contrary to common practice in a communication network. In a communication network, particularly in a wireless network, the existence of an edge between a pair of vertices almost invariably depends on the Euclidean distance between the vertices. Because of this lack of geometric component in the Erdős–Rényi connection model, the Erdős–Rényi connection model does not fit into the framework of this book and hence the treatment of the Erdős–Rényi (ER) connection model is left out. Readers are referred to [28, 111] for a complete treatment of the Erdős–Rényi connection model of random graphs and its more recent extensions.

1.1.2 Unit Disk Connection Model

In the unit disk connection model, there is a link between a pair of nodes, say nodes i and j, if and only if their Euclidean distance is smaller than or equal to a given threshold:

$$\left\| \boldsymbol{x}_i - \boldsymbol{x}_j \right\| \leq r \tag{1.1.1}$$

where $\boldsymbol{x}_i$ and $\boldsymbol{x}_j$ are the coordinates of nodes i and j respectively, the parameter r is known as the transmission range of the node and $\| \|$ denotes the Euclidean norm. When investigating how certain property of a network scales with an increase of the number of nodes in the network, parameter r is often written as $r(n)$ to emphasize the dependence on n where parameter n denotes number of nodes.

Compared with the Erdős–Rényi connection model, the unit disk connection model grossly captures the geometric aspect in a communication network that when a pair of nodes is closer to each other, it is easier to establish a link between them. Therefore, the unit disk connection model is more popularly used than the Erdős–Rényi connection model in studying communication networks.

The unit disk connection model is a simple connection model that may help to simplify the analysis. However its limitation is also quite obvious: no real antenna has an antenna pattern resembling that of the disk connection model. Therefore, the log-normal connection model, which is more realistic than the unit disk connection model, has accordingly been considered for modeling connections between nodes.

1.1.3 Log-Normal Connection Model

Let P_t be the transmission power of a transmitting node i in dB unit. As the signal propagates towards a receiving node j, the signal strength will attenuate with the propagation distance. Let $PL(\boldsymbol{x}_i, \boldsymbol{x}_j)$ be the path loss from node i to node j in dB unit where $\boldsymbol{x}_i$ and $\boldsymbol{x}_j$ are the coordinate vectors of node i and node j respectively. Thus the received signal strength at node j in dB unit, denoted by $P_r(\boldsymbol{x}_i \to \boldsymbol{x}_j)$, can be written as

$$P_r(\boldsymbol{x}_i \to \boldsymbol{x}_j) = P_t - PL(\boldsymbol{x}_i, \boldsymbol{x}_j) \tag{1.1.2}$$

The model in (1.1.2) does not consider the surrounding environmental obstacles which may have vastly different effects at two different location at the same Euclidean distance from the transmitting node. In reality, the measured received signal strength at the receiving node j can be very different from the average value predicted in (1.1.2). A random component, denoted by Z, has to be incorporated into (1.1.2) to account for the impact of shadowing caused by obstacles affecting the wave propagation, sometimes referred to as *shadow fading*. When shadow fading is considered, (1.1.2) becomes

$$P_r\left(\boldsymbol{x}_i \rightarrow \boldsymbol{x}_j\right) = P_t - PL\left(\boldsymbol{x}_i, \boldsymbol{x}_j\right) + Z \tag{1.1.3}$$

When the received signal strength $P_r\left(\boldsymbol{x}_i \rightarrow \boldsymbol{x}_j\right)$ at the receiving node j is greater than or equal to a given threshold P_{th}, we say that node j *is directly connected to* node i.

In the log-normal connection model, the path loss is assumed to follow a log-normal model [22, 153, 172]:

$$PL\left(\boldsymbol{x}_i, \boldsymbol{x}_j\right) = PL_0(d_0) + 10\alpha \log_{10} \frac{x_{ij}}{d_0} \tag{1.1.4}$$

where $PL_0(d_0)$ is the reference path loss in dB unit at a reference distance d_0, x_{ij} is the Euclidean distance between nodes i and j, and α is the path loss exponent which indicates the rate at which the received signal strength decreases with distance. The reference path loss $PL_0(d_0)$ is calculated using the free space Friis equation or obtained through field measurements at distance d_0 [172], and are usually assumed to be known constants [140, 162]. The value of α depends on the environment and terrain structure and can vary between 2 in free space and 6 in heavily built urban areas [172].

Considering the shadowing effect, the wireless received signal strength $P_r\left(\boldsymbol{x}_i \rightarrow \boldsymbol{x}_j\right)$ (in dB unit) at the receiving node j from the transmitting node i can be modeled by a log-normal shadowing model:

$$P_r\left(\boldsymbol{x}_i \rightarrow \boldsymbol{x}_j\right) = P_t - PL_0(d_0) - 10\alpha \log_{10} \frac{x_{ij}}{d_0} + Z_\sigma \tag{1.1.5}$$

where Z_σ is a zero-mean Gaussian random variable (in dB) with standard deviation σ (also in dB unit). The value of σ is usually larger than zero and can be as high as 12 *dB* [172]. Node j *is directly connected to* node i if and only if $P_r\left(\boldsymbol{x}_i \rightarrow \boldsymbol{x}_j\right) \geq P_{th}$.

Note that $P_r\left(\boldsymbol{x}_i \rightarrow \boldsymbol{x}_j\right)$ in (1.1.5) is a random variable. Using (1.1.5), the probability that node j is directly connected to node i, denoted as $\mathbb{P}(\boldsymbol{x}_i \rightarrow \boldsymbol{x}_j)$, is given by

$$\mathbb{P}(x_{ij}) = \Pr\left(P_r\left(\boldsymbol{x}_i \rightarrow \boldsymbol{x}_j\right) \geq P_{th}\right) = \int_{10\alpha \log_{10} \frac{x_{ij}}{r_0}}^{\infty} \frac{1}{\sqrt{2\pi}\sigma} e^{-\frac{z^2}{2\sigma^2}} dz, \tag{1.1.6}$$

where

$$r_0 = d_0 \times 10^{\frac{P_t - PL_0(d_0) - P_{th}}{10\alpha}} \tag{1.1.7}$$

has the meaning of being the transmission range in the absence of shadowing (i.e., $\sigma = 0$).

When the shadow fading Z_σ between a transmitter and a receiver is dependent of that of another distinct transmitter-receiver pair, the connections between distinct pairs of transmitter and receivers are also independent.

When $\sigma = 0$, the log-normal connection model reduces to a unit disk connection model with a transmission range equal to r_0. When $\sigma > 0$, the received power $P_r(\boldsymbol{x}_i \to \boldsymbol{x}_j)$ is determined by both the deterministic function of the Euclidean distance x_{ij} and the random shadowing component Z_σ. The transmission area of a node is no longer a circular area.

As a consequence of the reciprocity theorem of electromagnetic field, the antenna patterns of a node are the same whether the antenna is transmitting or receiving. Therefore, under the additional assumption that the propagation path from node i to node j and the propagation path from node j to node i are symmetric, it can be obtained that if node j is connected to node j when node j is transmitting at power P_t, node i is also connected to node j when node i is transmitting at the same power P_t. That is, the link between node i and node j becomes undirectional.

1.1.4 Random Connection Model

A more general model than the unit disk connection model and the log-normal connection model is the random connection model. Under the random connection model, a pair of nodes at $\boldsymbol{x}_i$ and $\boldsymbol{x}_j$ are directly connected, or equivalently a link exists between the two nodes, with probability $g(\boldsymbol{x}_i - \boldsymbol{x}_j)$, where $g : \Re^d \to [0, 1]$, independent of the event that another pair of nodes are directly connected.

The function g is usually required to satisfy the following properties of nonincreasing monotonicity, integral boundedness, and rotational invariance [71, 143, Chapter 6]:

$$g(\boldsymbol{x}) \le g(\boldsymbol{y}) \quad \text{whenever} \quad \|\boldsymbol{x}\| \ge \|\boldsymbol{y}\| \tag{1.1.8}$$

$$0 < \int_{\Re^d} g(\boldsymbol{x}) d\boldsymbol{x} < \infty \tag{1.1.9}$$

$$g(\boldsymbol{x}) = g(\boldsymbol{y}) \quad \text{whenever} \quad \|\boldsymbol{x}\| = \|\boldsymbol{y}\| \tag{1.1.10}$$

It is obvious that the random connection model incorporates the unit disk connection model and the log-normal connection model as its two special cases. When $g(\boldsymbol{x}) = 1$ for $\|\boldsymbol{x}\| \le r$ and $g(\boldsymbol{x}) = 0$ for $\|\boldsymbol{x}\| > r$, the random connection model reduces to the unit disk model; when

$$g(\boldsymbol{x}) = \int_{10\alpha \log_{10} \frac{\|\boldsymbol{x}\|}{r_0}}^{\infty} \frac{1}{\sqrt{2\pi}\sigma} e^{-\frac{z^2}{2\sigma^2}} dz$$

the random connection model becomes the log-normal connection model.

1.1.5 SINR Connection Model

In the four connection models introduced in the previous subsections, it is either explicitly or implicitly considered that the existence of a connection between a pair of nodes and the existence of a connection between another distinct pair of nodes are independent. This independent connection assumption greatly simplifies the analysis of the resulting network. The independence assumption can be justified in a wireless network with a small to medium amount of traffic load where the impact of interference on connections is often small and negligible. In a wireless network with a heavy traffic load and a large number of simultaneous transmissions, however, the impact of interference has to be considered and consequently, the connections may no longer be independent.

Specifically, due to the nature of wireless communications, signals transmitted at the same time and using the same frequency band will mutually interfere with each other. The SINR (signal to interference plus noise) connection model has been widely used to capture the impact of interference [57, 91, 94]. Under the SINR connection model, the existence of a directed link between a pair of nodes is determined by the strength of the received signal from the desired transmitter, the interference caused by other concurrent transmissions, and the background noise. Let P_i be the transmission power of node i in decimal unit and let $\boldsymbol{x}_k$, $k \in \Gamma$, be the location of node k, where Γ represents the set of indices of all nodes in the network. A node j can successfully receive the transmitted signal from a node i (i.e., node j *is directly connected to* node i) if and only if the SINR at $\boldsymbol{x}_j$, denoted by $\text{SINR}\left(\boldsymbol{x}_i \rightarrow \boldsymbol{x}_j\right)$, is above a prescribed threshold β, i.e.,

$$\text{SINR}\left(\boldsymbol{x}_i \rightarrow \boldsymbol{x}_j\right) = \frac{P_i \ell\left(\boldsymbol{x}_i, \boldsymbol{x}_j\right)}{N_0 + \gamma \sum_{k \in \mathcal{T}_i} P_k \ell\left(\boldsymbol{x}_k, \boldsymbol{x}_j\right)} \geq \beta \tag{1.1.11}$$

where $\ell\left(\boldsymbol{x}_i, \boldsymbol{x}_j\right)$ denotes the power attenuation from node i to node j and includes both the path loss and the shadow fading, $\mathcal{T}_i \subseteq \Gamma$ denotes the subset of nodes that are simultaneously transmitting as node i and using the same frequency band, N_0 is the background noise power. The coefficient $0 \leq \gamma \leq 1$ is the inverse of the processing gain of the system and it weighs the impact of interference. In a broadband system using CDMA (code division multiple access), γ depends on the orthogonality between codes used during concurrent transmissions and is smaller than 1; in a narrow-band system, γ is equal to 1 [57, 94].

Similarly, node i can receive from node j (i.e., node i is directly connected to node j) if and only if

$$\text{SINR}\left(\boldsymbol{x}_j \rightarrow \boldsymbol{x}_i\right) = \frac{P \ell\left(\boldsymbol{x}_j, \boldsymbol{x}_i\right)}{N_0 + \gamma \sum_{k \in \mathcal{T}_j} P_k \ell\left(\boldsymbol{x}_k, \boldsymbol{x}_i\right)} \geq \beta \tag{1.1.12}$$

Therefore, node i and node j are directly connected, i.e., an undirectional link exists between node i and node j, if and only if both (1.1.11) and (1.1.12) are satisfied.

It is evident from (1.1.11) and (1.1.12) that in the SINR connection model, the existence of a connection from a node to another node not only depends on the transmission power and the power attenuation of this node pair but also depends on the activity, transmission power, and power attenuation of other nodes. Therefore, connections between nodes are no longer independent.

1.2 Network Models

As revealed in the connection models introduced in the previous subsection, the establishment of links between a pair of nodes often depend on the location of nodes. In this subsection, we continue to discuss several network models that are widely used to describe the location of nodes.

It is natural to model nodes in the network by a point process where a node in the network is uniquely represented by a point. This abstraction allows us to omit the impact of some physical details such as the size and shape of network devices and focuses on the topological aspects of the network. We make use of two categories of point processes: deterministic point process and random point process. In the first category, points are deterministically placed in a given region, resembling the placement of network devices in a specific deployment area. In the second category, points are randomly placed in a given region. The second category is particularly useful to capture some irregular and random aspects of the network, such as location of mobile phone users, location of unplanned wireless access points, or irregular location of cellular base stations, that are better described by statistical laws. In the second category of the point process, it is of particular interest to study properties of the network when the scale of the network becomes sufficiently large. A large part of the book is dedicated to studying the random point process.

We now define some concepts that are used later in the book.

Definition 1 (Homogeneous Poisson Point Process) A random set of points X deployed in $A \subseteq \Re^2$ is said to be a homogeneous Poisson point process on A with density $\rho > 0$ if it satisfies the conditions

(1) for mutually disjoint domains of A, $A_1, \ldots, A_k$, the random variables $X(A_1), \ldots, X(A_k)$ are mutually independent, where $X(A_j)$ denotes the random number of points of X inside A_j, $j = 1, \ldots, k$;
(2) for any bounded area $D \subseteq A$, we have that for every $m \geq 0$

$$\Pr(X(D) = m) = e^{-\rho|D|} \frac{(\rho |D|)^m}{m!} \tag{1.2.1}$$

where $|D|$ denotes the size of the area D.

Note that for any area of unit size $D \subseteq A$, we have $E(X(D)) = \rho$ and the density of the homogeneous Poisson point process corresponds to the expected number of points of the process in the unit area.

A homogeneous Poisson point process on A can be constructed by first drawing a random number of points from a Poisson distribution with parameter $\rho\,|A|$ and then distributing the points uniformly and independently over A.

A point process is called a *stationary point process* if the distribution of points in a given region of the plane is invariant under any translation of the region to another location of the plane. A homogeneous Poisson point process is a stationary point process. In contrast, an inhomogeneous Poisson point process, which is defined below, is a nonstationary point process.

Definition 2 (Inhomogeneous Poisson Point Process) A random set of points X deployed in $A \subseteq \Re^2$ is said to be an inhomogeneous Poisson point process on A with density function $\rho(\boldsymbol{x}) : \boldsymbol{x} \in A \to [0, \infty)$ if it satisfies the conditions

(1) for mutually disjoint domains of A, $A_1, \ldots, A_k$, the random variables $X(A_1), \ldots, X(A_k)$ are mutually independent, where $X\left(A_j\right)$ denotes the random number of points of X inside A_j, $j = 1, \ldots, k$;
(2) for any bounded area $D \subseteq A$, we have that for every $m \geq 0$

$$\Pr(X(D) = m) = e^{-\int_D \rho(\boldsymbol{x})d\boldsymbol{x}} \frac{\left[\int_D \rho(\boldsymbol{x})\, d\boldsymbol{x}\right]^m}{m!} \tag{1.2.2}$$

The following theorem, known as the independent thinning theorem, relates an inhomogeneous Poisson point process to a homogeneous Poisson point process:

Theorem 3 *Consider a homogeneous Poisson point process X on $A \subseteq \Re^2$ with density ρ. Removing each point in X with probability $1-p(\boldsymbol{x})$, independently of other points in X, where $\boldsymbol{x} \in A$ is the location of the point, there results an inhomogeneous Poisson point process on $A \subseteq \Re^2$ with density $\rho p(\boldsymbol{x})$.*

Proof Denote the remaining point process after the removal of points in X by X'. First note that for arbitrary disjoint domains of A: $A_1, \ldots, A_k$, the independence of the random variables $X'(A_1), \ldots, X'(A_k)$ follows readily from the facts that X is a homogeneous point process and that each point in X is removed independently of other points in X.

Now we consider the distribution of points in X' in an arbitrary bounded area $D \subseteq A$. Since X is a homogeneous Poisson point process with density ρ, $X(D)$ is a Poissonly distributed random variable with mean $\rho\,|D|$. Further note that a randomly chosen point of X in D is removed with probability $\frac{\int_D 1-p(\boldsymbol{x})d\boldsymbol{x}}{|D|} = 1 - \frac{\int_D p(\boldsymbol{x})d\boldsymbol{x}}{|D|}$. As an easy consequence of the total probability theorem, it can be obtained that

$$\begin{aligned}\Pr\left(X'(D) = k\right) &= \sum_{m=k}^{\infty} \binom{m}{k} \left(\frac{\int_D p(\boldsymbol{x})\, d\boldsymbol{x}}{|D|}\right)^k \left(1 - \frac{\int_D p(\boldsymbol{x})\, d\boldsymbol{x}}{|D|}\right)^{m-k} \Pr(X(D) = m) \\ &= \sum_{m=k}^{\infty} \binom{m}{k} \left(\frac{\int_D p(\boldsymbol{x})\, d\boldsymbol{x}}{|D|}\right)^k \left(1 - \frac{\int_D p(\boldsymbol{x})\, d\boldsymbol{x}}{|D|}\right)^{m-k} e^{-\rho|D|} \frac{(\rho\,|D|)^m}{m!}\end{aligned}$$

$$= \frac{\left(\int_D \rho p(\mathbf{x})\, d\mathbf{x}\right)^k}{k!} e^{-\rho|D|} \sum_{m=k}^{\infty} \frac{\left(\rho\,|D| - \int_D \rho p(\mathbf{x})\, d\mathbf{x}\right)^{m-k}}{(m-k)!}$$

$$= \frac{\left(\int_D \rho p(\mathbf{x})\, d\mathbf{x}\right)^k}{k!} e^{-\int_D \rho p(\mathbf{x}) d\mathbf{x}}$$

That is, $X'(D)$ is a Poissonly distributed random variable with mean $\int_D \rho p(\mathbf{x})\, d\mathbf{x}$. This completes the proof. □

Using Theorem 3, an inhomogeneous Poisson point process on A with density function $\rho(\mathbf{x}) : \mathbf{x} \in A \to [0, \infty)$ can be constructed by first constructing a homogeneous Poisson point process with density $\max_{\mathbf{x}\in A} \rho(\mathbf{x})$ and then removing each point in the homogeneous Poisson point process independently with probability $1 - \frac{\rho(\mathbf{y})}{\max_{\mathbf{x}\in A} \rho(\mathbf{x})}$ where $\mathbf{y}$ is the location of the point.

A useful property of the Poisson point process is that conditioned on the event that there is a point at $\mathbf{x} \in \Re^2$, due to the independence property of the Poisson point process, the probabilistic structure of the conditioned process is identical to that of the original process, apart from the given point at $\mathbf{x}$. The distribution of an independent Poisson point process with a given point at $\mathbf{x}$ is called the *Palm distribution* of the Poisson point process.

Another property of the Poisson point process is that the superposition of two independent Poisson point processes with density ρ_1 and ρ_2 respectively is a Poisson point process with density $\rho_1 + \rho_2$.

Definition 4 (Uniform Point Process) A random set of points X with $|X| = n$ is said to be a uniform point process on a bounded area A with parameter n if these n points are uniformly and independently distributed on A.

From the above definitions of the Poisson point process and the uniform point process, we can find a close relationship between the two point process: a homogeneous Poisson point process X on A, conditional on $X(A) = n$, becomes a uniform point process on A with parameter n. In the following paragraph, we further demonstrate that the number of points of a uniform point process with parameter n on $A \subset \Re^2$ and located in a domain $D \subset A$ of fixed size converges in distribution to the number of points in the same domain of a homogeneous Poisson point process on A with constant density ρ, as A grows toward $\Re^2$ while keeping $\frac{n}{|A|} = \rho$.

Consider a uniform point process with parameter n on A as A grows towards $\Re^2$ and n grows with A such that $\frac{n}{|A|} = \rho$. We ignore the granularity caused by that $\rho\,|A|$ may not be an integer in our discussion, which is of minor consequence and can be readily handled. The probability mass function of the number of points $X(D)$ in a domain $D \subset A$ of a fixed size is given by

$$\Pr(X(D) = m) = \binom{n}{m} \left(\frac{|D|}{|A|}\right)^m \left(1 - \frac{|D|}{|A|}\right)^{n-m}$$

As A grows towards $\Re^2$ and using the equality $\frac{n}{|A|} = \rho$, it can be shown that for any fixed integer m,

$$
\begin{aligned}
\lim_{|A|\to\infty} \Pr\left(X\left(D\right) = m\right) &= \lim_{|A|\to\infty} \binom{n}{m} \left(\frac{|D|}{|A|}\right)^m \left(1 - \frac{|D|}{|A|}\right)^{n-m} \\
&= \lim_{n\to\infty} \frac{n\left(n-1\right)\cdots\left(n-m\right)}{m!} \left(\frac{\rho\left|D\right|}{n}\right)^m \left(1 - \frac{\rho\left|D\right|}{n}\right)^{n-m} \\
&= \lim_{n\to\infty} \frac{n\left(n-1\right)\cdots\left(n-m\right)}{m!n^m} \left(\rho\left|D\right|\right)^m \left(1 - \frac{\rho\left|D\right|}{n}\right)^{n-m} \\
&= \lim_{n\to\infty} \frac{1}{m!} \left(\rho\left|D\right|\right)^m \left(1 - \frac{\rho\left|D\right|}{n}\right)^n \left(1 - \frac{\rho\left|D\right|}{n}\right)^{-m} \\
&= \frac{\left(\rho\left|D\right|\right)^m}{m!} \lim_{n\to\infty} \left(1 - \frac{\rho\left|D\right|}{n}\right)^n \\
&= \frac{\left(\rho\left|D\right|\right)^m}{m!} e^{-\rho|D|}
\end{aligned}
$$

That is, as A grows towards $\Re^2$, the distribution of $X\left(D\right)$ converges to a Poisson distribution with parameter $\rho\left|D\right|$.

Because of the close connections between a uniform point process and a Poisson point process, in large networks, results established assuming a uniform point process are often applicable to a Poisson point process of the same density. However there are two important differences between the Poisson point process and the uniform point process. Firstly, for a Poisson point process, the random variables $X\left(A_1\right), \ldots, X\left(A_k\right)$ are mutually independent for mutually disjoint domains of A, $A_1, \ldots, A_k$, whereas the random variables $X\left(A_1\right), \ldots, X\left(A_k\right)$ are not independent for a uniform point process. Secondly, for a Poisson point process, $X\left(A\right)$ is a random variable with a Poisson distribution whereas for a uniform point process $X\left(A\right)$ is a constant. Therefore, extra caution needs to be taken when the aforementioned two differences have a significant impact on the network properties under investigation.

Now we introduce three related but logically distinct network models that have been widely used in studying properties of large networks. The first model, often referred to as the *dense network model,* considers that the network is deployed in a finite area with a sufficiently large node density. The second model, often referred to as the *extended network model*, considers that the node density is fixed and the network area is sufficiently large. The third model, referred to as the *infinite network model*, has its origin in continuum percolation theory [143]. It considers a network deployed in an infinite area, i.e., $\Re^2$ in two dimensions, and analyzes properties of the network as the node density becomes sufficiently large.

In the following, we give the formal definitions of the three network models in terms of the Poisson point process and the random connection model. These definitions can be easily amended to suit another point process and connection model.

Definition 5 (Dense Network Model) Let $\mathcal{G}\left(\mathcal{X}_\rho, g_{f(\rho)}, A\right)$ be a network with nodes Poissonly distributed on a unit square $A \triangleq \left[-\frac{1}{2}, \frac{1}{2}\right]^2$ with density ρ and a pair of nodes separated by a displacement $\boldsymbol{x}$ are directly connected with probability $g_{f(\rho)}(\boldsymbol{x}) \triangleq g\left(\frac{x}{f(\rho)}\right)$, independent of the event that another distinct pair of nodes are directly connected. Here, $\mathcal{X}_\rho$ denotes the vertex set in $\mathcal{G}\left(\mathcal{X}_\rho, g_{f(\rho)}, A\right)$ and $f(\rho)$ is a function of ρ.

Definition 6 (Extended Network Model) Let $\mathcal{G}\left(\mathcal{X}_1, g_{f(\rho)}, A_{\sqrt{\rho}}\right)$ be a network with nodes Poissonly distributed on a square $A_{\sqrt{\rho}} \triangleq \left[-\frac{\sqrt{\rho}}{2}, \frac{\sqrt{\rho}}{2}\right]^2$ with density 1 and a pair of nodes separated by a displacement $\boldsymbol{x}$ are directly connected with probability $g_{f(\rho)}(\boldsymbol{x})$, independent of the event that another distinct pair of nodes are directly connected. Here, $\mathcal{X}_1$ denotes the vertex set in $\mathcal{G}\left(\mathcal{X}_1, g_{f(\rho)}, A_{\sqrt{\rho}}\right)$.

Definition 7 (Infinite Network Model) Let $\mathcal{G}\left(\mathcal{X}_\rho, g, \Re^2\right)$ be a network with nodes Poissonly distributed on $\Re^2$ with density $\boldsymbol{\rho}$ and a pair of nodes separated by a Euclidean distance x are directly connected with probability $g(x)$, independent of the event that another distinct pair of nodes are directly connected. Here, $\mathcal{X}_\rho$ denotes the vertex set in $\mathcal{G}\left(\mathcal{X}_\rho, g, \Re^2\right)$.

The above three network models are often studied in the asymptotic regime as $\rho \to \infty$. The function $f(\rho)$ determines the scaling law that the connection function should follow in order for the network to acquire a certain property, e.g., connectivity, asymptotically as $\rho \to \infty$. For the special case of $g(\boldsymbol{x}) = 1$ for $\|\boldsymbol{x}\| \leq 1$ and $g(\boldsymbol{x}) = 0$ otherwise, $g_{f(\rho)}(\boldsymbol{x})$ becomes a unit disk connection model with a transmission range of $f(\rho)$. That is, a pair of nodes are directly connected if and only if their Euclidean distance is smaller than or equal to $f(\rho)$.

1.3 Graph Theoretic Tools for Connectivity Analysis

In this section, we introduce several graph theoretic tools and results established using these tools that are widely used in studying network connectivity.

1.3.1 Continuum Percolation Theory

Continuum percolation is an extension of its predecessor, discrete percolation, into continuous space. More specifically, discrete percolation studies point processes where the points are located on an integer lattice where continuum percolation studies point processes where the points are randomly placed in some continuous space. Two widely studied models in continuum percolation are the Poisson Boolean model and the Poisson random connection model.

A Poisson Boolean model, denoted by $\mathcal{G}\left(\mathcal{X}_\rho, Y, \Re^2\right)$, is defined by a Poisson point process of density ρ on $\Re^2$ and a nonnegative random variable Y. Unless otherwise specified, we assume that $E\left(Y^2\right) < \infty$. Each point of $\mathcal{X}_\rho$ is the center of a disk with a random radius and the radii corresponding to different points are independently and identically distributed following the distribution of Y. The radii are also independent of $\mathcal{X}_\rho$. The union of the disks forms the Poisson Boolean model. In this way, the plane $\Re^2$ is partitioned into two regions: the *occupied region*, which is the region covered by at least one disk, and the *vacant region*, which is the complement of the occupied region. A region is said to be *connected* if any two points in the region can be connected by a line entirely contained within the region. A maximal connected subregion of the occupied region is called a *connected occupied region*. Two points in $\mathcal{X}_\rho$ are said to be *connected* if they are in the same connected occupied region. A *component* is defined as a maximal subset of points in $\mathcal{X}_\rho$ such that all points in the subset are in the same connected occupied region. A main focus of continuum percolation is studying the conditions under which there exist components of infinite order. The order of a component is the number of points of $\mathcal{X}_\rho$ in the component or the cardinality of the component.

In the special case of $Y = c$, i.e., Y is a deterministic constant equal to c, $\mathcal{G}\left(\mathcal{X}_\rho, c, \Re^2\right)$ has the same connectivity property as a network defined by a Poisson point process $\mathcal{X}_\rho$ of density ρ on $\Re^2$ and a unit disk connection model with a transmission range of $2c$.

A Poisson random connection model, denoted by $\mathcal{G}\left(\mathcal{X}_\rho, g, \Re^2\right)$, is defined by a Poisson point process X of density ρ on $\Re^2$ and a connection function $g : \Re^2 \to [0, 1]$. Two points in $\mathcal{X}_\rho$ are said to be *connected* if there is a path connecting them. A *path* between a pair of points is a sequence of *distinct* edges where the first edge starts at one point, the last edge ends at the other point and the starting point of each intermediate edge is the ending point of its preceding edge. A component of $\mathcal{G}\left(\mathcal{X}_\rho, g, \Re^2\right)$ is a maximal set of points in $\mathcal{X}_\rho$ where there is a path between any point in the set to any other point in the set.

Next we introduce several important results in continuum percolation theory. We refer readers to [143] for proofs of the results.

Theorem 8 *Let $\mathcal{G}\left(\mathcal{X}_\rho, Y, \Re^2\right)$ be a Poisson Boolean model with Y satisfying $E\left(Y^2\right) < \infty$. There exists $\rho_0 > 0$ such that the expected number of disks in the connected occupied region W, which contains the origin, is finite whenever $0 < \rho < \rho_0$, if and only if*

$$E\left(Y^4\right) < \infty$$

The fact that only a connected occupied region containing the origin is considered here is of no importance. Due to the stationarity of the homogeneous Poisson point process and the independence of the radii, different points of the plane cannot be distinguished probabilistically.

Theorem 9 *For a Poisson Boolean model $\mathcal{G}\left(\mathcal{X}_\rho, Y, \Re^2\right)$, if $E\left(Y^3\right) < \infty$, there exists $0 < \rho_0$ such that for all $0 < \rho < \rho_0$, the probability that the number of disks in any connected occupied region is finite is equal to* 1.

The Poisson Boolean model $\mathcal{G}\left(\mathcal{X}_\rho, Y, \Re^2\right)$ is said to be in the *subcritical phase* if the connected occupied region containing the origin contains a finite number of points in $\mathcal{X}_\rho$ almost surely; and is said to be in the *supercritical phase* if the connected occupied region containing the origin contains an infinite number of Poisson points in $\mathcal{X}_\rho$ with a positive probability. An event ξ is said to occur *almost surely* if its probability is equal to one. An event ξ_z depending on z is said to occur *asymptotically almost surely* if its probability tends to one as $z \to \infty$. There are also two other phases in the Poisson Boolean model: the expected number of Poisson points in the connected occupied region W containing the origin is finite and the expected number of Poisson points in the connected occupied region W containing the origin is infinite. To formalize the phase transition between the phases, we define the *critical densities*:

$$\rho_N \triangleq \inf\left\{\rho : \Pr\left(\mathcal{X}_\rho\left(W\right) = \infty\right) > 0\right\} \tag{1.3.1}$$

and

$$\rho_E \triangleq \inf\left\{\rho : E\left(\mathcal{X}_\rho\left(W\right)\right) = \infty\right\} \tag{1.3.2}$$

Critical densities can also be defined in terms of the Lebesgue measure $\ell\left(W\right)$ of W:

$$\rho_H \triangleq \inf\left\{\rho : \Pr\left(\ell\left(W\right) = \infty\right) > 0\right\} \tag{1.3.3}$$

and

$$\rho_T \triangleq \inf\left\{\rho : E\left(\ell\left(W\right)\right) = \infty\right\} \tag{1.3.4}$$

It can be readily established that the critical densities ρ_N, ρ_E, ρ_H and ρ_T depend on the distribution of Y. To emphasize this dependence and avoid confusion, sometimes we also write $\rho_N\left(Y\right)$, $\rho_E\left(Y\right)$, $\rho_H\left(Y\right)$ and $\rho_T\left(Y\right)$ for ρ_N, ρ_E, ρ_H and ρ_T respectively.

The following theorem establishes that the notion of the size of the component or the Lebesgue measure of the connected occupied region does not affect the critical densities when Y is bounded:

Theorem 10 *For a Poisson Boolean model $\mathcal{G}\left(\mathcal{X}_\rho, Y, \Re^2\right)$ with $0 < \rho \leq R$, almost surely for some $R > 0$, we have*

$$\rho_N\left(Y\right) = \rho_H\left(Y\right)$$

and

$$\rho_E\left(Y\right) = \rho_T\left(Y\right)$$

The following theorem further establishes that the critical densities ρ_N and ρ_E are equal.

Theorem 11 *For a Poisson Boolean model* $\mathcal{G}\left(\mathcal{X}_\rho, Y, \Re^2\right)$ *with* $0 < \rho \leq R$*, almost surely for some* $R > 0$*, we have* $\rho_E(Y) = \rho_N(Y)$.

In the supercritical regime, i.e., when $\rho > \rho_N$, unbounded components, i.e., components with an unbounded order, exist with a positive probability and that the number of such components is an almost sure constant (infinity is allowed). The following theorem establishes a stronger result that in a Poisson Boolean model, there can be at most one unbounded component almost surely.

Theorem 12 *In a Poisson Boolean model* $\mathcal{G}\left(\mathcal{X}_\rho, Y, \Re^2\right)$*, there can be at most one unbounded component almost surely.*

The above result on the uniqueness of the unbounded component can be extended to Boolean models driven by arbitrary point processes.

Denote by $\theta_Y(\rho) = \theta(\rho)$ the probability that the origin is within an unbounded connected occupied region. The function θ is called the *percolation function*. It can be readily established using the coupling technique introduced in the next subsection that $\theta(\rho)$ is a nondecreasing function of ρ.

1.3.1.1 Coupling and Scaling Technique

Coupling and scaling are two important techniques that have been widely used in the theory of continuum percolation. It is however quite challenging to give a clear and concise definition of the two techniques. In this subsection, we use several examples to illustrate the use of the two techniques.

Scaling can be used to expand or shrink the distances between all pairs of nodes by a common factor such that the original graph and the new graph has the same graphical properties. For example, let $\mathcal{G}\left(\mathcal{X}_\rho, g, A_1\right)$ be a network with nodes Poissonly distributed on a unit square A_1 with density ρ where a pair of nodes are directly connected with probability $g(x)$, independent of other pairs of nodes, and x is the Euclidean distance between the two nodes. By expanding the Euclidean distance between all node pairs by a common factor $\sqrt{\rho}$, one obtains another network $\mathcal{G}\left(\mathcal{X}_1, g_{\sqrt{\rho}}, A_{\sqrt{\rho}}\right)$ where $\mathcal{G}\left(\mathcal{X}_1, g_{\sqrt{\rho}}, A_{\sqrt{\rho}}\right)$ is a network with nodes Poissonly distributed on a square of size ρ and a pair of nodes are directly connected with probability $g_{\sqrt{\rho}}(x) \triangleq g\left(\frac{x}{\sqrt{\rho}}\right)$ where x is the Euclidean distance between the nodes in the new network. Obviously, such scaling operation does not change the graphical properties of the network. Therefore, $\mathcal{G}\left(\mathcal{X}_\rho, g, A_1\right)$ and $\mathcal{G}\left(\mathcal{X}_1, g_{\sqrt{\rho}}, A_{\sqrt{\rho}}\right)$ have the same graphical properties. By using the scaling technique, one can apply results obtained from a network on a unit area to a network on an area of an arbitrary size.

Coupling is used to establish the connection between two networks such that if one network has a certain property, the second network also has the same property. It is often the case that one network is easier to analyze than the other network. By proving that the property of interest holds in the network easier to analyze and using the coupling technique, the conclusion follows that the property also holds in the network that is harder to analyze. In this way, the analysis can often be greatly simplified. In the following paragraphs, we give an example to illustrate the use of the coupling technique.

Let $\mathcal{G}\left(\mathcal{X}_\rho, g, A_1\right)$ be the same network as defined previously. Let $\mathcal{G}^T\left(\mathcal{X}_\rho, g, A_1^T\right)$ be a network with nodes Poissonly distributed on a unit *torus* A_1^T with density ρ where a pair of nodes are directly connected with probability $g\left(x^T\right)$, independent of other pair of nodes. Here x^T is the *toroidal distance* between the two nodes. The two networks are essentially the same except that one network is on a unit square and the other is on a unit torus. The torus that is commonly discussed in random geometric graph theory and continuum percolation theory is essentially the same as a square except that the distance between two points on a torus is defined by their *toroidal distance*, instead of Euclidean distance. For a unit torus $A_1^T = \left[-\frac{1}{2}, \frac{1}{2}\right]^2$, the toroidal distance between two points $\boldsymbol{x}_1, \boldsymbol{x}_2 \in A_1^T$ is given by [165, p. 13]:

$$\|\boldsymbol{x}_1 - \boldsymbol{x}_2\|^T \triangleq \min\left\{\|\boldsymbol{x}_1 + z - \boldsymbol{x}_2\| : z \in \mathbb{Z}^2\right\} \tag{1.3.5}$$

where $\mathbb{Z}^2$ is a two-dimensional integer set. The use of toroidal distance allows nodes located near the boundary to have the same number of connections *probabilistically* as a node located near the center. Therefore, it allows the removal of the boundary effect that is present in a square. The consideration of a torus often implies that there is no need to consider special cases occurring near the boundary of the region and that events inside the region do not depend on the particular location inside the region. This often simplifies the analysis.

Let W be the number of isolated nodes in $\mathcal{G}\left(\mathcal{X}_\rho, g, A_1\right)$ and W^T be the number of isolated nodes in $\mathcal{G}^T\left(\mathcal{X}_\rho, g, A_1^T\right)$. Both W and W^T are random variables. A node is called *an isolated node* if it is not directly connected to any other node in the network. Further note that the number of isolated nodes in a network is a nonincreasing function of the number of connections in the network. That is, given a network, increasing the number of connections in the network will result in the number of isolated nodes either decrease or remain the same. The following property regarding W and W^T can be established using the coupling technique:

Theorem 13 $W^T \leq_{st} W$.

Proof Given a random instance of $\mathcal{G}^T\left(\mathcal{X}_\rho, g, A_1^T\right)$, removing each connection in $\mathcal{G}^T\left(\mathcal{X}_\rho, g, A_1^T\right)$ independently with probability $\frac{g(x)}{g\left(x^T\right)}$ where x and x^T are the Euclidean distance and the toroidal distance between the corresponding pair of nodes respectively. Recall that, as mentioned in Sect. 1.1.4, g is a monotonically nonincreasing function of x. Further note that the toroidal distance between a pair of nodes is always less than or equal to their Euclidean distance. Therefore, $\frac{g(x)}{g\left(x^T\right)}$ is less than or equal to 1. In the resulting network, a pair of nodes are directly connected with probability $g(x)$. Thus an instance of $\mathcal{G}(\mathcal{X}_\rho, g, A_1)$ is obtained. Conversely, an instance of $\mathcal{G}^{\mathcal{T}}(\mathcal{X}_\rho, g, A_1^T)$ can also be obtained from an instance of $\mathcal{G}(\mathcal{X}_\rho, g, A_1)$ by adding connections to the corresponding instance of $\mathcal{G}(\mathcal{X}_\rho, g, A_1)$.

Further using the property that the number of isolated nodes is a nonincreasing function of the number of connections in the network, the conclusion follows. □

In Theorem 13, the symbol $\leq_{st}$ means less than or equal to in *stochastic order*. For two real random variables A and B, we say that $A \leq_{st} B$ if $\Pr(A < x) \leq \Pr(B < x)$ for all $x \in (-\infty, \infty)$. The following conclusion readily follows from Theorem 13.

Corollary 14 *If $\mathcal{G}^{\mathcal{T}}(\mathcal{X}_\rho, g, A_1^T)$ has no isolated nodes almost surely, $\mathcal{G}(\mathcal{X}_\rho, g, A_1)$ also has no isolated nodes almost surely.*

Recall that an event is said to occur almost surely if its probability is equal to 1.

Further note that $\mathcal{G}(\mathcal{X}_\rho, g, A_1)$ has no isolated nodes is a necessary condition for $\mathcal{G}(\mathcal{X}_\rho, g, A_1)$ to be a connected network. Therefore, by analyzing the conditions for the number of isolated nodes in $\mathcal{G}^{\mathcal{T}}(\mathcal{X}_\rho, g, A_1^T)$ equal 0, a necessary condition for $\mathcal{G}(\mathcal{X}_\rho, g, A_1)$ to be connected can be obtained. In this way, one can avoid the intricate analysis analyzing special situations that occur when nodes are located near the border of $\mathcal{G}(\mathcal{X}_\rho, g, A_1)$, which results in much simpler analysis.

Coupling and scaling techniques will be used a number of times later in the book.

1.3.2 Branching Process

Branching process is another important tool that has been widely used in graph theoretic analysis. Generally speaking, a branching process is a process that models the evolution of a population from generation to generation where each member in the earlier generation produces a random number of members in the next generation. The member that produces members in the next generation is called *parent* of these members in the next generation and the members produced are called its *children*. The parent is also a child of the member that produces it in the immediately preceding generation, and so on.

In its simplest form, the number of children produced by a parent is distributed according to a fixed probability distribution that does not vary from member to member. Furthermore, the number of children of a parent is independent of that of another parent in both the same generation and a different generation.

A central question in the theory of branching process is the probability of *extinction*, where no members exist after some finite number of generations. Let Z_n be the random number of members in the n-th generation and let $X_{i,n}$ be the random number of children in the n-th generation produced by the i-th parent in the $n-1$-th generation. Figure 1.1 gives an illustration of the branching process. It can be readily shown using the independence property of the number of children produced by each parent that starting with a member in generation zero, the expected number of members in generation n, conditioned on $Z_{n-1} = k$, is given by

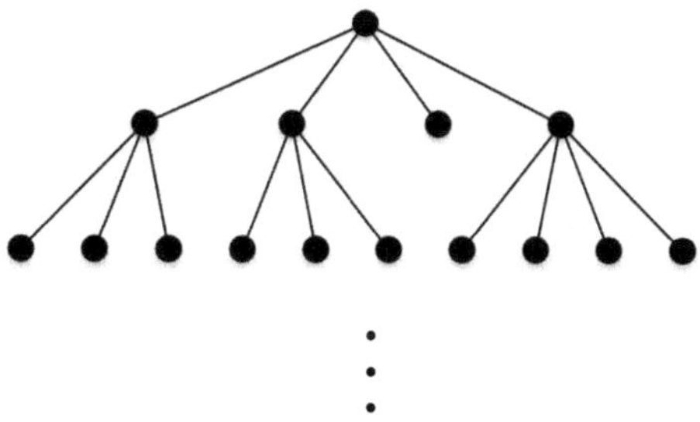

Fig. 1.1 An illustration of the branching process

$$E(Z_n | Z_{n-1} = k) = \sum_{i=1}^{k} E(X_{i,n}) = k\mu \tag{1.3.6}$$

where μ is the expected number of children of each parent. The member in generation zero is known as the *root* of the branching process. It can be further shown using Baye's formula that $E(Z_n) = \mu^n$. It follows that if $\mu < 1$, then the expected number of members in the n-th generation reduces rapidly to zero with the increase of n. Using Markov's inequality, this implies that the branching process becomes extinct with probability 1. If $\mu > 1$, the probability of extinction is less than 1 but no necessarily zero. For example, consider a branching process where each parent generates 0 or 10 children with an equal probability. It can be easily shown that with a nonzero probability, the branching process may become extinct in a finite number of generations. When $\mu = 1$, then extinction occurs with probability 1 unless each member always has exactly one child.

A well-known branching process is the Galton–Watson process. In the Galton–Watson process, $X_{i,n}$s are independently and identically distributed random non-negative integers for all $n \in Z^+$ and $i \in= \{1, \ldots, Z_{n-1}\}$, where Z^+ denotes the set of positive integers. Then, the recurrence equation describing the evolution of population of each generation is given by:

$$Z_{n+1} = \sum_{i=1}^{Z_n} X_{i,n} \text{ with } Z_0 = 1 \tag{1.3.7}$$

The ultimate extinction probability is given by

$$\lim_{n \to \infty} \Pr(Z_n = 0) \tag{1.3.8}$$

The process can be analyzed using the probability generating function. The probability generating function of a discrete random variable is a power series representation of the probability mass function of the random variable. Particularly, let p_k be the probability that a parent generates k children. When $p_0 = 0$, the process becomes trivial to analyze as the extinction probability can be easily shown to be 0.

Therefore, we focus on nontrivial cases where $p_0 > 0$. The probability generating function for p_i is given by

$$h(z) = p_0 + p_1 z + p_2 z^2 + \cdots$$

Let q_n be the probability that the process becomes extinct by the n-th generation. Obviously, $q_0 = 0$ and $0 = q_0 \leq q_1 \leq q_2 \cdots \leq 1$. If there are k members in the first generation, using the property that the number of children produced by each parent is independent, the probability that the process becomes extinct in the $n+1$-th generation is given by $(q_n)^k$. It follows that

$$q_{n+1} = p_0 + p_1 q_n + \cdots + p_k (q_n)^k + \cdots \tag{1.3.9}$$

Since q_n is bounded and nonincreasing in n, the limit $\lim_{n\to\infty} q_n$ exists. Letting $q \triangleq \lim_{n\to\infty} q_n$, the following equation regarding q must hold:

$$q = h(q) \tag{1.3.10}$$

By solving the above equation, the value of q can be obtained.

For the special case that $X_{i,n}$ are independently and identically distributed following a Poisson distribution with mean λ for all $n \in Z^+$ and $i \in \{1, \ldots, Z_{n-1}\}$, it can be shown that its probability generating function is given by

$$\begin{aligned} h(z) &= e^{-\lambda} + \lambda e^{-\lambda} z + \frac{\lambda^2}{2!} e^{-\lambda} z^2 + \cdots \\ &= \left(1 + \lambda z + \frac{\lambda^2}{2!} z^2 + \cdots\right) e^{-\lambda} \\ &= e^{(z-1)\lambda} \end{aligned} \tag{1.3.11}$$

The extinction probability is the solution to the following equation

$$p = e^{(p-1)\lambda} \tag{1.3.12}$$

and it can be further shown that

$$q = \frac{W\left(-\lambda e^{-\lambda}\right)}{-\lambda} \tag{1.3.13}$$

where $W(\cdot)$ is the Lambert W function [47].

The connection between the above branching process and continuum percolation theory can be readily established. Consider a random network on $\Re^2$. Place a node at the origin and designate that node as the root of the branching process. Nodes directly connected to the node at the origin form the members of the first generation of the branching process. Nodes directly connected to any node in the

n-th generation, excluding those nodes already included in the earlier generations, form the $n+1$-th generation. In this way, an associated branching process model can be established. It can be shown that the probability that the branching process does not become extinct, i.e., $1-q$, is equal to the probability that the origin is within an unbounded connected occupied region or equivalently an unbounded component, i.e., the *percolation probability*, which is a main metric of interest in continuum percolation theory. In the later chapters, we shall use the branching process to study network connectivity.

1.3.3 Algebraic Graph Theory

Algebraic graph theory studies the topological properties of graphs using algebraic methods and it provides another important class of tools widely used in graph theoretic analysis. Algebraic graph theory has been mainly used to study graphs of finite size.

Consider a network with a set of n nodes. Its properties can be studied using its *underlying graph* $G(V,E)$, where $V \triangleq \{v_1,\ldots,v_n\}$ denotes the vertex set and E denotes the edge set. The underlying graph is obtained by representing each node in the network uniquely using a vertex and the converse. An edge exists between two vertices if and only if there is a direct connection (or link) between the associated nodes. We limit our discussions to a *simple graph* (network) where there is at most one edge (link) between a pair of vertices (nodes) and an undirected graph whose edges are undirectional.

Define an *adjacency matrix* A_G of the graph $G(V,E)$ to be a symmetric $n\times n$ matrix whose $(i,j)^{th}, i \neq j$, entry is equal to one if there is an edge between v_i and v_j and is equal to zero otherwise. The diagonal entries of A_G are all equal to zero. The *eigenvalues of the graph* $G(V,E)$ are defined to be the eigenvalues of A_G. The network connectivity information, e.g., connectivity and k-connectivity, is entirely contained in its adjacency matrix. Many interesting connectivity and topological properties of the network can be obtained by studying the eigenvalues of its underlying graph. Furthermore, define the *Laplacian matrix* of a graph G to be

$$L_G \triangleq D - A_G \tag{1.3.14}$$

where D is a diagonal matrix with degrees of vertices in G on the diagonal where the *degree* of a vertex is the number of edges incident on it. Substantial results have been established in algebraic graph theory by studying the adjacency matrix and the Laplacian matrix of the graph. In the following paragraphs, we provide a brief summary of the main results related to network connectivity study.

We first give some formal definitions of the concepts and terms. Some of the definitions are not used in this chapter but it may be beneficial to put these definitions in a central place for later use in the book.

Definition 15 A *walk* of length ℓ between two vertices v_i and v_j is a sequence of ℓ edges (not necessarily distinct) where the first edge starts at v_i, the last edge ends at v_j and the starting vertex of each intermediate edge is the ending vertex of its preceding edge.

Definition 16 A *path* of length l is a walk of length l where all the edges are distinct.

Definition 17 A graph is *connected* if there is a path between any two distinct vertices.

Definition 18 The *distance* between two vertices is the length of the shortest path, measured by the number of hops, between them.

Definition 19 The *diameter* of a connected graph is the largest distance between any two vertices of the graph.

Definition 20 A *cycle* of length l is a path of length l where the starting vertex and the ending vertex are the same.

Definition 21 A *tree* is a connected graph with no cycles.

Definition 22 A graph $G'(V', E')$ is called a *subgraph* of $G(V, E)$ if $V' \subseteq V$ and $E' \subseteq E$ and $G(V, E)$ is called a *supergraph* of $G'(V', E')$.

Definition 23 If $G'(V', E')$ is a *subgraph* of $G(V, E)$ and E' contains all edges in E with endpoints in V', $G'(V', E')$ is called an *induced subgraph* of $G(V, E)$.

Definition 24 A spanning subgraph $G'(V', E')$ of a connected graph $G(V, E)$ is a subgraph where $V' = V$ and $G'(V', E')$ is a connected graph.

Definition 25 A *spanning tree* of a connected graph $G(V, E)$ is a spanning subgraph, which is a tree.

Definition 26 Two vertices are called *adjacent* if they are directly connected by an edge. Two graphs G and H are said to be *isomorphic*, if there is a one-to-one mapping, called an *isomorphism*, between the vertices of the two graphs such that two vertices are adjacent in G if and only if their corresponding vertices are adjacent in H.

Definition 27 A vertex is called *isolated* if there is no edge incident on it.

Definition 28 The complement of a graph $G(V, E)$, denoted by $\overline{G}\left(V, \bar{E}\right)$, is a graph with the same vertex set as $G(V, E)$ and an edge exists in $\bar{E}$ if and only if the edge is not in E.

Definition 29 A *bipartite graph* (or bigraph) $G(V, E)$ is a graph whose vertex set V can be partitioned into two disjoint sets V_1 and V_2 such that every edge in E is between a vertex in V_1 and a vertex in V_2.

Definition 30 A graph is called *k edge connected* if it remains connected whenever fewer than k edges are removed.

Definition 31 A graph is called *k vertex connected* if it remains connected whenever fewer than k vertices are removed. We often call a k vertex connected graph simply a k connected graph. A k connected graph is also k edge connected.

Definition 32 The *algebraic connectivity* of a graph $G(V, E)$ is the second-smallest eigenvalue of L_G.

Now we introduce some results established in algebraic graph theory that are most related to the connectivity analysis. For results that are easy to prove, we include their proof; otherwise we refer interested readers to the classical books of Biggs [25] and Godsil and Royle [80] for a proof.

Theorem 33 *The number of walks of length l between vertices v_i and v_j in $G(V, E)$ is equal to $\left(A_G^l\right)_{ij}$, i.e., the (i, j) entry of A_G^l.*

Proof We prove the theorem by recursion. When $l = 1$, the theorem is obviously correct. When $l = 2$,

$$\left(A_G^2\right)_{ij} = \sum_{m=1}^{n} a_{im} a_{mj}$$

where a_{im} is the (i, m) entry of A_G. Noting that a_{ii} is equal to 0 for $i \in \{1, \ldots, n\}$, $\left(A_G^2\right)_{ij}$ is therefore the total number of walks of length 2 between v_i and v_j via a third vertex.

Assuming that the theorem is correct for $l = k$, let us now consider the case when $l = k + 1$. It can be shown that

$$\left(A_G^{k+1}\right)_{ij} = \sum_{m=1}^{n} a_{im} \left(A_G^k\right)_{mj}$$

Since $\left(A_G^k\right)_{mj}$ is the total number of walks of length k between vertex v_j and a third vertex v_m and further noting that $a_{ii} = 0$ for $i \in \{1, \ldots, n\}$, it is easy to establish that $\left(A_G^{k+1}\right)_{ij}$ is the total number of walks of length $k + 1$ between v_i and v_j via a third distinct vertex v_m where $m \neq i, j$. The result can then be established. □

The following results can be established as an easy consequence of Theorem 33.

Corollary 34 *The number of walks of length smaller than or equal to l between vertices v_i and v_j in $G(V, E)$ is equal to $\left[(A_G + I)^l - I\right]_{ij}$, where $\left[(A_G + I)^l - I\right]_{ij}$ is the (i, j) entry of $\left[(A_G + I)^l - I\right]$ and I is the identity matrix.*

Corollary 35 *The graph $G(V, E)$ is a connected graph if and only if $(A_G + I)^n \geq 1$ where $n = |V|$, i.e., each entry of the matrix $(A_G + I)^n$ is greater than or equal to 1.*

Next we introduce some other important results closely related to the connectivity analysis.

Theorem 36 *The eigenvalues of a graph are real numbers.*

Theorem 37 *Let $\mu_1 \geq \ldots \geq \mu_n$ be the eigenvalues of a graph $G(V,E)$. If $\mu_1 = \mu_2$, then $G(V,E)$ is disconnected. If $\mu_1 = -\mu_n$ and $G(V,E)$ is not empty, then at least one connected component of $G(V,E)$ is nonempty and bipartite.*

Theorem 38 *If the number of distinct eigenvalues of $G(V,E)$ is r, then $G(V,E)$ has a diameter of at most $r-1$.*

Theorem 39 *The algebraic connectivity of a graph $G(V,E)$ is greater than 0 if and only if $G(V,E)$ is a connected graph.*

1.4 Notes and Further Readings

This chapter introduced several widely used connection models, network models, and graph theoretic tools for connectivity analysis, which lay the foundation for later studies. Several main results related to network connectivity study established using continuum percolation theory, branching process, and algebraic graph theory respectively were presented. For readers who are interested in a deeper investigation into these areas, we refer them to the classical book of Meester and Roy [143] for continuum percolation, to the book of Harris [98] for branching process, and to the books of Biggs [25] and Godsil and Royle [80] for algebraic graph theory. Furthermore, a concise summary of the main results established in graph theory can also be found in [86, 101] with the latter book [101] more focused on algebraic graph theory.

Part I
Connectivity of Large Static Networks

Chapter 2
Large Network Models and Their Implications

Abstract Connectivity of large wireless networks has been a primary concern for which asymptotic analysis is a useful tool. Three related but logically distinct network models are often considered in asymptotic analyses, viz., the dense network model, the extended network model, and the infinite network model, which considers respectively a network deployed in a fixed finite area with a sufficiently large node density, a network deployed in a sufficiently large area with a fixed node density, and a network deployed in an infinite plane with a sufficiently large node density. The infinite network model originated from continuum percolation theory and asymptotic results obtained from the infinite network model have often been applied to the dense and extended networks indiscriminately. In this chapter, through two case studies related to network connectivity on the expected number of isolated nodes and on the vanishing of components of finite order $k > 1$ respectively, we demonstrate some subtle but important differences between the infinite network model and the dense and extended network models. Therefore, extra scrutiny has to be used in order for the results obtained from the infinite network model to be applicable to the dense and extended network models. Asymptotic results are also established on the expected number of isolated nodes, the vanishingly small impact of the boundary effect on the number of isolated nodes and the vanishing of components of finite order $k > 1$ in the dense and extended network models using a general random connection model.

Connectivity of large networks as the number of nodes or node density in the network becomes sufficiently large has been a primary concern in the study of network connectivity. Asymptotic analysis, valid when the number of nodes in the network is large enough, is useful for understanding characteristics of these networks.

Three related but logically distinct network models, i.e., the dense network model, the extended network model, and the infinite network model defined in Sect. 1.2, have been widely used in the asymptotic analysis of large networks. The dense network model considers that the network is deployed in a finite area with a sufficiently large node density. The extended network model considers that the node density is fixed and the network area is sufficiently large. The infinite network model has its origin in continuum percolation theory [143]. It considers a network

G. Mao, *Connectivity of Communication Networks*,
DOI 10.1007/978-3-319-52989-9_2

deployed in an infinite area, i.e., $\Re^2$ in two dimensions, and analyzes properties of the network as the node density becomes sufficiently large. Due to the relatively longer history of research into continuum percolation theory, relatively abundant results in that area, and the close connections between the infinite network model and the dense and extended network models, results obtained in the infinite network model are often applied straightforwardly to the dense and extended models.

In this chapter, through two case studies on key events related to the network connectivity, i.e., the expected number of isolated nodes and the vanishing of components of fixed and finite order $k > 1$ (recall that the order of a component refers to the number of nodes in the component), and using a random connection model, we demonstrate some subtle but important differences between the infinite network model and the dense and extended network models due to the *truncation effect*, to be explained in the following paragraphs. Therefore, results obtained from an infinite network model *cannot be directly applied* to the dense and extended networks. Instead, some careful analysis of the impact of the truncation effect is required.

Here we give a detailed explanation of the above comments using a unit disk connection model as an example. In this chapter, we consider Poisson node distribution only and the conclusions obtained apply to uniform node distribution too. We refer readers to Sect. 1.1.1 for a discussion on the differences between the uniform node distribution and the Poisson node distribution. As introduced in Sect. 1.1, under the unit disk connection model, two nodes are directly connected if and only if their Euclidean distance is smaller than or equal to a given threshold $r(\rho)$, a parameter which is often taken as a function of a further parameter ρ, to be defined shortly, under the dense and extended network models. The parameter $r(\rho)$ is termed the *transmission range*. The dense and extended network models that are often considered assume respectively (a) nodes are Poissonly distributed in a unit area, say a square, with density ρ and $r(\rho) = \sqrt{\frac{\log \rho + c}{\pi \rho}}$ (the dense network model); (b) nodes are Poissonly distributed on a square $\sqrt{\rho} \times \sqrt{\rho}$ with density 1 and $r(\rho) = \sqrt{\frac{\log \rho + c}{\pi}}$ (the extended network model). The parameter c may be either a constant; or it can depend on ρ, in which case $c = o(\log \rho)$ (refer to Appendix for a definition of Landau's order notation). The corresponding infinite network model considers that nodes are Poissonly distributed in $\Re^2$ with density ρ and a pair of nodes are directly connected if and only if their Euclidean distance is smaller than or equal to r, which *does not* depend on ρ. The dense network model can be converted into the extended network model by using the scaling technique, more specifically scaling the Euclidean distances between *all pairs* of nodes by a common factor of $\sqrt{\rho}$ while maintaining their connections, and the converse. Therefore, the dense network model and the extended network model are equivalent in the analysis of connectivity. In the extended network model, as $\rho \to \infty$, the network area approaches $\Re^2$ and the *average node degree*, i.e., the average number of directly connected neighbors of a randomly chosen node, approaches infinity following $\Theta(\log \rho)$. That is, a node has more and more connections as $\rho \to \infty$. This resembles the situation that occurs in the infinite network model as $\rho \to \infty$. This close connection between the infinite network model and the dense and extended network models creates the *illusion* that as $\rho \to \infty$, results obtained in the infinite network model can be applied directly to the dense and extended models, e.g., those dealing with the vanishing of isolated nodes, the uniqueness of the component of infinite order, the vanishing of components of finite order $k > 1$.

Starting from the dense network model, however, if we scale the Euclidean distances between all pairs of nodes by a factor $1/\sqrt{\frac{\log \rho + c}{\pi \rho}}$, there results a network on a square $1/\sqrt{\frac{\log \rho + c}{\pi \rho}} \times 1/\sqrt{\frac{\log \rho + c}{\pi \rho}}$ with node density $\frac{\log \rho + c}{\pi}$, where $\frac{\log \rho + c}{\pi} \to \infty$ as $\rho \to \infty$, and a pair of nodes are directly connected if and only if their Euclidean distance is smaller than or equal to $r = 1$. Note that the transmission range r is equal to 1 and does not depend on the node density. This latter network model is also equivalent to the dense and extended network models in the connectivity analysis. On the other hand, this latter network can also be obtained from an infinite network on $\Re^2$ with node density $\frac{\log \rho + c}{\pi}$ and $r = 1$ by removing all nodes and the associated connections outside a square of $1/\sqrt{\frac{\log \rho + c}{\pi \rho}} \times 1/\sqrt{\frac{\log \rho + c}{\pi \rho}}$ in $\Re^2$. We term the effect associated with the above removal operation as the *truncation effect*. From the above discussion, it is clear that a prerequisite for the results obtained in the infinite network model to be applicable to the dense or extended network model is that the impact of the truncation effect on the property of interest must be vanishingly small as $\rho \to \infty$.

In this chapter, through two case studies, one on the expected number of isolated nodes and the other on the vanishing of components of fixed and finite order $k > 1$, using a random connection model, we show however that ensuring the impact of the truncation effect is vanishingly small either requires imposing a stronger requirement on the connection function or needs some nontrivial analysis to rule out the possibility of occurrence of some events associated with the truncation effect. Therefore, results obtained assuming an infinite network model *cannot* be applied directly to the dense and extended network models. In particular, we show that in order for the impact of the truncation effect on the number of isolated nodes to be vanishingly small, a stronger requirement on the connection function (than the usual requirements of rotational invariance, integral boundedness and nonincreasing monotonicity) needs to be imposed. We further show that some nontrivial analysis is required to rule out the possibility of occurrence of some events associated with the truncation effect in order to establish the result on the vanishing of components of components of fixed and finite order $k > 1$ in the dense and extended network models. For example, an infinite component in $\Re^2$ may, after truncation, yield multiple *components of extremely large order*, finite components of fixed order $k > 1$ and isolated nodes in a square of $1/\sqrt{\frac{\log \rho + c}{\pi \rho}} \times 1/\sqrt{\frac{\log \rho + c}{\pi \rho}}$, where these components are only connected via nodes and associated connections in the infinite component but outside the square of $1/\sqrt{\frac{\log \rho + c}{\pi \rho}} \times 1/\sqrt{\frac{\log \rho + c}{\pi \rho}}$. Thus the dense and extended networks may still possibly have finite components of order $k > 1$ even though the infinite network can be shown to *asymptotically almost surely* have no such finite components as $\rho \to \infty$. Recall that an event ξ is said to occur almost surely (*a.s.*) if its probability is equal to 1 and an event ξ_z depending on z is said to occur asymptotically almost surely (*a.a.s.*) if its probability tends to one as $z \to \infty$. Note that it is trivial to show that for any finite ρ, *almost surely* there is no infinite component in a network whose nodes are Poissonly distributed with density $\frac{\log \rho + c}{\pi}$ on a square of $1/\sqrt{\frac{\log \rho + c}{\pi \rho}} \times 1/\sqrt{\frac{\log \rho + c}{\pi \rho}}$. Therefore, we use the term *components of extremely large order* to refer to those components whose order may become *asymptotically infinite* as $\rho \to \infty$.

In this chapter, we also establish asymptotic results on the expected number of isolated nodes, the vanishingly small impact of the boundary effect on the number of isolated nodes and the vanishing of components of finite order $k > 1$ in the dense and extended network models using the random connection model. These results form key steps in establishing the sufficient and necessary condition for a large network to be asymptotically almost surely connected assuming the general random connection model, a problem to be tackled in the next chapter.

2.1 Comparative Outline of Three Large Network Models

In this section we intuitively explain the difference between the three widely used large network models: infinite network model, dense network model, and extended network model. From that, the concept of truncation effect will be defined and its different implications on analysis using the three models will be discussed.

Let $g : \Re^+ \to [0, 1]$ be a function satisfying the conditions of nonincreasing monotonicity and integral boundedness:

$$g(x) \le g(y) \quad \text{whenever} \quad x \ge y \tag{2.1.1}$$

$$0 < \int_{\Re^2} g(\|\boldsymbol{x}\|)\, d\boldsymbol{x} < \infty \tag{2.1.2}$$

where $\|\boldsymbol{x}\|$ denotes the Euclidean norm of $\boldsymbol{x}$. The function g is the connection function of the random connection model introduced in Sect. 1.1. Furthermore, the requirement of rotational invariance on the connection function in the random connection model has been met implicitly by letting g be a function of a scalar, typically representing the Euclidean distance between the two nodes being considered.

Using the integral boundedness condition on g and the nonincreasing monotonicity of g, it can be shown that

$$\int_{\Re^2} g(\|\boldsymbol{x}\|)\, d\boldsymbol{x} = \lim_{z \to \infty} \int_0^z 2\pi x g(x)\, dx$$

and

$$\lim_{z \to \infty} \int_z^\infty 2\pi x g(x)\, dx = 0$$

The above equation, together with the following derivations

$$\begin{aligned}
&\lim_{z\to\infty}\int_{z}^{\infty} 2\pi x g\left(x\right) dx \\
&\geq \lim_{z\to\infty}\int_{z}^{2z} 2\pi x g\left(x\right) dx \\
&\geq \lim_{z\to\infty}\int_{z}^{2z} 2\pi x g\left(2z\right) dx \\
&= \lim_{z\to\infty} 3\pi z^{2} g\left(2z\right)
\end{aligned}$$

allows us to conclude that

$$g\left(x\right) = o_x\left(\frac{1}{x^2}\right) \tag{2.1.3}$$

From time to time, we may require g to satisfy the more restrictive requirement that

$$g\left(x\right) = o_x\left(\frac{1}{x^2\log^2 x}\right) \tag{2.1.4}$$

and (2.1.1). We shall show later that the more restrictive requirement on g is required for the impact of the truncation effect to be vanishingly small as $\rho \to \infty$. When we do impose such additional constraint, we shall specify it clearly. It is obvious that conditions (2.1.1) and (2.1.2) imply (2.1.3) while condition (2.1.4) implies (2.1.2) and (2.1.3).

In the following analysis, we shall only use (2.1.1) and (2.1.4) (instead of (2.1.1) and (2.1.2)) when necessary. This helps to identify which part of the analysis relies on the more restrictive requirement on g. In our analysis, we assume that g has infinite support when necessary. Our results however apply to the situation when g has bounded support, which forms a special case and only makes the analysis easier.

Furthermore, define

$$r_\rho \triangleq \sqrt{\frac{\log\rho + b}{C\rho}} \tag{2.1.5}$$

for some nonnegative value ρ, where

$$0 < C = \int_{\Re^2} g\left(\|\boldsymbol{x}\|\right) d\boldsymbol{x} < \infty \tag{2.1.6}$$

and b is a constant ($+\infty$ is allowed).

Following the definitions of the dense network model, the extended network model and the infinite network model in Sect. 1.1, we denote the three corresponding models considered in this chapter by $\mathcal{G}\left(\mathcal{X}_\rho, g_{r_\rho}, A\right)$, $\mathcal{G}\left(\mathcal{X}_1, g_{\sqrt{\frac{\log\rho+b}{C}}}, A_{\sqrt{\rho}}\right)$ and $\mathcal{G}\left(\mathcal{X}_\rho, g, \Re^2\right)$. Recall that A and $A_{\sqrt{\rho}}$ are respectively a unit square $\left[-\frac{1}{2}, \frac{1}{2}\right]^2$ and a square of size ρ: $\left[-\frac{\sqrt{\rho}}{2}, \frac{\sqrt{\rho}}{2}\right]^2$. We further define a fourth network model in the following and the motivation for defining the new model appears shortly later.

Definition 40 Let $\mathcal{G}\left(\mathcal{X}_{\frac{\log\rho+b}{C}}, g, A_{\frac{1}{r_\rho}}\right)$ be a network with nodes Poissonly distributed on a square $A_{\frac{1}{r_\rho}} \triangleq \left[-\frac{1}{2r_\rho}, \frac{1}{2r_\rho}\right]^2$ with density $\frac{\log\rho+b}{C}$ and a pair of nodes separated by a Euclidean distance x are directly connected with probability $g(x)$, independent of the event that another distinct pair of nodes are directly connected. $\mathcal{X}_{\frac{\log\rho+b}{C}}$ denotes the vertex set in $\mathcal{G}\left(\mathcal{X}_{\frac{\log\rho+b}{C}}, g, A_{\frac{1}{r_\rho}}\right)$.

With minor abuse of the terminology, we use A (respectively $A_{\sqrt{\rho}}$, $A_{\frac{1}{r_\rho}}$) to denote both the square itself and the size of the square, and in the latter case, $A = 1$ (respectively $A_{\sqrt{\rho}} = \rho$, $A_{\frac{1}{r_\rho}} = \frac{1}{r_\rho^2}$).

The reason for choosing this particular form of r_ρ and the above network models is to avoid triviality in the analysis and to make the analysis compatible with those obtained assuming a unit disk connection model. Particularly, when g takes the form that $g(x) = 1$ for $x \leq 1$ and $g(x) = 0$ for $x > 1$, it can be shown that $\mathcal{G}\left(\mathcal{X}_\rho, g_{r_\rho}, A\right)$ reduces to the dense network model assuming a unit disk connection model discussed in [71, 92, 165] where $C = \pi$ and r_ρ corresponds to the critical transmission range for connectivity; $\mathcal{G}\left(\mathcal{X}_1, g_{\sqrt{\frac{\log\rho+b}{C}}}, A_{\sqrt{\rho}}\right)$ reduces to the extended network model assuming a unit disk connection model considered in [69, 71, Chapter 3.3.2]. Thus the above model easily incorporates the unit disk connection model as a special case. A similar conclusion can also be drawn for the log-normal connection model.

Now we establish the relationship between the three network models $\mathcal{G}\left(\mathcal{X}_\rho, g_{r_\rho}, A\right)$, $\mathcal{G}\left(\mathcal{X}_1, g_{\sqrt{\frac{\log\rho+b}{C}}}, A_{\sqrt{\rho}}\right)$, and $\mathcal{G}\left(\mathcal{X}_{\frac{\log\rho+b}{C}}, g, A_{\frac{1}{r_\rho}}\right)$ on finite and then asymptotically infinite regions respectively using the scaling and coupling technique. Given an instance of $\mathcal{G}\left(\mathcal{X}_\rho, g_{r_\rho}, A\right)$, if we scale the Euclidean distances between *all pairs* of nodes by a factor of $\sqrt{\rho}$ while maintaining their connections, there results a random network where nodes are Poissonly distributed on a square $A_{\sqrt{\rho}}$ with density 1 and a pair of nodes separated by a Euclidean distance x are directly connected with probability $g_{\sqrt{\frac{\log\rho+b}{\rho}}}(x)$, i.e., an instance of $\mathcal{G}\left(\mathcal{X}_1, g_{\sqrt{\frac{\log\rho+b}{C}}}, A_{\sqrt{\rho}}\right)$. All connectivity properties, e.g., connectivity, number of isolated nodes, number of components of a specified order, that hold in the instance of $\mathcal{G}\left(\mathcal{X}_\rho, g_{r_\rho}, A\right)$ are also valid for the associated instance in $\mathcal{G}\left(\mathcal{X}_1, g_{\sqrt{\frac{\log\rho+b}{C}}}, A_{\sqrt{\rho}}\right)$. To be more precise, the *underlying graphs* of these two network instances are *isomorphic*. Similarly, if we shrink the Euclidean distances between all pairs of nodes in a network, which is an instance of $\mathcal{G}\left(\mathcal{X}_1, g_{\sqrt{\frac{\log\rho+b}{C}}}, A_{\sqrt{\rho}}\right)$, by a factor of $\frac{1}{\sqrt{\rho}}$, there results an instance of $\mathcal{G}\left(\mathcal{X}_\rho, g_{r_\rho}, A\right)$ and the two networks again have the same connectivity property. Therefore, $\mathcal{G}\left(\mathcal{X}_\rho, g_{r_\rho}, A\right)$ and $\mathcal{G}\left(\mathcal{X}_1, g_{\sqrt{\frac{\log\rho+b}{C}}}, A_{\sqrt{\rho}}\right)$ are *equivalent* in that any connectivity property that holds in one model will necessarily hold in the other. Similarly, it can also be shown that $\mathcal{G}\left(\mathcal{X}_\rho, g_{r_\rho}, A\right)$ and $\mathcal{G}\left(\mathcal{X}_{\frac{\log\rho+b}{C}}, g, A_{\frac{1}{r_\rho}}\right)$ are equivalent in their connectivity properties. Therefore, in this chapter we only chose one model, i.e., $\mathcal{G}\left(\mathcal{X}_{\frac{\log\rho+b}{C}}, g, A_{\frac{1}{r_\rho}}\right)$, to discuss connectivity properties of finite and asymptotically infinite networks. The reason for choosing this network model is that under the model, a pair of nodes are directly connected following g, in the same way as nodes in the infinite network model $\mathcal{G}\left(\mathcal{X}_\rho, g, \Re^2\right)$ are directly connected. This facilitates the discussion and comparison between the finite (asymptotically infinite) network model and the infinite network model, which is a key focus of the chapter.

Furthermore, we point out that the above discussion on the equivalence of network models $\mathcal{G}\left(\mathcal{X}_\rho, g_{r_\rho}, A\right)$, $\mathcal{G}\left(\mathcal{X}_1, g_{\sqrt{\frac{\log\rho+b}{C}}}, A_{\sqrt{\rho}}\right)$ and $\mathcal{G}\left(\mathcal{X}_{\frac{\log\rho+b}{C}}, g, A_{\frac{1}{r_\rho}}\right)$ is only valid for the random connection model, including the unit disk connection model and the log-normal connection model as its two special cases. For the other widely used model, i.e., the SINR connection model defined in Sect. 1.1.5, under some special circumstances, e.g., the background noise is negligible and the attenuation function is a power law function, the three network models are equivalent; otherwise under more general conditions, *the three models are not equivalent*. However the key observation revealed in our analysis, i.e., results obtained from an infinite network model do not necessarily apply to the dense and extended network models, also holds for the SINR connection model.

2.2 Estimating the Number of Isolated Nodes

In this section we comparatively study the expected number of isolated nodes in $\mathcal{G}\left(\mathcal{X}_{\frac{\log\rho+b}{C}}, g, A_{\frac{1}{r_\rho}}\right)$ and the expected number of isolated nodes in its counterpart in an infinite network, i.e., a region with the same area as $A_{\frac{1}{r_\rho}}$ in an infinite network on $\Re^2$ and with the same node density $\frac{\log\rho+b}{C}$ and connection function g. Through the study, we show that under certain conditions, the impact of the truncation effect on the expected number of isolated nodes is non-negligible or may even be the dominant factor. The number of isolated nodes is a key parameter in the analysis of network connectivity. A necessary condition for a network to be connected is that the network has no isolated node. Such a necessary condition has been shown to be also a sufficient condition for a connected network as $\rho \to \infty$ under a unit disk connection model [165] and this is also true for a random connection model, as shown later in Chap. 3.

2.2.1 *Expected Number of Isolated Nodes in an Asymptotically Infinite Network*

In this subsection we analyze the expected number of isolated nodes in $\mathcal{G}\left(\mathcal{X}_{\frac{\log\rho+b}{C}}, g, A_{\frac{1}{r_\rho}}\right)$. For an arbitrary node in $\mathcal{G}\left(\mathcal{X}_{\frac{\log\rho+b}{C}}, g, A_{\frac{1}{r_\rho}}\right)$ at location $\boldsymbol{y} \in A_{\frac{1}{r_\rho}}$, it can be shown using the property that nodes are Poissonly distributed with density $\frac{\log\rho+b}{C}$ that the probability that a node at $\boldsymbol{y} \in A_{\frac{1}{r_\rho}}$, if exists, is isolated is given by (following the equation, explanations are given):

$$\begin{aligned}
\Pr\left(I_y = 1\right) &= \lim_{dA_x \to 0} \prod_{dA_x \subset A_{\frac{1}{r_\rho}}} \left[1 - \frac{\log\rho+b}{C} dA_x g\left(\|\boldsymbol{x}-\boldsymbol{y}\|\right)\right] \\
&= \lim_{dA_x \to 0} e^{\sum_{dA_x \subset A_{\frac{1}{r_\rho}}} \log\left(1 - \frac{\log\rho+b}{C} g(\|\boldsymbol{x}-\boldsymbol{y}\|) dA_x\right)} \\
&= \lim_{dA_x \to 0} e^{\sum_{dA_x \subset A_{\frac{1}{r_\rho}}} -\frac{\log\rho+b}{C} g(\|\boldsymbol{x}-\boldsymbol{y}\|) dA_x} \\
&= e^{-\int_{A_{\frac{1}{r_\rho}}} \frac{\log\rho+b}{C} g(\|\boldsymbol{x}-\boldsymbol{y}\|) d\boldsymbol{x}}
\end{aligned} \tag{2.2.1}$$

where I_y is an indicator random variable: $I_y = 1$ if the node at $\boldsymbol{y}$, if exists, is isolated and $I_y = 0$ otherwise. The term dA_x is a differential area in $A_{\frac{1}{r_\rho}}$ and $\boldsymbol{x} \in dA_x$. In the first step, $\frac{\log\rho+b}{C} dA_x$ is the probability that there is a node in the differential area $dA_x \subset A_{\frac{1}{r_\rho}}$. The probability that there are more than one nodes in dA_x is negligibly small. Since dA_x is a differential area, the actual location of the node in dA_x does not matter and is assumed to be $\boldsymbol{x}$. The term $\frac{\log\rho+b}{C} dA_x g\left(\|\boldsymbol{x}-\boldsymbol{y}\|\right)$ is the probability that there exists a node in $dA_x \subset A_{\frac{1}{r_\rho}}$ *and* that node is directly connected to the node at $\boldsymbol{y}$. With a little abuse of the terminology, we use dA_x to denote both the differential area itself and the size of the area. In the third step, the equality $\lim_{x\to 0} \frac{\log(1-x)}{-x} = 1$ is used.

Denote by W the number of isolated nodes in an instance of $\mathcal{G}\left(\mathcal{X}_{\frac{\log\rho+b}{C}}, g, A_{\frac{1}{r_\rho}}\right)$. It then follows that the expected number of isolated nodes in $\mathcal{G}\left(\mathcal{X}_{\frac{\log\rho+b}{C}}, g, A_{\frac{1}{r_\rho}}\right)$ is given by

$$E(W) = \int_{A_{\frac{1}{r_\rho}}} \frac{\log \rho + b}{C} e^{-\int_{A_{\frac{1}{r_\rho}}} \frac{\log \rho + b}{C} g(\|x-y\|)dx} dy \tag{2.2.2}$$

On the basis of (2.2.2), the following theorem can be obtained.

Theorem 41 *The expected number of isolated nodes in* $\mathcal{G}\left(\mathcal{X}_{\frac{\log \rho + b}{C}}, g, A_{\frac{1}{r_\rho}}\right)$ *is* $\int_{A_{\frac{1}{r_\rho}}} \frac{\log \rho + b}{C} e^{-\int_{A_{\frac{1}{r_\rho}}} \frac{\log \rho + b}{C} g(\|x-y\|)dx} dy$. *For g satisfying both (2.1.1) and (2.1.4), the expected number of isolated nodes in* $\mathcal{G}\left(\mathcal{X}_{\frac{\log \rho + b}{C}}, g, A_{\frac{1}{r_\rho}}\right)$ *converges asymptotically to* e^{-b} *as* $\rho \to \infty$.

The rest of this subsection is dedicated to the proof of Theorem 41.

We analyze $E(W)$ as $\rho \to \infty$. Denote by $D\left(y, r_\rho^{-\varepsilon}\right)$ a disk centered at $y \in A_{\frac{1}{r_\rho}}$ and with a radius $r_\rho^{-\varepsilon}$, where ε is a small positive constant and $\varepsilon < \frac{1}{4}$. Denote by $B\left(A_{\frac{1}{r_\rho}}\right) \subset A_{\frac{1}{r_\rho}}$ an area within $r_\rho^{-\varepsilon}$ of the border of $A_{\frac{1}{r_\rho}}$; denote by $\ell A_{\frac{1}{r_\rho}} \subset A_{\frac{1}{r_\rho}}$ a rectangular area of size $r_\rho^{-\varepsilon} \times \left(r_\rho^{-1} - 2r_\rho^{-\varepsilon}\right)$ within $r_\rho^{-\varepsilon}$ of one side of $A_{\frac{1}{r_\rho}}$, away from the corners of $A_{\frac{1}{r_\rho}}$ by $r_\rho^{-\varepsilon}$, and there are four such areas; let $\angle A_{\frac{1}{r_\rho}} \subset A_{\frac{1}{r_\rho}}$ denote a square of size $r_\rho^{-\varepsilon} \times r_\rho^{-\varepsilon}$ at the four corners of $A_{\frac{1}{r_\rho}}$. Figure 2.1 illustrates these areas.

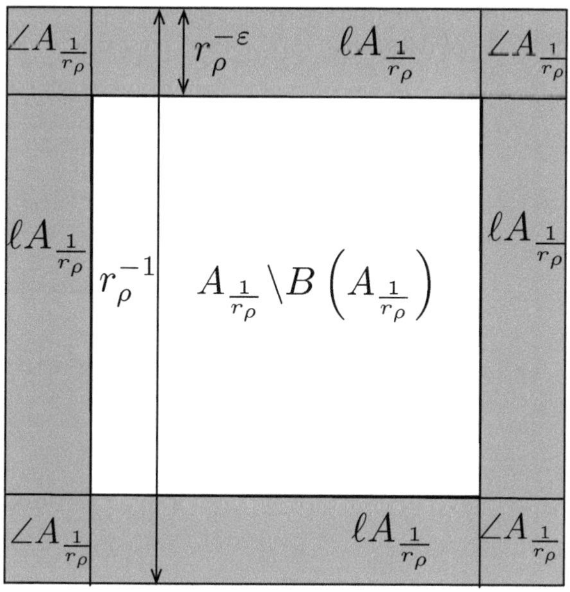

Fig. 2.1 An illustration of the boundary areas of $A_{\frac{1}{r_\rho}}$. The areas $\angle A_{\frac{1}{r_\rho}}$, $\ell A_{\frac{1}{r_\rho}}$ are self-explanatory and $B\left(A_{\frac{1}{r_\rho}}\right)$ is the shaded area in the figure

It follows from (2.2.2) that

$$
\begin{aligned}
&\lim_{\rho\to\infty} E(W) \\
&= \lim_{\rho\to\infty} \int_{A_{\frac{1}{r_\rho}}} \frac{\log\rho + b}{C} e^{-\int_{A_{\frac{1}{r_\rho}}} \frac{\log\rho+b}{C} g(\|\boldsymbol{x}-\boldsymbol{y}\|)d\boldsymbol{x}} d\boldsymbol{y} \\
&= \lim_{\rho\to\infty} \rho r_\rho^2 \int_{A_{\frac{1}{r_\rho}} \backslash B\left(A_{\frac{1}{r_\rho}}\right)} e^{-\rho r_\rho^2 \int_{A_{\frac{1}{r_\rho}}} g(\|\boldsymbol{x}-\boldsymbol{y}\|)d\boldsymbol{x}} d\boldsymbol{y} \\
&+ \lim_{\rho\to\infty} 4\rho r_\rho^2 \int_{\ell A_{\frac{1}{r_\rho}}} e^{-\rho r_\rho^2 \int_{A_{\frac{1}{r_\rho}}} g(\|\boldsymbol{x}-\boldsymbol{y}\|)d\boldsymbol{x}} d\boldsymbol{y} \\
&+ \lim_{\rho\to\infty} 4\rho r_\rho^2 \int_{\angle A_{\frac{1}{r_\rho}}} e^{-\rho r_\rho^2 \int_{A_{\frac{1}{r_\rho}}} g(\|\boldsymbol{x}-\boldsymbol{y}\|)d\boldsymbol{x}} d\boldsymbol{y} \qquad (2.2.3)
\end{aligned}
$$

The three summands in (2.2.3) represent respectively the expected number of isolated nodes in the central area $A_{\frac{1}{r_\rho}} \backslash B\left(A_{\frac{1}{r_\rho}}\right)$, in the boundary area along the four sides of $A_{\frac{1}{r_\rho}}$ and in the four corners of $A_{\frac{1}{r_\rho}}$. In the following analysis, we shall show that for g satisfying both (2.1.1) and (2.1.4), the first term approaches e^{-b} as $\rho \to \infty$, and the second and the third terms approach 0 as $\rho \to \infty$.

Consider the first summand in (2.2.3). Using the definition of r_ρ in (2.1.5), first it can be shown that for any $\boldsymbol{y} \in A_{\frac{1}{r_\rho}} \backslash B\left(A_{\frac{1}{r_\rho}}\right)$ (see Fig. 2.1 for an illustration of the region $A_{\frac{1}{r_\rho}} \backslash B\left(A_{\frac{1}{r_\rho}}\right)$):

$$
\begin{aligned}
&\lim_{\rho\to\infty} \rho e^{-\rho r_\rho^2 \int_{D\left(\boldsymbol{y}, r_\rho^{-\varepsilon}\right)} g(\|\boldsymbol{x}-\boldsymbol{y}\|)d\boldsymbol{x}} \\
&= \lim_{\rho\to\infty} \rho e^{-\rho r_\rho^2 \left(\int_{\Re^2} g(\|\boldsymbol{x}-\boldsymbol{y}\|)d\boldsymbol{x} - \int_{\Re^2 \backslash D\left(\boldsymbol{y}, r_\rho^{-\varepsilon}\right)} g(\|\boldsymbol{x}-\boldsymbol{y}\|)d\boldsymbol{x}\right)} \\
&= \lim_{\rho\to\infty} \rho e^{-\rho r_\rho^2 \left(C - \int_{\Re^2 \backslash D\left(\boldsymbol{y}, r_\rho^{-\varepsilon}\right)} g(\|\boldsymbol{x}-\boldsymbol{y}\|)d\boldsymbol{x}\right)} \\
&= \lim_{\rho\to\infty} e^{-b} e^{\rho r_\rho^2 \int_{\Re^2 \backslash D\left(\boldsymbol{y}, r_\rho^{-\varepsilon}\right)} g(\|\boldsymbol{x}-\boldsymbol{y}\|)d\boldsymbol{x}} \\
&= e^{-b} \lim_{\rho\to\infty} e^{\frac{\log\rho+b}{C} \int_{r_\rho^{-\varepsilon}}^{\infty} 2\pi r g(r)dr} \qquad (2.2.4)
\end{aligned}
$$

It can be shown further using (2.1.5) that (following the equation, detailed explanations are given):

$$
\begin{aligned}
&\lim_{\rho\to\infty} \frac{\log\rho+b}{C}\int_{r_\rho^{-\varepsilon}}^{\infty} 2\pi r g\left(r\right) dr \\
&= \lim_{\rho\to\infty} \frac{\int_{r_\rho^{-\varepsilon}}^{\infty} 2\pi r g\left(r\right) dr}{\frac{C}{\log\rho+b}} \\
&= \lim_{\rho\to\infty} \frac{-2\pi r_\rho^{-\varepsilon} g\left(r_\rho^{-\varepsilon}\right)\left(-\frac{\varepsilon}{2} r_\rho^{-\varepsilon-2}\frac{1-(\log\rho+b)}{C\rho^2}\right)}{-\frac{C}{\rho(\log\rho+b)^2}} \\
&= \lim_{\rho\to\infty} \pi\varepsilon\left(\log\rho+b\right)^2 r_\rho^{-2\varepsilon-2} g\left(r_\rho^{-\varepsilon}\right)\frac{\log\rho+b-1}{C\rho} \\
&= \lim_{\rho\to\infty} \pi\varepsilon\left(\log\rho+b\right)^2 r_\rho^{-2\varepsilon} g\left(r_\rho^{-\varepsilon}\right) && (2.2.5) \\
&= \lim_{\rho\to\infty} \pi\varepsilon\left(\log\rho+b\right)^2 r_\rho^{-2\varepsilon} o_\rho\left(\frac{1}{r_\rho^{-2\varepsilon}\log^2\left(r_\rho^{-2\varepsilon}\right)}\right) && (2.2.6) \\
&= \lim_{\rho\to\infty}\Bigg(\pi\varepsilon\left(\log\rho+b\right)^2 \\
&\quad o_\rho\left(\frac{1}{2\varepsilon^2\left(\log\left(\log\rho+b\right)-\log C-\log\rho\right)^2}\right)\Bigg) \\
&= 0 && (2.2.7)
\end{aligned}
$$

where L'Hospital's rule is used in the second step of the above equation, and $g\left(x\right) = o_x\left(\frac{1}{x^2\log^2 x}\right)$ is used from (2.2.5) to (2.2.6). As a result of (2.2.4) and (2.2.7)

$$
\lim_{\rho\to\infty} \rho e^{-\rho r_\rho^2 \int_{D\left(y, r_\rho^{-\varepsilon}\right)} g(\|\boldsymbol{x}-\boldsymbol{y}\|)d\boldsymbol{x}} = e^{-b} \tag{2.2.8}
$$

It follows that (see Fig. 2.1 for an illustration of the region $A_{\frac{1}{r_\rho}}\backslash B\left(A_{\frac{1}{r_\rho}}\right)$, which is unshaded in the figure.)

$$
\begin{aligned}
&\lim_{\rho\to\infty} \rho r_\rho^2 \int_{A_{\frac{1}{r_\rho}}\backslash B\left(A_{\frac{1}{r_\rho}}\right)} e^{-\rho r_\rho^2 \int_{A_{\frac{1}{r_\rho}}} g(\|\boldsymbol{x}-\boldsymbol{y}\|)d\boldsymbol{x}} d\boldsymbol{y} \\
&\le \lim_{\rho\to\infty} \rho r_\rho^2 \int_{A_{\frac{1}{r_\rho}}\backslash B\left(A_{\frac{1}{r_\rho}}\right)} e^{-\rho r_\rho^2 \int_{D\left(y, r_\rho^{-\varepsilon}\right)} g(\|\boldsymbol{x}-\boldsymbol{y}\|)d\boldsymbol{x}} d\boldsymbol{y}
\end{aligned}
$$

$$= \lim_{\rho\to\infty}\left(\rho e^{-\rho r_\rho^2 \int_{D\left(\mathbf{0}, r_\rho^{-\varepsilon}\right)} g(\|\mathbf{x}\|)d\mathbf{x}}\right)\left(r_\rho^2\int_{A_{\frac{1}{r_\rho}}\backslash B\left(A_{\frac{1}{r_\rho}}\right)} d\mathbf{y}\right)$$

$$= e^{-b}$$

and

$$\lim_{\rho\to\infty}\rho r_\rho^2\int_{A_{\frac{1}{r_\rho}}\backslash B\left(A_{\frac{1}{r_\rho}}\right)} e^{-\rho r_\rho^2\int_{A_{\frac{1}{r_\rho}}} g(\|\mathbf{x}-\mathbf{y}\|)d\mathbf{x}} d\mathbf{y}$$

$$\geq \lim_{\rho\to\infty}\rho r_\rho^2\int_{A_{\frac{1}{r_\rho}}\backslash B\left(A_{\frac{1}{r_\rho}}\right)} e^{-\rho r_\rho^2\int_{\Re^2} g(\|\mathbf{x}-\mathbf{y}\|)d\mathbf{x}} d\mathbf{y}$$

$$= e^{-b}$$

Therefore,

$$\lim_{\rho\to\infty}\rho r_\rho^2\int_{A_{\frac{1}{r_\rho}}/B\left(A_{\frac{1}{r_\rho}}\right)} e^{-\rho r_\rho^2\int_{A_{\frac{1}{r_\rho}}} g(\|\mathbf{x}-\mathbf{y}\|)d\mathbf{x}} d\mathbf{y} = e^{-b} \tag{2.2.9}$$

For the second term in (2.2.3), an illustration of the boundary area for $\mathbf{y} \in \ell A_{\frac{1}{r_\rho}}$ is shown in Fig. 2.2.

Define $L_y \triangleq \left(A_{\frac{1}{r_\rho}} \cap D\left(\mathbf{y}, r_\rho^{-\varepsilon}\right)\right)\backslash D_y$ (i.e., the shaded area in Fig. 2.2b). The symbols D_y, C_y, l_y, and R_y are defined in Fig. 2.2. It can be shown that

$$4\lim_{\rho\to\infty}\rho r_\rho^2\int_{\ell A_{\frac{1}{r_\rho}}} e^{-\rho r_\rho^2\int_{A_{\frac{1}{r_\rho}}} g(\|\mathbf{x}-\mathbf{y}\|)d\mathbf{x}} d\mathbf{y}$$

$$\leq 4\lim_{\rho\to\infty} r_\rho^2\int_{\ell A_{\frac{1}{r_\rho}}} \rho e^{-\rho r_\rho^2\int_{A_{\frac{1}{r_\rho}}\cap D\left(\mathbf{y}, r_\rho^{-\varepsilon}\right)} g(\|\mathbf{x}-\mathbf{y}\|)d\mathbf{x}} d\mathbf{y}$$

$$= 4\lim_{\rho\to\infty} r_\rho^2\int_{\ell A_{\frac{1}{r_\rho}}} \rho e^{-\rho r_\rho^2\left(\int_{D_y} g(\|\mathbf{x}-\mathbf{y}\|)d\mathbf{x}+\int_{L_y} g(\|\mathbf{x}-\mathbf{y}\|)d\mathbf{x}\right)} d\mathbf{y}$$

$$= 4\lim_{\rho\to\infty}\left(\left(\rho^{\frac{1}{2}} e^{-\frac{1}{2}\rho r_\rho^2\int_{D\left(\mathbf{0}, r_\rho^{-\varepsilon}\right)} g(\|\mathbf{x}\|)d\mathbf{x}}\right)\right.$$

$$\left.\left(\rho^{\frac{1}{2}} r_\rho^2\int_{\ell A_{\frac{1}{r_\rho}}} e^{-\rho r_\rho^2\int_{L_y} g(\|\mathbf{x}-\mathbf{y}\|)d\mathbf{x}} d\mathbf{y}\right)\right) \tag{2.2.10}$$

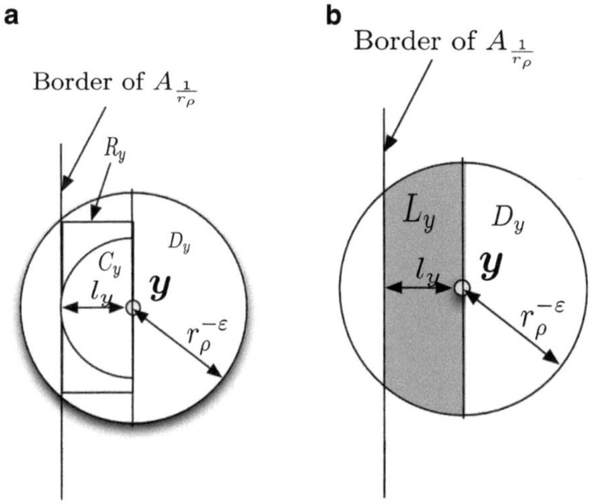

Fig. 2.2 An illustration of the boundary area for $\boldsymbol{y} \in \ell A_{\frac{1}{r_\rho}}$. The figure is drawn for $\boldsymbol{y}$ located near the left border of $A_{\frac{1}{r_\rho}}$. The situations for $\boldsymbol{y}$ near the top, bottom, and right borders of $A_{\frac{1}{r_\rho}}$ can be drawn analogously. $D\left(\boldsymbol{y}, r_\rho^{-\varepsilon}\right)$ is a disk centered at $\boldsymbol{y}$ and has a radius $r_\rho^{-\varepsilon}$. l_y is the distance between $\boldsymbol{y}$ and the border of $A_{\frac{1}{r_\rho}}$. D_y is a half disk centered at $\boldsymbol{y}$, with a radius $r_\rho^{-\varepsilon}$ and on the right side of $\boldsymbol{y}$. C_y is a half disk centered at $\boldsymbol{y}$, with a radius l_y and on the left side of $\boldsymbol{y}$. $R_y \subset A_{\frac{1}{r_\rho}} \cap D\left(\boldsymbol{y}, r_\rho^{-\varepsilon}\right)$ is a rectangle of $l_y \times 2\sqrt{r_\rho^{-2\varepsilon} - l_y^2}$ on the left side of $\boldsymbol{y}$. $L_y = \left(A_{\frac{1}{r_\rho}} \cap D\left(\boldsymbol{y}, r_\rho^{-\varepsilon}\right)\right) \backslash D_y$ is the shaded area in sub-figure b

For the first term $\rho^{\frac{1}{2}} e^{-\frac{1}{2}\rho r_\rho^2 \int_{D\left(\mathbf{0}, r_\rho^{-\varepsilon}\right)} g(\|\boldsymbol{x}\|)d\boldsymbol{x}}$ in (2.2.10), it can be shown that

$$
\begin{aligned}
&\lim_{\rho\to\infty} \rho^{\frac{1}{2}} e^{-\frac{1}{2}\rho r_\rho^2 \int_{D\left(\mathbf{0}, r_\rho^{-\varepsilon}\right)} g(\|\boldsymbol{x}\|)d\boldsymbol{x}} \\
&\quad = \lim_{\rho\to\infty} \rho^{\frac{1}{2}} e^{-\frac{1}{2}\rho r_\rho^2 \left(\int_{\Re^2} g(\|\boldsymbol{x}\|)d\boldsymbol{x} - \int_{\Re^2 \backslash D\left(\mathbf{0}, r_\rho^{-\varepsilon}\right)} g(\|\boldsymbol{x}\|)d\boldsymbol{x}\right)} \\
&\quad = \lim_{\rho\to\infty} \rho^{\frac{1}{2}} e^{-\frac{1}{2}\rho r_\rho^2 C} e^{\frac{1}{2}\rho r_\rho^2 \int_{\Re^2 \backslash D\left(\mathbf{0}, r_\rho^{-\varepsilon}\right)} g(\|\boldsymbol{x}\|)d\boldsymbol{x}} \\
&\quad = e^{-\frac{b}{2}}
\end{aligned}
\tag{2.2.11}
$$

where (2.2.7) is used in reaching (2.2.11).

Let γ be a positive constant and $\frac{1}{2} > \gamma > \frac{\varepsilon}{2}$. Let $\triangle$ be a positive constant such that

$$
\int_0^{\triangle} 2\pi x g(x)\, dx = \gamma 2C \tag{2.2.12}
$$

The existence of such a positive constant $\triangle$ can be validated by using (2.1.6) and noting that $2\gamma < 1$. Using the nonincreasing monotonicity of g, it can also be shown that $g(\triangle) > 0$; otherwise it can be shown that $\int_0^{\triangle} 2\pi x g(x)\,dx = C$ which implies $\gamma = \frac{1}{2}$. This constitutes a contradiction with the requirement that $\frac{1}{2} > \gamma > \frac{\varepsilon}{2}$. Therefore, $g(\triangle) > 0$. In the following analysis, it is assumed that ρ is sufficiently large such that $r_\rho^{-\varepsilon} \geq 2\triangle$.

For the second term in (2.2.10), it can be shown that

$$\begin{aligned}
&\lim_{\rho\to\infty} \rho^{\frac{1}{2}} r_\rho^2 \int_{\ell A_{\frac{1}{r_\rho}}} e^{-\rho r_\rho^2 \int_{L_{\boldsymbol{y}}} g(\|\boldsymbol{x}-\boldsymbol{y}\|)d\boldsymbol{x}} d\boldsymbol{y} \\
&= \lim_{\rho\to\infty} \left(\rho^{\frac{1}{2}} r_\rho^2 \left(r_\rho^{-1} - 2r_\rho^{-\varepsilon}\right)\right. \\
&\quad \times \left.\int_0^{r_\rho^{-\varepsilon}} e^{-\rho r_\rho^2 \int_{L_{\boldsymbol{y}}} g(\|\boldsymbol{x}-\boldsymbol{y}\|)d\boldsymbol{x}} dy\right) \qquad (2.2.13) \\
&\leq \lim_{\rho\to\infty} \rho^{\frac{1}{2}} r_\rho \int_0^{r_\rho^{-\varepsilon}} e^{-\rho r_\rho^2 \int_{L_{\boldsymbol{y}}} g(\|\boldsymbol{x}-\boldsymbol{y}\|)d\boldsymbol{x}} dy \\
&= \lim_{\rho\to\infty} \sqrt{\frac{\log\rho + b}{C}} \int_0^{r_\rho^{-\varepsilon}} e^{-\rho r_\rho^2 \int_{L_{\boldsymbol{y}}} g(\|\boldsymbol{x}-\boldsymbol{y}\|)d\boldsymbol{x}} dy \\
&= \lim_{\rho\to\infty} \sqrt{\frac{\log\rho + b}{C}} \left(\int_0^{\triangle} e^{-\rho r_\rho^2 \int_{L_{\boldsymbol{y}}} g(\|\boldsymbol{x}-\boldsymbol{y}\|)d\boldsymbol{x}} dy\right. \\
&\quad + \left.\int_{\triangle}^{r_\rho^{-\varepsilon}} e^{-\rho r_\rho^2 \int_{L_{\boldsymbol{y}}} g(\|\boldsymbol{x}-\boldsymbol{y}\|)d\boldsymbol{x}} dy\right) \qquad (2.2.14)
\end{aligned}$$

where in (2.2.13) $\boldsymbol{y}$ represents a (any) point in ℓA_ρ at a Euclidean distance $y \in [0, r_\rho^{-\varepsilon}]$ apart from the border of A_ρ. Define $\lambda \triangleq \frac{\log\rho+b}{C}$ for convenience, it can be further shown that in (2.2.14)

$$\begin{aligned}
&\lim_{\rho\to\infty} \sqrt{\lambda} \int_{\triangle}^{r_\rho^{-\varepsilon}} e^{-\rho r_\rho^2 \int_{L_{\boldsymbol{y}}} g(\|\boldsymbol{x}-\boldsymbol{y}\|)d\boldsymbol{x}} dy \\
&\leq \lim_{\rho\to\infty} \sqrt{\lambda} \int_{\triangle}^{r_\rho^{-\varepsilon}} e^{-\rho r_\rho^2 \int_{C_{\boldsymbol{y}}} g(\|\boldsymbol{x}-\boldsymbol{y}\|)d\boldsymbol{x}} dy \\
&= \lim_{\rho\to\infty} \sqrt{\lambda} \int_{\triangle}^{r_\rho^{-\varepsilon}} e^{-\frac{1}{2}\rho r_\rho^2 \int_0^y 2\pi x g(x)dx} dy \\
&= \lim_{\rho\to\infty} \sqrt{\lambda} \int_{\triangle}^{r_\rho^{-\varepsilon}} e^{-\frac{1}{2}\rho r_\rho^2 \left(\int_0^{\triangle} 2\pi x g(x)dx + \int_{\triangle}^{y} 2\pi x g(x)dx\right)} dy
\end{aligned}$$

$$\leq \lim_{\rho\to\infty} \sqrt{\lambda} \int_{\Delta}^{r_\rho^{-\varepsilon}} e^{-\frac{1}{2}\rho r_\rho^2 \int_0^\Delta 2\pi x g(x)dx} dy$$

$$= \lim_{\rho\to\infty} \sqrt{\lambda} \int_{\Delta}^{r_\rho^{-\varepsilon}} e^{-\gamma(\log\rho+b)} dy \tag{2.2.15}$$

$$= \lim_{\rho\to\infty} \sqrt{\lambda} \left(e^{-\gamma b} \rho^{-\gamma} \left(\left(\frac{\log\rho+b}{C\rho} \right)^{-\frac{\varepsilon}{2}} - \Delta \right) \right)$$

$$= 0 \tag{2.2.16}$$

where (2.2.12) is used in reaching (2.2.15), and $\gamma > \frac{\varepsilon}{2}$ is used in reaching (2.2.16). It can also be shown that for the other term in (2.2.14),

$$\lim_{\rho\to\infty} \sqrt{\frac{\log\rho+b}{C}} \int_0^\Delta e^{-\rho r_\rho^2 \int_{L_y} g(\|\mathbf{x}-\mathbf{y}\|)d\mathbf{x}} dy$$

$$\leq \lim_{\rho\to\infty} \sqrt{\lambda} \int_0^\Delta e^{-\rho r_\rho^2 \int_{R_y} g(\|\mathbf{x}-\mathbf{y}\|)d\mathbf{x}} dy \tag{2.2.17}$$

$$= \lim_{\rho\to\infty} \sqrt{\lambda} \int_0^\Delta e^{-\rho r_\rho^2 2 \int_0^y \int_0^{\sqrt{r_\rho^{-2\varepsilon}-x^2}} g(\sqrt{x^2+z^2})dzdx} dy$$

$$\leq \lim_{\rho\to\infty} \sqrt{\lambda} \int_0^\Delta e^{-\rho r_\rho^2 2 \int_0^y \int_0^{r_\rho^{-\varepsilon}-x} g(\sqrt{x^2+z^2})dzdx} dy \tag{2.2.18}$$

$$\leq \lim_{\rho\to\infty} \sqrt{\lambda} \int_0^\Delta e^{-\rho r_\rho^2 2 \int_0^y \int_0^{r_\rho^{-\varepsilon}-\Delta} g(\sqrt{x^2+z^2})dzdx} dy \tag{2.2.19}$$

$$\leq \lim_{\rho\to\infty} \sqrt{\lambda} \int_0^\Delta e^{-\rho r_\rho^2 2 \int_0^y \int_0^{r_\rho^{-\varepsilon}-\Delta} g(z+\Delta)dzdx} dy \tag{2.2.20}$$

$$= \lim_{\rho\to\infty} \sqrt{\lambda} \int_0^\Delta e^{-\rho r_\rho^2 2y \int_0^{r_\rho^{-\varepsilon}-\Delta} g(z+\Delta)dz} dy \tag{2.2.21}$$

where (2.2.18) is obtained by noting that $r_\rho^{-2\varepsilon} - x^2 \geq \left(r_\rho^{-\varepsilon} - x\right)^2$ for $r_\rho^{-\varepsilon} \geq x$ (note that for ρ sufficiently large, $r_\rho^{-\varepsilon} > \Delta \geq y \geq x$), (2.2.19) is obtained by noting that $x \leq \Delta$ and (2.2.20) is obtained by noting that $y \leq \Delta$ and the nonincreasing monotonicity of g.

Let ρ be sufficiently large such that $r_\rho^{-\varepsilon} \geq 2\Delta$ and also note that $g(\Delta) > 0$. Therefore, $\beta \triangleq \int_0^\Delta g(z+\Delta)\,dz$ is a positive constant and $\beta > 0$. It then follows from (2.2.21) that

$$
\begin{aligned}
&\lim_{\rho\to\infty}\sqrt{\frac{\log\rho+b}{C}}\int_0^{\Delta} e^{-\rho r_\rho^2\int_{L_y} g(\|x-y\|)dx}dy\\
&\le \lim_{\rho\to\infty}\sqrt{\frac{\log\rho+b}{C}}\int_0^{\Delta} e^{-\rho r_\rho^2 2\beta y}dy\\
&= \lim_{\rho\to\infty}\sqrt{\frac{\log\rho+b}{C}}\times\frac{1-e^{-\rho r_\rho^2 2\beta\Delta}}{\rho r_\rho^2 2\beta}\\
&= \lim_{\rho\to\infty}\sqrt{\frac{\log\rho+b}{C}}\times\frac{1-e^{-2\beta\Delta\frac{\log\rho+b}{C}}}{2\beta\frac{\log\rho+b}{C}}\\
&=0
\end{aligned}
\tag{2.2.22}
$$

As a result of (2.2.16) and (2.2.22), both terms on the right-hand side of (2.2.14) go to zero. It follows that

$$
\lim_{\rho\to\infty}\rho^{\frac{1}{2}}r_\rho^2\int_{\ell A_\rho} e^{-\rho r_\rho^2\int_{L_y} g(\|x-y\|)dx}dy=0
$$

The above equation, together with (2.2.10) and (2.2.11), leads to the conclusion that

$$
4\lim_{\rho\to\infty}\rho r_\rho^2\int_{\ell A_\rho} e^{-\rho r_\rho^2\int_{A_\rho} g(\|x-y\|)dx}dy=0 \tag{2.2.23}
$$

i.e., the second term in (2.2.3) approaches 0 as $\rho\to\infty$.

For the third term in (2.2.3), it can be shown that

$$
\begin{aligned}
&4\lim_{\rho\to\infty}\rho r_\rho^2\int_{\angle A_{\frac{1}{r_\rho}}} e^{-\rho r_\rho^2\int_{A_{\frac{1}{r_\rho}}} g(\|x-y\|)dx}dy\\
&\le 4\lim_{\rho\to\infty}\rho r_\rho^2\int_{\angle A_{\frac{1}{r_\rho}}} e^{-\rho r_\rho^2\int_{A_{\frac{1}{r_\rho}}\cap D\left(y,r_\rho^{-\varepsilon}\right)} g(\|x-y\|)dx}dy\\
&\le 4\lim_{\rho\to\infty}\rho r_\rho^2\int_{\angle A_{\frac{1}{r_\rho}}} e^{-\frac{1}{4}\rho r_\rho^2\int_{D\left(y,r_\rho^{-\varepsilon}\right)} g(\|x-y\|)dx}dy\\
&=4\lim_{\rho\to\infty}\left(r_\rho^{-\varepsilon}\right)^2 r_\rho^2\rho e^{-\frac{1}{4}\rho r_\rho^2\int_{D\left(y,r_\rho^{-\varepsilon}\right)} g(\|x-y\|)dx}\\
&=4\lim_{\rho\to\infty} r_\rho^{2-2\varepsilon}\rho e^{-\frac{1}{4}\rho r_\rho^2\left(C-\int_{\Re^2\backslash D\left(y,r_\rho^{-\varepsilon}\right)} g(\|x-y\|)dx\right)}\\
&=4\lim_{\rho\to\infty}\left(\frac{\log\rho+b}{C\rho}\right)^{1-\varepsilon}\rho e^{-\frac{1}{4}(\log\rho+b)} && (2.2.24)\\
&=4C^{-1+\varepsilon}e^{-\frac{1}{4}b}\lim_{\rho\to\infty}\frac{(\log\rho+b)^{1-\varepsilon}}{\rho^{\frac{1}{4}-\varepsilon}}\\
&=0 && (2.2.25)
\end{aligned}
$$

where the second step results by noting that for any $\boldsymbol{y} \in \angle A_{\frac{1}{r_\rho}}$, $A_\rho \cap D\left(\boldsymbol{y}, r_\rho^{-\varepsilon}\right)$ covers at least one quarter of $D\left(\boldsymbol{y}, r_\rho^{-\varepsilon}\right)$, (2.2.7) is used in reaching (2.2.24), and $\varepsilon < \frac{1}{4}$ is used in the final step.

As a result of (2.2.3), (2.2.9), (2.2.23), and (2.2.25):

$$\lim_{\rho\to\infty} E\left(W\right) = e^{-b} \tag{2.2.26}$$

Hence Theorem 41 is proved.

2.2.2 *Impact of Boundary Effect on the Number of Isolated Nodes*

Before we proceed to the comparison of the expected number of isolated nodes in $\mathcal{G}\left(\mathcal{X}_{\frac{\log\rho+b}{C}}, g, A_{\frac{1}{r_\rho}}\right)$ and the expected number in its counterpart in an infinite network, we first examine the impact of boundary effect on the number of isolated nodes in $\mathcal{G}\left(\mathcal{X}_{\frac{\log\rho+b}{C}}, g, A_{\frac{1}{r_\rho}}\right)$. Boundary effect is a common concern in the analysis of network connectivity. The analysis of the impact of the boundary effect is done by comparing the number of isolated nodes in $\mathcal{G}\left(\mathcal{X}_{\frac{\log\rho+b}{C}}, g, A_{\frac{1}{r_\rho}}\right)$ and that in a network with nodes Poissonly distributed on a torus $A^T_{\frac{1}{r_\rho}} \triangleq \left[-\frac{1}{2r_\rho}, \frac{1}{2r_\rho}\right]^2$ with node density $\frac{\log\rho+b}{C}$ and where a pair of nodes separated by a *toroidal distance* x^T are directly connected with probability $g\left(x^T\right)$, independent of the event that another distinct pair of nodes are directly connected. Denote the network on a torus by $\mathcal{G}^T\left(\mathcal{X}_{\frac{\log\rho+b}{C}}, g, A^T_{\frac{1}{r_\rho}}\right)$. The following lemma can be established.

Lemma 42 *The expected number of isolated nodes in* $\mathcal{G}^T\left(\mathcal{X}_{\frac{\log\rho+b}{C}}, g, A^T_{\frac{1}{r_\rho}}\right)$ *is* $\rho e^{-\int_{A_{\frac{1}{r_\rho}}} \frac{\log\rho+b}{C} g(\|\boldsymbol{x}\|)d\boldsymbol{x}}$. *For g satisfying both (2.1.1) and (2.1.4), the expected number of isolated nodes in* $\mathcal{G}^T\left(\mathcal{X}_{\frac{\log\rho+b}{C}}, g, A_{\frac{1}{r_\rho}}\right)$ *converges to* e^{-b} *as* $\rho \to \infty$.

Proof As discussed in Sect. 1.3.1, the torus that is commonly discussed in random geometric graph theory is essentially the same as a square except that the distance between two points on a torus is defined by their *toroidal distance*, instead of Euclidean distance. From now on, whenever the difference between a torus and a square affects the parameter being discussed, we use superscript ^{T} to mark the parameter in a torus.

We note the following relation between toroidal distance and Euclidean distance on a square area centered at the origin:

$$\|\boldsymbol{x}_1-\boldsymbol{x}_2\|^T \le \|\boldsymbol{x}_1-\boldsymbol{x}_2\| \tag{2.2.27}$$

$$\|\boldsymbol{x}\|^T = \|\boldsymbol{x}\| \tag{2.2.28}$$

which will be used in the later analysis.

It can then be shown that for an arbitrary node in $\mathcal{G}^T\left(\mathcal{X}_{\frac{\log\rho+b}{C}}, g, A^T_{\frac{1}{r_\rho}}\right)$ at location $\boldsymbol{y}$, the probability that the node is isolated is given by:

$$\begin{aligned}\Pr\left(I_y^T=1\right) &= e^{-\int_{A^T_{\frac{1}{r_\rho}}}\frac{\log\rho+b}{C}g\left(\|\boldsymbol{x}-\boldsymbol{y}\|^T\right)d\boldsymbol{x}}\\ &= e^{-\int_{A^T_{\frac{1}{r_\rho}}}\frac{\log\rho+b}{C}g\left(\|\boldsymbol{x}\|^T\right)d\boldsymbol{x}}\\ &= e^{-\int_{A_{\frac{1}{r_\rho}}}\frac{\log\rho+b}{C}g(\|\boldsymbol{x}\|)d\boldsymbol{x}}\end{aligned}$$

where in the second step, the property of a torus that the probability that an arbitrary node at location $\boldsymbol{y}$ is isolated is equal to the probability that a node at the origin is isolated is used; in the third step (2.2.28) is used.

Thus the expected number of isolated nodes in $\mathcal{G}^T\left(\mathcal{X}_{\frac{\log\rho+b}{C}}, g, A^T_{\frac{1}{r_\rho}}\right)$ is given by

$$E\left(W^T\right)$$

$$= \int_{A_{\frac{1}{r_\rho}}}\frac{\log\rho+b}{C}e^{-\int_{A_{\frac{1}{r_\rho}}}\frac{\log\rho+b}{C}g(\|\boldsymbol{x}\|)d\boldsymbol{x}}d\boldsymbol{y} \tag{2.2.29}$$

$$= \frac{1}{r_\rho^2}\frac{\log\rho+b}{C}e^{-\int_{A_{\frac{1}{r_\rho}}}\frac{\log\rho+b}{C}g(\|\boldsymbol{x}\|)d\boldsymbol{x}} \tag{2.2.30}$$

$$= \rho e^{-\int_{A_{\frac{1}{r_\rho}}}\frac{\log\rho+b}{C}g(\|\boldsymbol{x}\|)d\boldsymbol{x}} \tag{2.2.31}$$

First, it can be shown using (2.1.6) that for g satisfying (2.1.4)

$$\begin{aligned}&\lim_{\rho\to\infty}\rho e^{-\int_{D\left(\boldsymbol{0},r_\rho^{-\varepsilon}\right)}\frac{\log\rho+b}{C}g(\|\boldsymbol{x}\|)d\boldsymbol{x}}\\ &\quad= \lim_{\rho\to\infty}\rho e^{-\frac{\log\rho+b}{C}\left(C-\int_{\Re^2\backslash D\left(\boldsymbol{0},r_\rho^{-\varepsilon}\right)}g(\|\boldsymbol{x}\|)d\boldsymbol{x}\right)}\end{aligned}$$

$$= e^{-b} \lim_{\rho\to\infty} e^{\frac{\log\rho+b}{C}\int_{r_\rho^{-\varepsilon}}^{\infty} 2\pi x g(x)dx}$$

$$= e^{-b} \tag{2.2.32}$$

where $D(\mathbf{0}, x)$ denotes a disk centered at the origin and with a radius x, ε is a small positive constant, and the last step results because

$$\begin{aligned}
&\lim_{\rho\to\infty} \frac{\int_{r_\rho^{-\varepsilon}}^{\infty} 2\pi x g(x)\,dx}{\frac{1}{\log\rho+b}} \\
&= \lim_{\rho\to\infty} \frac{\pi\varepsilon r_\rho^{-\varepsilon} g\left(r_\rho^{-\varepsilon}\right) r_\rho^{-\varepsilon-2} \frac{\log\rho+b-1}{C\rho^2}}{\frac{1}{\rho(\log\rho+b)^2}} \qquad (2.2.33)\\
&= \lim_{\rho\to\infty} \pi\varepsilon(\log\rho+b)^2 r_\rho^{-2\varepsilon} o_\rho\left(\frac{1}{r_\rho^{-2\varepsilon}\log^2\left(r_\rho^{-2\varepsilon}\right)}\right) \\
&= 0
\end{aligned}$$

where L'Hospital's rule is used in reaching (2.2.33) and in the third step $g(x) = o_x\left(\frac{1}{x^2\log^2 x}\right)$ is used. Note that by the definition of C in (2.1.6),

$$\rho e^{-\int_{\Re^2}\frac{\log\rho+b}{C}g(\|\boldsymbol{x}\|)d\boldsymbol{x}} = e^{-b} \tag{2.2.34}$$

and

$$\begin{aligned}
&\rho e^{-\int_{\Re^2}\frac{\log\rho+b}{C}g(\|\boldsymbol{x}\|)d\boldsymbol{x}} \\
&\le \rho e^{-\int_{A_{\frac{1}{r_\rho}}}\frac{\log\rho+b}{C}g(\|\boldsymbol{x}\|)d\boldsymbol{x}} \\
&\le \rho e^{-\int_{D\left(\mathbf{0},r_\rho^{-\varepsilon}\right)}\frac{\log\rho+b}{C}g(\|\boldsymbol{x}\|)d\boldsymbol{x}} \qquad (2.2.35)
\end{aligned}$$

As a result of (2.2.29), (2.2.32), (2.2.34), and (2.2.35)

$$\lim_{\rho\to\infty} E\left(W^T\right) = e^{-b} \tag{2.2.36}$$

□

On the basis of Theorem 41 and Lemma 42, and using the coupling technique, the following lemma can be obtained.

Lemma 43 *For g satisfying both (2.1.1) and (2.1.4), the number of isolated nodes in* $\mathcal{G}\left(\mathcal{X}_{\frac{\log\rho+b}{C}}, g, A_{\frac{1}{r_\rho}}\right)$ *due to the boundary effect is asymptotically almost surely* 0 *as* $\rho \to \infty$.

Proof Comparing Theorem 41 and Lemma 42, it is noted that the expected numbers of isolated nodes on a torus and on a square respectively asymptotically converge to the same *nonzero finite constant* e^{-b} as $\rho \to \infty$. Now we use the coupling technique [143] to construct the connection between W and W^T, the number of isolated nodes in the corresponding instance of $\mathcal{G}^T\left(\mathcal{X}_{\frac{\log\rho+b}{C}}, g, A_{\frac{1}{r_\rho}}\right)$. Consider an instance of $\mathcal{G}^T\left(\mathcal{X}_{\frac{\log\rho+b}{C}}, g, A^T_{\frac{1}{r_\rho}}\right)$. The number of isolated nodes in that network is W^T, which depends on ρ. Remove each connection of the above network with probability $1 - \frac{g(x)}{g(x^T)}$, independent of the event that another connection is removed, where x is the Euclidean distance between the two endpoints of the connection and x^T is the corresponding toroidal distance. Due to $x^T \le x$ (see (2.2.27)) and the nonincreasing property of g, $0 \le 1 - \frac{g(x)}{g(x^T)} \le 1$. Further note that only connections between nodes near the boundary with $x^T < x$ will be affected, i.e., when $x = x^T$ the removal probability is zero. Denote the number of *newly* appearing isolated nodes by W^E. W^E has the meaning of being *the number of isolated nodes due to the boundary effect*. It is straightforward to show that W^E is a *nonnegative random integer*, depending on ρ. Furthermore, such a connection removal process results a random network with nodes Poissonly distributed with density $\frac{\log\rho+b}{C}$ where a pair of nodes separated by a *Euclidean* distance x are directly connected with probability $g(x)$, i.e., a random network on a square with the boundary effect included. The following equation results as a consequence of the above discussion:

$$W = W^E + W^T$$

Using Theorem 41, Lemma 42, and the above equation, it can be shown that

$$\lim_{\rho\to\infty} E\left(W^E\right) = \lim_{\rho\to\infty} E\left(W - W^T\right) = 0$$

Due to the nonnegativity of W^E:

$$\lim_{\rho\to\infty} \Pr\left(W^E = 0\right) = 1$$

□

Remark 44 Note that for g not satisfying (2.1.4), $E(W)$ and $E\left(W^T\right)$ are not necessarily convergent as $\rho \to \infty$. Particularly, using the same analysis in the previous two subsections (see also (2.2.42) in Sect. 2.2.4 below), it can be shown that when $g(x) = \omega_x\left(\frac{1}{x^2\log^2 x}\right)$, both $\lim_{\rho\to\infty} E(W)$ and $\lim_{\rho\to\infty} E\left(W^T\right)$

are unbounded. When $g(x) = \Theta_x\left(\frac{1}{x^2\log^2 x}\right)$, $\lim_{\rho\to\infty} E(W)$ and $\lim_{\rho\to\infty} E\left(W^T\right)$ start to depend on the asymptotic behavior of g and is only convergent when $\lim_{x\to\infty} g(x)\, x^2 \log^2 x = a$, where $0 < a < \infty$ is a positive constant. In that case, it can be shown that $\lim_{\rho\to\infty} E(W)$ and $\lim_{\rho\to\infty} E\left(W^T\right)$ converge to $e^{-b+\frac{4\pi}{C}a}$. For $\lim_{\rho\to\infty} E\left(W^T\right)$ the above result can be established by first choosing a small positive constant $\Delta\varepsilon$ and then letting ρ be sufficiently large such that $D\left(\mathbf{0}, \frac{1}{2}r_\rho^{-1-\Delta\varepsilon}\right)$ contains $A_{\frac{1}{r_\rho}}$, where $D(\boldsymbol{x}, r)$ denotes a disk centered at $\boldsymbol{x}$ and with a radius r. An upper and lower bound on $E\left(W^T\right)$ can then be established by noting that

$$\begin{aligned}
&\lim_{\rho\to\infty} \rho e^{-\int_{D\left(\mathbf{0},\frac{1}{2}r_\rho^{-1-\Delta\varepsilon}\right)} \frac{\log\rho+b}{C} g(\|\boldsymbol{x}\|)dx} \\
&\leq \lim_{\rho\to\infty} E\left(W^T\right) = \rho e^{-\int_{A_{\frac{1}{r_\rho}}} \frac{\log\rho+b}{C} g(\|\boldsymbol{x}\|)dx} \\
&\leq \lim_{\rho\to\infty} \rho e^{-\int_{D\left(\mathbf{0},\frac{1}{2}r_\rho^{-1}\right)} \frac{\log\rho+b}{C} g(\|\boldsymbol{x}\|)dx}
\end{aligned}$$

Following the exactly same procedure as that in (2.2.32) and (2.2.33) and finally letting $\Delta\varepsilon \to 0$, the result for $\lim_{\rho\to\infty} E\left(W^T\right)$ can be obtained. The result for $\lim_{\rho\to\infty} E(W)$ can be obtained following a similar procedure as that in Sect. 2.2.1.

2.2.3 The Number of Isolated Nodes in a Region $A_{\frac{1}{r_\rho}}$ of an Infinite Network with Node Density $\frac{\log\rho+b}{C}$

In this section, we consider the number of isolated nodes in the counterpart of $\mathcal{G}\left(\mathcal{X}_{\frac{\log\rho+b}{C}}, g, A_{\frac{1}{r_\rho}}\right)$ in an infinite network. Specifically, for a meaningful comparison with the number of isolated nodes in $\mathcal{G}\left(\mathcal{X}_{\frac{\log\rho+b}{C}}, g, A_{\frac{1}{r_\rho}}\right)$, we consider the number of isolated nodes, denoted by W^∞ (with superscript $^\infty$ marking the parameter in an infinite network), in a square $A_{\frac{1}{r_\rho}}$ of an infinite network on $\Re^2$ with Poissonly distributed node at density $\frac{\log\rho+b}{C}$. Denote the infinite network by $\mathcal{G}\left(\mathcal{X}_{\frac{\log\rho+b}{C}}, g, \Re^2\right)$. For g satisfying (2.1.2), a randomly chosen node in $\mathcal{G}\left(\mathcal{X}_{\frac{\log\rho+b}{C}}, g, \Re^2\right)$, at location $\boldsymbol{y} \in A_{\frac{1}{r_\rho}}$, is isolated with probability

$$\Pr\left(I_y^\infty = 1\right) = e^{-\int_{\Re^2} \frac{\log\rho+b}{C} g(\|\boldsymbol{x}-\boldsymbol{y}\|)dx} = \frac{1}{\rho}e^{-b} \tag{2.2.37}$$

where (2.1.2) is used in the above equation. Therefore,

$$\begin{aligned} E\left(W^{\infty}\right) &= \int_{A_{\frac{1}{r_\rho}}} \frac{\log\rho + b}{C} \times \frac{1}{\rho} e^{-b} dy \\ &= \frac{\log\rho + b}{C} \times \frac{1}{\rho} e^{-b} \times \left(\frac{1}{r_\rho}\right)^2 \\ &= e^{-b} \end{aligned} \tag{2.2.38}$$

The last line follows by (2.1.5).

The above result is summarized in the following lemma:

Lemma 45 *For g satisfying (2.1.2), the expected number of isolated nodes in a region $A_{\frac{1}{r_\rho}}$ of $\mathcal{G}\left(\mathcal{X}_{\frac{\log\rho+b}{C}}, g, \Re^2\right)$ is e^{-b}.*

Remark 46 Indeed only the integral boundedness requirement on g in (2.1.2) is required for Lemma 45. However such difference does not affect the latter discussion on the key focus of the chapter, i.e., the truncation effect may have a non-negligible impact on the number of isolated nodes. Therefore, we choose to omit such difference to focus on the main idea.

2.2.4 A Comparison of the Expected Number of Isolated Nodes in $\mathcal{G}\left(\mathcal{X}_{\frac{\log\rho+b}{C}}, g, A_{\frac{1}{r_\rho}}\right)$ and in Its Counterpart in an Infinite Network

Comparing Theorem 41 and Lemma 45, we note that:

(1) The expected number of isolated nodes in $\mathcal{G}\left(\mathcal{X}_{\frac{\log\rho+b}{C}}, g, A_{\frac{1}{r_\rho}}\right)$ only *converges asymptotically* to e^{-b} as $\rho \to \infty$ whereas the expected number of isolated nodes in an area of the same size in $\mathcal{G}\left(\mathcal{X}_{\frac{\log\rho+b}{C}}, g, \Re^2\right)$ *is always* e^{-b} no matter which value ρ takes.

(2) The expected number of isolated nodes in $\mathcal{G}\left(\mathcal{X}_{\frac{\log\rho+b}{C}}, g, A_{\frac{1}{r_\rho}}\right)$ converges asymptotically to e^{-b} for g *satisfying both (2.1.1) and (2.1.4)* whereas the expected number of isolated nodes in an area of the same size in $\mathcal{G}\left(\mathcal{X}_{\frac{\log\rho+b}{C}}, g, \Re^2\right)$ is e^{-b} for g *satisfying (2.1.2) only*.

In the following we examine the reason behind the differences.

Using (2.2.1), (2.2.2), (2.2.37), and (2.2.38), it can be shown that

$$
\begin{aligned}
&\frac{E(W)}{E(W^{\infty})} \\
&= e^{b} \int_{A_{\frac{1}{r_\rho}}} \frac{\log \rho + b}{C} e^{-\int_{A_{\frac{1}{r_\rho}}} \frac{\log \rho + b}{C} g(\|x-y\|)dx} dy \\
&= e^{b} \int_{A_{\frac{1}{r_\rho}}} \frac{\log \rho + b}{C} \exp\left(-\int_{\Re^2} \frac{\log \rho + b}{C} g(\|x-y\|)\,dx\right. \\
&\quad \left. + \int_{\Re^2 \setminus A_{\frac{1}{r_\rho}}} \frac{\log \rho + b}{C} g(\|x-y\|)\,dx\right) dy \\
&= \int_{A_{\frac{1}{r_\rho}}} \frac{\log \rho + b}{C\rho} e^{\int_{\Re^2 \setminus A_{\frac{1}{r_\rho}}} \frac{\log \rho + b}{C} g(\|x-y\|)dx} dy
\end{aligned}
\tag{2.2.39}
$$

It is trivial to show that the value in (2.2.39) is always greater than 1 for g with infinite support. That is, for any g with infinite support, the expected number of isolated nodes in $\mathcal{G}\left(\mathcal{X}_{\frac{\log\rho+b}{C}}, g, A_{\frac{1}{r_\rho}}\right)$ is strictly larger than the expected number of isolated nodes in an area $A_{\frac{1}{r_\rho}}$ of $\mathcal{G}\left(\mathcal{X}_{\frac{\log\rho+b}{C}}, g, \Re^2\right)$. Furthermore, it can be shown that the value in (2.2.39) accounts for the cumulative effect of nodes outside $A_{\frac{1}{r_\rho}}$ in $\mathcal{G}\left(\mathcal{X}_{\frac{\log\rho+b}{C}}, g, \Re^2\right)$ and the associated connections between these nodes and nodes inside $A_{\frac{1}{r_\rho}}$ on reducing the expected number of isolated nodes in $A_{\frac{1}{r_\rho}}$. Because $\mathcal{G}\left(\mathcal{X}_{\frac{\log\rho+b}{C}}, g, A_{\frac{1}{r_\rho}}\right)$ can be obtained from $\mathcal{G}\left(\mathcal{X}_{\frac{\log\rho+b}{C}}, g, \Re^2\right)$ by removing all these nodes and associated connections outside an area of $A_{\frac{1}{r_\rho}}$ in $\mathcal{G}\left(\mathcal{X}_{\frac{\log\rho+b}{C}}, g, \Re^2\right)$, we term this distinction *the truncation effect*. Theorem 41 and Lemma 45 show that when g satisfies both (2.1.1) and (2.1.4) (i.e., g has to decrease fast enough), the impact of the truncation effect on isolated nodes becomes vanishingly small as $\rho \to \infty$.

Based on the above discussion, the following theorem can be established:

Theorem 47 *For g satisfying (2.1.2), the expected number of isolated nodes in an area of* $A_{\frac{1}{r_\rho}}$ *in* $\mathcal{G}\left(\mathcal{X}_{\frac{\log\rho+b}{C}}, g, \Re^2\right)$ *is* e^{-b}. *Removing all nodes of* $\mathcal{G}\left(\mathcal{X}_{\frac{\log\rho+b}{C}}, g, \Re^2\right)$ *outside* $A_{\frac{1}{r_\rho}}$ *and the associated connections, there results* $\mathcal{G}\left(\mathcal{X}_{\frac{\log\rho+b}{C}}, g, A_{\frac{1}{r_\rho}}\right)$. *The expected number of isolated nodes in* $\mathcal{G}\left(\mathcal{X}_{\frac{\log\rho+b}{C}}, g, A_{\frac{1}{r_\rho}}\right)$ *converges to* e^{-b} *if g satisfies both (2.1.1) and (2.1.4). The more restrictive requirement on g is a sufficient condition for the impact of the* truncation effect *associated with the above removal operations on the number of isolated nodes in* $\mathcal{G}\left(\mathcal{X}_{\frac{\log\rho+b}{C}}, g, A_{\frac{1}{r_\rho}}\right)$ *to be vanishingly small as* $\rho \to \infty$.

In the following, we show that the more restrictive requirement on g in (2.1.4) (compared with (2.1.1) and (2.1.2)) is also necessary for the impact of the truncation effect to become vanishingly small as $\rho \to \infty$. Specifically, consider the case when (2.1.4) is not satisfied. Let

$$f(x) \triangleq g(x)\, x^2 \log^2 x \tag{2.2.40}$$

Condition (2.1.4) not being satisfied means

$$\lim_{x\to\infty} f(x) \neq 0 \tag{2.2.41}$$

i.e., $\lim_{x\to\infty} f(x)$ may equal to a positive constant, ∞, or does not exist (e.g., $f(x)$ is a periodic function of x).

It can be shown that (following the equation, detailed explanations are given and see also (2.2.29))

$$\begin{aligned}
&\lim_{\rho\to\infty} E(W)\\
&\quad\geq \lim_{\rho\to\infty} E\left(W^T\right)\\
&\quad= \lim_{\rho\to\infty} \rho e^{-\int_{A_{\frac{1}{r_\rho}}} \frac{\log\rho+b}{C} g(\|\boldsymbol{x}\|)d\boldsymbol{x}}\\
&\quad\geq \lim_{\rho\to\infty} \rho e^{-\int_{D\left(\mathbf{0},\frac{1}{2}r_\rho^{-1}\right)} \frac{\log\rho+b}{C} g(\|\boldsymbol{x}\|)d\boldsymbol{x}}\\
&\quad= e^{-b} \lim_{\rho\to\infty} e^{\int_{\Re^2\setminus D\left(\mathbf{0},\frac{1}{2}r_\rho^{-1}\right)} \frac{\log\rho+b}{C} g(\|\boldsymbol{x}\|)d\boldsymbol{x}}\\
&\quad= e^{-b+\frac{4\pi}{C}\lim_{x\to\infty} f(x)}
\end{aligned} \tag{2.2.42}$$

where the last step results because of the following derivations:

$$\begin{aligned}
&\int_{\Re^2\setminus D\left(\mathbf{0},\frac{1}{2}r_\rho^{-1}\right)} \frac{\log\rho+b}{C} g(\|\boldsymbol{x}\|)\, d\boldsymbol{x}\\
&\quad= \lim_{\rho\to\infty} \int_{\frac{1}{2}r_\rho^{-1}}^{\infty} \frac{\log\rho+b}{C} 2\pi x g(x)\, dx\\
&\quad= \lim_{\rho\to\infty} \frac{\frac{\pi}{2} r_\rho^{-4} g\left(\frac{1}{2}r_\rho^{-1}\right) \frac{\log\rho+b-1}{C\rho^2}}{\frac{C}{\rho(\log\rho+b)^2}}\\
&\quad= \lim_{\rho\to\infty} \frac{\pi}{2C} (\log\rho+b)^2\, r_\rho^{-2} g\left(\frac{1}{2}r_\rho^{-1}\right)
\end{aligned}$$

$$= \lim_{\rho\to\infty} \frac{\pi}{2C} (\log\rho + b)^2 r_\rho^{-2} \frac{f\left(\frac{1}{2} r_\rho^{-1}\right)}{\frac{1}{4} r_\rho^{-2} \log^2 \left(\frac{1}{2} r_\rho^{-1}\right)}$$

$$= \lim_{\rho\to\infty} \frac{2\pi (\log\rho + b)^2 f\left(\frac{1}{2} r_\rho^{-1}\right)}{C\left(\log\frac{1}{2} - \frac{1}{2}\log(\log\rho + b) + \frac{1}{2}\log\rho + \frac{1}{2}\log C\right)^2}$$

$$= \frac{4\pi}{C} \lim_{\rho\to\infty} f\left(\frac{1}{2} r_\rho^{-1}\right)$$

$$= \frac{4\pi}{C} \lim_{x\to\infty} f(x)$$

where in the second step, L'Hospital's rule with $\frac{C}{\log\rho + b}$ being the denominator and $\int_{\frac{1}{2} r_\rho^{-1}}^{\infty} 2\pi x g(x)\, dx$ being the numerator is used; in the third step, (2.2.40) is used.

Remark 48 Equation (2.2.42) shows also that $\lim_{\rho\to\infty} E\left(W^T\right) \geq e^{-b+\frac{4\pi}{C}\lim_{x\to\infty} f(x)}$ where $E\left(W^T\right)$ is the expected number of isolated nodes on a torus, which does not include the contribution of the boundary effect to the number of isolated nodes. Note also that the expected number of isolated nodes in an area of $A_{\frac{1}{r_\rho}}$ in $\mathcal{G}\left(\mathcal{X}_{\frac{\log\rho + b}{C}}, g, \Re^2\right)$ is e^{-b}. Therefore, the term $e^{\frac{4\pi}{C}\lim_{x\to\infty} f(x)}$ is entirely attributable to the truncation effect.

Note that $f(x)$ is a nonnegative function for $x > 1$. It is obvious from (2.2.42) that *unless* $\lim_{x\to\infty} f(x) = 0$, i.e., (2.1.4) is satisfied, the expected number of isolated node in $\mathcal{G}\left(\mathcal{X}_{\frac{\log\rho + b}{C}}, g, A_{\frac{1}{r_\rho}}\right)$ will be larger than the expected number of isolated nodes in an area of $A_{\frac{1}{r_\rho}}$ in $\mathcal{G}\left(\mathcal{X}_{\frac{\log\rho + b}{C}}, g, \Re^2\right)$. That is, the impact of the *truncation effect* on the number of isolated nodes in $\mathcal{G}\left(\mathcal{X}_{\frac{\log\rho + b}{C}}, g, A_{\frac{1}{r_\rho}}\right)$ will *not be* vanishingly small as $\rho \to \infty$. In particular, it can be shown that for $g(x) = \Theta_x\left(\frac{1}{x^2\log^2 x}\right)$, the impact of the truncation effect is non-negligible or even dominant in determining the number of isolated nodes in $\mathcal{G}\left(\mathcal{X}_{\frac{\log\rho + b}{C}}, g, A_{\frac{1}{r_\rho}}\right)$. Using (2.2.42), it can also be shown that for $g(x) = \omega_x\left(\frac{1}{x^2\log^2 x}\right)$, $\lim_{\rho\to\infty} E(W)$ is unbounded, i.e., connectivity cannot be achieved for $g(x) = \omega_x\left(\frac{1}{x^2\log^2 x}\right)$ even if (2.1.1) and (2.1.2) are satisfied.

The above discussion leads to the following conclusion:

Theorem 49 *The more restrictive requirement on g that it satisfies (2.1.4) is a necessary condition for the impact of the truncation effect on the number of isolated nodes in* $\mathcal{G}\left(\mathcal{X}_{\frac{\log\rho + b}{C}}, g, A_{\frac{1}{r_\rho}}\right)$ *to be vanishingly small as* $\rho \to \infty$. *Furthermore, for* $g(x) = \Theta_x\left(\frac{1}{x^2\log^2 x}\right)$, *the impact of the truncation effect is non-negligible or even* *dominant in determining the number of isolated nodes in* $\mathcal{G}\left(\mathcal{X}_{\frac{\log\rho+b}{C}}, g, A_{\frac{1}{r_\rho}}\right)$*; and for* $g(x) = \omega_x\left(\frac{1}{x^2\log^2 x}\right)$*, the truncation effect is the dominant factor in determining the number of isolated nodes in* $\mathcal{G}\left(\mathcal{X}_{\frac{\log\rho+b}{C}}, g, A_{\frac{1}{r_\rho}}\right)$.

Noting that the number of isolated nodes in a network is a nonnegative integer, the following result can be obtained as an easy consequence of Theorem 47 (see also [139]). Note that in formulating this result, we drop the assumption that b, originally introduced in (2.1.5), is a constant, and allow it instead to be ρ-dependent.

Corollary 50 *For* g *satisfying both (2.1.1) and (2.1.4), a necessary condition for* $\mathcal{G}\left(\mathcal{X}_{\frac{\log\rho+b}{C}}, g, A_{\frac{1}{r_\rho}}\right)$ *to be* asymptotically almost surely *(as* $\rho \to \infty$*) connected is* $b \to \infty$.

Remark 51 As pointed out in [143, p. 151], the three requirements on g in the random connection model, i.e., rotational invariance, nonincreasing monotonicity, and integral boundedness, are not equally important. Particularly, rotational invariance and nonincreasing monotonicity are required only to simplify the analysis such that "the notation and formulae will be somewhat simpler." Similarly, we expect the results obtained in this section and in the next section requiring nonincreasing monotonicity in (2.1.1) are also valid when the condition in (2.1.1) is removed. These however require more complicated handling of $g(x)$, particularly when x is sufficiently large.

2.3 Vanishing of Finite Components with More Than One Nodes

In this section we consider the events of the asymptotic vanishing of components of fixed and finite order $k > 1$ in the infinite network $\mathcal{G}\left(\mathcal{X}_{\frac{\log\rho+b}{C}}, g, \Re^2\right)$ and in $\mathcal{G}\left(\mathcal{X}_{\frac{\log\rho+b}{C}}, g, A_{\frac{1}{r_\rho}}\right)$ respectively as $\rho \to \infty$.

In [143, Theorem 6.4], it was shown that as $\rho \to \infty$ (and $\frac{\log\rho+b}{C} \to \infty$) the probability for a node to be isolated given that its component is finite converges to 1. In other words, as $\rho \to \infty$, asymptotically almost surely $\mathcal{G}\left(\mathcal{X}_{\frac{\log\rho+b}{C}}, g, \Re^2\right)$ has only isolated nodes and components of infinite order, and components of fixed and finite order $k > 1$ asymptotically vanish. In the following we show that due to the truncation effect, the above result obtained in $\mathcal{G}\left(\mathcal{X}_{\frac{\log\rho+b}{C}}, g, \Re^2\right)$ does *not* carry over to the conclusion that as $\rho \to \infty$, asymptotically almost surely $\mathcal{G}\left(\mathcal{X}_{\frac{\log\rho+b}{C}}, g, A_{\frac{1}{r_\rho}}\right)$ has only isolated nodes and infinite components too, without further analysis on the impact of the truncation effect. Specifically, an infinite component in $\mathcal{G}\left(\mathcal{X}_{\frac{\log\rho+b}{C}}, g, \Re^2\right)$ may *possibly* consist of components of extremely large order, components of fixed and finite order $k > 1$ and isolated nodes involving nodes and connections entirely contained inside $A_{\frac{1}{r_\rho}}$, where these components are only connected to each other via nodes and connections outside $A_{\frac{1}{r_\rho}}$. Note that for any finite ρ, almost surely there is no infinite component in $\mathcal{G}\left(\mathcal{X}_{\frac{\log\rho+b}{C}}, g, A_{\frac{1}{r_\rho}}\right)$. Therefore, we use the term *component of extremely large order* to refer to a component whose order may become asymptotically infinite as $\rho \to \infty$. As the nodes and associated connections outside $A_{\frac{1}{r_\rho}}$ are removed, the infinite component in $\Re^2$ may *possibly* leave components of extremely large order, components of finite order $k > 1$ and isolated nodes in $A_{\frac{1}{r_\rho}}$. As such, vanishing of components of finite order $k > 1$ in $\mathcal{G}\left(\mathcal{X}_{\frac{\log\rho+b}{C}}, g, \Re^2\right)$ as $\rho \to \infty$ does not *necessarily* carry the conclusion that components of finite order $k > 1$ in $\mathcal{G}\left(\mathcal{X}_{\frac{\log\rho+b}{C}}, g, A_{\frac{1}{r_\rho}}\right)$ also vanish as $\rho \to \infty$, even when $A_{\frac{1}{r_\rho}}$ approaches $\Re^2$ as $\rho \to \infty$. An example is illustrated in Fig. 2.3.

We further point out that many other topologies, particularly under a random connection model where even a pair of nodes separated by a large distance may have a nonzero probability to be directly connected, can be drawn for an infinite component in $\Re^2$, where after removing all nodes and associated connections of the infinite component outside $A_{\frac{1}{r_\rho}}$, the infinite component leaves components of finite order $k > 1$ inside $A_{\frac{1}{r_\rho}}$, even when $A_{\frac{1}{r_\rho}}$ grows as $\rho \to \infty$. We emphasize that we are not hinting that the topology of the infinite component shown in Fig. 2.3 is likely to occur in $\mathcal{G}\left(\mathcal{X}_{\frac{\log\rho+b}{C}}, g, \Re^2\right)$ as $\rho \to \infty$, but neither can such a possibility be precluded. Therefore, a conclusion cannot be drawn straightforwardly from [143, Theorem 6.4] that asymptotically almost surely components of finite order $k > 1$ in $\mathcal{G}\left(\mathcal{X}_{\frac{\log\rho+b}{C}}, g, A_{\frac{1}{r_\rho}}\right)$ vanish as $\rho \to \infty$. Instead, some sophisticated analysis is required to establish such a conclusion in $\mathcal{G}\left(\mathcal{X}_{\frac{\log\rho+b}{C}}, g, A_{\frac{1}{r_\rho}}\right)$.

We present such a result for the vanishing of components of finite order $k > 1$ in $\mathcal{G}\left(\mathcal{X}_{\frac{\log\rho+b}{C}}, g, A_{\frac{1}{r_\rho}}\right)$ as $\rho \to \infty$ to fill this theoretical gap:

Theorem 52 *For g satisfying (2.1.1) and (2.1.4), asymptotically almost surely there is no component of finite order $k > 1$ in $\mathcal{G}\left(\mathcal{X}_{\frac{\log\rho+b}{C}}, g, A_{\frac{1}{r_\rho}}\right)$.*

The rest of this section is dedicated to the proof of Theorem 52.

For convenience, let λ be the node density in $\mathcal{G}\left(\mathcal{X}_{\frac{\log\rho+b}{C}}, g, A_{\frac{1}{r_\rho}}\right)$ where $\lambda \triangleq \rho r_\rho^2 = \frac{\log\rho+b}{C}$. Using the above notations, $\mathcal{G}\left(\mathcal{X}_{\frac{\log\rho+b}{C}}, g, A_{\frac{1}{r_\rho}}\right)$ can be written as $\mathcal{G}\left(\mathcal{X}_\lambda, g, A_{\frac{1}{r_\rho}}\right)$.

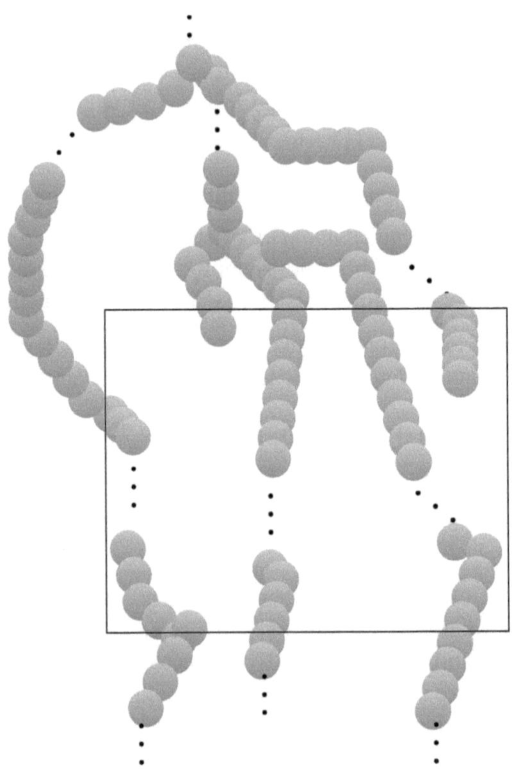

Fig. 2.3 An illustration that an infinite component in $\Re^2$ may leave components of extremely large order, components of finite order $k > 1$, and isolated nodes in a finite (or asymptotically infinite) region in $\Re^2$ when nodes and connections outside the finite (asymptotically infinite) region is removed. The figure uses the unit disk connection model as a special case for easy illustration. Each ball has a radius of half of the transmission range and is centered at a node. Two adjacent balls overlap if and only if the associated nodes are directly connected. The figure shows an infinite component with nodes organized in a tree structure. The *square area* represents the finite (asymptotically infinite) region. Even as the square grows to include more and more nodes of the infinite component, it is still possible for the square to have components of finite order $k > 1$ when nodes and connections outside the square are removed

Note that for any finite ρ the total number of nodes in $\mathcal{G}\left(\mathcal{X}_\lambda, g, A_{\frac{1}{r_\rho}}\right)$, hence the total number of components in $\mathcal{G}\left(\mathcal{X}_\lambda, g, A_{\frac{1}{r_\rho}}\right)$, is almost surely finite. Denote by ξ_k the (random) number of components of order k in an instance of $\mathcal{G}\left(\mathcal{X}_\lambda, g, A_{\frac{1}{r_\rho}}\right)$. It then suffices to show that for an arbitrarily large positive integer M:

$$\lim_{\rho\to\infty} \Pr\left(\sum_{k=2}^{M} \xi_k = 0\right) = 1 \tag{2.3.1}$$

The following symbols and notations are used:

Denote by $g_1(\boldsymbol{x}_1, \boldsymbol{x}_2, \ldots, \boldsymbol{x}_k)$ the probability that a set of k nodes at nonrandom positions $\boldsymbol{x}_1, \boldsymbol{x}_2, \ldots, \boldsymbol{x}_k \in A_{\frac{1}{r_\rho}}$ forms a connected component.

Denote by $g_2(\boldsymbol{y}; \boldsymbol{x}_1, \boldsymbol{x}_2, \ldots, \boldsymbol{x}_k)$ the probability that a node at nonrandom position $\boldsymbol{y}$ is connected to at least one node in $\{\boldsymbol{x}_1, \boldsymbol{x}_2, \ldots, \boldsymbol{x}_k\}$. It can be shown that

$$g_2(\boldsymbol{y}; \boldsymbol{x}_1, \boldsymbol{x}_2, \ldots, \boldsymbol{x}_k) = 1 - \prod_{i=1}^{k} \left(1 - g\left(\|\boldsymbol{y} - \boldsymbol{x}_i\|\right)\right) \tag{2.3.2}$$

and

$$g_2(\boldsymbol{y}; \boldsymbol{x}_1, \boldsymbol{x}_2, \ldots, \boldsymbol{x}_k) \geq g_2(\boldsymbol{y}; \boldsymbol{x}_1, \boldsymbol{x}_2, \ldots, \boldsymbol{x}_i) \text{ for } 1 \leq i \leq k \tag{2.3.3}$$

As an easy consequence of the union bound,

$$g_2(\boldsymbol{y}; \boldsymbol{x}_1, \boldsymbol{x}_2, \ldots, \boldsymbol{x}_k) \leq \sum_{i=1}^{k} g\left(\|\boldsymbol{y} - \boldsymbol{x}_i\|\right) \tag{2.3.4}$$

Using the monotonicity and positive integral boundedness properties of g in (2.1.1) and (2.1.4), it can be shown that there exists a positive constant r such that $g(r^-)\left(1 - g\left(r^+\right)\right) > 0$ where $g(r^-) \triangleq \lim_{x \to r^-} g(x)$ and $g\left(r^+\right) \triangleq \lim_{x \to r^+} g(x)$. If g is a continuous function, then $g(r^-) = g\left(r^+\right)$; if g is a discontinuous function, e.g., a unit disk connection model, by choosing r to be the transmission range, $g(r^-)\left(1 - g\left(r^+\right)\right) = 1$. For convenience in notations, we use β for $g(r^-)\left(1 - g\left(r^+\right)\right)$, i.e.,

$$\beta \triangleq g(r^-)\left(1 - g\left(r^+\right)\right) \tag{2.3.5}$$

Denote by $\partial A_{\frac{1}{r_\rho}}$ the border of $A_{\frac{1}{r_\rho}}$. Denote by $\ell A_{\frac{1}{r_\rho}} \subset A_{\frac{1}{r_\rho}}$ a rectangular area of size $\left(\frac{1}{r_\rho} - 2r\right) \times r$ along one side of the border of $A_{\frac{1}{r_\rho}}$, within a distance r of the border and away from the four corners of $A_{\frac{1}{r_\rho}}$ by at least r. There are four such areas in $A_{\frac{1}{r_\rho}}$. Denote by $\angle A_{\frac{1}{r_\rho}} \subset A_{\frac{1}{r_\rho}}$ a square area of size $r \times r$ located at a corner of $A_{\frac{1}{r_\rho}}$. There are four such corner squares in $A_{\frac{1}{r_\rho}}$. Denote by $B_d\left(A_{\frac{1}{r_\rho}}\right) \subset A_{\frac{1}{r_\rho}}$ a boundary area within a distance d of the border of $A_{\frac{1}{r_\rho}}$. Note the difference of the definitions of those symbols from those used Sect. 2.2 and particularly Fig. 2.1.

Let $D(\boldsymbol{x}, d) \subset \Re^2$ represents a disk centered at $\boldsymbol{x} \in A_{\frac{1}{r_\rho}}$ and with a radius d.

We first establish some preliminary results that will be used in the proof.

Lemma 53 *In $\mathcal{G}\left(\mathcal{X}_\lambda, g, A_{\frac{1}{r_\rho}}\right)$, the expected number of components of order k is given by*

$$E(\xi_k) = \frac{\lambda^k}{k!} \int_{\left(A_{\frac{1}{r_\rho}}\right)^k} g_1(\boldsymbol{x}_1, \boldsymbol{x}_2, \ldots, \boldsymbol{x}_k)\, e^{-\lambda \int_{A_{\frac{1}{r_\rho}}} g_2(\boldsymbol{y}; \boldsymbol{x}_1, \boldsymbol{x}_2, \ldots, \boldsymbol{x}_k) d\boldsymbol{y}}\, d(\boldsymbol{x}_1 \cdots \boldsymbol{x}_k) \tag{2.3.6}$$

Proof It can be shown that for any $n \geq k$:

$$E\left(\xi_k \,||\mathcal{X}_\lambda| = n\right) = \frac{\binom{n}{k}}{\left(A_{\frac{1}{r_\rho}}\right)^n} \int_{\left(A_{\frac{1}{r_\rho}}\right)^n} g_1\left(\boldsymbol{x}_1, \boldsymbol{x}_2, \ldots, \boldsymbol{x}_k\right) \prod_{i=k+1}^{n} \left(1 - g_2\left(\boldsymbol{x}_i; \boldsymbol{x}_1, \boldsymbol{x}_2, \ldots, \boldsymbol{x}_k\right)\right) d\left(\boldsymbol{x}_1 \cdots \boldsymbol{x}_n\right) \tag{2.3.7}$$

In (2.3.7), $\binom{n}{k}$ is the number of distinct sets of k nodes drawn from a total of n nodes and the rest term represents the probability of the event that a *randomly chosen* set of k nodes forms a component of order k. From (2.3.7), it follows that

$$\begin{aligned}
&E\left(\xi_k\right) \\
&= \sum_{n=k}^{\infty} E\left(\xi_k \,||\mathcal{X}_\lambda| = n\right) \frac{\left(\lambda A_{\frac{1}{r_\rho}}\right)^n}{n!} e^{-\lambda A_{\frac{1}{r_\rho}}} \\
&= \sum_{n=k}^{\infty} \frac{\left(\lambda A_{\frac{1}{r_\rho}}\right)^n}{n!} e^{-\lambda A_{\frac{1}{r_\rho}}} \frac{\binom{n}{k}}{\left(A_{\frac{1}{r_\rho}}\right)^n} \int_{\left(A_{\frac{1}{r_\rho}}\right)^n} g_1\left(\boldsymbol{x}_1, \boldsymbol{x}_2, \ldots, \boldsymbol{x}_k\right) \prod_{i=k+1}^{n} \left(1 - g_2\left(\boldsymbol{x}_i; \boldsymbol{x}_1, \boldsymbol{x}_2, \ldots, \boldsymbol{x}_k\right)\right) \\
&\quad \times d\left(\boldsymbol{x}_1 \cdots \boldsymbol{x}_n\right) \\
&= \sum_{n=k}^{\infty} \frac{\lambda^n}{n!} e^{-\lambda A_{\frac{1}{r_\rho}}} \binom{n}{k} \int_{\left(A_{\frac{1}{r_\rho}}\right)^k} g_1\left(\boldsymbol{x}_1, \boldsymbol{x}_2, \ldots, \boldsymbol{x}_k\right) \left(\int_{A_{\frac{1}{r_\rho}}} 1 - g_2\left(\boldsymbol{y}; \boldsymbol{x}_1, \boldsymbol{x}_2, \ldots, \boldsymbol{x}_k\right) dy\right)^{n-k} \\
&\quad \times d\left(\boldsymbol{x}_1 \cdots \boldsymbol{x}_k\right) \\
&= \int_{\left(A_{\frac{1}{r_\rho}}\right)^k} g_1\left(\boldsymbol{x}_1, \boldsymbol{x}_2, \ldots, \boldsymbol{x}_k\right) \left(\sum_{n=k}^{\infty} \frac{\lambda^n}{n!} e^{-\lambda A_\rho} \binom{n}{k} \left(\int_{A_{\frac{1}{r_\rho}}} 1 - g_2\left(\boldsymbol{y}; \boldsymbol{x}_1, \boldsymbol{x}_2, \ldots, \boldsymbol{x}_k\right) dy\right)^{n-k}\right) \\
&\quad \times d\left(\boldsymbol{x}_1 \cdots \boldsymbol{x}_k\right) \\
&= \frac{\lambda^k}{k!} \int_{\left(A_{\frac{1}{r_\rho}}\right)^k} g_1\left(\boldsymbol{x}_1, \boldsymbol{x}_2, \ldots, \boldsymbol{x}_k\right) \left(\sum_{n=k}^{\infty} \frac{\left(\lambda\left(\int_{A_{\frac{1}{r_\rho}}} 1 - g_2\left(\boldsymbol{y}; \boldsymbol{x}_1, \boldsymbol{x}_2, \ldots, \boldsymbol{x}_k\right) dy\right)\right)^{n-k}}{(n-k)!} e^{-\lambda A_{\frac{1}{r_\rho}}}\right) \\
&\quad \times d\left(\boldsymbol{x}_1 \cdots \boldsymbol{x}_k\right) \\
&= \frac{\lambda^k}{k!} \int_{\left(A_{\frac{1}{r_\rho}}\right)^k} g_1\left(\boldsymbol{x}_1, \boldsymbol{x}_2, \ldots, \boldsymbol{x}_k\right) e^{-\lambda \int_{A_{\frac{1}{r_\rho}}} g_2\left(\boldsymbol{y}; \boldsymbol{x}_1, \boldsymbol{x}_2, \ldots, \boldsymbol{x}_k\right) dy} d\left(\boldsymbol{x}_1 \cdots \boldsymbol{x}_k\right)
\end{aligned}$$

□

A similar technique as that used in the proof of Proposition 6.2 in [143], originally due to Penrose [163], was used in the proof of Lemma 53 .

The following lemma is also used in the analysis of $E\left(\xi_k\right)$.

Lemma 54 *A sufficient and necessary condition for a given set of nodes to form a single connected component is that there exists an ordering of the nodes, which can start from any node in the set, such that each node appearing later in the order is connected to at least one node appearing earlier in the order.*

The proof is trivial and omitted.

Using Lemma 54, the following result can be established:

Lemma 55 *Let Γ_k denote the set $\{1, \ldots, k\}$. The function $g_1(\boldsymbol{x}_1, \boldsymbol{x}_2, \ldots, \boldsymbol{x}_k)$ satisfies the following inequality*

$$
\begin{aligned}
& g_1(\boldsymbol{x}_1, \boldsymbol{x}_2, \ldots, \boldsymbol{x}_k) \\
& \leq \sum_{i_2 \in \Gamma_k \setminus \{1\}, i_3 \in \Gamma_k \setminus \{1, i_2\}, \cdots, i_k \in \Gamma_k \setminus \{1, i_2, \ldots, i_{k-1}\}} g_2(\boldsymbol{x}_{i_2}; \boldsymbol{x}_1)\, g_2(\boldsymbol{x}_{i_3}; \boldsymbol{x}_1, \boldsymbol{x}_{i_2}) \cdots \\
& \quad g_2(\boldsymbol{x}_{i_k}; \boldsymbol{x}_1, \boldsymbol{x}_{i_2}, \ldots, \boldsymbol{x}_{i_{k-1}})
\end{aligned}
$$

Proof Without loss of generality, we assume that such ordering described in Lemma 54 starts from $\boldsymbol{x}_1 \in \{\boldsymbol{x}_1, \boldsymbol{x}_2, \ldots, \boldsymbol{x}_k\}$. Denote by $\xi_{(1, i_2, \ldots, i_k)}$ the event that $(\boldsymbol{x}_1, \boldsymbol{x}_{i_2}, \ldots, \boldsymbol{x}_{i_k})$ is one of such an ordering described in Lemma 54, where $i_2 \in \Gamma_k \setminus \{1\}, i_3 \in \Gamma_k \setminus \{1, i_2\}, \cdots, i_k \in \Gamma_k \setminus \{1, i_2, \ldots, i_{k-1}\}$. Using Lemma 54, it can be shown that

$$
\Pr\left(\xi_{(1, i_2, \ldots, i_{k-1})}\right) = g_2(\boldsymbol{x}_{i_2}; \boldsymbol{x}_1)\, g_2(\boldsymbol{x}_{i_3}; \boldsymbol{x}_1, \boldsymbol{x}_{i_2}) \cdots g_2(\boldsymbol{x}_{i_k}; \boldsymbol{x}_1, \boldsymbol{x}_{i_2}, \ldots, \boldsymbol{x}_{i_{k-1}})
$$

Then, it follows that

$$
g_1(\boldsymbol{x}_1, \boldsymbol{x}_2, \ldots, \boldsymbol{x}_k) = \Pr\left(\cup_{i_2 \in \Gamma_k \setminus \{1\}, i_3 \in \Gamma_k \setminus \{1, i_2\}, \cdots, i_k \in \Gamma_k \setminus \{1, i_2, \ldots, i_{k-1}\}} \xi_{(1, i_2, \ldots, i_k)}\right)
$$

As an easy consequence of the above equation and the union bound:

$$
\begin{aligned}
& g_1(\boldsymbol{x}_1, \boldsymbol{x}_2, \ldots, \boldsymbol{x}_k) \\
& \leq \sum_{i_2 \in \Gamma_k \setminus \{1\}, i_3 \in \Gamma_k \setminus \{1, i_2\}, \cdots, i_k \in \Gamma_k \setminus \{1, i_2, \ldots, i_{k-1}\}} g_2(\boldsymbol{x}_{i_2}; \boldsymbol{x}_1)\, g_2(\boldsymbol{x}_{i_3}; \boldsymbol{x}_1, \boldsymbol{x}_{i_2}) \cdots \\
& \quad g_2(\boldsymbol{x}_{i_k}; \boldsymbol{x}_1, \boldsymbol{x}_{i_2}, \ldots, \boldsymbol{x}_{i_{k-1}})
\end{aligned}
$$

□

The following geometric results are also used in the proof of Theorem 52.

Lemma 56 *Consider two points $\boldsymbol{x}_1, \boldsymbol{x}_2 \in A_{\frac{1}{r_\rho}}$ and let $z \triangleq \|\boldsymbol{x}_2 - \boldsymbol{x}_1\|$. For a positive constant $c_1 = \sqrt{3}r$ and $z \leq r$*

$$
|D(\boldsymbol{x}_1, r) \setminus D(\boldsymbol{x}_2, r)| \geq c_1 z
$$

where $|D(\boldsymbol{x}_1, r) \setminus D(\boldsymbol{x}_2, r)|$ denotes the area of $D(\boldsymbol{x}_1, r) \setminus D(\boldsymbol{x}_2, r)$.

Proof First it can be shown that for $z \geq 2r$

$$|D(\mathbf{x}_1, r) \backslash D(\mathbf{x}_2, r)| = \pi r^2$$

and for $z < 2r$

$$\begin{aligned} & f(z) \\ & \triangleq |D(\mathbf{x}_2, r) \backslash D(\mathbf{x}_1, r)| \\ & = \pi r^2 - 2r^2 \arcsin\left(\sqrt{1 - \frac{z^2}{4r^2}}\right) + zr\sqrt{1 - \frac{z^2}{4r^2}} \end{aligned}$$

Furthermore, it can be shown that

$$\frac{df(z)}{dz} = 2r\sqrt{1 - \frac{z^2}{4r^2}}$$

Therefore, $f(z)$ is an increasing function of z for $z < 2r$ and $\frac{df(z)}{dz} \geq \sqrt{3}r$ for $z \leq r$. It then follows from $f(0) = 0$ that $f(z) \geq \sqrt{3}rz$ for $z \leq r$. □

Lemma 57 *Consider two points* $\mathbf{x}_1 \in \ell A_{\frac{1}{r\rho}}$ *and* $\mathbf{x}_2 \in A_{\frac{1}{r\rho}} \cap D(\mathbf{x}_1, r)$ *and let* $z \triangleq \|\mathbf{x}_2 - \mathbf{x}_1\|$. *When* $\gamma(\mathbf{x}_2) \leq \gamma(\mathbf{x}_1)$,

$$\left|A_{\frac{1}{r\rho}} \cap D(\mathbf{x}_1, r) \backslash D(\mathbf{x}_2, r)\right| \geq \frac{c_1}{2} z$$

When $\gamma(\mathbf{x}_2) > \gamma(\mathbf{x}_1)$, *for any positive constant* c_2, *there exists a positive constant* $z_0 < r$ *such that for all* $z \leq z_0$

$$\left|A_{\frac{1}{r\rho}} \cap D(\mathbf{x}_1, r) \backslash D(\mathbf{x}_2, r)\right| \geq (r - c_2) z - r \times |\gamma(\mathbf{x}_2) - \gamma(\mathbf{x}_1)|$$

where $\gamma(\mathbf{x}_1)$ $(\gamma(\mathbf{x}_2))$ *represents the shortest Euclidean distance between* $\mathbf{x}_1$ $(\mathbf{x}_2)$ *and a border of* $A_{\frac{1}{r\rho}}$ *that is adjacent to* $\ell A_{\frac{1}{r\rho}}$ *(i.e.,* $\partial A_{\frac{1}{r\rho}} \cap \ell A_{\frac{1}{r\rho}}$, *see Fig. 2.4 for an illustration of* $\gamma(\mathbf{x}_2)$ *where* $\gamma(\mathbf{x}_1) = 0$ *in the figure).*

Proof The first part of the lemma can be easily proved by noting that when $\gamma(\mathbf{x}_2) \leq \gamma(\mathbf{x}_1)$,

$$\left|A_{\frac{1}{r\rho}} \cap D(\mathbf{x}_1, r) \backslash D(\mathbf{x}_2, r)\right| \geq \frac{1}{2} |D(\mathbf{x}_1, r) \backslash D(\mathbf{x}_2, r)|$$

and the lemma can then be proved using Lemma 56.

Now let us focus on the situation when $\gamma(\mathbf{x}_2) > \gamma(\mathbf{x}_1)$. It can be easily shown (see also Fig. 2.4) that when changing the value of $\gamma(\mathbf{x}_1)$ while keeping $\mathbf{x}_2 - \mathbf{x}_1$ *fixed* (i.e., $\mathbf{x}_2$ has the same displacement as $\mathbf{x}_1$), $\left|A_{\frac{1}{r\rho}} \cap D(\mathbf{x}_1, r) \backslash D(\mathbf{x}_2, r)\right|$ is minimized as $\gamma(\mathbf{x}_1) = 0$ (i.e., $\mathbf{x}_1 \in \ell A_{\frac{1}{r\rho}} \cap \partial A_{\frac{1}{r\rho}}$) and $(r - c_2) z - r \times |\gamma(\mathbf{x}_2) - \gamma(\mathbf{x}_1)|$ remains constant. Therefore, we focus on the worst case when $\mathbf{x}_1 \in \ell A_{\frac{1}{r\rho}} \cap \partial A_{\frac{1}{r\rho}}$. When $\mathbf{x}_1 \in \ell A_{\frac{1}{r\rho}} \cap \partial A_{\frac{1}{r\rho}}$, $|\gamma(\mathbf{x}_2) - \gamma(\mathbf{x}_1)| = \gamma(\mathbf{x}_2)$.

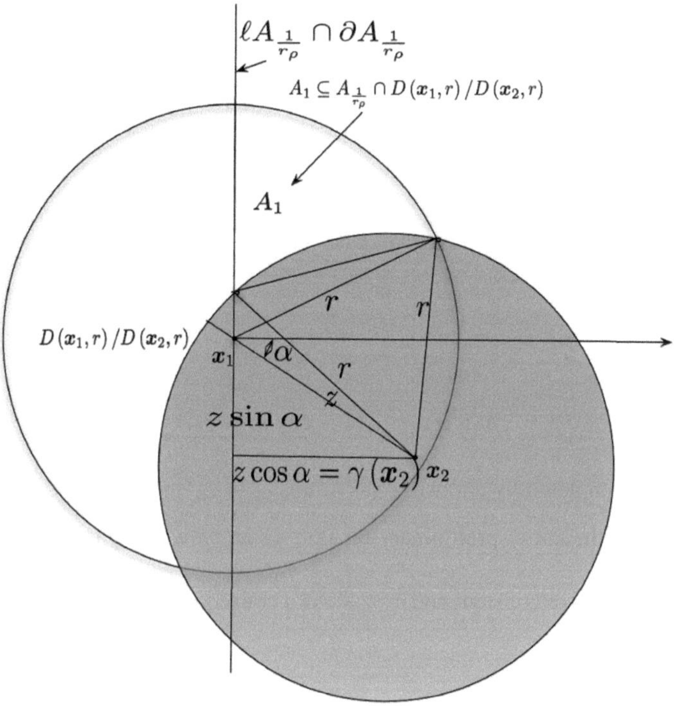

Fig. 2.4 An illustration of $\left|A_{\frac{1}{r\rho}} \cap D(\boldsymbol{x}_2, r)/D(\boldsymbol{x}_1, r)\right|$ for $\boldsymbol{x}_1 \in \ell A_{\frac{1}{r\rho}} \cap \partial A_{\frac{1}{r\rho}}$ and $\boldsymbol{x}_2 \in A_{\frac{1}{r\rho}} \cap D(\boldsymbol{x}_1, r)$. Note that A_1 is the upper part of $A_\rho \cap D(\boldsymbol{x}_1, r) \setminus D(\boldsymbol{x}_2, r)$ above the line connecting $\boldsymbol{x}_1$ and $\boldsymbol{x}_2$. Depending on the relative positions of $\boldsymbol{x}_1$ and $\boldsymbol{x}_2$, $A_\rho \cap D(\boldsymbol{x}_1, r) \setminus D(\boldsymbol{x}_2, r)$ may also contain a nonempty region below the line connecting $\boldsymbol{x}_1$ and $\boldsymbol{x}_2$

Figure 2.4 shows $A_{\frac{1}{r\rho}} \cap D(\boldsymbol{x}_1, r) \setminus D(\boldsymbol{x}_2, r)$ for $\boldsymbol{x}_1 \in \ell A_{\frac{1}{r\rho}} \cap \partial A_{\frac{1}{r\rho}}$ and $\boldsymbol{x}_2 \in A_{\frac{1}{r\rho}} \cap D(\boldsymbol{x}_1, r)$. It can be shown that under the above conditions for $\boldsymbol{x}_1$ and $\boldsymbol{x}_2$ (see Fig. 2.4 for definitions of α and A_1 and some detailed but straightforward geometric analysis omitted in the following equation)

$$
\begin{aligned}
&\left|A_\rho \cap D(\boldsymbol{x}_1, r) \setminus D(\boldsymbol{x}_2, r)\right| \\
&\geq |A_1| \\
&= h(z, \alpha) \\
&\triangleq \frac{\pi r^2}{4} - r^2 \arccos\frac{z}{2r} + \frac{1}{2} zr\sqrt{1 - \frac{z^2}{4r^2}} + \frac{r^2}{2}\arccos\frac{z\cos\alpha}{r} \\
&\quad - \frac{1}{2} zr\sqrt{1 - \frac{z^2}{r^2}\cos^2\alpha}\cos\alpha + \frac{1}{2} z^2 \sin\alpha\cos\alpha
\end{aligned}
$$

Note that $h(0,\alpha) = 0$,

$$\left.\frac{\partial h(z,\alpha)}{\partial z}\right|_{z=0} = r(1-\cos\alpha)$$

and $\cos\alpha = \frac{\gamma(\boldsymbol{x}_2)}{z}$. Therefore,

$$\lim_{z\to 0+} \frac{h(z,\alpha) - h(0,\alpha)}{z} = r(1-\cos\alpha)$$

i.e., for a given positive constant c_2, there exists $z_\alpha > 0$ depending on α such that for all $0 \le z \le z_\alpha$

$$h(z,\alpha) \ge (r(1-\cos\alpha) - c_2)\,z$$

The proof is completed by choosing $z_0 = \min_{0\le\alpha\le\frac{\pi}{2}} z_\alpha$ and using $\cos\alpha = \frac{\gamma(\boldsymbol{x}_2)}{z}$. $\square$

On the basis of the above preliminary results, we are now ready to start the proof of Theorem 52.

Let δ be a positive constant and $\delta \le \frac{r}{2}$. As a consequence of Lemma 53, it can be shown that

$$\begin{aligned}
&E(\xi_k)\\
&= \frac{\lambda^k}{k!}\int_{\left(A_{\frac{1}{r\rho}}\right)^{k-1}}\int_{A_{\frac{1}{r\rho}}} g_1(\boldsymbol{x}_1,\boldsymbol{x}_2,\ldots,\boldsymbol{x}_k)\, e^{-\lambda\int_{A_{\frac{1}{r\rho}}} g_2(\boldsymbol{y};\boldsymbol{x}_1,\boldsymbol{x}_2,\ldots,\boldsymbol{x}_k)d\boldsymbol{y}}\\
&\quad d\boldsymbol{x}_1 d(\boldsymbol{x}_2\cdots\boldsymbol{x}_k)\\
&= \frac{\lambda^k}{k!}\int_{A_{\frac{1}{r\rho}}}\int_{\left(A_{\frac{1}{r\rho}}\right)^{k-1}\cap(D(\boldsymbol{x}_1,\delta))^{k-1}} g_1(\boldsymbol{x}_1,\boldsymbol{x}_2,\ldots,\boldsymbol{x}_k)\, e^{-\lambda\int_{A_{\frac{1}{r\rho}}} g_2(\boldsymbol{y};\boldsymbol{x}_1,\boldsymbol{x}_2,\ldots,\boldsymbol{x}_k)d\boldsymbol{y}}\\
&\quad d(\boldsymbol{x}_2\cdots\boldsymbol{x}_k)\,d\boldsymbol{x}_1\\
&+ \frac{\lambda^k}{k!}\int_{A_{\frac{1}{r\rho}}}\int_{\left(A_{\frac{1}{r\rho}}\right)^{k-1}\backslash(D(\boldsymbol{x}_1,\delta))^{k-1}} g_1(\boldsymbol{x}_1,\boldsymbol{x}_2,\ldots,\boldsymbol{x}_k)\, e^{-\lambda\int_{A_{\frac{1}{r\rho}}} g_2(\boldsymbol{y};\boldsymbol{x}_1,\boldsymbol{x}_2,\ldots,\boldsymbol{x}_k)d\boldsymbol{y}}\\
&\quad d(\boldsymbol{x}_2\cdots\boldsymbol{x}_k)\,d\boldsymbol{x}_1
\end{aligned} \tag{2.3.8}$$

Denote by $E(\xi_{k,1})$ and $E(\xi_{k,2})$ the two summands in (2.3.8) respectively. In the following analysis, we shall show that by choosing δ to be sufficiently small, $\lim_{\rho\to\infty}\sum_{k=2}^{\infty} E(\xi_{k.1}) = 0$ and $\lim_{\rho\to\infty}\sum_{k=2}^{M} E(\xi_{k,2}) = 0$. $\sum_{k=2}^{\infty} E(\xi_{k.1})$ has the meaning of being the expected total number of components of finite orders $\infty >$

$k > 1$, where all other nodes of the component are located within a δ neighborhood of a randomly designated node (i.e., $\boldsymbol{x}_1$ in (2.3.8)). $\lim_{\rho\to\infty}\sum_{k=2}^{\infty} E(\xi_{k.1}) = 0$ implies $\lim_{\rho\to\infty}\sum_{k=2}^{M} E(\xi_{k.1}) = 0$. $\sum_{k=2}^{M} E(\xi_{k,2}) = 0$ has the meaning of being the expected total number of components of finite orders $M \geq k > 1$ where at least one of the nodes forming the component is located outside a δ neighborhood of a randomly designated node (i.e., $\boldsymbol{x}_1$ in (2.3.8)) in the component.

An Analysis of the First Term in (2.3.8)

Denote by $D_\delta^i \subset \left(A_{\frac{1}{r_\rho}}\right)^{k-1}$ the set $\left\{(\boldsymbol{x}_2,\ldots,\boldsymbol{x}_k) \in \left(A_{\frac{1}{r_\rho}}\right)^{k-1} \cap (D(\boldsymbol{x}_1,\delta))^{k-1} : \|\boldsymbol{x}_i-\boldsymbol{x}_1\| \geq \max_{j\in\{2,\ldots k\},j\neq i}\left\|\boldsymbol{x}_j-\boldsymbol{x}_1\right\|\right\}$, $i \in \{2,\ldots k\}$. Using (2.3.3) and the definition of D_δ^i, it can be shown that

$$
\begin{aligned}
&E(\xi_{k,1})\\
&\triangleq \frac{\lambda^k}{k!}\int_{A_\rho}\int_{\left(A_{\frac{1}{r_\rho}}\right)^{k-1}\cap(D(\boldsymbol{x}_1,\delta))^{k-1}} g_1(\boldsymbol{x}_1,\boldsymbol{x}_2,\ldots,\boldsymbol{x}_k)\, e^{-\lambda\int_{A_{\frac{1}{r_\rho}}} g_2(\boldsymbol{y};\boldsymbol{x}_1,\boldsymbol{x}_2,\ldots,\boldsymbol{x}_k)d\boldsymbol{y}}\\
&\quad d(\boldsymbol{x}_2\cdots\boldsymbol{x}_k)\,d\boldsymbol{x}_1\\
&=\sum_{i=2}^{k}\frac{\lambda^k}{k!}\int_{A_{\frac{1}{r_\rho}}}\int_{D_\delta^i} g_1(\boldsymbol{x}_1,\boldsymbol{x}_2,\ldots,\boldsymbol{x}_k)\, e^{-\lambda\int_{A_{\frac{1}{r_\rho}}} g_2(\boldsymbol{y};\boldsymbol{x}_1,\boldsymbol{x}_2,\ldots,\boldsymbol{x}_k)d\boldsymbol{y}}\\
&\quad d(\boldsymbol{x}_2\cdots\boldsymbol{x}_k)\,d\boldsymbol{x}_1\\
&\leq \frac{\lambda^k}{(k-2)!k}\int_{A_{\frac{1}{r_\rho}}}\int_{D_\delta^2} g_1(\boldsymbol{x}_1,\boldsymbol{x}_2,\ldots,\boldsymbol{x}_k)\, e^{-\lambda\int_{A_{\frac{1}{r_\rho}}} g_2(\boldsymbol{y};\boldsymbol{x}_1,\boldsymbol{x}_2)d\boldsymbol{y}}\\
&\quad d(\boldsymbol{x}_2\cdots\boldsymbol{x}_k)\,d\boldsymbol{x}_1\\
&\leq \frac{\lambda^k}{(k-2)!k}\int_{A_{\frac{1}{r_\rho}}}\int_{D_\delta^2} e^{-\lambda\int_{A_{\frac{1}{r_\rho}}} g_2(\boldsymbol{y};\boldsymbol{x}_1,\boldsymbol{x}_2)d\boldsymbol{y}}\\
&\quad d(\boldsymbol{x}_2\cdots\boldsymbol{x}_k)\,d\boldsymbol{x}_1\\
&\leq \frac{\lambda^k}{(k-2)!k}\int_{A_{\frac{1}{r_\rho}}}\int_{A_{\frac{1}{r_\rho}}\cap D(\boldsymbol{x}_1,\delta)}\left(\pi\left\|\boldsymbol{x}_2-\boldsymbol{x}_1\right\|^2\right)^{k-2} e^{-\lambda\int_{A_{\frac{1}{r_\rho}}} g_2(\boldsymbol{y};\boldsymbol{x}_1,\boldsymbol{x}_2)d\boldsymbol{y}}\\
&\quad d\boldsymbol{x}_2 d\boldsymbol{x}_1
\end{aligned}
\tag{2.3.9}
$$

As a result of the following inequality:

$$\begin{aligned}
&\sum_{k=2}^{\infty} \frac{\lambda^k \left(\pi \|\boldsymbol{x}_2 - \boldsymbol{x}_1\|^2\right)^{k-2}}{(k-2)!k} \\
&= \lambda^2 \left(\sum_{k=0}^{\infty} \frac{\lambda^k \left(\pi \|\boldsymbol{x}_2 - \boldsymbol{x}_1\|^2\right)^{k}}{k!\,(k+2)} \right) \\
&\le \lambda^2 \left(\sum_{k=0}^{\infty} \frac{\lambda^k \left(\pi \|\boldsymbol{x}_2 - \boldsymbol{x}_1\|^2\right)^{k}}{k!} e^{-\lambda\pi \|\boldsymbol{x}_2 - \boldsymbol{x}_1\|^2} \right) e^{\lambda\pi \|\boldsymbol{x}_2 - \boldsymbol{x}_1\|^2} \\
&= \lambda^2 e^{\lambda\pi \|\boldsymbol{x}_2 - \boldsymbol{x}_1\|^2}
\end{aligned}$$

it follows from (2.3.9) that

$$\begin{aligned}
&\sum_{k=2}^{\infty} E\left(\xi_{k,1}\right) \\
&\le \lambda^2 \int_{A_{\frac{1}{r_\rho}}} \int_{A_{\frac{1}{r_\rho}} \cap D(\boldsymbol{x}_1,\delta)} e^{-\lambda\left(\int_{A_{\frac{1}{r_\rho}}} g_2(\boldsymbol{y};\boldsymbol{x}_1,\boldsymbol{x}_2)d\boldsymbol{y} - \pi\|\boldsymbol{x}_2-\boldsymbol{x}_1\|^2\right)} d\boldsymbol{x}_2 d\boldsymbol{x}_1 \\
&= \lambda^2 \int_{A_{\frac{1}{r_\rho}}} \int_{A_{\frac{1}{r_\rho}} \cap D(\boldsymbol{x}_1,\delta)} e^{-\lambda\left(\int_{A_{\frac{1}{r_\rho}}} g(\|\boldsymbol{y}-\boldsymbol{x}_2\|)d\boldsymbol{y} + \int_{A_{\frac{1}{r_\rho}}} g(\|\boldsymbol{y}-\boldsymbol{x}_1\|)(1-g(\|\boldsymbol{y}-\boldsymbol{x}_2\|))d\boldsymbol{y} - \pi\|\boldsymbol{x}_2-\boldsymbol{x}_1\|^2\right)} d\boldsymbol{x}_2 d\boldsymbol{x}_1 \\
&\le \lambda^2 \int_{A_{\frac{1}{r_\rho}}} \int_{A_{\frac{1}{r_\rho}} \cap D(\boldsymbol{x}_1,\delta)} \\
&\quad e^{-\lambda\left(\int_{A_{\frac{1}{r_\rho}}} g(\|\boldsymbol{y}-\boldsymbol{x}_2\|)d\boldsymbol{y} + \int_{A_{\frac{1}{r_\rho}} \cap D(\boldsymbol{x}_1,r)\backslash D(\boldsymbol{x}_2,r)} g(\|\boldsymbol{y}-\boldsymbol{x}_1\|)(1-g(\|\boldsymbol{y}-\boldsymbol{x}_2\|))d\boldsymbol{y} - \pi\|\boldsymbol{x}_2-\boldsymbol{x}_1\|^2\right)} d\boldsymbol{x}_2 d\boldsymbol{x}_1 \quad (2.3.10) \\
&\le \lambda^2 \int_{A_{\frac{1}{r_\rho}}} \int_{A_{\frac{1}{r_\rho}} \cap D(\boldsymbol{x}_1,\delta)} \\
&\quad e^{-\lambda\left(\int_{A_{\frac{1}{r_\rho}}} g(\|\boldsymbol{y}-\boldsymbol{x}_2\|)d\boldsymbol{y} + g(r^-)(1-g(r^+))\left|A_{\frac{1}{r_\rho}} \cap D(\boldsymbol{x}_1,r)\backslash D(\boldsymbol{x}_2,r)\right| - \pi\|\boldsymbol{x}_2-\boldsymbol{x}_1\|^2\right)} d\boldsymbol{x}_2 d\boldsymbol{x}_1 \quad (2.3.11)
\end{aligned}$$

where in (2.3.10) the parameter $r > 0$ is chosen such that $g\left(r^-\right)\left(1-g\left(r^+\right)\right) > 0$. For convenience, use β for $g\left(r^-\right)\left(1-g\left(r^+\right)\right)$ as defined in (2.3.5). It follows from (2.3.11) that

$$\sum_{k=2}^{\infty} E\left(\xi_{k,1}\right)$$

$$\leq \lambda^2 \int_{B_r\left(A_{\frac{1}{r\rho}}\right)} \int_{A_{\frac{1}{r\rho}} \cap D(\boldsymbol{x}_1,\delta)} e^{-\lambda\left(\int_{A_{\frac{1}{r\rho}}} g(\|\boldsymbol{y}-\boldsymbol{x}_2\|)d\boldsymbol{y}+\beta\left|A_{\frac{1}{r\rho}} \cap D(\boldsymbol{x}_1,r)\setminus D(\boldsymbol{x}_2,r)\right|-\pi\|\boldsymbol{x}_2-\boldsymbol{x}_1\|^2\right)}$$

$$d\boldsymbol{x}_2 d\boldsymbol{x}_1$$

$$+ \lambda^2 \int_{A_{\frac{1}{r\rho}} \setminus B_r\left(A_{\frac{1}{r\rho}}\right)} \int_{A_{\frac{1}{r\rho}} \cap D(\boldsymbol{x}_1,\delta)} e^{-\lambda\left(\int_{A_{\frac{1}{r\rho}}} g(\|\boldsymbol{y}-\boldsymbol{x}_2\|)d\boldsymbol{y}+\beta\left|A_{\frac{1}{r\rho}} \cap D(\boldsymbol{x}_1,r)\setminus D(\boldsymbol{x}_2,r)\right|-\pi\|\boldsymbol{x}_2-\boldsymbol{x}_1\|^2\right)}$$

$$d\boldsymbol{x}_2 d\boldsymbol{x}_1 \tag{2.3.12}$$

For the first summand in the above equation, it can be shown that

$$\lambda^2 \int_{B_r\left(A_{\frac{1}{r\rho}}\right)} \int_{A_{\frac{1}{r\rho}} \cap D(\boldsymbol{x}_1,\delta)} e^{-\lambda\left(\int_{A_{\frac{1}{r\rho}}} g(\|\boldsymbol{y}-\boldsymbol{x}_2\|)d\boldsymbol{y}+\beta\left|A_{\frac{1}{r\rho}} \cap D(\boldsymbol{x}_1,r)\setminus D(\boldsymbol{x}_2,r)\right|-\pi\|\boldsymbol{x}_2-\boldsymbol{x}_1\|^2\right)} d\boldsymbol{x}_2 d\boldsymbol{x}_1$$

$$= 4\lambda^2 \int_{\ell A_{\frac{1}{r\rho}}} \int_{A_{\frac{1}{r\rho}} \cap D(\boldsymbol{x}_1,\delta)} e^{-\lambda\left(\int_{A_{\frac{1}{r\rho}}} g(\|\boldsymbol{y}-\boldsymbol{x}_2\|)d\boldsymbol{y}+\beta\left|A_{\frac{1}{r\rho}} \cap D(\boldsymbol{x}_1,r)\setminus D(\boldsymbol{x}_2,r)\right|-\pi\|\boldsymbol{x}_2-\boldsymbol{x}_1\|^2\right)} d\boldsymbol{x}_2 d\boldsymbol{x}_1$$

$$+ 4\lambda^2 \int_{\angle A_{\frac{1}{r\rho}}} \int_{A_{\frac{1}{r\rho}} \cap D(\boldsymbol{x}_1,\delta)} e^{-\lambda\left(\int_{A_{\frac{1}{r\rho}}} g(\|\boldsymbol{y}-\boldsymbol{x}_2\|)d\boldsymbol{y}+\beta\left|A_{\frac{1}{r\rho}} \cap D(\boldsymbol{x}_1,r)\setminus D(\boldsymbol{x}_2,r)\right|-\pi\|\boldsymbol{x}_2-\boldsymbol{x}_1\|^2\right)} d\boldsymbol{x}_2 d\boldsymbol{x}_1$$

$$\tag{2.3.13}$$

Denote by $\gamma(\boldsymbol{x})$ the shortest Euclidean distance between a point $\boldsymbol{x} \in \ell_{r+\delta} A_{\frac{1}{r\rho}}$ and a border of $A_{\frac{1}{r\rho}}$ adjacent to $\ell_{r+\delta} A_{\frac{1}{r\rho}}$ (i.e., $\partial A_{\frac{1}{r\rho}} \cap \ell_{r+\delta} A_{\frac{1}{r\rho}}$), where $\ell_{r+\delta} A_{\frac{1}{r\rho}}$ denotes a boundary rectangular area of size $\left(r_\rho^{-1} - 2(r-\delta)\right) \times (r+\delta)$ within $r+\delta$ of the border of $A_{\frac{1}{r\rho}}$ and away from the corners of $A_{\frac{1}{r\rho}}$ by at least $r-\delta$. Denote by $B_{\gamma(\boldsymbol{x}_2)}\left(A_{\frac{1}{r\rho}}\right) \subset A_\rho$ a boundary area $\left\{\boldsymbol{x} \in \ell_{r+\delta} A_{\frac{1}{r\rho}} : \gamma(\boldsymbol{x}) \leq \gamma(\boldsymbol{x}_2)\right\}$, and denote by $R(\boldsymbol{x}_2, 2r)$ a rectangular area of size $2r \times \gamma(\boldsymbol{x}_2)$ located between $\boldsymbol{x}_2$ and $\partial A_{\frac{1}{r\rho}}$ with $\boldsymbol{x}_2$ at the center of one side of $R(\boldsymbol{x}_2, 2r)$. See Fig. 2.5 for an illustration of the areas defined above.

We first evaluate the term: $\int_{B_{\gamma(\boldsymbol{x}_2)}\left(A_{\frac{1}{r\rho}}\right)} g(\|\boldsymbol{y}-\boldsymbol{x}_2\|)\, d\boldsymbol{y} + \beta\left|A_{\frac{1}{r\rho}} \cap D(\boldsymbol{x}_1, r) \setminus D(\boldsymbol{x}_2, r)\right|$. It can be shown that for $\boldsymbol{x}_1 \in \ell A_{\frac{1}{r\rho}}$ and $\boldsymbol{x}_2 \in A_{\frac{1}{r\rho}} \cap D(\boldsymbol{x}_1, \delta)$, when

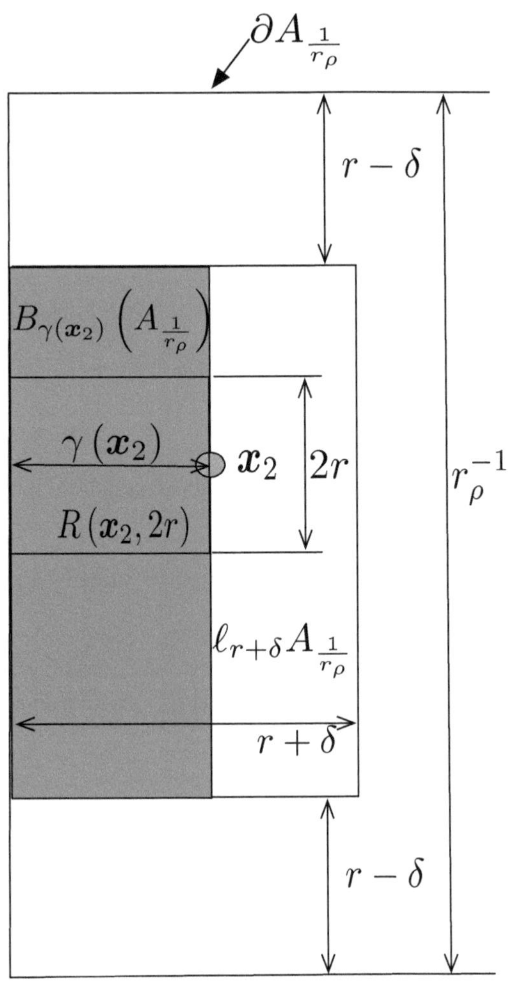

Fig. 2.5 An illustration of the areas $\ell_{r+\delta}A_{\frac{1}{r\rho}}$, $B_{\gamma(\boldsymbol{x}_2)}\left(A_{\frac{1}{r\rho}}\right)$ and $R(\boldsymbol{x}_2, 2r)$. The *shaded area* is $B_{\gamma(\boldsymbol{x}_2)}\left(A_{\frac{1}{r\rho}}\right)$

$\gamma(\boldsymbol{x}_2) \geq \gamma(\boldsymbol{x}_1)$:

$$\begin{aligned}
&\int_{B_{\gamma(\boldsymbol{x}_2)}\left(A_{\frac{1}{r\rho}}\right)} g\left(\|\boldsymbol{y}-\boldsymbol{x}_2\|\right) d\boldsymbol{y} \\
&\geq \int_{B_{\gamma(\boldsymbol{x}_2)}\left(A_{\frac{1}{r\rho}}\right)\cap R(\boldsymbol{x}_2,2r)} g\left(\|\boldsymbol{y}-\boldsymbol{x}_2\|\right) d\boldsymbol{y} \\
&\geq \int_0^r \int_0^{\gamma(\boldsymbol{x}_2)} g\left(\sqrt{x^2+y^2}\right) dxdy
\end{aligned}$$

$$
\begin{aligned}
&\geq \int_0^r \int_0^{|\gamma(\boldsymbol{x}_2)-\gamma(\boldsymbol{x}_1)|} g\left(\sqrt{x^2+y^2}\right) dxdy \\
&\geq \int_0^r \int_0^{|\gamma(\boldsymbol{x}_2)-\gamma(\boldsymbol{x}_1)|} g\left(\sqrt{\left(\frac{r}{2}\right)^2+y^2}\right) dxdy \qquad (2.3.14) \\
&= c_3 \left|\gamma\left(\boldsymbol{x}_2\right)-\gamma\left(\boldsymbol{x}_1\right)\right|
\end{aligned}
$$

where in (2.3.14), the nonincreasing monotonicity condition on g and that $|\gamma(\boldsymbol{x}_2)-\gamma(\boldsymbol{x}_1)| \leq \|\boldsymbol{x}_2-\boldsymbol{x}_1\| \leq \delta \leq \frac{r}{2}$ is used, and $c_3 \triangleq \int_0^r g\left(\sqrt{\left(\frac{r}{2}\right)^2+y^2}\right) dy$. Since $g\left(r^-\right)\left(1-g\left(r^+\right)\right) > 0$, it follows from the nonincreasing monotonicity condition on g that $g\left(\frac{r}{2}\right) > 0$ and c_3 is a positive constant, i.e., $c_3 > 0$.

Choose c_2 to be sufficiently small such that $c_4 \triangleq \frac{c_3}{\beta} - c_2 > 0$ and choose δ to be sufficiently small such that $\delta \leq z_0$. Using (2.3.14) and Lemma 57, it follows that

$$
\begin{aligned}
&\int_{B_{\gamma(\boldsymbol{x}_2)}\left(A_{\frac{1}{r\rho}}\right)} g\left(\|\boldsymbol{y}-\boldsymbol{x}_2\|\right) d\boldsymbol{y} + \beta\left|A_{\frac{1}{r\rho}} \cap D\left(\boldsymbol{x}_1, r\right) \backslash D\left(\boldsymbol{x}_2, r\right)\right| \\
&\geq c_3\left|\gamma\left(\boldsymbol{x}_2\right)-\gamma\left(\boldsymbol{x}_1\right)\right| + \beta\left(\left(r-c_2\right)\|\boldsymbol{x}_2-\boldsymbol{x}_1\| - r \times \left|\gamma\left(\boldsymbol{x}_2\right)-\gamma\left(\boldsymbol{x}_1\right)\right|\right) \\
&= \beta\left(\left(r-c_2\right) - \left(r-\frac{c_3}{\beta}\right) \times \frac{\left|\gamma\left(\boldsymbol{x}_2\right)-\gamma\left(\boldsymbol{x}_1\right)\right|}{\|\boldsymbol{x}_2-\boldsymbol{x}_1\|}\right)\|\boldsymbol{x}_2-\boldsymbol{x}_1\|
\end{aligned}
$$

Note that $\frac{|\gamma(\boldsymbol{x}_2)-\gamma(\boldsymbol{x}_1)|}{\|\boldsymbol{x}_2-\boldsymbol{x}_1\|} \leq 1$, therefore,

$$
\left(r-c_2\right) - \left(r-\frac{c_3}{\beta}\right) \times \frac{\left|\gamma\left(\boldsymbol{x}_2\right)-\gamma\left(\boldsymbol{x}_1\right)\right|}{\|\boldsymbol{x}_2-\boldsymbol{x}_1\|} \geq \frac{c_3}{\beta} - c_2
$$

$$
\begin{aligned}
&\int_{B_{\gamma(\boldsymbol{x}_2)}\left(A_{\frac{1}{r\rho}}\right)} g\left(\|\boldsymbol{y}-\boldsymbol{x}_2\|\right) d\boldsymbol{y} + \beta\left|A_{\frac{1}{r\rho}} \cap D\left(\boldsymbol{x}_1, r\right) \backslash D\left(\boldsymbol{x}_2, r\right)\right| \\
&\geq \beta c_4\|\boldsymbol{x}_2-\boldsymbol{x}_1\| \qquad (2.3.15)
\end{aligned}
$$

When $\gamma\left(\boldsymbol{x}_2\right) < \gamma\left(\boldsymbol{x}_1\right)$, using Lemma 57

$$
\begin{aligned}
&\int_{B_{\gamma(\boldsymbol{x}_2)}\left(A_{\frac{1}{r\rho}}\right)} g\left(\|\boldsymbol{y}-\boldsymbol{x}_2\|\right) d\boldsymbol{y} + \beta\left|A_{\frac{1}{r\rho}} \cap D\left(\boldsymbol{x}_1, r\right) \backslash D\left(\boldsymbol{x}_2, r\right)\right| \\
&\geq \beta\left|A_{\frac{1}{r\rho}} \cap D\left(\boldsymbol{x}_1, r\right) \backslash D\left(\boldsymbol{x}_2, r\right)\right| \\
&\geq \beta\frac{c_1}{2}\|\boldsymbol{x}_2-\boldsymbol{x}_1\| \qquad (2.3.16)
\end{aligned}
$$

Let $c_5 \triangleq \min\left\{\frac{c_1}{2}, c_4\right\}$. It follows from (2.3.15) and (2.3.16) that

$$\int_{B_{\gamma(\boldsymbol{x}_2)}\left(A_{\frac{1}{r_\rho}}\right)} g\left(\|\boldsymbol{y}-\boldsymbol{x}_2\|\right) d\boldsymbol{y} + \beta \left|A_{\frac{1}{r_\rho}} \cap D\left(\boldsymbol{x}_1, r\right) \backslash D\left(\boldsymbol{x}_2, r\right)\right|$$
$$\geq \beta c_5 \left\|\boldsymbol{x}_2 - \boldsymbol{x}_1\right\|$$

Choose δ to be sufficiently small such that $\pi\delta \leq \frac{1}{2}c_5\beta$ and also $\delta \leq z_0$. Note also that for $\boldsymbol{x}_2 \in A_{\frac{1}{r_\rho}} \cap D\left(\boldsymbol{x}_1, \delta\right)$, $\|\boldsymbol{x}_2 - \boldsymbol{x}_1\| \leq \delta$. Then, it follows that

$$4\lambda^2 \int_{\ell A_{\frac{1}{r_\rho}}} \int_{A_{\frac{1}{r_\rho}} \cap D(\boldsymbol{x}_1,\delta)} e^{-\lambda\left(\int_{A_{\frac{1}{r_\rho}}} g(\|\boldsymbol{y}-\boldsymbol{x}_2\|)d\boldsymbol{y} + \beta\left|A_{\frac{1}{r_\rho}} \cap D(\boldsymbol{x}_1,r)\backslash D(\boldsymbol{x}_2,r)\right| - \pi\|\boldsymbol{x}_2-\boldsymbol{x}_1\|^2\right)} d\boldsymbol{x}_2 d\boldsymbol{x}_1$$

$$\leq 4\lambda^2 \int_{\ell A_{\frac{1}{r_\rho}}} \int_{A_{\frac{1}{r_\rho}} \cap D(\boldsymbol{x}_1,\delta)} e^{-\lambda\left(\int_{A_{\frac{1}{r_\rho}} \backslash B_{\gamma(\boldsymbol{x}_2)}\left(A_{\frac{1}{r_\rho}}\right)} g(\|\boldsymbol{y}-\boldsymbol{x}_2\|)d\boldsymbol{y} + \frac{1}{2}c_5\beta\|\boldsymbol{x}_2-\boldsymbol{x}_1\|\right)} d\boldsymbol{x}_2 d\boldsymbol{x}_1$$

$$\leq 4\lambda^2 \int_{\ell_{r+\delta} A_{\frac{1}{r_\rho}}} \int_{A_{\frac{1}{r_\rho}} \cap D(\boldsymbol{x}_2,\delta)} e^{-\lambda\left(\int_{A_{\frac{1}{r_\rho}} \backslash B_{\gamma(\boldsymbol{x}_2)}\left(A_{\frac{1}{r_\rho}}\right)} g(\|\boldsymbol{y}-\boldsymbol{x}_2\|)d\boldsymbol{y} + \frac{1}{2}c_5\beta\|\boldsymbol{x}_2-\boldsymbol{x}_1\|\right)} d\boldsymbol{x}_1 d\boldsymbol{x}_2$$

$$\leq 4\lambda^2 \int_0^\delta e^{-\lambda\frac{1}{2}c_5\beta x} 2\pi x dx \int_{\ell_{r+\delta} A_{\frac{1}{r_\rho}}} e^{-\lambda \int_{A_{\frac{1}{r_\rho}} \backslash B_{\gamma(\boldsymbol{x}_2)}\left(A_{\frac{1}{r_\rho}}\right)} g(\|\boldsymbol{y}-\boldsymbol{x}_2\|)d\boldsymbol{y}} d\boldsymbol{x}_2$$

$$= 32\pi \frac{1 - e^{-\lambda\frac{1}{2}c_5\beta\delta}\left(1 + \lambda\frac{1}{2}c_5\beta\delta\right)}{\left(c_5\beta\right)^2} \int_{\ell_{r+\delta} A_{\frac{1}{r_\rho}}} e^{-\lambda \int_{A_{\frac{1}{r_\rho}} \backslash B_{\gamma(\boldsymbol{x}_2)}\left(A_{\frac{1}{r_\rho}}\right)} g(\|\boldsymbol{y}-\boldsymbol{x}_2\|)d\boldsymbol{y}} d\boldsymbol{x}_2$$

We further divide $\ell_{r+\delta} A_{\frac{1}{r_\rho}}$ into two parts: one rectangular area of size $\left(r_\rho^{-1} - 2r_\rho^{-\varepsilon}\right) \times (r+\delta)$ in the center of $\ell_{r+\delta} A_{\frac{1}{r_\rho}}$, denoted by $\ell^1_{r+\delta} A_{\frac{1}{r_\rho}}$, and the other area $\ell^2_{r+\delta} A_\rho = \ell_{r+\delta} A_{\frac{1}{r_\rho}} \backslash \ell^1_{r+\delta} A_{\frac{1}{r_\rho}}$. It can be shown that

$$\lim_{\rho\to\infty} \int_{\ell^1_{r+\delta} A_{\frac{1}{r_\rho}}} e^{-\lambda \int_{A_{\frac{1}{r_\rho}} \backslash B_{\gamma(\boldsymbol{x}_2)}\left(A_{\frac{1}{r_\rho}}\right)} g(\|\boldsymbol{y}-\boldsymbol{x}_2\|)d\boldsymbol{y}} d\boldsymbol{x}_2$$

$$\leq \lim_{\rho\to\infty} \int_{\ell^1_{r+\delta} A_{\frac{1}{r_\rho}}} e^{-\lambda \int_{A_{\frac{1}{r_\rho}} \backslash B_{\gamma(\boldsymbol{x}_2)}\left(A_{\frac{1}{r_\rho}}\right) \cap D\left(\boldsymbol{x}_2, r_\rho^{-\varepsilon}\right)} g(\|\boldsymbol{y}-\boldsymbol{x}_2\|)d\boldsymbol{y}} d\boldsymbol{x}_2$$

$$= \lim_{\rho\to\infty} \left(r_\rho^{-1} - 2r_\rho^{-\varepsilon}\right) \times (r+\delta)\, e^{-\frac{1}{2}\lambda \int_{D\left(\mathbf{0},r_\rho^{-\varepsilon}\right)} g(\|\mathbf{y}\|)d\mathbf{y}}$$

$$= 0 \tag{2.3.17}$$

where the last step results due to (2.2.11), which showed that for g satisfying both (2.1.1) and (2.1.4), $\lim_{\rho\to\infty} \rho^{\frac{1}{2}} e^{-\frac{1}{2}\rho r_\rho^2 \int_{D\left(\mathbf{0},r_\rho^{-\varepsilon}\right)} g(\|\mathbf{x}\|)d\mathbf{x}} = e^{-\frac{b}{2}}$. Then, the result follows readily from the definition of r_ρ in (2.1.5). Note that the result in (2.3.17) cannot be obtained for g satisfying (1.1.8) and (1.1.9) only.

Using similar steps that resulted in (2.2.25), it can be shown that

$$\lim_{\rho\to\infty} \int_{\ell^2_{r+\delta} A_{\frac{1}{r_\rho}}} e^{-\lambda \int_{A_{\frac{1}{r_\rho}} \backslash B_\gamma(\mathbf{x}_2)\left(A_{\frac{1}{r_\rho}}\right)} g(\|\mathbf{y}-\mathbf{x}_2\|)d\mathbf{y}} \, d\mathbf{x}_2 = 0$$

The above equation, together with (2.3.17), allows us to conclude that the first term in (2.3.13) converges to 0 as $\rho \to \infty$:

$$\lim_{\rho\to\infty} 4\lambda^2 \int_{\ell A_{\frac{1}{r_\rho}}} \int_{A_{\frac{1}{r_\rho}} \cap D(\mathbf{x}_1,\delta)} e^{-\lambda\left(\int_{A_{\frac{1}{r_\rho}}} g(\|\mathbf{y}-\mathbf{x}_2\|)d\mathbf{y} + \beta\left|A_{\frac{1}{r_\rho}} \cap D(\mathbf{x}_1,r)\backslash D(\mathbf{x}_2,r)\right| - \pi\|\mathbf{x}_2-\mathbf{x}_1\|^2\right)}$$

$$d\mathbf{x}_2 d\mathbf{x}_1 = 0 \tag{2.3.18}$$

Now let us consider the second term in (2.3.13). First it can be shown that

$$\int_{A_{\frac{1}{r_\rho}}} g\left(\|\mathbf{y}-\mathbf{x}_2\|\right) d\mathbf{y} \geq \int_{A_{\frac{1}{r_\rho}} \cap D\left(\mathbf{x}_2, r_\rho^{-\varepsilon}\right)} g\left(\|\mathbf{y}-\mathbf{x}_2\|\right) d\mathbf{y}$$

and for any $\mathbf{x}_2 \in A_{\frac{1}{r_\rho}}$, $A_{\frac{1}{r_\rho}} \cap D\left(\mathbf{x}_2, r_\rho^{-\varepsilon}\right)$ contains at least one quarter of $D\left(\mathbf{x}_2, r_\rho^{-\varepsilon}\right)$. Furthermore, since

$$\lim_{\rho\to\infty} \int_{D\left(\mathbf{x}_2, r_\rho^{-\varepsilon}\right)} g\left(\|\mathbf{y}-\mathbf{x}_2\|\right) d\mathbf{y} = C$$

there exists a ρ_0 such that for $\rho \geq \rho_0$ and any positive constant $\gamma < 1$

$$\int_{D\left(\mathbf{x}_2, r_\rho^{-\varepsilon}\right)} g\left(\|\mathbf{y}-\mathbf{x}_2\|\right) d\mathbf{y} \geq \gamma C$$

As a result of the above discussions, it can be shown that for sufficiently large $\rho \geq \rho_0$

$$4\lambda^2 \int_{\angle A_{\frac{1}{r\rho}}} \int_{A_{\frac{1}{r\rho}} \cap D(x_1,\delta)} e^{-\lambda\left(\int_{A_{\frac{1}{r\rho}}} g(\|y-x_2\|)dy+\beta\left|A_{\frac{1}{r\rho}} \cap D(x_1,r)\backslash D(x_2,r)\right|-\pi\|x_2-x_1\|^2\right)} dx_2 dx_1$$

$$\leq 4\lambda^2 \int_{\angle A_{\frac{1}{r\rho}}} \int_{A_{\frac{1}{r\rho}} \cap D(x_1,\delta)} e^{-\lambda\left(\frac{1}{4}\gamma C-\pi\|x_2-x_1\|^2\right)} dx_2 dx_1$$

$$\leq 4\lambda^2 \pi \delta^2 r^2 e^{-\lambda\left(\frac{1}{4}\gamma C-\pi\delta^2\right)} \tag{2.3.19}$$

where by choosing $\delta < \frac{1}{4\pi}\gamma C$, the above equation can be easily shown as converging to 0 as $\rho \to \infty$.

In summary, using (2.3.13), (2.3.18), and (2.3.19), it can be shown that for $\delta < \min\left\{\frac{1}{4\pi}\gamma C, \frac{r}{2}, \frac{1}{2\pi}c_5\beta, z_0\right\}$, for the first term in (2.3.12), we have

$$\lim_{\rho\to\infty} \lambda^2 \int_{B_r\left(A_{\frac{1}{r\rho}}\right)} \int_{A_{\frac{1}{r\rho}} \cap D(x_1,\delta)} e^{-\lambda\left(\int_{A_{\frac{1}{r\rho}}} g(\|y-x_2\|)dy+\beta\left|A_{\frac{1}{r\rho}} \cap D(x_1,r)\backslash D(x_2,r)\right|-\pi\|x_2-x_1\|^2\right)}$$
$$dx_2 dx_1 = 0 \tag{2.3.20}$$

For the second term in (2.3.12), using Lemma 56, it can be shown that

$$\lambda^2 \int_{A_{\frac{1}{r\rho}} \backslash B_r\left(A_{\frac{1}{r\rho}}\right)} \int_{A_{\frac{1}{r\rho}} \cap D(x_1,\delta)} e^{-\lambda\left(\int_{A_{\frac{1}{r\rho}}} g(\|y-x_2\|)dy+\beta\left|A_{\frac{1}{r\rho}} \cap D(x_1,r)\backslash D(x_2,r)\right|-\pi\|x_2-x_1\|^2\right)}$$
$$dx_2 dx_1$$

$$= \lambda^2 \int_{A_{\frac{1}{r\rho}} \backslash B_r\left(A_{\frac{1}{r\rho}}\right)} \int_{A_{\frac{1}{r\rho}} \cap D(x_1,\delta)} e^{-\lambda\left(\int_{A_{\frac{1}{r\rho}}} g(\|y-x_2\|)dy+\beta|D(x_1,r)\backslash D(x_2,r)|-\pi\|x_2-x_1\|^2\right)} dx_2 dx_1$$

$$\leq \lambda^2 \int_{A_{\frac{1}{r\rho}} \backslash B_r\left(A_{\frac{1}{r\rho}}\right)} \int_{A_{\frac{1}{r\rho}} \cap D(x_1,\delta)} e^{-\lambda\left(\int_{A_{\frac{1}{r\rho}}} g(\|y-x_2\|)dy+\left(\beta\sqrt{3}r-\pi\delta\right)\|x_2-x_1\|\right)} dx_2 dx_1$$

$$\leq \lambda^2 \int_{A_{\frac{1}{r\rho}}} \int_{A_{\frac{1}{r\rho}} \cap D(x_1,\delta)} e^{-\lambda\left(\int_{A_{\frac{1}{r\rho}}} g(\|y-x_2\|)dy+\left(\beta\sqrt{3}r-\pi\delta\right)\|x_2-x_1\|\right)} dx_2 dx_1$$

$$= \lambda^2 \int_{A_{\frac{1}{r_\rho}}} \int_{A_{\frac{1}{r_\rho}} \cap D(\boldsymbol{x}_2,\delta)} e^{-\lambda\left(\beta\sqrt{3}r-\pi\delta\right)\|\boldsymbol{x}_2-\boldsymbol{x}_1\|} d\boldsymbol{x}_1 e^{-\lambda \int_{A_{\frac{1}{r_\rho}}} g(\|\boldsymbol{y}-\boldsymbol{x}_2\|)d\boldsymbol{y}} d\boldsymbol{x}_2$$

$$\leq \frac{1-e^{-\lambda\left(\beta\sqrt{3}r-\pi\delta\right)\delta}\left(1+\lambda\left(\beta\sqrt{3}r-\pi\delta\right)\delta\right)}{\lambda\left(\beta\sqrt{3}r-\pi\delta\right)^2} \lambda \int_{A_{\frac{1}{r_\rho}}} e^{-\lambda \int_{A_{\frac{1}{r_\rho}}} g(\|\boldsymbol{y}-\boldsymbol{x}_2\|)d\boldsymbol{y}} d\boldsymbol{x}_2$$

In Theorem 41, we have established that

$$\lim_{\rho\to\infty} \lambda \int_{A_{\frac{1}{r_\rho}}} e^{-\lambda \int_{A_{\frac{1}{r_\rho}}} g(\|\boldsymbol{y}-\boldsymbol{x}_2\|)d\boldsymbol{y}} d\boldsymbol{x}_2 = e^{-b}$$

Therefore, it follows straightforwardly that for $\delta < \beta\sqrt{3}r/\pi$

$$\lim_{\rho\to\infty} \lambda^2 \int_{A_{\frac{1}{r_\rho}} \setminus B_r\left(A_{\frac{1}{r_\rho}}\right)} \int_{A_{\frac{1}{r_\rho}} \cap D(\boldsymbol{x}_1,\delta)} e^{-\lambda\left(\int_{A_{\frac{1}{r_\rho}}} g(\|\boldsymbol{y}-\boldsymbol{x}_2\|)d\boldsymbol{y} + \beta\left|A_{\frac{1}{r_\rho}} \cap D(\boldsymbol{x}_1,r)\setminus D(\boldsymbol{x}_2,r)\right| - \pi\|\boldsymbol{x}_2-\boldsymbol{x}_1\|^2\right)}$$
$$d\boldsymbol{x}_2 d\boldsymbol{x}_1 = 0 \tag{2.3.21}$$

Using (2.3.11), (2.3.12), (2.3.20), and (2.3.21), we are able to conclude that by choosing δ to be a positive constant such that

$$\delta < \min\left\{\frac{1}{4\pi}\gamma C, \frac{r}{2}, \frac{1}{2\pi}c_5\beta, \beta\sqrt{3}r/\pi, z_0\right\}$$

$$\lim_{\rho\to\infty} \sum_{k=2}^{\infty} E\left(\xi_{k,1}\right) = 0 \tag{2.3.22}$$

An Analysis of the Second Term in (2.3.8)

Now let us consider the second term in (2.3.8), i.e.,

$$E\left(\xi_{k,2}\right)$$
$$= \frac{\lambda^k}{k!} \int_{A_{\frac{1}{r_\rho}}} \int_{\left(A_{\frac{1}{r_\rho}}\right)^{k-1} \setminus (D(\boldsymbol{x}_1,\delta))^{k-1}} g_1\left(\boldsymbol{x}_1, \boldsymbol{x}_2, \ldots, \boldsymbol{x}_k\right) e^{-\lambda \int_{A_{\frac{1}{r_\rho}}} g_2(\boldsymbol{y};\boldsymbol{x}_1,\boldsymbol{x}_2,\ldots,\boldsymbol{x}_k)d\boldsymbol{y}}$$
$$d\left(\boldsymbol{x}_2 \cdots \boldsymbol{x}_k\right) d\boldsymbol{x}_1$$

For $(\boldsymbol{x}_2 \cdots \boldsymbol{x}_k) \in \left(A_{\frac{1}{r\rho}}\right)^{k-1} \setminus (D(\boldsymbol{x}_1, \delta))^{k-1}$, there is at least one node in $\{\boldsymbol{x}_2 \cdots \boldsymbol{x}_k\}$ outside a Euclidean distance δ of $\boldsymbol{x}_1$ and belonging to $A_{\frac{1}{r\rho}} \backslash D(\boldsymbol{x}_1, \delta)$. Without losing generality, let that node be $\boldsymbol{x}_j \in A_{\frac{1}{r\rho}} \backslash D(\boldsymbol{x}_1, \delta)$, where $j \in \Gamma_k / \{1\}$.

Let $\angle A_{\frac{1}{r\rho}} \subset A_{\frac{1}{r\rho}}$ be a square area of size $r \times r$ located at a corner of $A_{\frac{1}{r\rho}}$ and let $\overline{\angle A_{\frac{1}{r\rho}}} \subset A_{\frac{1}{r\rho}}$ be an area in $A_{\frac{1}{r\rho}}$ excluding the four corner squares $\angle A_{\frac{1}{r\rho}}$. It is straightforward from the proofs of Lemmas 57 and 56 that for $\boldsymbol{x}_1 \in \overline{\angle A_{\frac{1}{r\rho}}}$ and $\boldsymbol{x}_j \in A_{\frac{1}{r\rho}} \backslash D(\boldsymbol{x}_1, \delta)$, i.e., $\left\|\boldsymbol{x}_j - \boldsymbol{x}_1\right\| \geq \delta$, there exists a positive constant $c_6 > 0$, depending on δ, such that

$$\left|A_{\frac{1}{r\rho}} \cap D(\boldsymbol{x}_1, r) \backslash D(\boldsymbol{x}_2, r)\right| \geq c_6$$

Using the above inequality and (2.3.3), it follows that

$$\begin{aligned}
&\int_{A_{\frac{1}{r\rho}}} g_2(\boldsymbol{y}; \boldsymbol{x}_1, \boldsymbol{x}_2, \ldots, \boldsymbol{x}_k)\, d\boldsymbol{y} \\
&\geq \int_{A_{\frac{1}{r\rho}}} g_2(\boldsymbol{y}; \boldsymbol{x}_1, \boldsymbol{x}_j)\, d\boldsymbol{y} \\
&= \int_{A_{\frac{1}{r\rho}}} g\left(\left\|\boldsymbol{y} - \boldsymbol{x}_j\right\|\right) + g\left(\left\|\boldsymbol{y} - \boldsymbol{x}_1\right\|\right)\left(1 - g\left(\left\|\boldsymbol{y} - \boldsymbol{x}_j\right\|\right)\right) d\boldsymbol{y} \\
&\geq \int_{A_{\frac{1}{r\rho}}} g\left(\left\|\boldsymbol{y} - \boldsymbol{x}_j\right\|\right) d\boldsymbol{y} + \beta \left|A_{\frac{1}{r\rho}} \cap D(\boldsymbol{x}_1, r) \backslash D(\boldsymbol{x}_j, r)\right| \\
&\geq \int_{A_{\frac{1}{r\rho}}} g\left(\left\|\boldsymbol{y} - \boldsymbol{x}_j\right\|\right) d\boldsymbol{y} + \beta c_6
\end{aligned}$$

Therefore,

$$\begin{aligned}
&\frac{\lambda^k}{k!} \int_{\overline{\angle A_{\frac{1}{r\rho}}}} \int_{\left(A_{\frac{1}{r\rho}}\right)^{k-1} \backslash (D(\boldsymbol{x}_1, \delta))^{k-1}} g_1(\boldsymbol{x}_1, \boldsymbol{x}_2, \ldots, \boldsymbol{x}_k)\, e^{-\lambda \int_{A_{\frac{1}{r\rho}}} g_2(\boldsymbol{y}; \boldsymbol{x}_1, \boldsymbol{x}_2, \ldots, \boldsymbol{x}_k) d\boldsymbol{y}} \\
&\quad d(\boldsymbol{x}_2 \cdots \boldsymbol{x}_k)\, d\boldsymbol{x}_1 \\
&\leq \frac{\lambda^k}{k!} \int_{\overline{\angle A_{\frac{1}{r\rho}}}} \int_{\left(A_{\frac{1}{r\rho}}\right)^{k-1} \backslash (D(\boldsymbol{x}_1, \delta))^{k-1}} g_1(\boldsymbol{x}_1, \boldsymbol{x}_2, \ldots, \boldsymbol{x}_k)\, e^{-\lambda \int_{A_{\frac{1}{r\rho}}} g\left(\left\|\boldsymbol{y} - \boldsymbol{x}_j\right\|\right) d\boldsymbol{y} - \lambda \beta c_6} \\
&\quad d(\boldsymbol{x}_2 \cdots \boldsymbol{x}_k)\, d\boldsymbol{x}_1
\end{aligned}$$

$$\leq \frac{\lambda^k}{k!} \int_{\left(A_{\frac{1}{r_\rho}}\right)^k} g_1\left(\boldsymbol{x}_1, \boldsymbol{x}_2, \ldots, \boldsymbol{x}_k\right) e^{-\lambda \int_{A_{\frac{1}{r_\rho}}} g\left(\left\|\boldsymbol{y}-\boldsymbol{x}_j\right\|\right) d\boldsymbol{y} - \lambda \beta c_6} d\left(\boldsymbol{x}_1 \boldsymbol{x}_2 \cdots \boldsymbol{x}_k\right)$$

$$= \frac{\lambda^k}{k!} \int_{\left(A_{\frac{1}{r_\rho}}\right)^k} g_1\left(\boldsymbol{x}_1, \boldsymbol{x}_2, \ldots, \boldsymbol{x}_k\right) e^{-\lambda \int_{A_{\frac{1}{r_\rho}}} g\left(\left\|\boldsymbol{y}-\boldsymbol{x}_1\right\|\right) d\boldsymbol{y} - \lambda \beta c_6} d\left(\boldsymbol{x}_1 \boldsymbol{x}_2 \cdots \boldsymbol{x}_k\right) \tag{2.3.23}$$

where a renumbering of the nodes occurred in the last step of the above equation. First using Lemma 55, and then using (2.3.4) and the inequality that $\int_{A_{\frac{1}{r_\rho}}} g\left(\left\|\boldsymbol{x}_j - \boldsymbol{x}_i\right\|\right) d\boldsymbol{x}_i \leq C$, it can be shown that

$$\frac{\lambda^k}{k!} \int_{\left(A_{\frac{1}{r_\rho}}\right)^k} g_1\left(\boldsymbol{x}_1, \boldsymbol{x}_2, \ldots, \boldsymbol{x}_k\right) e^{-\lambda \int_{A_{\frac{1}{r_\rho}}} g\left(\left\|\boldsymbol{y}-\boldsymbol{x}_1\right\|\right) d\boldsymbol{y} - \lambda \beta c_6} d\left(\boldsymbol{x}_1 \boldsymbol{x}_2 \cdots \boldsymbol{x}_k\right)$$

$$\leq \frac{\lambda^k}{k!} \int_{\left(A_{\frac{1}{r_\rho}}\right)^k} \sum_{i_2 \in \Gamma_k \backslash \{1\}, \cdots, i_k \in \Gamma_k \backslash \{1, i_2, \ldots, i_{k-1}\}} g_2\left(\boldsymbol{x}_{i_2}; \boldsymbol{x}_1\right) \cdots g_2\left(\boldsymbol{x}_{i_k}; \boldsymbol{x}_1, \boldsymbol{x}_{i_2}, \ldots, \boldsymbol{x}_{i_{k-1}}\right)$$

$$\times e^{-\lambda \int_{A_{\frac{1}{r_\rho}}} g\left(\left\|\boldsymbol{y}-\boldsymbol{x}_1\right\|\right) d\boldsymbol{y} - \lambda \beta c_6} d\left(\boldsymbol{x}_{i_k} \cdots \boldsymbol{x}_{i_2} \boldsymbol{x}_1\right)$$

$$\leq \frac{\lambda^k C^{k-1}}{k!} (k-1)!\,(k-1)! \int_{A_{\frac{1}{r_\rho}}} e^{-\lambda \int_{A_{\frac{1}{r_\rho}}} g\left(\left\|\boldsymbol{y}-\boldsymbol{x}_1\right\|\right) d\boldsymbol{y} - \lambda \beta c_6} d\boldsymbol{x}_1$$

$$= \frac{(k-1)!}{k} e^{-\frac{b\beta c_6}{C}} \times \frac{(\log \rho + b)^{k-1}}{\rho^{\frac{\beta c_6}{C}}} \times \lambda \int_{A_{\frac{1}{r_\rho}}} e^{-\lambda \int_{A_\rho} g\left(\left\|\boldsymbol{y}-\boldsymbol{x}_1\right\|\right) d\boldsymbol{y}} d\boldsymbol{x}_1 \tag{2.3.24}$$

Using Theorem 41, (2.3.23), and (2.3.24), it follows that

$$\lim_{\rho \to \infty} \frac{\lambda^k}{k!} \int_{\angle A_{\frac{1}{r_\rho}}} \int_{\left(A_{\frac{1}{r_\rho}}\right)^{k-1} \backslash (D(\boldsymbol{x}_1, \delta))^{k-1}} g_1\left(\boldsymbol{x}_1, \boldsymbol{x}_2, \ldots, \boldsymbol{x}_k\right) e^{-\lambda \int_{A_{\frac{1}{r_\rho}}} g_2\left(\boldsymbol{y}; \boldsymbol{x}_1, \boldsymbol{x}_2, \ldots, \boldsymbol{x}_k\right) d\boldsymbol{y}}$$

$$d\left(\boldsymbol{x}_2 \cdots \boldsymbol{x}_k\right) d\boldsymbol{x}_1 = 0 \tag{2.3.25}$$

Using similar steps as those leading to (2.3.24), it can be shown that

$$\frac{\lambda^k}{k!} \int_{\angle A_{\frac{1}{r_\rho}}} \int_{\left(A_{\frac{1}{r_\rho}}\right)^{k-1} \backslash (D(\boldsymbol{x}_1, \delta))^{k-1}} g_1\left(\boldsymbol{x}_1, \boldsymbol{x}_2, \ldots, \boldsymbol{x}_k\right) e^{-\lambda \int_{A_{\frac{1}{r_\rho}}} g_2\left(\boldsymbol{y}; \boldsymbol{x}_1, \boldsymbol{x}_2, \ldots, \boldsymbol{x}_k\right) d\boldsymbol{y}}$$

$$d\left(\boldsymbol{x}_2 \cdots \boldsymbol{x}_k\right) d\boldsymbol{x}_1$$

$$\leq \frac{\lambda^k}{k!} \int_{\angle A_{\frac{1}{r_\rho}}} \int_{\left(A_{\frac{1}{r_\rho}}\right)^{k-1}} g_1\left(\boldsymbol{x}_1, \boldsymbol{x}_2, \ldots, \boldsymbol{x}_k\right) e^{-\lambda \int_{A_{\frac{1}{r_\rho}}} g_2\left(\left\|\boldsymbol{y}-\boldsymbol{x}_1\right\|\right) d\boldsymbol{y}} d\left(\boldsymbol{x}_2 \cdots \boldsymbol{x}_k\right) d\boldsymbol{x}_1$$

$$\leq \frac{\lambda^k C^{k-1}}{k} (k-1)! \int_{\angle A_{\frac{1}{r_\rho}}} e^{-\lambda \int_{A_{\frac{1}{r_\rho}}} g\left(\left\|\boldsymbol{y}-\boldsymbol{x}_1\right\|\right) d\boldsymbol{y}} d\boldsymbol{x}_1$$

Using similar steps that resulted in (2.3.19), it can be shown that

$$\begin{aligned}
&\lim_{\rho\to\infty}\frac{\lambda^k C^{k-1}}{k}(k-1)!\int_{\angle A_{\frac{1}{r_\rho}}} e^{-\lambda\int_{A_{\frac{1}{r_\rho}}} g(\|y-x_1\|)dy-\lambda\beta c_6}\, d\boldsymbol{x}_1\\
&\leq \lim_{\rho\to\infty}\frac{\lambda^k C^{k-1}}{k}(k-1)!\delta^2 e^{-\frac{1}{4}\lambda\gamma C}\\
&=0
\end{aligned}\tag{2.3.26}$$

The combination of (2.3.25) and (2.3.26) allows us to conclude that

$$\lim_{\rho\to\infty} E\left(\xi_{k,2}\right)=0$$

It follows that for any fixed but arbitrarily large integer M

$$\lim_{\rho\to\infty}\sum_{k=2}^{M} E\left(\xi_{k,2}\right)=0\tag{2.3.27}$$

Finally, from (2.3.22) and (2.3.27), we conclude that

$$\lim_{\rho\to\infty}\left(\sum_{k=2}^{M} E\left(\xi_k\right)=0\right)=0$$

Noting that ξ_k is a nonnegative integer, therefore,

$$\lim_{\rho\to\infty}\Pr\left(\sum_{k=2}^{M}\xi_k=0\right)=1$$

Hence Theorem 52 is proved.

Remark 58 Theorem 52 gives a sufficient condition on g required for the number of components of fixed and finite order $k>1$ in $\mathcal{G}\left(\mathcal{X}_{\frac{\log\rho+b}{C}}, g, A_{\frac{1}{r_\rho}}\right)$ to be vanishingly small as $\rho\to\infty$. It is also interesting to obtain a necessary condition on g required for the number of components of fixed and finite order $k>1$ in $\mathcal{G}\left(\mathcal{X}_{\frac{\log\rho+b}{C}}, g, A_{\frac{1}{r_\rho}}\right)$ to be vanishingly small. The technique used in the proof of Theorem 52 however cannot answer the above question on a necessary condition on g. More specifically, denote by ξ_k the (random) number of components of order k in an instance of $\mathcal{G}\left(\mathcal{X}_\lambda, g, A_{\frac{1}{r_\rho}}\right)$ and let M be an arbitrarily large positive integer M. The proof of Theorem 52 is based on an analysis of $E\left(\sum_{k=2}^{M}\xi_k\right)$. By showing that $\lim_{\rho\to\infty} E\left(\sum_{k=2}^{M}\xi_k\right)=0$, it follows that

$\lim_{\rho\to\infty}\Pr\left(\sum_{k=2}^{M}\xi_k=0\right)=1$. However $\lim_{\rho\to\infty}E\left(\sum_{k=2}^{M}\xi_k\right)=0$ is only a sufficient condition for $\lim_{\rho\to\infty}\Pr\left(\sum_{k=2}^{M}\xi_k=0\right)=1$, *not* a necessary condition. In the next chapter, we shall develop a technique to obtain a tight necessary condition on g required for the number of components of fixed and finite order $k>1$ in $\mathcal{G}\left(\mathcal{X}_{\frac{\log\rho+b}{C}},g,A_{\frac{1}{r_\rho}}\right)$ to be vanishingly small.

2.4 Notes and Further Readings

In this chapter, we discussed three widely used network models, including the dense network model $\mathcal{G}\left(\mathcal{X}_\rho,g_{r_\rho},A\right)$, the extended network model $\mathcal{G}\left(\mathcal{X}_1,g_{\sqrt{\frac{\log\rho+b}{C}}},A_{\sqrt{\rho}}\right)$, and the infinite network model $\mathcal{G}\left(\mathcal{X}_\rho,g,\Re^2\right)$, and their implications on the connectivity analysis. Using the scaling and coupling technique, it is shown that the dense network model and the extended network model are equivalent in their connectivity properties and they are also equivalent to the network model $\mathcal{G}\left(\mathcal{X}_{\frac{\log\rho+b}{C}},g,A_{\frac{1}{r_\rho}}\right)$, which can be obtained from the infinite network model $\mathcal{G}\left(\mathcal{X}_{\frac{\log\rho+b}{C}},g,\Re^2\right)$ by removing all nodes and associated connections outside the area $A_{\frac{1}{r_\rho}}$ of $\mathcal{G}\left(\mathcal{X}_\rho,g,\Re^2\right)$. Define the effect associated with the above removal operation as the truncation effect. A prerequisite for any (asymptotic) conclusion obtained in the infinite network model to be applicable to the dense and extended network models is that the impact of the truncation effect must be vanishingly small on the parameter concerned as $\rho\to\infty$—a conclusion that often needs nontrivial analysis to establish. We then conducted two case studies using a random connection model, on the expected number of isolated nodes and on the vanishing of components of fixed and finite order $k>1$ respectively, with a focus on examining the impact of the truncation effect and showed that the connection function g has to decrease sufficiently fast in order for the truncation effect to have a vanishingly small impact.

Extensive research has been done on connectivity problems using the well-known random geometric graph and the unit disk connection model [164, 165]. In [167], Philips et al. proved that the average node degree must grow logarithmically with the area of the network to ensure that the network is connected, where nodes are placed randomly on a square according to a Poisson point process with a known density in $\Re^2$. This result by Philips et al. actually provides a necessary condition on the average node degree required for connectivity. In [193], Xue et al. showed that in a network with a total of n nodes randomly and uniformly distributed on a unit square in $\Re^2$, if each node is connected to $c\log n$ nearest neighbors with $c\le 0.074$ then the resulting random network is asymptotically almost surely disconnected as $n\to\infty$; and if each node is connected to $c\log n$ nearest neighbors with $c\ge 5.1774$ then the network is asymptotically almost surely connected as $n\to\infty$. In [12], Balister et al. advanced the results in [193] and improved the lower and upper bounds to

$0.3043 \log n$ and $0.5139 \log n$ respectively. In a more recent work [13], Balister et al. achieved much improved results by showing that there exists a constant c_{crit} such that if each node is connected to $\lfloor c \log n \rfloor$ nearest neighbors with $c < c_{crit}$ then the network is asymptotically almost surely disconnected as $n \to \infty$, and if each node is connected to $\lfloor c \log n \rfloor$ nearest neighbors with $c > c_{crit}$ then the network is asymptotically almost surely connected as $n \to \infty$. Here, $\lfloor x \rfloor$ means the largest integer that is smaller than or equal to x. In both [12] and [13], the authors considered nodes randomly distributed following a Poisson point process of density one on a square of area n in $\Re^2$. In [173], Ravelomanana investigated the critical transmission range for connectivity in three-dimensional wireless sensor networks and derived similar results to the two-dimensional results in [92].

The log-normal connection model, which is more realistic than the unit disk connection model, has also been considered for investigating network connectivity in [20, 21, 100, 145, 146, 160]. In [20, 21, 100, 145, 146, 160], the authors investigated from different perspectives the necessary condition for a network with nodes uniformly or Poissonly distributed in a bounded area in $\Re^2$ and a pair of nodes are directly connected following the log-normal connection model to be connected.

Other work in the area include [57, 81, 119, 129], which studied from the percolation perspective, the impact of mutual interference caused by simultaneous transmissions, the impact of physical layer cooperative transmissions, the impact of directional antennas, and the impact of unreliable links on connectivity respectively.

Chapter 3
Connectivity of Large Wireless Networks: Sufficient and Necessary Conditions

Abstract This chapter studies the sufficient and necessary condition for a large wireless network to be asymptotically almost surely connected. Consider a dense network model, we show that as the node density approaches infinity, (a) the distribution of the number of isolated nodes converges to a Poisson distribution; (b) asymptotically almost surely (*a.a.s.*) there is no component of fixed and finite order $k > 1$; (c) *a.a.s.* the number of components with an unbounded order is one. Therefore, as the node density approaches infinity, the network *a.a.s.* contains a unique unbounded component and isolated nodes only; a sufficient and necessary condition for the network to be *a.a.s.* connected is that there is no isolated node in the network. These results, established assuming a general random connection model, readily incorporate existing results established assuming the unit disk connection model and the fewer results assuming the log-normal connection model as its special cases.

In this chapter, we investigate the sufficient and necessary condition for a large wireless network to be asymptotically almost surely (*a.a.s.*) connected.

More specifically, we consider a network where all nodes are distributed on a unit square $A \triangleq [-\frac{1}{2}, \frac{1}{2}]^2$ following a Poisson distribution with known density ρ and a pair of nodes are directly connected following a random connection model, viz., a pair of nodes separated by a Euclidean distance x are directly connected with probability $g_{r_\rho}(x) \triangleq g(x/r_\rho)$, where $g : [0, \infty) \to [0, 1]$, independent of the event that another pair of nodes are directly connected. Here

$$r_\rho = \sqrt{\frac{\log \rho + b}{C\rho}} \tag{3.0.1}$$

and b is a constant. The reason for choosing this particular form of r_ρ is that the analysis becomes nontrivial when b is a constant. Other forms of r_ρ can be accommodated by dropping the assumption that b is constant, i.e., b becomes a function of ρ, and allowing $b \to \infty$ or $b \to -\infty$ as $\rho \to \infty$. The results are rapidly attainable, and we discuss these situations separately in Sects. 3.2 and 3.2.2.

G. Mao, *Connectivity of Communication Networks*,
DOI 10.1007/978-3-319-52989-9_3

Same as in Chap. 2, the connection function g is required to satisfy the following properties of monotonicity, integral boundedness and rotational invariance:

$$g(x) \le g(y) \qquad whenever\ \ x \ge y \tag{3.0.2}$$

$$0 < C \triangleq \int_{\Re^2} g(\|\boldsymbol{x}\|)d\boldsymbol{x} < \infty \tag{3.0.3}$$

As shown in Sect. 2.1, (3.0.2) and (3.0.3) allow us to conclude that

$$g(x) = o_x(1/x^2) \tag{3.0.4}$$

However, we require g to satisfy the more restrictive requirement that

$$g(x) = o_x(1/(x^2 \log^2 x)) \tag{3.0.5}$$

As discussed in Chap. 2, the more restrictive requirement is required to ensure that the impact of the *truncation effect* on connectivity is asymptotically vanishingly small as $\rho \to \infty$.

For convenience we also assume that g has infinite support when necessary. Our results however apply to the situation when g has bounded support, which forms a special case and actually makes the analysis easier.

Denote the above network by $\mathcal{G}\left(\mathcal{X}_\rho, g_{r_\rho}, A\right)$. The results established for $\mathcal{G}\left(\mathcal{X}_\rho, g_{r_\rho}, A\right)$ on a unit square can be readily extended to networks of other size using the coupling and scaling technique. The results are also applicable for networks employing other connection models such as the unit disk connection model and the log-normal connection model, which form special cases of the random connection model.

Several results will be established when we investigate the sufficient and necessary condition for $\mathcal{G}\left(\mathcal{X}_\rho, g_{r_\rho}, A\right)$ to be asymptotically almost surely connected as $\rho \to \infty$. First, using the Chen-Stein technique [7, 15], to be introduced later in the chapter, we show that the distribution of the number of isolated nodes in $\mathcal{G}\left(\mathcal{X}_\rho, g_{r_\rho}, A\right)$ asymptotically converges to a Poisson distribution as $\rho \to \infty$. This result readily leads to a necessary condition for $\mathcal{G}\left(\mathcal{X}_\rho, g_{r_\rho}, A\right)$ to be asymptotically almost surely connected as $\rho \to \infty$. Secondly, we show that as $\rho \to \infty$, the number of components in $\mathcal{G}\left(\mathcal{X}_\rho, g_{r_\rho}, A\right)$ of unbounded order converges to one. This result, together with the result in Chap. 2 that the number of components of finite order $k > 1$ in $\mathcal{G}\left(\mathcal{X}_\rho, g_{r_\rho}, A\right)$ asymptotically vanishes as $\rho \to \infty$, allows us to conclude that as $\rho \to \infty$, asymptotically almost surely there are only a unique unbounded component and isolated nodes in $\mathcal{G}\left(\mathcal{X}_\rho, g_{r_\rho}, A\right)$. The above results allow us to establish that the sufficient and necessary condition for $\mathcal{G}\left(\mathcal{X}_\rho, g_{r_\rho}, A\right)$ to be asymptotically almost surely connected is that there is no isolated node in the network. On that basis, we obtain the asymptotic probability that $\mathcal{G}\left(\mathcal{X}_\rho, g_{r_\rho}, A\right)$ forms a connected network as $\rho \to \infty$ and the sufficient and necessary condition for $\mathcal{G}\left(\mathcal{X}_\rho, g_{r_\rho}, A\right)$ to be asymptotically almost surely connected.

3.1 Connectivity of Large Wireless Networks: The Unit Disk Connection Model

Before we delve into the study of connectivity of $\mathcal{G}\left(\mathcal{X}_\rho, g_{r_\rho}, A\right)$, in this section, we first give an introduction to the work of Gupta and Kumar [92] on connectivity of large wireless networks assuming the unit disk connection model. Specifically, in [92] Gupta and Kumar considered a network with n nodes independently and identically distributed on a disk of unit area, denoted by D, following a uniform distribution, where a pair of nodes are directly connected if and only if their Euclidean distance is smaller than or equal to $r(n)$, known as the transmission range. Let $r(n) = \sqrt{\frac{\log(n)+c(n)}{\pi n}}$. They showed that as $n \to \infty$, the above network is asymptotically almost surely connected if and only if $c(n) \to \infty$. The work of Gupta and Kumar [92] is among the earliest works that studied connectivity of large wireless networks. The techniques used in their work have influenced many later studies including ours. Therefore, in this section, we introduce the main techniques used in their study.

In this section, we ignore the impact of the boundary effect which arises when nodes are located near the boundary of D to focus on the main ideas. For readers interested in techniques to handle the boundary effect, we refer to Appendix of [92] and Chap. 2.

Denote the network studied by Gupta and Kumar [92] by $G(n, r(n))$. In [92], they first studied a lower bound on the probability that there is at least one isolated node in $G(n, r(n))$. A necessary condition on $r(n)$ for $G(n, r(n))$ to be asymptotically almost surely connected is that the probability of having at least one isolated node approaches 0. In this way, a necessary condition on $r(n)$ for $G(n, r(n))$ to be asymptotically almost surely connected is obtained. Secondly, they studied an upper bound on the probability that a network with nodes Poissonly distributed on D with node density n and transmission range $r(n)$ has no isolated node. Denote the network with Poisson node distribution by $G^{Poisson}(n, r(n))$. In continuum percolation theory, it has been established that in an infinite network with node density n and transmission range $r(n)$, as $n \to \infty$, the infinite network asymptotically almost surely has only a giant component of an infinite order and isolated nodes. For a finite (asymptotically infinite) network assuming *a unit disk connection model*, a similar result has also been established in [165, Chapter 10]. Applying the above results and showing that the difference between $G(n, r(n))$ and $G^{Poisson}(n, r(n))$ is negligible for large n, it was concluded that as $n \to \infty$, $G(n, r(n))$ has only a component of extremely large order and isolated nodes. Therefore, when the isolated nodes vanish, the network becomes connected. On that basis, a sufficient condition on $r(n)$ for $G(n, r(n))$ to be asymptotically almost surely connected results.

In the rest of this section, we introduce in greater details the techniques used in [92] to establish the key results summarized in the previous paragraphs.

We start with the analysis of the probability that there is at least one isolated node in $G(n, r(n))$. For brevity, we use $P_{iso}(G)$ to denote the probability that the graph G

has at least one isolated node and use $P_d(G)$ to denote the probability that the graph G is disconnected. Denote by v_i, $i \in \{1, \ldots, n\}$ the i-th node in $G(n, r(n))$. It can be established that

$$\begin{aligned} &P_{iso}(G(n, r(n))) \\ &\quad = \Pr\left(\cup_{i=1}^{n}\{v_i \text{ is an isolated node}\}\right) \\ &\quad \geq \sum_{i=1}^{n}\Pr(\{v_i \text{ is an isolated node}\}) - \sum_{i=1}^{n}\sum_{j\neq i}\Pr\left(\{v_i \text{ and } v_j \text{ are isolated}\}\right) \end{aligned} \tag{3.1.1}$$

For the two terms in (3.1.1), neglecting the boundary effect and noting that the nodes are on a disk of unit area, it can be further established using some straightforward geometric analysis taking into account that v_i may be at a distance between $r(n)$ and $2r(n)$ from v_j:

$$\Pr(\{v_i \text{ is an isolated node}\}) = \left(1 - \pi r^2(n)\right)^{n-1} \tag{3.1.2}$$

$$\begin{aligned} &\Pr\left(\{v_i \text{ and } v_j \text{ are isolated}\}\right) \\ &\quad = \left(4\pi r^2(n) - \pi r^2(n)\right)\left(1 - \frac{5}{4}\pi r^2(n)\right)^{n-2} \\ &\quad + \left(1 - 4\pi r^2(n)\right)\left(1 - 2\pi r^2(n)\right)^{n-2} \end{aligned} \tag{3.1.3}$$

Putting (3.1.2) and (3.1.3) into (3.1.1), there results

$$\begin{aligned} &P_{iso}(G(n, r(n))) \\ &\quad \geq n\left(1 - \pi r^2(n)\right)^{n-1} - n(n-1) \\ &\quad \times \left[3\pi r^2(n)\left(1 - \frac{5}{4}\pi r^2(n)\right)^{n-2} + \left(1 - 4\pi r^2(n)\right)\left(1 - 2\pi r^2(n)\right)^{n-2}\right] \end{aligned} \tag{3.1.4}$$

Further using the following two bounds

Lemma 59 ([92, Lemma 2.1]) *For any $p \in [0, 1]$, $(1 - p) \leq e^{-p}$. For any given $\theta \geq 1$, there exists $p_0 \in [0, 1]$, such that*

$$e^{-\theta p} \leq 1 - p, \text{ for all } 0 \leq p \leq p_0$$

If $\theta > 1$, then $p_0 > 0$.

Lemma 60 ([92, Lemma 2.2]) *If $\pi r^2(n) = \frac{\log n + c(n)}{n}$, then for any fixed $\theta < 1$ and for all sufficiently large n*

$$n\left(1 - \pi r^2(n)\right)^{n-1} \geq \theta e^{-c(n)}$$

and noting that $\pi r^2(n) = \frac{\log n + c(n)}{n}$, it can be obtained that for any fixed $\theta < 1$ and $\varepsilon > 0$,

$$\begin{aligned}
& P_{iso}(G(n, r(n))) \\
& \geq \theta e^{-c(n)} - n(n-1)\left(3\pi r^2(n)\, e^{-\frac{5}{4}(n-2)\pi r^2(n)} + e^{-2(n-2)\pi r^2(n)}\right) \\
& \geq \theta e^{-c(n)} - (1+\varepsilon)\, e^{-2c(n)}
\end{aligned}$$

It follows that as $n \to \infty$, a necessary condition for $P_{iso}(G(n, r(n))) \to 0$ is that $c(n) \to \infty$.

The above discussion can be summarized into the following theorem.

Theorem 61 *As $n \to \infty$, $G(n, r(n))$ is asymptotically almost surely connected only if $c(n) \to \infty$.*

We now move on to study the sufficient condition for $G(n, r(n))$ to be asymptotically almost surely connected. The following result can be first established:

Lemma 62 *If $\pi r^2(n) = \frac{\log n + c(n)}{n}$, then*

$$\limsup_{n\to\infty} P_{iso}\left(G^{Poisson}(n, r(n))\right) \leq e^{-c}$$

where $c = \lim_{n\to\infty} c(n)$.

Proof The number of nodes in $G^{Poisson}(n, r(n))$ is Poisson distributed with mean n. Let $W(k)$ be the random number of isolated nodes in $G(k, r(n))$, it follows that

$$\begin{aligned}
& P_{iso}\left(G^{Poisson}(n, r(n))\right) \\
& = \sum_{k=1}^{\infty} \frac{n^k}{k!} e^{-n} P_{iso}(G(k, r(n))) \\
& \leq \sum_{k=1}^{\infty} \frac{n^k}{k!} e^{-n} E[W(k)] \\
& = \sum_{k=1}^{\infty} \frac{n^k}{k!} e^{-n} k\left(1 - \pi r^2(n)\right)^{k-1} \\
& = n \sum_{k=0}^{\infty} \frac{n^k}{k!} e^{-n}\left(1 - \pi r^2(n)\right)^{k} \\
& = n e^{-n\pi r^2(n)} \\
& = e^{-c(n)}
\end{aligned}$$

The result follows. □

Using Lemma 62, together with the result in continuum percolation theory and random geometric graph theory that in an infinite network with node density n and transmission range $r(n)$, as $n \to \infty$, the infinite network asymptotically almost surely has only a giant component of an infinite order and isolated nodes, it was concluded in [92] that for any $\varepsilon > 0$ and for any sufficiently large n,

$$P_d\left(G^{Poisson}(n, r(n))\right) \leq (1+\varepsilon) P_{iso}\left(G^{Poisson}(n, r(n))\right) \tag{3.1.5}$$

Noting the connection that

$$P_d\left(G^{Poisson}(n, r(n))\right) = \sum_{k=1}^{\infty} P_d(G(k, r(n))) \frac{n^k}{k!} e^{-n} \tag{3.1.6}$$

When a new node, say v_k, is added into $G(k-1, r(n))$, the new node can be either a) an isolated node; b) or falling into the transmission range of a node in $G(k-1, r(n))$, which potentially converts $G(k-1, r(n))$ from a disconnected network to a connected network. Therefore, the following inequality can be established:

$$\begin{aligned} & P_d(G(k, r(n))) \\ & \quad \leq \Pr(\{v_k \text{ is isolated in } G(k, r(n))\}) + P_d(G(k-1, r(n))) \end{aligned} \tag{3.1.7}$$

By recursion, it can be further obtained that

$$\begin{aligned} & P_d(G(n, r(n))) \\ & \quad \leq \sum_{k=j+1}^{n} \Pr(\{v_k \text{ is isolated in } G(k, r(n))\}) + P_d(G(j, r(n))) \\ & \quad = \sum_{k=j+1}^{n} \left(1 - \pi r^2(n)\right)^{k-1} + P_d(G(j, r(n))) \\ & \quad \leq \frac{\left(1 - \pi r^2(n)\right)^j}{\pi r^2(n)} + P_d(G(j, r(n))) \end{aligned} \tag{3.1.8}$$

Substituting (3.1.8) into (3.1.6) and using the relationship that, for all $\varepsilon > 0$ and sufficiently large n, $\sum_{j=1}^{n} e^{-n} \frac{n^j}{j!} \geq \left(\frac{1}{2} - \varepsilon\right)$ in the last step, it can be obtained that

$$\begin{aligned} & P_d\left(G^{Poisson}(n, r(n))\right) \\ & \quad \geq P_d(G(n, r(n))) \sum_{j=1}^{n} e^{-n} \frac{n^j}{j!} - \sum_{j=1}^{n-1} \frac{\left(1 - \pi r^2(n)\right)^{j-1}}{\pi r^2(n)} e^{-n} \frac{n^j}{j!} \\ & \quad \geq P_d(G(n, r(n))) \left(\frac{1}{2} - \varepsilon\right) - \frac{e^{-n\pi r^2(n)}}{\pi r^2(n)} \end{aligned}$$

Further using (3.1.7), it can be obtained that

$$P_d\left(G\left(n,r\left(n\right)\right)\right)\leq 2\left(1+4\varepsilon\right)\left[P_{iso}\left(G^{Poisson}\left(n,r\left(n\right)\right)\right)+\frac{e^{-n\pi r^2(n)}}{\pi r^2\left(n\right)}\right]$$

Using the relationship that $\pi r^2\left(n\right)=\frac{\log n+c(n)}{n}$ and Lemma 62, it can be shown that for any fixed $\varepsilon>0$, the following inequality holds for all sufficiently large n

$$P_d\left(G\left(n,r\left(n\right)\right)\right)\leq 2\left(1+4\varepsilon\right)\left[e^{-c(n)}+\frac{e^{-c(n)}}{\log n+c\left(n\right)}\right]$$

Since $\varepsilon>0$ is arbitrary, it follows that

$$\limsup_{n\to\infty}P_d\left(G\left(n,r\left(n\right)\right)\right)\leq 2e^{-c}$$

As an easy consequence of the above inequality,

$$\liminf_{n\to\infty}\Pr\left(\left\{G\left(n,r\left(n\right)\right)\text{ is connected}\right\}\right)\geq 1-2e^{-c}$$

Thus, a sufficient condition for $G\left(n,r\left(n\right)\right)$ to be asymptotically almost surely connected results. The above discussion can be formally summarized into the following theorem.

Theorem 63 *As $n\to\infty$, $G\left(n,r\left(n\right)\right)$ is asymptotically almost surely connected for $\pi r^2\left(n\right)=\frac{\log n+c(n)}{n}$ if $c\left(n\right)\to\infty$.*

Combining Theorems 61 and 63, a sufficient and necessary condition for $G\left(n,r\left(n\right)\right)$ to be asymptotically almost surely connected is obtained:

Theorem 64 *Graph $G\left(n,r\left(n\right)\right)$ with $\pi r^2\left(n\right)=\frac{\log n+c(n)}{n}$ is asymptotically almost surely connected as $n\to\infty$ if and only if $c\left(n\right)\to\infty$.*

3.2 Connectivity of Large Networks: The Random Connection Model

In the last section, we introduced the work of Gupta and Kumar [92] on the sufficient and necessary condition on the transmission range required for a network with uniformly distributed nodes and assuming the unit disk connection model to be asymptotically almost surely connected. Now we return to the main focus of this chapter and study the sufficient and necessary condition for the network $\mathcal{G}\left(\mathcal{X}_\rho,g_{r_\rho},A\right)$ to be asymptotically almost surely connected as $\rho\to\infty$. We start with the necessary condition.

3.2.1 Necessary Condition for Almost Sure Connectivity

As an intermediate step to obtaining the main result, we first and temporarily consider a network with the same node distribution and connection model as $\mathcal{G}\left(\mathcal{X}_\rho, g_{r_\rho}, A\right)$ however with nodes deployed on a unit torus $A^T \triangleq [-\frac{1}{2}, \frac{1}{2}]^2$. Denote the network on the torus by $\mathcal{G}^T\left(\mathcal{X}_\rho, g_{r_\rho}, A\right)$. We show that as $\rho \to \infty$, the distribution of the number of isolated nodes in $\mathcal{G}^T\left(\mathcal{X}_\rho, g_{r_\rho}, A\right)$, denoted by W^T, asymptotically converges to a Poisson distribution with mean e^{-b}. We then extend the above result to $\mathcal{G}\left(\mathcal{X}_\rho, g_{r_\rho}, A\right)$. On that basis, we obtain a necessary condition for $\mathcal{G}\left(\mathcal{X}_\rho, g_{r_\rho}, A\right)$ to be asymptotically almost surely connected as $\rho \to \infty$.

3.2.1.1 Distribution of the Number of Isolated Nodes on a Torus

In this subsection, we analyze the distribution of the number of isolated nodes in $\mathcal{G}^T\left(\mathcal{X}_\rho, g_{r_\rho}, A\right)$.

The use of a toroidal rather than planar region as a tool in analyzing network properties is well known and has also been used previously in Chaps. 1 and 2. In this section, whenever the difference between a torus and a square affects the parameter being discussed, we use superscript ^{T} to mark the parameter in a torus while the unmarked parameter is associated with a square.

We note the following relation between toroidal distance and Euclidean distance on a square area centered at the origin:

$$\|\boldsymbol{x}_1 - \boldsymbol{x}_2\|^T \leq \|\boldsymbol{x}_1 - \boldsymbol{x}_2\| \quad \text{and} \quad \|\boldsymbol{x}\|^T = \|\boldsymbol{x}\| \tag{3.2.1}$$

which will be used in the later analysis.

The main result of this subsection is given in Theorem 65.

Theorem 65 *The distribution of the number of isolated nodes in* $\mathcal{G}^T\left(\mathcal{X}_\rho, g_{r_\rho}, A\right)$ *converges to a Poisson distribution with mean* e^{-b} *as* $\rho \to \infty$.

The rest of this subsection is dedicated to the proof of Theorem 65.

Our proof relies on the use of the Chen-Stein bound [7, 15]. We first establish some preliminary results that allow us to use the Chen-Stein bound for the analysis of number of isolated nodes in $\mathcal{G}^T\left(\mathcal{X}_\rho, g_{r_\rho}, A\right)$.

Divide the unit torus into m^2 nonoverlapping squares each with size $\frac{1}{m^2}$. Denote the i_m^{th} square by A_{i_m}. Define two sets of indicator random variables $J_{i_m}^T$ and $I_{i_m}^T$ with $i_m \in \Gamma_m \triangleq \{1, \ldots m^2\}$, where $J_{i_m}^T = 1$ if and only if there exists exactly one node in A_{i_m}, otherwise $J_{i_m}^T = 0$; $I_{i_m}^T = 1$ if and only if there is exactly one node in A_{i_m} *and* that node is isolated, $I_{i_m}^T = 0$ otherwise. Obviously $J_{i_m}^T$ is independent of $J_{j_m}^T, j_m \in \Gamma_m \backslash \{i_m\}$. Denote the center of $A_{i_m}^T$ by $\boldsymbol{x}_{i_m}$ and without loss of generality we assume that when $J_{i_m}^T = 1$, the associated node in A_{i_m} is at $\boldsymbol{x}_{i_m}$. As we are mainly concerned with the case that $m \to \infty$, i.e., the size of the square is vanishingly small, the actual position of the node in the square is not important. Observe that for any fixed m, the values of $\Pr\left(I_{i_m}^T = 1\right)$ and $\Pr\left(J_{i_m}^T = 1\right)$ do not depend on the particular index i_m on a torus. However both the set of indices Γ_m and a particular index i_m depend on m. As m changes, the square associated with $I_{i_m}^T$ and $J_{i_m}^T$ also changes.

Remark 66 Here we are only interested in the limiting values of various parameters associated with a sub-square as $m \to \infty$. Also because of the consideration of a torus, the value of a particular index i_m does not affect the discussion of the associated parameters, i.e., these parameters $I_{i_m}^T$ and $J_{i_m}^T$ do not depend on i_m. Therefore, in the following, we omit some straightforward discussions on the convergence of various parameters, e.g., i_m, $\boldsymbol{x}_{i_m}$, $I_{i_m}^T$ and $J_{i_m}^T$, as $m \to \infty$.

Without causing ambiguity, we drop the explicit dependence on m in our notations for convenience. As an easy consequence of the Poisson node distribution, $\Pr(J_i^T = 1) \sim_m \rho/m^2$. Using [143, Proposition 1.3], $\Pr(I_i^T = 1) = \Pr(I_i^T = 1| J_i^T = 1)\Pr(J_i^T = 1)$ and the property of a torus (see also Lemma 42), it can be shown that

$$\begin{aligned}\Pr(I_i^T = 1) &\sim_m \frac{\rho}{m^2} e^{-\int_A \rho g(\frac{\|\boldsymbol{x}-\boldsymbol{x}_i\|^T}{r_\rho})d\boldsymbol{x}}\\ &= \frac{\rho}{m^2} e^{-\int_A \rho g(\frac{\|\boldsymbol{x}\|^T}{r_\rho})d\boldsymbol{x}}\end{aligned} \tag{3.2.2}$$

Now consider the event $I_i^T I_j^T = 1, i \neq j$, conditioned on the event that $J_i^T J_j^T = 1$, meaning that both nodes having been placed inside A_i and A_j, respectively, are isolated. Following the same steps leading to (3.2.2), it can be shown that

$$\begin{aligned}&\lim_{m\to\infty} \Pr(I_i^T I_j^T = 1|J_i^T J_j^T = 1)\\ &= (1 - g(\frac{\|\boldsymbol{x}_i - \boldsymbol{x}_j\|^T}{r_\rho}))\exp\left[-\int_A \rho(g(\frac{\|\boldsymbol{x}-\boldsymbol{x}_i\|^T}{r_\rho})\right.\\ &\quad\left. + g(\frac{\|\boldsymbol{x}-\boldsymbol{x}_j\|^T}{r_\rho}) - g(\frac{\|\boldsymbol{x}-\boldsymbol{x}_i\|^T}{r_\rho})g(\frac{\|\boldsymbol{x}-\boldsymbol{x}_j\|^T}{r_\rho})\right) d\boldsymbol{x}\right]\end{aligned} \tag{3.2.3}$$

where the term $(1-g(\frac{\|\boldsymbol{x}_i-\boldsymbol{x}_j\|^T}{r_\rho}))$ is due to the requirement that the two nodes located inside A_i and A_j cannot be directly connected given that they are both isolated nodes. Observe also that $\Pr(I_i^T I_j^T = 1) = \Pr(J_i^T J_j^T = 1)\Pr(I_j^T I_j^T = 1|J_i^T J_j^T = 1)$. Using the above equation, (3.2.2) and (3.2.3), it can be established that

$$\begin{aligned}&\frac{\Pr(I_i^T I_j^T = 1)}{\Pr(I_i^T = 1)\Pr(I_j^T = 1)}\\ &\sim_m (1 - g(\frac{\|\boldsymbol{x}_i - \boldsymbol{x}_j\|^T}{r_\rho}))e^{\int_A \rho g(\frac{\|\boldsymbol{x}-\boldsymbol{x}_i\|^T}{r_\rho})g(\frac{\|\boldsymbol{x}-\boldsymbol{x}_j\|^T}{r_\rho})d\boldsymbol{x}}\end{aligned} \tag{3.2.4}$$

Now we are ready to use the Chen-Stein bound to prove Theorem 65. Particularly, we shall show using the Chen-Stein bound that

$$W^T = \lim_{m\to\infty} \sum_{i\in\Gamma_m} I_i^T \tag{3.2.5}$$

asymptotically converges to a Poisson distribution with mean e^{-b} as $\rho \to \infty$.

The following theorem gives a formal statement of the Chen-Stein bound:

Theorem 67 ([15, Theorem 1.A]) *For a set of indicator random variables I_i, $i \in \Gamma$, define $W \triangleq \sum_{i\in\Gamma} I_i$, $p_i \triangleq E(I_i)$ and $\eta \triangleq E(W)$. For any choice of the index set $\Gamma_{s,i} \subset \Gamma$ and $\Gamma_{s,i} \cap \{i\} = \{\emptyset\}$,*

$$\begin{aligned} & d_{TV}(\mathcal{L}(W), Po(\eta)) \\ & \le \sum_{i\in\Gamma} [(p_i^2 + p_i E(\sum_{j\in\Gamma_{s,i}} I_j))] \min(1, \frac{1}{\eta}) \\ & + \sum_{i\in\Gamma} E(I_i \sum_{j\in\Gamma_{s,i}} I_j) \min(1, \frac{1}{\eta}) \\ & + \sum_{i\in\Gamma} E|E\{I_i|(I_j, j\in\Gamma_{w,i})\} - p_i| \min(1, \frac{1}{\eta}) \end{aligned}$$

where $\mathcal{L}(W)$ denotes the distribution of W, $Po(\eta)$ denotes a Poisson distribution with mean η, $\Gamma_{w,i} = \Gamma \backslash \{\Gamma_{s,i} \cup \{i\}\}$ and d_{TV} denotes the total variation distance. The total variation distance between two probability distributions α and β on $\mathbb{Z}^+$ is given by $d_{TV}(\alpha, \beta) \triangleq \sup\left\{|\alpha(A) - \beta(A)| : A \subset \mathbb{Z}^+\right\}$.

For convenience, we separate the bound in Theorem 67 into three terms $b_1 \min(1, \frac{1}{\eta})$, $b_2 \min(1, \frac{1}{\eta})$ and $b_3 \min(1, \frac{1}{\eta})$ where

$$b_1 \triangleq \sum_{i\in\Gamma} \left[(p_i^2 + p_i E(\sum_{j\in\Gamma_{s,i}} I_j)) \right] \tag{3.2.6}$$

$$b_2 \triangleq \sum_{i\in\Gamma} E\left(I_i \sum_{j\in\Gamma_{s,i}} I_j \right) \tag{3.2.7}$$

$$b_3 \triangleq \sum_{i\in\Gamma} E\left|E\left\{I_i | \left(I_j, j\in\Gamma_{w,i}\right)\right\} - p_i\right| \tag{3.2.8}$$

The set of indices $\Gamma_{s,i}$ is often chosen to contain all those j, other than i, for which I_j is "strongly" dependent on I_i, and the set $\Gamma_{w,i}$ often contains all other indices apart from i for which I_j is at most "weakly" dependent on I_i.

Remark 68 A main challenge in using the Chen-Stein bound to prove Theorem 65 is that under the random connection model, the two events I_i and I_j may be correlated even when $\boldsymbol{x}_i$ and $\boldsymbol{x}_j$ are separated by a very large Euclidean distance. Therefore, the dependence structure is global, which significantly increases the complexity of the analysis. In comparison, in applications where the dependence structure is local, by a suitable choice of $\Gamma_{s,i}$, the b_3 term can be easily made to be 0 and the evaluation of the b_1 and b_2 terms involves the computation of the first two moments of W only, which can often be achieved relatively easily. An example is a random geometric network under the unit disk model. If $\Gamma_{s,i}$ is chosen to be a neighborhood of i containing indices of all nodes whose Euclidean distance to node i is less than or equal to twice the transmission range, the b_3 term is easily shown to be 0. It can then be readily shown that the b_1 and b_2 terms approach 0 as the neighborhood size of a node becomes vanishingly small compared to the overall network size as $\rho \to \infty$ [70]. However this is certainly not the case for the random connection model.

Remark 69 The key idea involved using the Chen-Stein bound to prove Theorem 65 is constructing a neighborhood of a node, i.e., $\Gamma_{s,i}$ in Theorem 67, such that a) the size of the neighborhood becomes vanishingly small compared with A as $\rho \to \infty$. This is required for the b_1 and b_2 terms to approach 0 as $\rho \to \infty$; b) asymptotically almost surely the neighborhood contains all nodes that may have a direct connection with the node. This is required for the b_3 term to approach 0 as $\rho \to \infty$. Such a neighborhood is defined in the next paragraph.

Let $D^T(\boldsymbol{x}_i, r) \triangleq \{\boldsymbol{x} \in A : \|\boldsymbol{x} - \boldsymbol{x}_i\|^T \leq r\}$ and when $\boldsymbol{x}_i$ is not within distance r of the border of A, $D^T(\boldsymbol{x}_i, r)$ becomes the same as $D(\boldsymbol{x}_i, r)$ where $D(\boldsymbol{x}_i, r) \triangleq \{\boldsymbol{x} \in A : \|\boldsymbol{x} - \boldsymbol{x}_i\| \leq r\}$. Further define the neighborhood of an index $i \in \Gamma$ as $\Gamma_{s,i} \triangleq \{j : \boldsymbol{x}_j \in D^T\left(\boldsymbol{x}_i, 2r_\rho^{1-\epsilon}\right)\}\backslash\{i\}$ and define the non-neighborhood of the index i as $\Gamma_{w,i} \triangleq \{j : \boldsymbol{x}_j \notin D^T(\boldsymbol{x}_i, 2r_\rho^{1-\epsilon})\}$ where ϵ is a small positive constant and $\epsilon \in (0, \frac{1}{2})$. It can be shown that

$$|\Gamma_{s,i}| = m^2 4\pi r_\rho^{2-2\epsilon} + o_m(m^2 4\pi r_\rho^{2-2\epsilon}) \tag{3.2.9}$$

Note that in Theorem 67, $p_i = E(I_i^T)$ and $E(I_i^T)$ has been given in (3.2.2). Furthermore, as an easy consequence of (3.2.5) and Lemma 42 which showed that

$$\lim_{\rho\to\infty} E(W^T) = \lim_{\rho\to\infty} \rho e^{-\int_A \rho g\left(\frac{\|\boldsymbol{x}\|^T}{r_\rho}\right) d\boldsymbol{x}} = e^{-b} \tag{3.2.10}$$

$\lim_{\rho\to\infty} \lim_{m\to\infty} \eta = e^{-b}$.

Using (3.2.2), $p_i = E(I_i^T)$ and (3.2.10), it follows that

$$\lim_{m\to\infty} m^2 p_i = \rho e^{-\int_A \rho g(\frac{\|x\|^T}{r_\rho})dx} \tag{3.2.11}$$

$$\lim_{\rho\to\infty}\lim_{m\to\infty} m^2 p_i = e^{-b} \tag{3.2.12}$$

Next we shall evaluate the b_1, b_2, and b_3 terms in the following three subsections separately and show that all three terms converge to 0 as $\rho \to \infty$.

3.2.1.2 An Evaluation of the b_1 Term

It can be shown that (following the equation, detailed explanations are given)

$$\begin{aligned}
&\lim_{\rho\to\infty}\lim_{m\to\infty} \sum_{i\in\Gamma}(p_i^2 + p_i E(\sum_{j\in\Gamma_{s,i}} I_j^T)) \\
&= \lim_{\rho\to\infty}\lim_{m\to\infty} m^2 p_i E\left(\sum_{j\in\Gamma_{s,i}\cup\{i\}} I_j^T\right) \\
&= \lim_{\rho\to\infty}\lim_{m\to\infty} (m^2 p_i)^2 4\pi r_\rho^{2-2\epsilon} && (3.2.13) \\
&= \lim_{\rho\to\infty} 4\pi \left(\rho e^{-\int_A \rho g(\frac{\|x-x_i\|^T}{r_\rho})dx}\right)^2 \left(\frac{\log\rho + b}{C\rho}\right)^{1-\epsilon} && (3.2.14) \\
&= 4\pi e^{-2b} \lim_{\rho\to\infty} (\frac{\log\rho + b}{C\rho})^{1-\epsilon} = 0 && (3.2.15)
\end{aligned}$$

where (3.2.9) is used in obtaining (3.2.13); (3.0.1) and (3.2.11) are used in obtaining (3.2.14); and (3.2.10) and (3.2.12) are used in obtaining (3.2.15). Therefore, $\lim_{\rho\to\infty}\lim_{m\to\infty} b_1 = 0$.

3.2.1.3 An Evaluation of the b_2 Term

For the b_2 term, assume that ρ is sufficiently large such that $\frac{1}{2r_\rho} >> 2r_\rho^{-\epsilon}$ and let $A_{\frac{1}{r_\rho}} = [-\frac{1}{2r_\rho}, \frac{1}{2r_\rho}]^2$. Using (3.2.3) in the first step; and first using some translation and scaling operations and then using (3.2.1) in the last step, (3.2.16) can be obtained:

$$\begin{aligned}
&\lim_{m\to\infty} \sum_{i\in\Gamma} E\left(I_i^T \sum_{j\in\Gamma_{s,i}} I_j^T\right) \\
&= \lim_{m\to\infty} \frac{\rho^2}{m^2} \sum_{j\in\Gamma_{s,i}} \{(1 - g\left(\left\|\frac{x_i - x_j}{r_\rho}\right\|^T\right) \exp[-\int_A \rho(g\left(\left\|\frac{x - x_i}{r_\rho}\right\|^T\right)
\end{aligned}$$

$$+ g(\left\|\frac{\boldsymbol{x}-\boldsymbol{x}_j}{r_\rho}\right\|^T) - g(\left\|\frac{\boldsymbol{x}-\boldsymbol{x}_i}{r_\rho}\right\|^T)g(\left\|\frac{\boldsymbol{x}-\boldsymbol{x}_j}{r_\rho}\right\|^T))])\}d\boldsymbol{x}$$

$$= \rho^2 \int_{D^T(\boldsymbol{x}_i, 2r_\rho^{1-\epsilon})} \{(1-g(\frac{\|\boldsymbol{x}_i-\boldsymbol{y}\|^T}{r_\rho})) \exp[-\int_A \rho(g\left(\left\|\frac{\boldsymbol{x}-\boldsymbol{x}_i}{r_\rho}\right\|^T\right)$$

$$+ g(\left\|\frac{\boldsymbol{x}-\boldsymbol{y}}{r_\rho}\right\|^T) - g(\left\|\frac{\boldsymbol{x}-\boldsymbol{x}_i}{r_\rho}\right\|^T)g(\left\|\frac{\boldsymbol{x}-\boldsymbol{y}}{r_\rho}\right\|^T))d\boldsymbol{x}]\}d\boldsymbol{y}$$

$$= \rho^2 r_\rho^2 \int_{D(\boldsymbol{0}, 2r_\rho^{-\epsilon})} \{(1-g(\|\boldsymbol{y}\|^T)) \exp[-\rho r_\rho^2 \int_{A_{\frac{1}{r_\rho}}} (g(\|\boldsymbol{x}\|^T)$$

$$+ g(\|\boldsymbol{x}-\boldsymbol{y}\|^T) - g(\|\boldsymbol{x}\|^T)g(\|\boldsymbol{x}-\boldsymbol{y}\|^T))d\boldsymbol{x}]\}d\boldsymbol{y} \tag{3.2.16}$$

Letting $\lambda \triangleq \frac{\log \rho + b}{C}$ for convenience, noting that (using (3.2.1) and (3.0.3))

$$\lim_{\rho\to\infty} \int_{A_{\frac{1}{r_\rho}}} g(\|\boldsymbol{x}\|^T)d\boldsymbol{x} = \lim_{\rho\to\infty} \int_{A_{\frac{1}{r_\rho}}} g(\|\boldsymbol{x}-\boldsymbol{y}\|^T)d\boldsymbol{x} = C$$

and that $1-g(\|\boldsymbol{y}\|^T) \le 1$, it can further be shown following (3.2.16) that as $\rho \to \infty$,

$$\lim_{\rho\to\infty} \lim_{m\to\infty} \sum_{i\in\Gamma} E(I_i^T \sum_{j\in\Gamma_{s,i}} I_j^T)$$

$$\le e^{-2b} \lim_{\rho\to\infty} \frac{\lambda}{\rho} \int_{D(\boldsymbol{0}, 2r_\rho^{-\epsilon})} e^{\lambda \int_{A_{\frac{1}{r_\rho}}} g(\|\boldsymbol{x}\|^T)g(\|\boldsymbol{x}-\boldsymbol{y}\|^T)d\boldsymbol{x}} d\boldsymbol{y} \tag{3.2.17}$$

In the following paragraphs, we shall show that the right-hand side of (3.2.17) converges to 0 as $\rho \to \infty$. Using (3.0.2) and (3.0.3), we assert that there exists a positive constant r such that $g(r^-)(1-g(r^+)) > 0$ where $g(r^-) \triangleq \lim_{x\to r^-} g(x)$ and $g(r^+) \triangleq \lim_{x\to r^+} g(x)$. Indeed if g is a continuous function, any positive constant r with $g(r) > 0$ satisfies the requirement; if g is a discontinuous function, e.g., a unit disk model, by choosing r to be the transmission range, $g(r^-)(1-g(r^+)) = 1$.

In the following discussion we assume that ρ is sufficiently large such that $\frac{1}{2r_\rho} >> 2r_\rho^{-\epsilon} >> r$. It can be shown using (3.0.2), (3.0.3) and (3.2.1) that for $\boldsymbol{y} \in D(\boldsymbol{0}, 2r_\rho^{-\epsilon})$,

$$\int_{A_{\frac{1}{r_\rho}}} g\left(\|\boldsymbol{x}\|^T\right) g\left(\|\boldsymbol{x}-\boldsymbol{y}\|^T\right) d\boldsymbol{x}$$

$$\le \int_{\Re^2} g(\|\boldsymbol{x}\|) g(\|\boldsymbol{x}-\boldsymbol{y}\|) d\boldsymbol{x}$$

$$= C - \int_{\Re^2} g(\|\boldsymbol{x}\|)(1 - g(\|\boldsymbol{x}-\boldsymbol{y}\|))d\boldsymbol{x}$$

$$\leq C - \int_{D(\boldsymbol{0},r)\backslash D(\boldsymbol{y},r)} g(\|\boldsymbol{x}\|)(1 - g(\|\boldsymbol{x}-\boldsymbol{y}\|))\, d\boldsymbol{x}$$

$$\leq C - g\left(r^-\right)\left(1 - g\left(r^+\right)\right)|D(\boldsymbol{0},r)\backslash D(\boldsymbol{y},r)| \tag{3.2.18}$$

Let $f(x) \triangleq \pi r^2 - 2r^2 \arcsin(\sqrt{1-x^2/(4r^2)}) + rx\sqrt{1-x^2/(4r^2)}$. Using some straightforward geometric analysis, it can be shown that

- when $\|\boldsymbol{y}\| > 2r$, $|D(\boldsymbol{0},r)\backslash D(\boldsymbol{y},r)| = \pi r^2$; and
- when $\|\boldsymbol{y}\| \leq 2r$, $|D(\boldsymbol{0},r)\backslash D(\boldsymbol{y},r)| = f(\|\boldsymbol{y}\|)$.

Furthermore, using the definition of $f(x)$, it can be shown that

- when $\|\boldsymbol{y}\| \leq r$, $|D(\boldsymbol{0},r)\backslash D(\boldsymbol{y},r)| \geq \sqrt{3}r\,\|\boldsymbol{y}\|$; and
- when $\|\boldsymbol{y}\| > r$, $|D(\boldsymbol{0},r)\backslash D(\boldsymbol{y},r)| \geq (\frac{\pi}{3} + \frac{\sqrt{3}}{2})r^2$.

For convenience, let $c_1 \triangleq g\left(r^-\right)\left(1 - g\left(r^+\right)\right)\sqrt{3}r$ and $c_2 \triangleq g\left(r^-\right)\left(1 - g\left(r^+\right)\right)\left(\frac{\pi}{3} + \frac{\sqrt{3}}{2}\right)r^2$. Noting that $g\left(r^-\right)\left(1 - g\left(r^+\right)\right) > 0$, c_1 and c_2 are positive constants, independent of both $\boldsymbol{y}$ and ρ.

As a result of (3.2.18) and the above inequalities on $|D(\boldsymbol{0},r)\backslash D(\boldsymbol{y},r)|$, it follows that

$$\lim_{\rho\to\infty} \frac{\lambda}{\rho} \int_{D(\boldsymbol{0},2r_\rho^{-\epsilon})} e^{\lambda \int_{A_{\frac{1}{r_\rho}}} g\left(\|\boldsymbol{x}\|^T\right) g\left(\|\boldsymbol{x}-\boldsymbol{y}\|^T\right) d\boldsymbol{x}} d\boldsymbol{y}$$

$$\leq \lim_{\rho\to\infty} \frac{\lambda}{\rho} \int_{D(\boldsymbol{0},r)} e^{\lambda(C - c_1\|\boldsymbol{y}\|)} d\boldsymbol{y}$$

$$+ \lim_{\rho\to\infty} \frac{\lambda}{\rho} \int_{D\left(\boldsymbol{0},2r_\rho^{-\epsilon}\right)\backslash D(\boldsymbol{0},r)} e^{\lambda(C-c_2)} d\boldsymbol{y} \tag{3.2.19}$$

For the first summand in the above equation, it can be shown that:

$$\lim_{\rho\to\infty} \frac{\lambda}{\rho} \int_{D(\boldsymbol{0},r)} e^{\lambda(C - c_1\|\boldsymbol{y}\|)} d\boldsymbol{y}$$

$$= \lim_{\rho\to\infty} \frac{\log\rho + b}{C\rho} \int_0^r 2\pi y e^{\frac{\log\rho+b}{C}(C - c_1 y)} dy \tag{3.2.20}$$

$$= 0 \tag{3.2.21}$$

For the second summand in (3.2.19), by choosing $\varepsilon < \frac{c_2}{C}$ and using (3.0.1), it follows that

$$\lim_{\rho\to\infty}\frac{\lambda}{\rho}\int_{D(\mathbf{0},2r_\rho^{-\epsilon})\setminus D(\mathbf{0},r)} e^{\lambda(C-c_2)}dy$$

$$= \lim_{\rho\to\infty}\frac{e^{b(1-\frac{c_2}{C})}}{C}\times\frac{\log\rho+b}{\rho^{\frac{c_2}{C}}}\times\pi(4r_\rho^{-2\epsilon}-r^2) \quad (3.2.22)$$

$$= 0 \quad (3.2.23)$$

Combining (3.2.19), (3.2.20), and (3.2.22), it follows that

$$\lim_{\rho\to\infty}\frac{\lambda}{\rho}\int_{D(\mathbf{0},2r_\rho^{-\epsilon})} e^{\lambda\int_{A_{\frac{1}{r_\rho}}} g(\|x\|^T)g(\|x-y\|^T)dx}dy = 0 \quad (3.2.24)$$

As a result of (3.2.17) and the above equation: $\lim_{\rho\to\infty}\lim_{m\to\infty} b_2 = 0$.

3.2.1.4 An Evaluation of the b_3 Term

We first obtain an analytical expression of the term $E\{I_i|(I_j, j\in\Gamma_{w,i})\}$ in b_3. Using the same procedure that results in (3.2.4), it can be obtained that (for convenience we use g_i for $g(\frac{\|x-x_i\|^T}{r_\rho})$ and use g_{ij} for $g(\frac{\|x_i-x_j\|^T}{r_\rho})$ in the following equation):

$$\lim_{m\to\infty}\frac{Pr(I_i^T=1, I_j^T=1, I_k^T=0)}{Pr(I_i^T=1)Pr(I_j^T=1, I_k^T=0)}$$

$$= \lim_{m\to\infty}\frac{Pr(I_i^T=1, I_j^T=1)-Pr(I_i^T=1, I_j^T=1, I_k^T=1)}{Pr(I_i^T=1)(Pr(I_j^T=1)-Pr(I_j^T=1, I_k^T=1))}$$

$$\sim_m \left(1-g_{ij}\right)e^{\int_A \rho g_i g_j dx}$$

$$\times\frac{1-\frac{\rho}{m^2}(1-g_{ik})(1-g_{kj})e^{-\int_A \rho(g_k-g_ig_k-g_kg_j+g_ig_jg_k)dx}}{1-\frac{\rho}{m^2}(1-g_{kj})e^{-\int_A \rho(g_k-g_kg_j)dx}} \quad (3.2.25)$$

Using (3.0.2), (3.0.3), (3.0.4), and (3.2.1), it can be shown that when $j\in\Gamma_{w,i}$ (or equivalently $\|x_i-x_j\|^T > 2r_\rho^{1-\varepsilon}$), the integrals of some higher order terms inside the exponential function in (3.2.25) satisfy:

$$\int_A \rho g(\frac{\|x-x_i\|^T}{r_\rho})g(\frac{\|x-x_j\|^T}{r_\rho})dx$$

$$= \int_{D^T(x_i, r_\rho^{1-\varepsilon})}\rho g(\frac{\|x-x_i\|^T}{r_\rho})g(\frac{\|x-x_j\|^T}{r_\rho})dx$$

$$+\int_{A\backslash D^T(\boldsymbol{x}_i, r_\rho^{1-\varepsilon})} \rho g(\frac{\|\boldsymbol{x}-\boldsymbol{x}_i\|^T}{r_\rho}) g(\frac{\|\boldsymbol{x}-\boldsymbol{x}_j\|^T}{r_\rho}) d\boldsymbol{x}$$

$$\leq 2C\rho r_\rho^2 g(r_\rho^{-\varepsilon}) \sim_\rho o_\rho(1)$$

Note also that $g_{ik} = g(\frac{\|\boldsymbol{x}_i-\boldsymbol{x}_k\|^T}{r_\rho}) = o_\rho(1)$ for $k \in \Gamma_{w,i}$. Using the above equations and (3.2.4), it can be further shown following (3.2.25) that when $j, k \in \Gamma_{w,i}$.

$$\lim_{\rho\to\infty}\lim_{m\to\infty} \frac{Pr(I_i^T = 1, I_j^T = 1, I_k^T = 0)}{Pr(I_i^T = 1)Pr(I_j^T = 1, I_k^T = 0)}$$

$$= \lim_{\rho\to\infty}\lim_{m\to\infty} \frac{Pr(I_i^T = 1, I_j^T = 1)}{Pr(I_i^T = 1)Pr(I_j^T = 1)} \quad (3.2.26)$$

Equation (3.2.26) shows that the impact of those events, whose associated indicator random variables $I_k^T = 0, k \in \Gamma_{w,i}$, on the event $I_i^T = 1$ is asymptotically vanishingly small, hence can be ignored. Denote by Γ_i a random set of indices containing all indices j where $j \in \Gamma_{w,i}$ *and* $I_j = 1$, i.e., the node in question is also isolated. Denote by γ_i an instance of Γ_i. Define $n \triangleq |\gamma_i|$. Following the same procedure that results in (3.2.26), it can be established that (with some verbose but straightforward discussions omitted)

$$\lim_{\rho\to\infty}\lim_{m\to\infty} \frac{E\{I_i^T|(I_j^T, j \in \Gamma_{w,i})\}}{\frac{\rho}{m^2}}$$

$$= \lim_{\rho\to\infty}\lim_{m\to\infty} \frac{E\{I_i^T|(I_j^T = 1, j \in \gamma_i)\}}{\frac{\rho}{m^2}}$$

$$= \lim_{\rho\to\infty} E[e^{-\int_A \rho g(\frac{\|\boldsymbol{x}-\boldsymbol{x}_i\|^T}{r_\rho})\prod_{j\in\gamma_i}(1-g(\frac{\|\boldsymbol{x}-\boldsymbol{x}_j\|^T}{r_\rho}))d\boldsymbol{x}}$$

$$\times \prod_{j\in\gamma_i}(1-g(\frac{\|\boldsymbol{x}_i-\boldsymbol{x}_j\|^T}{r_\rho}))] \quad (3.2.27)$$

Equation (3.2.27) gives an analytical expression of the term $E\{I_i^T|(I_j^T, j \in \Gamma_{w,i})\}$. To solve the challenges associated with handling the absolute value term in b_3, viz., $\left|E\{I_i^T|(I_j^T, j \in \Gamma_{w,i})\} - p_i\right|$, we further obtain an upper and a lower bound of $I_i^T|(I_j^T, j \in \Gamma_{w,i})$, which allows us to remove the absolute value sign in the further analysis of b_3.

Note that $\boldsymbol{x}_i$ and $\boldsymbol{x}_j, j \in \Gamma_{w,i}$ are separated by a distance not smaller than $2r_\rho^{-\epsilon}$. Using (3.0.2), a lower bound on the value inside the expectation operator in (3.2.27) is given by

$$B_{L,i} \triangleq (1 - g(2r_\rho^{-\epsilon}))^n e^{-\int_A \rho g(\frac{\|\boldsymbol{x}-\boldsymbol{x}_i\|^T}{r_\rho})d\boldsymbol{x}} \quad (3.2.28)$$

An upper bound on the value inside the expectation operator in (3.2.27) is given by

$$B_{U,i} \triangleq e^{-\int_A \rho g\left(\frac{\|x-x_i\|^T}{r_\rho}\right) \prod_{j \in \gamma_i}\left(1-g\left(\frac{\|x-x_j\|^T}{r_\rho}\right)\right) dx} \tag{3.2.29}$$

Using $p_i = E(I_i^T)$ and (3.2.2), it can be shown that

$$B_{U,i} \geq \lim_{m\to\infty} \frac{m^2 p_i}{\rho} \geq B_{L,i} \tag{3.2.30}$$

Let us consider $E|E\{I_i|(I_j, j \in \Gamma_{w,i})\} - p_i|$ now. From (3.2.27), (3.2.28), (3.2.29), and (3.2.30), it is clear that

$$\begin{aligned}
&\lim_{\rho\to\infty}\lim_{m\to\infty} \sum_{i\in\Gamma} E\left|E\{I_i^T|(I_j^T, j \in \Gamma_{w,i})\} - p_i\right| \\
&\in \left[0, \max\left\{\lim_{\rho\to\infty}\lim_{m\to\infty} m^2 p_i - \rho E(B_{L,i}),\right.\right. \\
&\quad \left.\left.\lim_{\rho\to\infty}\lim_{m\to\infty} \rho E(B_{U,i}) - m^2 p_i\right\}\right]
\end{aligned} \tag{3.2.31}$$

In the following, we shall show that both terms $\lim_{m\to\infty} m^2 p_i - \rho E(B_{L,i})$ and $\lim_{m\to\infty} \rho E(B_{U,i}) - m^2 p_i$ in (3.2.31) approach 0 as $\rho \to \infty$. First it can be shown following (3.2.28) that

$$\begin{aligned}
&\lim_{m\to\infty} \rho E(B_{L,i}) \\
&\geq \lim_{m\to\infty} \rho E\left[(1 - ng(2r_\rho^{-\epsilon}))e^{-\int_A \rho g\left(\frac{\|x-x_i\|^T}{r_\rho}\right) dx}\right] \\
&= \lim_{m\to\infty} \rho(1 - E(n)\, g(2r_\rho^{-\epsilon}))e^{-\int_A \rho g\left(\frac{\|x-x_i\|^T}{r_\rho}\right) dx}
\end{aligned} \tag{3.2.32}$$

where $\lim_{m\to\infty} E(n)$ is the expected number of isolated nodes in $A\backslash D(\boldsymbol{x}_i, 2r_\rho^{1-\epsilon})$. In the first step of the above equation, the inequality $(1-x)^n \geq 1 - nx$ for $0 \leq x \leq 1$ and $n \geq 0$ is used. When $\rho \to \infty$, $r_\rho^{1-\epsilon} \to 0$ and $r_\rho^{-\epsilon} \to \infty$, $\lim_{\rho\to\infty}\lim_{m\to\infty} E(n) = \lim_{\rho\to\infty} E(W^T) = e^{-b}$ is a bounded value and $\lim_{\rho\to\infty}\lim_{m\to\infty} g\left(2r_\rho^{-\epsilon}\right) \to 0$, which is an immediate outcome of (3.0.4). Using (3.2.11), it then follows that

$$\lim_{\rho\to\infty}\lim_{m\to\infty} \frac{\rho E(B_{L,i})}{m^2 p_i} \geq \lim_{\rho\to\infty}\lim_{m\to\infty} (1 - E(n)\, g(2r_\rho^{-\epsilon})) = 1$$

Together with (3.2.12) and (3.2.30), we conclude that

$$\lim_{\rho\to\infty}\lim_{m\to\infty} m^2 p_i - \rho E\left(B_{L,i}\right) = 0 \tag{3.2.33}$$

Now let us consider the second term $\lim_{m\to\infty}\rho E\left(B_{U,i}\right) - m^2 p_i$, it can be observed that

$$\begin{aligned}
&\lim_{m\to\infty} E\left(B_{U,i}\right)\\
&\le E\left[e^{-\int_{D(\boldsymbol{x}_i,r_\rho^{1-\epsilon})}\rho g(\frac{\|\boldsymbol{x}-\boldsymbol{x}_i\|^T}{r_\rho})\prod_{j\in\mathcal{Y}_i}(1-g(\frac{\|\boldsymbol{x}-\boldsymbol{x}_j\|^T}{r_\rho}))d\boldsymbol{x}}\right]\\
&\le \lim_{m\to\infty} E\left[e^{-\int_{D(\boldsymbol{x}_i,r_\rho^{1-\epsilon})}\rho g(\frac{\|\boldsymbol{x}-\boldsymbol{x}_i\|^T}{r_\rho})\prod_{j\in\mathcal{Y}_i}(1-g(\frac{r_\rho^{1-\epsilon}}{r_\rho}))d\boldsymbol{x}}\right]\\
&= \lim_{m\to\infty} E\left[e^{-(1-g(r_\rho^{-\epsilon}))^n\int_{D(\boldsymbol{x}_i,r_\rho^{1-\epsilon})}\rho g(\frac{\|\boldsymbol{x}-\boldsymbol{x}_i\|^T}{r_\rho})d\boldsymbol{x}}\right]\\
&\le \lim_{m\to\infty} E\left[e^{-(1-ng(r_\rho^{-\epsilon}))\int_{D(\boldsymbol{x}_i,r_\rho^{1-\epsilon})}\rho g(\frac{\|\boldsymbol{x}-\boldsymbol{x}_i\|^T}{r_\rho})d\boldsymbol{x}}\right]
\end{aligned}\tag{3.2.34}$$

where in the second step, the nonincreasing monotonicity of g, and the fact that $\boldsymbol{x}_j$ is located in $A\backslash D(\boldsymbol{x}_i, 2r_\rho^{1-\epsilon})$ and $\boldsymbol{x}$ is located in $D(\boldsymbol{x}_i, r_\rho^{1-\epsilon})$, therefore $\left\|\boldsymbol{x}-\boldsymbol{x}_j\right\|^T \ge r_\rho^{1-\epsilon}$ is used. It can be further demonstrated that the term $\int_{D(\boldsymbol{x}_i,r_\rho^{1-\epsilon})}\rho g(\frac{\|\boldsymbol{x}-\boldsymbol{x}_i\|}{r_\rho})d\boldsymbol{x}$ in (3.2.34) has the following property:

$$\begin{aligned}
\eta\left(\varepsilon,\rho\right) &\triangleq \int_{D(\boldsymbol{x}_i,r_\rho^{1-\epsilon})}\rho g(\frac{\left\|\boldsymbol{x}-\boldsymbol{x}_i\right\|^T}{r_\rho})d\boldsymbol{x}\\
&= \rho r_\rho^2\int_{D(\frac{\boldsymbol{x}_i}{r_\rho},r_\rho^{-\epsilon})} g(\left\|\boldsymbol{x}-\boldsymbol{x}_i/r_\rho\right\|^T)d\boldsymbol{x}\\
&\le C\rho r_\rho^2 = \log\rho + b
\end{aligned}\tag{3.2.35}$$

For the other term $ng(r_\rho^{-\epsilon})$ in (3.2.34), choosing a positive constant $\delta < 2\epsilon$ and using Markov's inequality, it can be shown that $\Pr\left(n \ge r_\rho^{-\delta}\right) \leqslant r_\rho^{\delta} E\left(n\right)$. Therefore,

$$\begin{aligned}
&\lim_{\rho\to\infty}\lim_{m\to\infty}\Pr\left(ng(r_\rho^{-\epsilon})\eta(\varepsilon,\rho) \ge r_\rho^{-\delta}g(r_\rho^{-\epsilon})\eta(\varepsilon,\rho)\right)\\
&\le \lim_{\rho\to\infty}\lim_{m\to\infty} r_\rho^{\delta}E\left(n\right)
\end{aligned}$$

where $\lim_{\rho\to\infty} r_\rho^{-\delta}g(r_\rho^{-\epsilon})\eta(\varepsilon,\rho) = 0$ due to (3.0.4), (3.2.35) and $\delta < 2\epsilon$, $\lim_{\rho\to\infty} r_\rho^{B} = 0$ for any positive constant B, and $\lim_{\rho\to\infty}\lim_{m\to\infty} r_\rho^{\delta}E\left(n\right) = 0$

due to that $\lim_{\rho\to\infty}\lim_{m\to\infty} E(n) = \lim_{\rho\to\infty} E(W^T) = e^{-b}$ is a bounded value and that $\lim_{\rho\to\infty} r_\rho^\delta = 0$. Therefore,

$$\lim_{\rho\to\infty}\lim_{m\to\infty} \Pr\left(ng(r_\rho^{-\epsilon})\eta(\varepsilon,\rho) = 0\right) = 1 \tag{3.2.36}$$

As a result of (3.2.1), (3.2.34), (3.2.35), and (3.2.36):

$$\begin{aligned}
&\lim_{\rho\to\infty}\lim_{m\to\infty} \rho E(B_{U,i}) \\
&\le \lim_{\rho\to\infty}\lim_{m\to\infty} \rho E\left[e^{-\int_{D(x_i,r_\rho^{1-\epsilon})}\rho g\left(\frac{\|x-x_i\|^T}{r_\rho}\right)dx}\right] \\
&= \lim_{\rho\to\infty} \rho e^{-\int_{D(\mathbf{0},r_\rho^{1-\varepsilon})}\rho g\left(\frac{\|x\|}{r_\rho}\right)dx} \\
&= \lim_{\rho\to\infty} \rho e^{-\rho r_\rho^2\left(C-\int_{\Re^2\setminus D(\mathbf{0},r_\rho^{-\varepsilon})} g(\|x\|)dx\right)} \\
&= e^{-b}\lim_{\rho\to\infty} e^{\rho r_\rho^2\int_{\Re^2\setminus D(\mathbf{0},r_\rho^{-\varepsilon})} g(\|x\|)dx} \\
&= e^{-b}
\end{aligned} \tag{3.2.37}$$

where the last step results because

$$\begin{aligned}
&\lim_{\rho\to\infty} \rho r_\rho^2 \int_{\Re^2\setminus D(\mathbf{0},r_\rho^{-\varepsilon})} g(\|x\|)\,dx \\
&= \lim_{\rho\to\infty} \frac{\int_{r_\rho^{-\varepsilon}}^{\infty} 2\pi x g(x)\,dx}{\frac{C}{\log\rho+b}} \\
&= \lim_{\rho\to\infty} \frac{\pi\varepsilon r_\rho^{-\varepsilon} g(r_\rho^{-\varepsilon}) r_\rho^{-\varepsilon-2}\frac{\log\rho+b-1}{C\rho^2}}{\frac{C}{\rho(\log\rho+b)^2}} && (3.2.38) \\
&= \lim_{\rho\to\infty} \frac{\pi\varepsilon}{C}(\log\rho+b)^2 r_\rho^{-2\varepsilon} o_\rho\left(\frac{1}{r_\rho^{-2\varepsilon}\log^2(r_\rho^{-2\varepsilon})}\right) \\
&= 0 && (3.2.39)
\end{aligned}$$

where L'Hospital's rule is used in reaching (3.2.38) and in the third step (3.0.5) is used. Using (3.2.12), (3.2.30), and (3.2.37), it can be shown that

$$\lim_{\rho\to\infty}\lim_{m\to\infty} \rho E(B_{U,i}) - m^2 p_i = 0 \tag{3.2.40}$$

As a result of (3.2.31), (3.2.33), and (3.2.40), $\lim_{\rho\to\infty}\lim_{m\to\infty} b_3 = 0$.

A combination of the analysis in Sects. 3.2.1.2, 3.2.1.3, and 3.2.1.4 completes this proof.

3.2.1.5 Distribution of the Number of Isolated Nodes on a Square

We now consider the asymptotic distribution of the number of isolated nodes in $\mathcal{G}\left(\mathcal{X}_\rho, g_{r_\rho}, A\right)$.

Let W be the number of isolated nodes in $\mathcal{G}\left(\mathcal{X}_\rho, g_{r_\rho}, A\right)$ and W^E be the number of isolated nodes in $\mathcal{G}\left(\mathcal{X}_\rho, g_{r_\rho}, A\right)$ due to the boundary effect. It has been shown in Chap. 2 using the coupling technique that $W = W^E + W^T$. Using the above equation, Theorem 65, Lemma 43, which showed that $\lim_{\rho\to\infty} \Pr(W^E = 0) = 1$, and Slutsky's theorem [84] which establishes the convergence of W, the following result on the asymptotic distribution of W can be readily obtained. Note that let $\mathcal{G}\left(\mathcal{X}_\lambda, g, A_{\frac{1}{r_\rho}}\right)$ be a network with nodes Poissonly distributed on a square $A_{\frac{1}{r_\rho}} = [-\frac{1}{2r_\rho}, \frac{1}{2r_\rho}]^2$ with density $\lambda = (\log\rho + b)/C$ and a pair of nodes separated by a Euclidean distance x are directly connected with probability $g(x)$, independent of other connections. Results in Chap. 2 are derived for $\mathcal{G}\left(\mathcal{X}_\lambda, g, A_{\frac{1}{r_\rho}}\right)$. By proper scaling, it is straightforward to extend the results for $\mathcal{G}\left(\mathcal{X}_\lambda, g, A_{\frac{1}{r_\rho}}\right)$ to $\mathcal{G}\left(\mathcal{X}_\rho, g_{r_\rho}, A\right)$. Therefore, we ignore the difference.

Theorem 70 *The distribution of the number of isolated nodes in* $\mathcal{G}\left(\mathcal{X}_\rho, g_{r_\rho}, A\right)$ *converges to a Poisson distribution with mean* e^{-b} *as* $\rho \to \infty$.

Corollary 71 follows immediately from Theorem 70.

Corollary 71 *As* $\rho \to \infty$, *the probability that there is no isolated node in* $\mathcal{G}\left(\mathcal{X}_\rho, g_{r_\rho}, A\right)$ *converges to* $e^{-e^{-b}}$.

Now we relax requirement that b is a constant to obtain a necessary condition for $\mathcal{G}\left(\mathcal{X}_\rho, g_{r_\rho}, A\right)$ to be asymptotically almost surely connected. Specifically, consider the situation when $b \to -\infty$ or $b \to \infty$ as $\rho \to \infty$. Note that the property that the network $\mathcal{G}\left(\mathcal{X}_\rho, g_{r_\rho}, A\right)$ has no isolated node is an *increasing* property. For an arbitrary network, a particular property is termed *increasing* if the property is preserved when more connections (edges) are added into the network. Using a coupling technique similar to that used in [71, Chapter 2] and with a few simple steps (omitted), the following theorem and corollary can be obtained:

Theorem 72 *In* $\mathcal{G}\left(\mathcal{X}_\rho, g_{r_\rho}, A\right)$, *if* $b \to \infty$ *as* $\rho \to \infty$, *asymptotically almost surely there is no isolated node in the network; if* $b \to -\infty$ *as* $\rho \to \infty$, *asymptotically almost surely the network has at least one isolated node.*

Corollary 73 $b \to \infty$ *is a necessary condition for* $\mathcal{G}\left(\mathcal{X}_\rho, g_{r_\rho}, A\right)$ *to be asymptotically almost surely connected as* $\rho \to \infty$.

3.2.2 Sufficient Condition for Almost Sure Connectivity

In this subsection, we continue to investigate the sufficient condition for $\mathcal{G}\left(\mathcal{X}_\rho, g_{r_\rho}, A\right)$ to be asymptotically almost surely connected. In Chap. 2, we showed that vanishing of components of finite order $k > 1$ in $\mathcal{G}\left(\mathcal{X}_\rho, g, \Re^2\right)$ as $\rho \to \infty$ (as shown in [143, Theorems 6.3]) does not *necessarily* carry over to the conclusion that components of finite order $k > 1$ in $\mathcal{G}\left(\mathcal{X}_\rho, g_{r_\rho}, A\right)$ also vanish as $\rho \to \infty$, contrary perhaps to intuition. Then, we presented a result for the vanishing of components of finite order $k > 1$ in $\mathcal{G}\left(\mathcal{X}_\rho, g_{r_\rho}, A\right)$ as $\rho \to \infty$ to fill this theoretical gap. On the basis of the above results, we shall further demonstrate in this subsection that asymptotically almost surely the number of unbounded components in $\mathcal{G}\left(\mathcal{X}_\rho, g_{r_\rho}, A\right)$ is one as $\rho \to \infty$. A sufficient condition for $\mathcal{G}\left(\mathcal{X}_\rho, g_{r_\rho}, A\right)$ to be asymptotically almost surely connected readily follows.

In [143, Theorem 6.3], it was shown that there can be at most one unbounded component in $\mathcal{G}\left(\mathcal{X}_\rho, g, \Re^2\right)$. However due to the truncation effect introduced in Chap. 2, it appears difficult to establish such a conclusion using [143, Theorem 6.3]. Indeed, differently from $\mathcal{G}\left(\mathcal{X}_\rho, g, \Re^2\right)$ in which an unbounded component may exist for a finite ρ, it can be easily shown that for any finite ρ, $\Pr\left(\left|\mathcal{X}_\rho\right| < \infty\right) = 1$, i.e., the total number of nodes in $\mathcal{G}\left(\mathcal{X}_\rho, g_{r_\rho}, A\right)$ is almost surely finite. It then follows that for any finite ρ almost surely there is no unbounded component in $\mathcal{G}\left(\mathcal{X}_\rho, g_{r_\rho}, A\right)$.

In this chapter, we solve the above conceptual difficulty involving use of the term "unbounded component" by considering the number of components in $\mathcal{G}\left(\mathcal{X}_\rho, g_{r_\rho}, A\right)$ of order greater than M, denoted by $\xi_{>M}$, where M is an arbitrarily large positive integer. We then show that $\lim_{M\to\infty}\lim_{\rho\to\infty}\Pr(\xi_{>M} = 1) = 1$. The analytical result is summarized in the following theorem:

Theorem 74 *As $\rho \to \infty$, asymptotically almost surely the number of unbounded components in $\mathcal{G}\left(\mathcal{X}_\rho, g_{r_\rho}, A\right)$ is one.*

In the following paragraphs, we prove Theorem 74. For notational convenience, we prove the result for $\mathcal{G}\left(\mathcal{X}_\lambda, g, A_{\frac{1}{r_\rho}}\right)$ and the result is equally valid for $\mathcal{G}\left(\mathcal{X}_\rho, g_{r_\rho}, A\right)$. The proof is based on analyzing the number of components in $\mathcal{G}\left(\mathcal{X}_\lambda, g, A_{\frac{1}{r_\rho}}\right)$ of order greater than some integer M as $\rho \to \infty$. Specifically we shall show that $\lim_{M\to\infty}\lim_{\rho\to\infty}\Pr(\xi_{>M} = 1) = 1$.

A direct analysis of $\Pr(\xi_{>M} = 1)$ can be difficult. We first analyze $E(\xi_{>M})$ and then use the result on $E(\xi_{>M})$ to establish the desired asymptotic result on $\Pr(\xi_{>M} = 1)$.

Denote by $g_1\left(\boldsymbol{x}_1, \ldots, \boldsymbol{x}_k\right)$ the probability that a set of k nodes at nonrandom positions $\boldsymbol{x}_1, \ldots, \boldsymbol{x}_k \in A_{\frac{1}{r_\rho}}$ forms a connected component where nodes are connected randomly and independently following the connection function g. Denote by $g_2\left(\boldsymbol{y}; \boldsymbol{x}_1, \boldsymbol{x}_2, \ldots, \boldsymbol{x}_k\right)$ the probability that a node at nonrandom position $\boldsymbol{y}$ is connected to at least one node in $\{\boldsymbol{x}_1, \boldsymbol{x}_2, \ldots, \boldsymbol{x}_k\}$. As an easy consequence of Lemma 53, which showed that the expected number of components of order k, denoted by ξ_k, in

$\mathcal{G}\left(\mathcal{X}_\lambda, g, A_{\frac{1}{r_\rho}}\right)$ is given by

$$E\left(\xi_k\right)=\frac{\lambda^k}{k!}\int_{(A_{\frac{1}{r_\rho}})^k} g_1(\boldsymbol{x}_1,\ldots,\boldsymbol{x}_k)e^{-\lambda\int_{A_{\frac{1}{r_\rho}}} g_2(\boldsymbol{y};\boldsymbol{x}_1,\ldots,\boldsymbol{x}_k)d\boldsymbol{y}} d(\boldsymbol{x}_1\cdots\boldsymbol{x}_k)$$

it follows that

$$\begin{aligned}
&E\left(\xi_{>M}\right)\\
&=\sum_{k=M+1}^{\infty}\frac{\lambda^k}{k!}\int_{(A_{\frac{1}{r_\rho}})^k}\left(g_1(\boldsymbol{x}_1,\ldots,\boldsymbol{x}_k)e^{-\lambda\int_{A_{\frac{1}{r_\rho}}} g_2(\boldsymbol{y};\boldsymbol{x}_1,\ldots,\boldsymbol{x}_k)d\boldsymbol{y}}\right)d(\boldsymbol{x}_1\cdots\boldsymbol{x}_k)\\
&\le\sum_{k=M+1}^{\infty}\frac{\lambda^k}{k!}\int_{(A_{\frac{1}{r_\rho}})^k} e^{-\lambda\int_{A_{\frac{1}{r_\rho}}} g_2(\boldsymbol{y};\boldsymbol{x}_1,\ldots,\boldsymbol{x}_k)d\boldsymbol{y}} d(\boldsymbol{x}_1\cdots\boldsymbol{x}_k)\\
&=\sum_{k=1}^{\infty}\frac{\lambda^k}{k!}\int_{(A_{\frac{1}{r_\rho}})^k} e^{-\lambda\int_{A_{\frac{1}{r_\rho}}} g_2(\boldsymbol{y};\boldsymbol{x}_1,\ldots,\boldsymbol{x}_k)d\boldsymbol{y}} d(\boldsymbol{x}_1\cdots\boldsymbol{x}_k)\\
&\quad-\sum_{k=1}^{M}\frac{\lambda^k}{k!}\int_{(A_{\frac{1}{r_\rho}})^k} e^{-\lambda\int_{A_{\frac{1}{r_\rho}}} g_2(\boldsymbol{y};\boldsymbol{x}_1,\ldots,\boldsymbol{x}_k)d\boldsymbol{y}} d(\boldsymbol{x}_1\cdots\boldsymbol{x}_k)
\end{aligned}\tag{3.2.41}$$

In the following we shall show that as $\rho\to\infty$, the first term in (3.2.41) converges to $e^{e^{-b}}$, and the second term in (3.2.41) after the "–" sign is lower-bounded by $\sum_{k=1}^{M}\frac{(e^{-b})^k}{k!}$. The conclusion then follows that $E(\xi_{>M})$ converge to 1 as $\rho\to\infty$ and $M\to\infty$.

Let us consider the first term in (3.2.41) now. Let

$$\Phi\triangleq\lambda\int_{A_{\frac{1}{r_\rho}}}\left(1-g_2(\boldsymbol{y};\boldsymbol{x}_1,\ldots\boldsymbol{x}_k)\right)d\boldsymbol{y}\tag{3.2.42}$$

for convenience. It can be shown that

$$\begin{aligned}
&\lim_{\rho\to\infty}\sum_{k=1}^{\infty}\frac{\lambda^k}{k!}\int_{(A_{\frac{1}{r_\rho}})^k} e^{-\lambda\int_{A_{\frac{1}{r_\rho}}} g_2(\boldsymbol{y};\boldsymbol{x}_1,\ldots\boldsymbol{x}_k)d\boldsymbol{y}} d(\boldsymbol{x}_1,\ldots\boldsymbol{x}_k)\\
&=\lim_{\rho\to\infty}\sum_{k=1}^{\infty}\frac{\lambda^k}{k!}e^{-\rho}\int_{(A_{\frac{1}{r_\rho}})^k} e^{\Phi}d(\boldsymbol{x}_1,\ldots\boldsymbol{x}_k)\\
&=\lim_{\rho\to\infty}\sum_{k=1}^{\infty}\frac{\lambda^k}{k!}e^{-\rho}\int_{(A_{\frac{1}{r_\rho}})^k}\sum_{n=0}^{\infty}\frac{\Phi^n}{n!}d(\boldsymbol{x}_1,\ldots\boldsymbol{x}_k)\\
&=\sum_{n=0}^{\infty}\frac{1}{n!}\lim_{\rho\to\infty}\sum_{k=1}^{\infty}\frac{\lambda^k}{k!}e^{-\rho}\int_{(A_{\frac{1}{r_\rho}})^k}\Phi^n d(\boldsymbol{x}_1,\ldots\boldsymbol{x}_k)
\end{aligned}\tag{3.2.43}$$

Next we shall show that in (3.2.43), $\lim_{\rho\to\infty}\sum_{k=1}^{\infty}\frac{\lambda^k}{k!}e^{-\rho}\int_{(A_{\frac{1}{r_\rho}})^k}\Phi^n d(\boldsymbol{x}_1,\ldots\boldsymbol{x}_k) = (e^{-b})^n$. Given this result, conclusion readily follows from (3.2.43) that the first term in (3.2.41) converges to $e^{e^{-b}}$.

A direct computation of the term $\lim_{\rho\to\infty}\sum_{k=1}^{\infty}\frac{\lambda^k}{k!}e^{-\rho}\int_{(A_{\frac{1}{r_\rho}})^k}\Phi^n d(\boldsymbol{x}_1,\ldots\boldsymbol{x}_k)$ turns out to be very difficult. To resolve the difficulty, we construct a random integer X, depending on ρ, such that on one hand, the probability mass function of X has an analytical form that can be easily related to the term $\sum_{k=1}^{\infty}\frac{\lambda^k}{k!}e^{-\rho}\int_{(A_{\frac{1}{r_\rho}})^k}\Phi^n d(\boldsymbol{x}_1,\ldots\boldsymbol{x}_k)$; and on the other hand using the Chen-Stein bound we are familiar with, the probability mass function can be shown to converge to a Poisson distribution as $\rho\to\infty$. In this way, we are able to compute the above term using the intermediate random integer X. In the following, we give details of the analysis.

We first construct the random integer X described in the last paragraph and demonstrate its properties related to our analysis.

Consider an additional *independent* Poisson point process $\mathcal{X}'_\lambda$ with nodes Poissonly distributed on $A_{\frac{1}{r_\rho}}$ and with density λ, being added to $\mathcal{G}\left(\mathcal{X}_\lambda, g, A_{\frac{1}{r_\rho}}\right)$. Furthermore, nodes in $\mathcal{X}'_\lambda$ are connected with nodes in $\mathcal{X}_\lambda$ following g independently, i.e., a node in $\mathcal{X}'_\lambda$ and a node in $\mathcal{X}_\lambda$ separated by a Euclidean distance x are connected with probability $g(x)$, independent of any other connection.

Let X be the number of nodes in $\mathcal{X}'_\lambda$ that are *not* directly connected to any node in $\mathcal{X}_\lambda$. It is evident that, conditioned on $\mathcal{X}_\lambda = (\boldsymbol{x}_1,\ldots\boldsymbol{x}_k)$ where $\boldsymbol{x}_1,\ldots\boldsymbol{x}_k \in A_{\frac{1}{r_\rho}}$ and $|\mathcal{X}_\lambda| > 0$, a randomly chosen node in $\mathcal{X}'_\lambda$ at location $\boldsymbol{y}$ is *not* directly connected to any node in $\mathcal{X}_\lambda$ with probability $1 - g_2(\boldsymbol{y};\boldsymbol{x}_1,\ldots,\boldsymbol{x}_k)$, which is determined by its location only. It readily follows that the conditional distribution of X, i.e., $X|\mathcal{X}_\lambda = (\boldsymbol{x}_1,\ldots\boldsymbol{x}_k)$, is Poisson with mean $\lambda\int_{A_{\frac{1}{r_\rho}}}[1 - g_2(\boldsymbol{y};\boldsymbol{x}_1,\ldots,\boldsymbol{x}_k)]d\boldsymbol{y}$. As a result of the above discussion:

$$\Pr\left(X = m|\mathcal{X}_\lambda = (\boldsymbol{x}_1,\ldots\boldsymbol{x}_k)\right) = \frac{\Phi^m}{m!}e^{-\Phi} \tag{3.2.44}$$

Obviously when $\mathcal{X}_\lambda = \emptyset$, $\Pr(X = m|\mathcal{X}_\lambda = \emptyset) = \Pr(|\mathcal{X}'_\lambda| = m)$. Therefore, the unconditional distribution of X is given by:

$$\Pr(X = m)$$
$$= \sum_{k=1}^{\infty}\frac{\lambda^k}{k!}e^{-\rho}\int_{(A_{\frac{1}{r_\rho}})^k}\frac{\Phi^m}{m!}e^{-\Phi}d(\boldsymbol{x}_1,\ldots\boldsymbol{x}_k) + \frac{\rho^m}{m!}e^{-2\rho} \tag{3.2.45}$$

Note that as $\rho\to\infty$, the term $\frac{\rho^m}{m!}e^{-2\rho}$ in (3.2.45), which is associated with $\mathcal{X}_\lambda = \emptyset$, becomes vanishingly small. Further note that $\sum_{m=0}^{\infty}\frac{\rho^m}{m!}e^{-2\rho} = e^{-\rho}\to 0$ as $\rho\to\infty$, i.e., as $\rho\to\infty$ even the cumulative contribution of X is negligibly small.

If we define $g_2(\boldsymbol{y};\emptyset) \triangleq 0$ for completeness, we can also write (3.2.45) as

$$\Pr(X=m)=\sum_{k=0}^{\infty}\frac{\lambda^k}{k!}e^{-\rho}\int_{(A_{\frac{1}{r_\rho}})^k}\frac{\Phi^m}{m!}e^{-\Phi}d(\boldsymbol{x}_1,\ldots\boldsymbol{x}_k) \tag{3.2.46}$$

Using (3.2.46), it can be readily shown that

$$\begin{aligned}
&E(X)\\
&=\sum_{m=0}^{\infty}m\Pr(X=m) && (3.2.47)\\
&=\sum_{k=0}^{\infty}\frac{\lambda^k}{k!}e^{-\rho}\int_{(A_{\frac{1}{r_\rho}})^k}\Phi d(\boldsymbol{x}_1,\ldots\boldsymbol{x}_k)\\
&=\sum_{k=0}^{\infty}\frac{\lambda^k}{k!}e^{-\rho}\{\lambda\int_{A_{\frac{1}{r_\rho}}}\{\int_{A_{\frac{1}{r_\rho}}}[1-g(\|\boldsymbol{x}-\boldsymbol{y}\|)]d\boldsymbol{x}\}^k d\boldsymbol{y}\}\\
&=\lambda\int_{A_{\frac{1}{r_\rho}}}e^{-\lambda\int_{A_{\frac{1}{r_\rho}}}g(\|\boldsymbol{x}-\boldsymbol{y}\|)d\boldsymbol{x}}d\boldsymbol{y} && (3.2.48)
\end{aligned}$$

Comparing the above equation with Theorem 41, the conclusion follows that the above value is equal to the expected number of isolated nodes in $\mathcal{G}\left(\mathcal{X}_\lambda, g, A_{\frac{1}{r_\rho}}\right)$, denoted by W. It then follows from Theorem 41 that $\lim_{\rho\to\infty}E(X)=e^{-b}$. In fact, a stronger result that the distributions of X and W converge to the same Poisson distribution as $\rho\to\infty$ can be established:

Lemma 75 *As $\rho\to\infty$, the distribution of X converges to a Poisson distribution with mean e^{-b}, i.e., the total variation distance between the distribution of X and a Poisson distribution with mean e^{-b} reduces to 0 as $\rho\to\infty$.*

Lemma 75 can be proved using exactly the same steps as those used in proving Theorem 70. Therefore, the proof is omitted.

As a result of Lemma 75, for an arbitrary set of nonnegative integers, denoted by Γ,

$$\lim_{\rho\to\infty}\sum_{m\in\Gamma}\Pr(X=m)=\sum_{m\in\Gamma}\frac{(e^{-b})^m}{m!}e^{-e^{-b}} \tag{3.2.49}$$

Now we are ready to continue our analysis on $\lim_{\rho\to\infty}\sum_{k=1}^{\infty}\frac{\lambda^k}{k!}e^{-\rho}\int_{(A_{\frac{1}{r_\rho}})^k}\Phi^n d(\boldsymbol{x}_1,\ldots\boldsymbol{x}_k)$. Using (3.2.45) first and then using (3.2.49), it can be shown that for any positive integer n:

$$\begin{aligned}
&\lim_{\rho\to\infty}\sum_{k=1}^{\infty}\frac{\lambda^k}{k!}e^{-\rho}\int_{(A_{\frac{1}{r_\rho}})^k}\Phi^n d(\boldsymbol{x}_1,\ldots\boldsymbol{x}_k)\\
&=\lim_{\rho\to\infty}\sum_{k=1}^{\infty}\frac{\lambda^k}{k!}e^{-\rho}\int_{(A_{\frac{1}{r_\rho}})^k}\sum_{m=0}^{\infty}\Phi^n\frac{\Phi^m}{m!}e^{-\Phi}d(\boldsymbol{x}_1,\ldots\boldsymbol{x}_k)\\
&=\sum_{m=0}^{\infty}\frac{1}{m!}\lim_{\rho\to\infty}\sum_{k=1}^{\infty}\frac{\lambda^k}{k!}e^{-\rho}\int_{(A_{\frac{1}{r_\rho}})^k}\Phi^{n+m}e^{-\Phi}d(\boldsymbol{x}_1,\ldots\boldsymbol{x}_k)\\
&=\sum_{m=0}^{\infty}\frac{1}{m!}\lim_{\rho\to\infty}(\Pr(X=n+m)-\frac{\rho^{(n+m)}}{(n+m)!}e^{-2\rho})(n+m)!\\
&=\sum_{m=0}^{\infty}\frac{(e^{-b})^{n+m}}{m!}e^{-e^{-b}}\\
&=(e^{-b})^n
\end{aligned}$$

Using the above equation, it follows from (3.2.43) that

$$\begin{aligned}
&\lim_{\rho\to\infty}\sum_{k=1}^{\infty}\frac{\lambda^k}{k!}\int_{(A_{\frac{1}{r_\rho}})^k}e^{-\lambda\int_{A_{\frac{1}{r_\rho}}}g_2(\boldsymbol{y};\boldsymbol{x}_1,\ldots\boldsymbol{x}_k)d\boldsymbol{y}}d(\boldsymbol{x}_1,\ldots\boldsymbol{x}_k)\\
&=\sum_{n=0}^{\infty}\frac{(e^{-b})^n}{n!} && (3.2.50)\\
&=e^{e^{-b}} && (3.2.51)
\end{aligned}$$

This deals with the first term on the right of (3.2.41). Now we continue with the analysis of the second term in (3.2.41). As an easy consequence of the union bound, $g_2(\boldsymbol{y};\boldsymbol{x}_1,\boldsymbol{x}_2,\ldots,\boldsymbol{x}_k)\le\sum_{i=1}^{k}g(\|\boldsymbol{y}-\boldsymbol{x}_i\|)$, it can then be shown that

$$\begin{aligned}
&\frac{\lambda^k}{k!}\int_{(A_{\frac{1}{r_\rho}})^k}e^{-\lambda\int_{A_{\frac{1}{r_\rho}}}g_2(\boldsymbol{y};\boldsymbol{x}_1,\ldots\boldsymbol{x}_k)d\boldsymbol{y}}d(\boldsymbol{x}_1\cdots\boldsymbol{x}_k)\\
&\ge\frac{\lambda^k}{k!}\int_{(A_{\frac{1}{r_\rho}})^k}e^{-\lambda\int_{A_{\frac{1}{r_\rho}}}\sum_{i=1}^{k}g(\|\boldsymbol{y}-\boldsymbol{x}_i\|)d\boldsymbol{y}}d(\boldsymbol{x}_1\cdots\boldsymbol{x}_k)\\
&=\frac{(\lambda\int_{A_{\frac{1}{r_\rho}}}e^{-\lambda\int_{A_{\frac{1}{r_\rho}}}g(\|\boldsymbol{x}-\boldsymbol{y}\|)d\boldsymbol{y}}d\boldsymbol{x})^k}{k!} && (3.2.52)
\end{aligned}$$

Using Theorem 41, it can be further shown that

$$\lim_{\rho\to\infty} \frac{\lambda^k}{k!} \int_{(A_{\frac{1}{r_\rho}})^k} e^{-\lambda \int_{A_{\frac{1}{r_\rho}}} g_2(\mathbf{y};\mathbf{x}_1,\ldots\mathbf{x}_k)d\mathbf{y}} d(\mathbf{x}_1 \cdots \mathbf{x}_k)$$

$$\geq \frac{(e^{-b})^k}{k!} \tag{3.2.53}$$

Note that (3.2.53) can also be obtained from Jensen's inequality.
Combining (3.2.41), (3.2.50), and (3.2.53), it follows that

$$\lim_{\rho\to\infty} E(\xi_{>M}) \leq e^{e^{-b}} - \sum_{k=1}^{M} \frac{(e^{-b})^k}{k!} = 1 + \frac{(\eta_M)^{M+1}}{(M+1)!} \tag{3.2.54}$$

where in the last step Taylor's theorem is used, η_M is a number depending on M and $0 \leq \eta_M \leq e^{-b}$.

In Theorems 70 and 52, we have established, respectively, that the asymptotic distribution of the number of isolated nodes in $\mathcal{G}\left(\mathcal{X}_\lambda, g, A_{\frac{1}{r_\rho}}\right)$ is Poisson with mean e^{-b} and the number of components in $\mathcal{G}\left(\mathcal{X}_\lambda, g, A_{\frac{1}{r_\rho}}\right)$ of order within $[2, M]$ vanishes as $\rho \to \infty$. As a consequence of the above two results,

$$\lim_{\rho\to\infty} \Pr(\xi_{>M} \geq 1) = 1 \quad \text{and} \quad \lim_{\rho\to\infty} \Pr(\xi_{>M} = 0) = 0 \tag{3.2.55}$$

Further note that

$$E(\xi_{>M}) = \sum_{m=1}^{\infty} m \Pr(\xi_{>M} = m)$$

$$\geq \Pr(\xi_{>M} = 1) + 2\sum_{m=2}^{\infty} \Pr(\xi_{>M} = m)$$

$$= \Pr(\xi_{>M} = 1) + 2\,(1 - \Pr(\xi_{>M} = 1) - \Pr(\xi_{>M} = 0)) \tag{3.2.56}$$

Combining the three equations (3.2.54), (3.2.55), and (3.2.56):

$$\lim_{\rho\to\infty} \Pr(\xi_{>M} = 1) \geq 1 - \frac{(\eta_M)^{M+1}}{(M+1)!} \tag{3.2.57}$$

As an easy consequence of the above equation:

$$\lim_{M\to\infty} \lim_{\rho\to\infty} \Pr\left(\xi_{>M} = 1\right) = 1$$

This completes the proof of Theorem 74.

Remark 76 Proof of the type of results in Theorem 74 usually requires some complicated geometric analysis. Particularly the proof of Lemma 75, which forms a foundation of the proof of Theorem 74, needs sophisticated geometric analysis. In this subsection, we omitted the proof of Lemma 75 because the proof is exactly the same as the proof of Theorem 70, which in turn relies on some results established in Chap. 2. We refer interested readers to the proof of Theorem 41 for techniques on handling geometric obstacles involved in analyzing the boundary effect and to the proof of Theorem 52 for techniques on handling geometric obstacles involved in analyzing the number of components in $\mathcal{G}\left(\mathcal{X}_\rho, g_{r_\rho}, A\right)$.

An implication of Theorem 74 is that for an arbitrarily small positive constant ε, there exists large positive constants M_0 and ρ_0 such that for all $M > M_0$ and $\rho > \rho_0$, $\Pr(\xi_{>M} = 1) > 1 - \varepsilon$. From (3.2.57), it can further be concluded that for a fixed positive integer M and an arbitrarily small positive constant ε, there exists ρ_0 such that for all $\rho > \rho_0$,

$$\Pr(\xi_{>M} = 1) > 1 - \frac{e^{-(M+1)b}}{(M+1)!} - \varepsilon \tag{3.2.58}$$

The following corollary can therefore be obtained from Theorems 52 and 74:

Corollary 77 *As $\rho \to \infty$, asymptotically almost surely $\mathcal{G}\left(\mathcal{X}_\rho, g_{r_\rho}, A\right)$ forms a connected network if and only if there is no isolated node in it.*

Proof Let ξ be the total number of components in $\mathcal{G}\left(\mathcal{X}_\rho, g_{r_\rho}, A\right)$. It is clear that $\xi = \xi_1 + \sum_{k=2}^{M} \xi_k + \xi_{>M}$, where ξ_k is the number of components of order k. Noting that $\xi = 1$ if and only if $\mathcal{G}\left(\mathcal{X}_\rho, g_{r_\rho}, A\right)$ forms a connected network, it suffices to show that $\lim_{\rho\to\infty} \Pr\left(\xi = 1 | \xi_1 = 0\right) = 1$. We observe that

$$\begin{aligned}
&\Pr\left(\xi = 1, \xi_1 = 0\right) \\
&\geq \Pr(\xi_1 = 0, \sum_{k=2}^{M} \xi_k = 0, \xi_{>M} = 1) \\
&= \Pr(\xi_1 = 0) - (\Pr(\overline{\sum_{k=2}^{M} \xi_k = 0}) + \Pr(\overline{\xi_{>M} = 1}))
\end{aligned} \tag{3.2.59}$$

where in (3.2.59) $\overline{\xi_{>M} = 1}$ represents the complement of the event $\xi_{>M} = 1$ and (3.2.59) results as a consequence of the union bound. Further note that (3.2.59) is valid for *any* value of M and that $\Pr\left(\xi_1 = 0\right)$ converges to a *nonzero* constant $e^{-e^{-b}}$ as $\rho \to \infty$ (Theorem 70). Using the above results, Theorem 52 which showed that $\lim_{\rho\to\infty} \Pr(\sum_{k=2}^{M} \xi_k = 0) = 1$, and (3.2.58), and following a few simple steps (omitted), it can be shown that for an arbitrarily small positive constant ε, by choosing M to be sufficiently large, there exists ρ_0 such that for all $\rho > \rho_0$, $\Pr\left(\xi = 1 | \xi_1 = 0\right) > 1 - \varepsilon$. □

As an easy consequence of Theorem 70 and Corollary 77, the following theorem can be established:

Theorem 78 *As* $\rho \to \infty$, *the probability that* $\mathcal{G}\left(\mathcal{X}_\rho, g_{r_\rho}, A\right)$ *forms a connected network converges to* $e^{-e^{-b}}$.

Using the above theorem and a similar analysis as that leading to Theorem 72 and Corollary 73, the following theorem on the sufficient and necessary condition for $\mathcal{G}\left(\mathcal{X}_\rho, g_{r_\rho}, A\right)$ to be asymptotically almost surely connected can be obtained:

Theorem 79 *As* $\rho \to \infty$, $\mathcal{G}\left(\mathcal{X}_\rho, g_{r_\rho}, A\right)$ *is asymptotically almost surely connected if and only if* $b \to \infty$; $\mathcal{G}\left(\mathcal{X}_\rho, g_{r_\rho}, A\right)$ *is asymptotically almost surely disconnected if and only if* $b \to -\infty$.

3.3 Special Cases of the Network Model and the Random Connection Model

In the previous section, we established the sufficient and necessary condition required for $\mathcal{G}\left(\mathcal{X}_\rho, g_{r_\rho}, A\right)$ to be asymptotically almost surely connected as $\rho \to \infty$. The results are established for a network with nodes Poissonly distributed on a unit square following a general random connection model. Using the scaling and coupling technique, the results can be readily extended to a network of any other size. Furthermore, as discussed in Sect. 1.2, although uniform point process and Poisson point process are two conceptually distinct point processes, their difference is often negligible for a sufficient large value of ρ. Therefore, the results obtained using $\mathcal{G}\left(\mathcal{X}_\rho, g_{r_\rho}, A\right)$ with Poisson node distribution can be applied to a network with a total of $\rho\,|A|$ nodes independently, identically and uniformly distributed on A where a pair of nodes are directly connected following the random connection function g_{r_ρ}, independent of other pairs of nodes. Techniques to extend results obtained assuming a Poisson point process to a uniform point process, and the converse, are well known and have also been briefly discussed in Sect. 1.2.

The results in the previous section are established for a unit square. However they should also be applicable for a network area of any other shape, e.g., a disk of unit area, although some complicated geometrical analysis may be required to handle the boundary effect when nodes are located in a network area of a different shape. Generally, we expect the results to be valid for a network area of any other shape as long as the ratio of the size of the boundary area to the total network area becomes vanishingly small as $\rho \to \infty$. For nodes directly connected following the random connection function g_{r_ρ}, the size of the boundary area is given by the product of the length of the border and r_ρ.

The random connection model includes many other models such as the unit disk connection model and the log-normal connection model as its special cases.

Particularly, letting g_{r_ρ} be such that

$$g_{r_\rho}(x) = g\left(\frac{x}{r_\rho}\right) = \begin{cases} 1 \text{ if } x \le r_\rho \\ 0 \text{ if } x > r_\rho \end{cases}$$

$\mathcal{G}\left(\mathcal{X}_\rho, g_{r_\rho}, A\right)$ reduces to the same network considered in [92] assuming the unit disk connection model with a transmission range of r_ρ. On the other hand, letting

$$g(x) = \int_{10\alpha \log_{10} \frac{x}{r_0}}^{\infty} \frac{1}{\sqrt{2\pi}\sigma} e^{-\frac{z^2}{2\sigma^2}} dz \tag{3.3.1}$$

where r_0 and σ are the parameters of the log-normal connection model introduced in Sect. 1.1.3, the results established for $\mathcal{G}\left(\mathcal{X}_\rho, g_{r_\rho}, A\right)$ can be applied to the log-normal connection model too. There are also other interesting extensions of the connection model, e.g., those taking into account Rayleigh or Rician fading, those considering propagation models other than the log-normal model, and those considering the directionality of the antenna pattern [172] etc., the random connection model is general enough to include these models as its special cases.

However, the results on $\mathcal{G}\left(\mathcal{X}_\rho, g_{r_\rho}, A\right)$ however critically rely on the assumption that the random events underpinning generation of a connection are independent, which is not necessarily true in some real networks due to channel correlation and interference. Particularly, the presence of interference makes the connection between a pair of nodes dependent on the location and activities of other nearby nodes. The connection model that captures the impact of interference is known as the SINR connection model, which has been introduced earlier in Sect. 1.1.5. In the latter part of this book, we shall introduce techniques for connectivity analysis assuming the SINR connection model. The major obstacle in dealing with the impact of channel correlation is that there is no widely accepted model in the wireless communication community capturing the impact of channel correlation on connections. Thus, the results obtained assuming the random connection model cannot be readily extended to the SINR connection model or to a model where the connections are correlated.

3.4 Notes and Further Readings

Following the seminal work of Penrose [164, 165] and Gupta and Kumar [92] on the asymptotic connectivity of large random networks with Poisson node distribution and assuming the unit disk connection model, there is general expectation that there is a range of connection models to which the above results [92, 164, 165] obtained assuming the unit disk connection model can carry over. However, for quite a long time, both the asymptotic laws that the network should follow and the conditions on the connection function required for the network to be asymptotically almost surely connected under a more general setting have been unknown. In this chapter, we filled in the gap by providing the sufficient and necessary condition for a network with nodes Poissonly distributed on a unit square and following a general random connection model to be asymptotically almost surely connected as $\rho \to \infty$. The conditions on the connection function required in order for the above network to be *asymptotically almost surely* connected were also provided. Therefore, the results bring models addressed by theoretical research closer to reality.

Some work exists on the analysis of the asymptotic distribution of the number of isolated nodes [70, 71, 165, 197] assuming the unit disk connection model. In [197], Yi et al. considered a total of n nodes distributed independently and uniformly on a unit-area disk and each node may be active independently with some probability p. A node is considered to be isolated if it is not directly connected to any of the active nodes. Using some complicated geometric analysis, they showed that if all nodes have a maximum transmission range $r(n) = \sqrt{(\log n + \xi)/(\pi p n)}$ for some constant ξ, the total number of isolated nodes is asymptotically Poissonly distributed with mean $e^{-\xi}$. In [70, 71], Franceschetti et al. derived essentially the same result using the Chen-Stein technique. A similar result can also be found in the earlier work of Penrose [165] in a continuum percolation setting.

Chapter 4
Giant Component in Large Wireless Networks

Abstract In this chapter, we study the giant component, the largest component containing a nonvanishing fraction of nodes, in large wireless networks in both one-dimensional and two-dimensional space. Specifically, consider a network with a total of n nodes randomly, independently, and uniformly distributed in a unit interval or unit square. For one-dimensional networks, we derive a closed-form analytical formula for calculating the probability of having a giant component of order above pn with any fixed $0.5 < p \leq 1$. The asymptotic behavior of one-dimensional network having a giant component is then investigated, which is distinctly different from its two-dimensional counterpart. For two-dimensional networks, we derive an asymptotic analytical upper bound on the minimum transmission range at which the probability of having a giant component of order above qn for any fixed $0 < q < 1$ tends to one as $n \to \infty$. Using the result, we show that significant energy savings can be achieved if we only require a large percentage of nodes (e.g., 95%, 99%) to be connected rather than requiring all nodes to be connected. That is, with a slightly relaxed requirement on connectivity, substantial energy savings can be achieved for a large network. For one-dimensional space, we study the network assuming the unit disk connection model. For two-dimensional space, we first consider the unit disk connection model and then discuss how results obtained assuming the unit disk connection model can be extended to more general connection models.

In the last chapter, we studied connectivity of $\mathcal{G}\left(\mathcal{X}_\rho, g_{r_\rho}, A\right)$ as $\rho \to \infty$ and showed that the network is asymptotically almost surely connected if and only if $b \to \infty$, where $g_{r_\rho}(x) = g(x/r_\rho)$, $r_\rho = \sqrt{(\log \rho + b)/(C\rho)}$ and $C = \int_{\Re^2} g(\|\boldsymbol{x}\|)d\boldsymbol{x}$. The *average node degree* of $\mathcal{G}\left(\mathcal{X}_\rho, g_{r_\rho}, A\right)$, i.e., the average number of directly connected neighbors per node, is given by:

$$\begin{aligned}
&\int_A \rho g_{r_\rho}(\|\boldsymbol{x}\|)d\boldsymbol{x} \\
&= \int_A \rho g(\frac{\|\boldsymbol{x}\|}{r_\rho})d\boldsymbol{x} \\
&= \int_{A\frac{1}{r_\rho}} \rho g(\|\boldsymbol{y}\|)\left(r_\rho\right)^2 d\boldsymbol{y}
\end{aligned}$$

G. Mao, *Connectivity of Communication Networks*,
DOI 10.1007/978-3-319-52989-9_4

where $A_{\frac{1}{r_\rho}}$ is a square of size $\left(\frac{1}{r_\rho}\right)^2$. Noting that $b = o(\log \rho)$, it readily follows from the above equation that as $\rho \to \infty$

$$\lim_{\rho \to \infty} \frac{\int_A \rho g_{r_\rho}(\|\boldsymbol{x}\|)d\boldsymbol{x}}{\log \rho} = 1 \tag{4.0.1}$$

Equation (4.0.1) implies that as the network becomes larger and larger (to be more precise, the average number of nodes ρ in the network increases as $\rho \to \infty$), the average node degree has to increase following $\sim \log \rho$, i.e., every node has to directly connect to more and more other nodes, in order to maintain the network connected. With the increased requirement on the average node degree, every node has to transmit at a higher and higher transmission power, which in turn causes more and more interference in the network. Therefore, connectivity of a large network is a very demanding requirement.

The above observation motivates us to inquire whether in some applications, we can tolerate a (small) fraction of nodes to be disconnected from other nodes in the network; and whether there is any benefit in terms of reduced requirement on transmission range or power if we allow a (small) fraction of nodes to be disconnected?

For the first query, we note that in many applications, it is unnecessary for all nodes to be connected to each other all the time [26]. Examples of such applications include a wireless sensor network used for habitat monitoring [104, 112], or environmental monitoring [105, 161], and a mobile ad hoc network in which users can tolerate short off-service intervals [117, 199], etc.

In environmental monitoring, there are scenarios where the size of the monitored phenomenon is very large (e.g., rain clouds), or the parameters (e.g., temperature, humidity) being monitored change slowly both in space and in time. When the number of nodes used for monitoring the phenomenon or measuring the parameters is very large, having a few disconnected nodes will not cause statistically significant change in the monitored parameters. An example of such applications is a wireless sensor network deployed underneath the Briksdalsbreen glacier in Norway to monitor the pressure, humidity, and temperature of ice in order to understand glacier dynamics in response to climate change [161].

In habitat monitoring, there are scenarios where the number of objects (e.g., zebras, cane toads) being monitored is large and these objects are distributed randomly and almost independently in the surveyed region. Having a few nodes disconnected or lost may not significantly affect the accuracy of the monitored parameter, e.g., the size or the density of the population. Examples of such applications include the experiment reported in [104] where a wireless acoustic sensor network was used to monitor the population distribution of invasive cane toads in northern Australia. In this application, having a few nodes disconnected has little impact on the accuracy of the estimated population distribution.

Furthermore, in many mobile ad hoc networks, having a number of nodes temporarily disconnected is also not critical as long as users can tolerate short off-

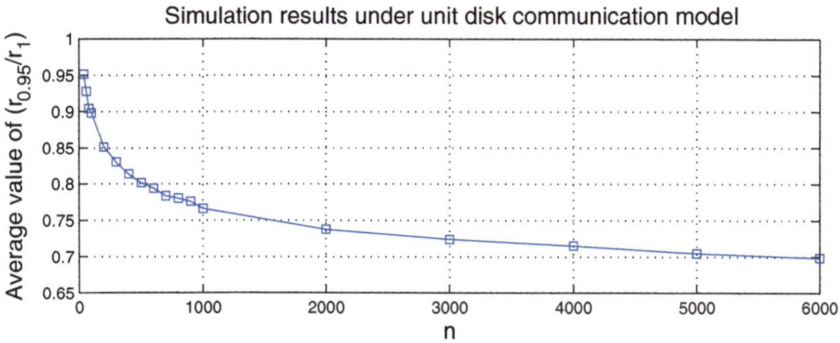

Fig. 4.1 Simulation results for the average value of the ratio $r_{0.95}/r_1$ as a function of the number of nodes n in the network where r_1 is the transmission range required for a connected network and $r_{0.95}$ is the transmission range required for 95% of nodes to be connected. The ratio shown is the average value and each average value is obtained over 2000 random topologies where a total of n nodes are uniformly, randomly, and independently distributed on a unit square

service intervals. For example, in a campus-wide wireless network, students and staff can share information using wireless devices (e.g., laptops, smartphones, and iPads) around campus [199]. When a wireless device loses its wireless connection temporarily, it can store the data and complete the work after getting the connection later.

For the second query, we conduct simulations which show that by allowing a small percentage of nodes to be disconnected (versus requiring all nodes to be connected), much less transmission power or range is required. As an example, Fig. 4.1 shows simulation results comparing the transmission range required for all nodes to be connected with the transmission range required for 95% of nodes to be connected in a network of n nodes randomly, independently, and uniformly distributed on a unit square assuming the unit disk connection model.

As shown in Fig. 4.1, the ratio of the transmission range required for 95% nodes to be connected to that for a connected network decreases with an increase in the total number of nodes in the network. When the number of nodes is 1000, the transmission range required for 95% nodes to be connected is 24% less than that required for a connected network. Based on a conservative estimate that the required transmission power increases with the square of the required transmission range (see Sect. 1.1.3 for a discussion on how power attenuates with distance), this means a power saving of at least 42%. In addition, the ratio decreases further as the total number of nodes n increases. As we shall show later in Sect. 4.2, the ratio will go to zero when $n \to \infty$. This means that the power saving is even more significant in a network with a larger number of nodes. Since many real applications do not require all nodes to be connected, it is appropriate to consider slightly relaxing the connectivity requirement, i.e., requiring most nodes (e.g., 95%) to be connected rather than requiring all nodes to be connected, in order to achieve significant savings in power consumption. It is therefore both important and interesting to investigate the largest connected component containing a nonvanishing fraction of nodes, termed the *giant component* [100, 165].

Motivated by the above observations and simulation studies, in this chapter, we investigate the conditions for having a giant component of order above a given fraction of nodes in both one- and two-dimensional spaces. For two-dimensional space, we first consider the unit disk connection model and then discuss how results obtained assuming the unit disk connection model can be extended to more general connection models.

4.1 Giant Component in One-Dimensional Networks

In this section, we investigate the probability of having a giant component of order above pn in a network of n nodes on a unit interval for any fixed $0.5 < p \leq 1$ in one-dimensional space assuming a unit disk connection model with a transmission range r. Denote the probability by $P(n, r, p)$. We first analyze the situation of finite n and then extend to the case of infinite n.

4.1.1 Giant Component in a Finite Network

The main result of this subsection on the probability $P(n, r, p)$ for a finite network of n nodes is given in the following theorem.

Theorem 80 *Consider a one-dimensional random network $G_1(\mathcal{X}_n, r)$ with a total of n nodes deployed on a unit interval $[0, 1]$ and assuming a unit disk connection model with a transmission range of r. Denote the network by $G_1(\mathcal{X}_n, r)$. Let p be a fixed real number in $(0.5, 1]$. Let $P(n, r, p)$ be the probability that $G_1(\mathcal{X}_n, r)$ has a giant component of order above pn. Then,*

$$\begin{aligned}
&P(n,r,p)\\
&= \sum_{i=\lceil np\rceil}^{n-1}\left[2\sum_{j=0}^{\min\{i,\lfloor\frac{1}{r}\rfloor-1\}}\binom{i}{j}(-1)^j(1-(j+1)r)^n\right.\\
&\left.\quad+(n-i+1)\sum_{j=0}^{\min\{i-1,\lfloor\frac{1}{r}\rfloor-2\}}\binom{i-1}{j}(-1)^j(1-(j+2)r)^n\right]\\
&\quad+\sum_{j=0}^{\min\{n-1,\lfloor\frac{1}{r}\rfloor\}}\binom{n-1}{j}(-1)^j(1-jr)^n
\end{aligned} \tag{4.1.1}$$

The rest of this subsection is dedicated to the proof of Theorem 80. The following three lemmas, viz., Lemma 81, Lemma 82, and Lemma 83, are required for the proof. Lemma 82 and Lemma 83 are used to prove Theorem 80; Lemma 81 is used to derive Lemma 82 and Lemma 83.

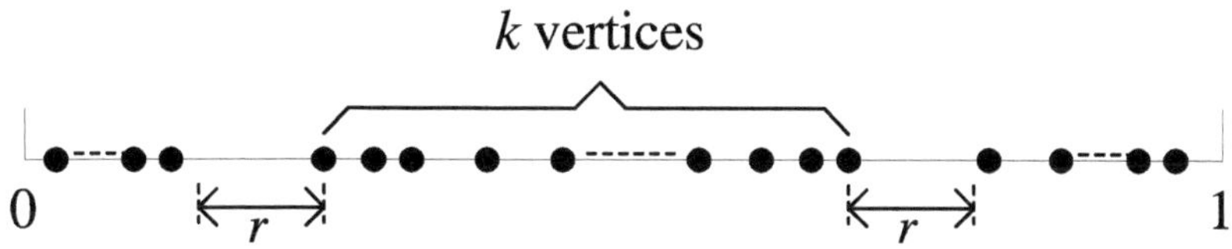

Fig. 4.2 An illustration of the event $\mathcal{F}_k^1$

Lemma 81 (**[79, Lemma 1]**) *Let* $[x, x + y]$ *be a subinterval of length y entirely contained within a unit interval* $[0, 1]$. *Let two of k given vertices be placed on the borders of this subinterval. Let* $P(k, y, r)$ *be the probability that the remaining* $k - 2$ *vertices placed independently, randomly, and uniformly on* $[0, 1]$ *are inside* $[x, x+y]$ *and the k vertices form a connected subgraph of length y. Then,*

$$P(k, y, r) = \sum_{j=0}^{\min\{k-1, \lfloor y/r \rfloor\}} \binom{k-1}{j} (-1)^j (y - jr)^{k-2} \tag{4.1.2}$$

Lemma 82 *Denote by* $\mathcal{F}_k^1$ *the event that there exists a connected subgraph with exactly k* $(k < n)$ *vertices in* $G_1(\mathcal{X}_n, r)$ *and both endpoints of this subgraph are not within distance r from the borders of the unit interval, and none of the remaining* $n - k$ *vertices is connected to this subgraph (refer to Fig. 4.2 for an illustration of the event.). Then*

$$\Pr\left(\mathcal{F}_k^1\right) = (n - k + 1) \sum_{j=0}^{\min\{k-1, \lfloor \frac{1}{r} \rfloor - 2\}} \binom{k-1}{j} (-1)^j (1 - (j + 2)r)^n \tag{4.1.3}$$

Proof There are $\binom{n}{k}$ distinct combinations for selecting k vertices from a total of n vertices. Consider a subinterval $[x, x + y]$, where x and $x + y$ are the positions of the left border and the right border, respectively. For any given k vertices, there are $\binom{k}{2}$ different combinations for selecting 2 vertices as endpoints, and two permutations of each selection in placing them on the borders of $[x, x+y]$. The probability that the remaining $k-2$ vertices placed randomly and uniformly on $[0, 1]$ are inside $[x, x+y]$ *and* the k vertices form a connected subgraph is given by (4.1.2). Therefore,

$$\begin{aligned}\Pr\left(\mathcal{F}_k^1\right) &= 2\binom{n}{k} \times \binom{k}{2} \int_0^{1-2r} \left(\int_r^{1-r-y} dx\right) P(k, y, r)(1 - y - 2r)^{n-k} dy \\ &= 2\binom{n}{k} \times \binom{k}{2} \int_0^{1-2r} P(k, y, r)(1 - y - 2r)^{n-k+1} dy \end{aligned} \tag{4.1.4}$$

Partitioning the integration interval $[0, 1-2r]$ into subintervals, i.e., $[0, r)$, $[r, 2r)$,..., and using Lemma 81, (4.1.4) becomes

$$\Pr\left(\mathcal{F}_k^1\right) = 2\binom{n}{k} \times \binom{k}{2} \sum_{i=0}^{\lfloor \frac{1}{r} \rfloor - 3} \int_{ir}^{(i+1)r} (1 - y - 2r)^{n-k+1}$$

$$\times \left[\sum_{j=0}^{L(i)} \binom{k-1}{j} (-1)^j (y - jr)^{k-2} \right] dy$$

$$+ 2\binom{n}{k} \times \binom{k}{2} \int_{\lfloor \frac{1}{r} \rfloor r - 2r}^{1-2r} (1 - y - 2r)^{n-k+1}$$

$$\times \left[\sum_{j=0}^{L(i)} \binom{k-1}{j} (-1)^j (y - jr)^{k-2} \right] dy \tag{4.1.5}$$

where $L(i) = \min\{k-1, i\}$. Taking the inner sums outside the integrals and letting $L' = \min\{k-1, \lfloor 1/r \rfloor - 2\}$, (4.1.5) becomes

$$\Pr\left(\mathcal{F}_k^1\right) = 2\binom{n}{k} \times \binom{k}{2} \sum_{j=0}^{L'} \binom{k-1}{j} (-1)^j$$

$$\times \left[\int_{jr}^{1-2r} (y - jr)^{k-2} (1 - y - 2r)^{n-k+1} dy \right]$$

$$= 2\binom{n}{k} \times \binom{k}{2} \sum_{j=0}^{L'} \binom{k-1}{j} (-1)^j$$

$$\times (1 - jr - 2r)^n \int_0^1 t^{k-2} (1 - t)^{n-k+1} dt \tag{4.1.6}$$

Note that the integral on the right-hand side of (4.1.6) is the *Beta Function*. Therefore, it follows

$$\int_0^1 t^{k-2} (1 - t)^{n-k+1} dt = \frac{(k-2)!(n-k+1)!}{n!} \tag{4.1.7}$$

Inserting (4.1.7) into (4.1.6), (4.1.3) can be readily obtained. □

Lemma 83 *Let $\mathcal{F}_k^2$ be the event that there exists a connected subgraph with exactly k $(k < n)$ vertices in $G_1(\mathcal{X}_n, r)$* and *the leftmost vertex of the subgraph is located within distance r from the left border of the unit interval* and *the remaining $n-k$ vertices are all located on the right side of this subgraph* and *none of them is*

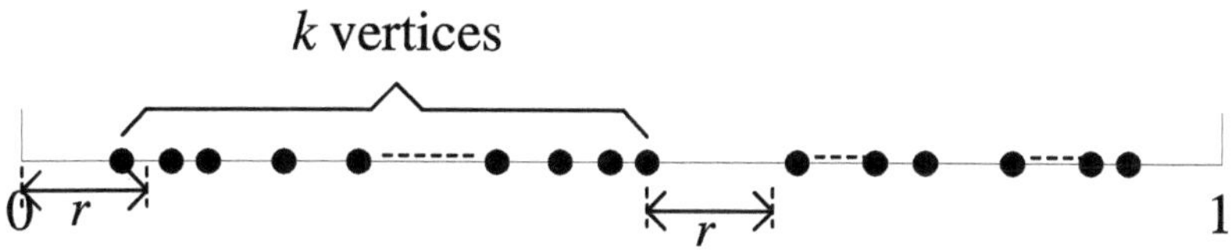

Fig. 4.3 An illustration of the event $\mathcal{F}_k^2$

connected to this subgraph (refer to Fig. 4.3). Then,

$$\Pr\left(\mathcal{F}_k^2\right) = \sum_{j=0}^{\min\{k,\lfloor\frac{1}{r}\rfloor-1\}} \binom{k}{j} (-1)^j \left[1-(j+1)r\right]^n \tag{4.1.8}$$

Proof Similar to the proof of Lemma 82, we have

$$\Pr\left(\mathcal{F}_k^2\right) = 2\binom{n}{k} \times \binom{k}{2} \int_0^{1-r} \left[\int_0^X P(k,y,r)(1-y-x-r)^{n-k}dx\right]dy \tag{4.1.9}$$

where $X = \min\{r, 1-r-y\}$. After some simplifications, we have

$$\begin{aligned}\Pr\left(\mathcal{F}_k^2\right) &= 2\binom{n}{k} \times \binom{k}{2} \int_0^{1-r} P(k,y,r)\frac{(1-r-y)^{n-k+1}}{n-k+1}dy \\ &\quad - 2\binom{n}{k} \times \binom{k}{2} \int_0^{1-2r} P(k,y,r)\frac{(1-2r-y)^{n-k+1}}{n-k+1}dy\end{aligned}$$

Partitioning the integration interval into subintervals, i.e., $[0,r)$, $[r,2r)$, ..., and using the same method as that used in the proof of Lemma 82, we have

$$\begin{aligned}\Pr\left(\mathcal{F}_k^2\right) = &\sum_{j=0}^{\min\{k-1,\lfloor\frac{1}{r}\rfloor-1\}} \binom{k-1}{j}(-1)^j(1-(j+1)r)^n \\ &- \sum_{j=0}^{\min\{k-1,\lfloor\frac{1}{r}\rfloor-2\}} \binom{k-1}{j}(-1)^j(1-(j+2)r)^n\end{aligned} \tag{4.1.10}$$

Combining the two sums in (4.1.10), the result follows. □

Now we are able to prove Theorem 80 using results established in Lemmas 82 and 83.

Recall that $L(G_1(\mathcal{X}_n, r))$ is the order of the giant component in $G_1(\mathcal{X}_n, r)$. It is clear that

$$P(n,r,p) = \sum_{i=\lceil np \rceil}^{n} \Pr\left(L(G_1(\mathcal{X}_n, r)) = i\right)$$
$$= \sum_{i=\lceil np \rceil}^{n-1} \Pr\left(L(G_1(\mathcal{X}_n, r)) = i\right) + \Pr\left(L(G_1(\mathcal{X}_n, r)) = n\right)$$

$P(n,r,p)$ can then be obtained by combining the following two calculations:

(1) $\Pr\left(L(G_1(\mathcal{X}_n, r)) = n\right)$. This probability is actually the probability that the network $G_1(\mathcal{X}_n, r)$ is connected, denoted as P_{con}. It is given by Corollary 1 in [79] as

$$P_{con} = \sum_{j=0}^{\min\{n-1, \lfloor 1/r \rfloor\}} \binom{n-1}{j} (-1)^j (1-jr)^n \tag{4.1.11}$$

(2) $\Pr\left(L(G_1(\mathcal{X}_n, r)) = i\right)$ for $\lceil np \rceil \leq i < n$. This probability is equal to the probability that there exists a connected subgraph with exactly i, $\lceil np \rceil \leq i < n$, vertices and none of the remaining $n - i$ vertices is connected to this subgraph. There are three different sub-cases in which this event may happen.

(a) Both endpoints of this subgraph are not within a distance r from the borders of the unit interval. Lemma 82 provides the probability for this case.
(b) The left (right) endpoint of this subgraph is within a distance r from the left (right) border of the unit interval. Lemma 83 provides the probability for this case.
(c) Both endpoints of this subgraph are within distance r from the borders of the unit interval. This can only happen when $i = n$, but here we require $i < n$. Hence, the probability of this case is zero.

Note that in Theorem 80, it is required that $p > 0.5$. In Lemmas 82 and 83, the connected subgraph with exactly i vertices is not necessarily the largest connected subgraph in which we are interested. To ensure that the connected subgraph with exactly i vertices is the largest connected subgraph, it suffices that we restrict $p > 0.5$.

Finally, the formula for the probability $P(n,r,p)$ as a key result of this subsection can be readily derived as

$$P(n,r,p) = \sum_{i=\lceil np \rceil}^{n-1} \left[2\Pr\{\mathcal{F}_i^2\} + \Pr\{\mathcal{F}_i^1\}\right] + P_{con} \tag{4.1.12}$$

Substituting (4.1.3), (4.1.8), and (4.1.11) into (4.1.12), we can obtain (4.1.1).

This completes the proof of Theorem 80.

4.1.2 Giant Component in Asymptotically Infinite Networks

In this subsection, we investigate the asymptotic behavior of the giant component by studying the probability $P(n, r, p)$ as $n \to \infty$. We assume that the transmission range r is a function of n. We investigate the property of a one-dimensional network having a giant component when r falls into three different related ranges: 1) when $n(1-r)^n \to e^{-c}$ as $n \to \infty$, 2) when $n(1-2r)^n \to 0$ and $n(1-r)^n \to \infty$ as $n \to \infty$, and 3) when $n(1-2r)^n \to e^{-c}$ and $n(1-r)^n \to \infty$ as $n \to \infty$.

4.1.2.1 When $n(1-r)^n \to e^{-c}$ as $n \to \infty$

The following three lemmas are used in our analysis.

Lemma 84 *If $n(1-r)^n \to e^{-c}$ as $n \to \infty$, then*

$$r \to 0, \quad rn \to \infty, \quad (1-r)^n \to e^{-rn}, \quad as \;\; n \to \infty$$

Proof Note that when $n(1-r)^n \to e^{-c}$, we have $r \to 0$ as $n \to \infty$. Otherwise $n(1-r)^n$ will tend to zero as $n \to \infty$.

Since $r \to 0$ as $n \to \infty$, we have $\frac{1}{r} \to \infty$. It follows that

$$(1-r)^n = (1-r)^{\frac{1}{r} \cdot r \cdot n} \sim e^{-rn}, \quad as \;\; n \to \infty$$

Using the above equation, it can be readily obtained that when $n(1-r)^n \to e^{-c}$ as $n \to \infty$, $ne^{-rn} \to e^{-c}$ as $n \to \infty$. Consequently,

$$rn \to \log n + c \to \infty, \quad as \;\; n \to \infty$$

The result follows. □

Lemma 85 *For any $y > 0$ and any fixed $b > 0$,*

$$\lim_{y \to 0} \frac{y}{1-e^{-y}} = 1 \; and \;\; \lim_{y \to \infty} \frac{y}{e^{by}} = 0$$

Lemma 85 can be readily proved using L'Hospital's rule. Thus the proof is omitted.

Lemma 86 *Given a geometric progression $A_a = a, a^2, a^3, \ldots, a^i, \ldots$ with $a > 0$ and $a \neq 1$. For any two fixed integers m, l $(0 \leq m < l)$,*

$$S(A_a, m, l) \triangleq \sum_{i=m}^{l} ia^i = \frac{ma^m - la^{l+1}}{1-a} + \frac{a^{m+1}(1-a^{l-m})}{(1-a)^2}$$

Proof Using the definition of $S(A_a, m, l)$, we have

$$a \times S(A_a, m, l) = \sum_{i=m}^{l} ia^{i+1}.$$

Since $a \neq 1$, we have

$$S(A_a, m, l) - a \times S(A_a, m, l) = ma^m - la^{l+1} + \sum_{i=m+1}^{l} a^i \tag{4.1.13}$$

Solving (4.1.13), the result follows. □

Using Theorem 80, Lemmas 84, 85, and 86, the following theorem on $P(n, r, p)$ as $n \to \infty$ can be established.

Theorem 87 *Adopt the same assumptions as in Theorem 80. If* $n(1-r)^n \to e^{-c}$ *as* $n \to \infty$, *or equivalently if* $r = \frac{\log n + c + o(1)}{n}$,

$$\lim_{n\to\infty} P(n, r, p) = e^{-pe^{-c}} + (1-p)e^{-c}e^{-pe^{-c}} \tag{4.1.14}$$

In addition, asymptotically almost surely there are only components whose orders are $\Theta(n)$.

Proof In (4.1.1), for any integer i satisfying $\lceil np \rceil \le i \le n-1$, we have $i = \Theta(n)$. Further using Lemma 84, we have $r \to 0$ and $nr \to \infty$ as $n \to \infty$. Hence, $\lfloor \frac{1}{r} \rfloor \to \infty$ and $\lfloor \frac{1}{r} \rfloor = o(n)$ as $n \to \infty$. Thus, we have $\lfloor \frac{1}{r} \rfloor = o(i)$, and for a sufficiently large n, $\min\{i, \lfloor \frac{1}{r} \rfloor - 1\} = \lfloor \frac{1}{r} \rfloor - 1$ and $\min\{i-1, \lfloor \frac{1}{r} \rfloor - 2\} = \lfloor \frac{1}{r} \rfloor - 2$.

Define $S_1(i)$, $S_2(i)$ as the first and the second sub-sum in the first sum of (4.1.1), respectively, i.e.,

$$S_1(i) \triangleq 2 \sum_{j=0}^{\lfloor \frac{1}{r} \rfloor - 1} \binom{i}{j} (-1)^j (1-(j+1)r)^n \tag{4.1.15}$$

$$S_2(i) \triangleq (n-i+1) \sum_{j=0}^{\lfloor \frac{1}{r} \rfloor - 2} \binom{i-1}{j} (-1)^j (1-(j+2)r)^n \tag{4.1.16}$$

First, let us consider the term $S_1(i)$. It is clear that for any integer j satisfying $0 \le j \le \lfloor \frac{1}{r} \rfloor - 1$, $j = o(i)$ and $0 \le 1-(j+1)r < 1$.

When $(j+1)r \to 0$ as $n \to \infty$, e.g., $j = 0, 1, 2$, we have

$$\begin{aligned}
&\binom{i}{j}(-1)^j(1-(j+1)r)^n \\
&= \frac{i \cdot (i-1) \dots (i-j+1)}{j!}(-1)^j(1-(j+1)r)^n \\
&\sim \frac{i^j}{j!}(-1)^j e^{-(j+1)rn} = e^{-rn}\frac{(-1)^j(ie^{-rn})^j}{j!}, \text{ as } n \to \infty
\end{aligned}$$

Define an integer set J_1 as

$$J_1 \triangleq \left\{ j : j \in \mathbb{N} \cup \{0\},\ 0 \le j \le \lfloor \frac{1}{r} \rfloor - 1,\ (j+1)r \to 0 \right\}$$

where $\mathbb{N}$ denotes the set of natural numbers. It follows that

$$\begin{aligned}
&2\sum_{j\in J_1}\binom{i}{j}(-1)^j(1-(j+1)r)^n \\
&\quad\sim 2\sum_{j\in J_1} e^{-rn}\frac{(-1)^j(ie^{-rn})^j}{j!} \\
&\quad= 2e^{-rn}\sum_{j\in J_1}\frac{(-1)^j(ie^{-rn})^j}{j!} \\
&\quad\sim 2e^{-rn}\times\exp\left(-ie^{-rn}\right),\quad as\ \ n\to\infty
\end{aligned} \tag{4.1.17}$$

When $(j+1)r \nrightarrow 0$ as $n \to \infty$, e.g., $j = \lfloor \frac{1}{r} \rfloor - 1, j = \lfloor \frac{1}{r} \rfloor - 2$, define $t = t(j)$ such that $\frac{1}{t} = 1 - (j+1)r$. It can be shown that $t > 1$ and $t \nrightarrow 1$ as $n \to \infty$. Hence, we have

$$\begin{aligned}
&n\times i\times\binom{i}{j}(-1)^j(1-(j+1)r)^n \\
&\sim n\times i\times\frac{i^j}{j!}(-1)^j\left(\frac{1}{t}\right)^n = o(1),\quad as\ \ n\to\infty
\end{aligned}$$

Define an integer set J_2 as

$$J_2 \triangleq \left\{ j : j \in \mathbb{N} \cup \{0\},\ 0 \le j \le \lfloor \frac{1}{r} \rfloor - 1,\ (j+1)r \nrightarrow 0 \right\}$$

It can be shown that

$$\lim_{n\to\infty}\left(2n\sum_{j\in J_2}\binom{i}{j}(-1)^j(1-(j+1)r)^n\right)$$

It follows from the above equation that

$$2\sum_{j\in J_2}\binom{i}{j}(-1)^j(1-(j+1)r)^n = o\left(\frac{1}{n}\right),\ as\ n\to\infty \tag{4.1.18}$$

Using (4.1.17) and (4.1.18), the first sub-sum $S_1(i)$ can be calculated as

$$S_1(i) = 2\sum_{j\in J_1}\binom{i}{j}(-1)^j(1-(j+1)r)^n$$
$$+2\sum_{j\in J_2}\binom{i}{j}(-1)^j(1-(j+1)r)^n$$
$$\sim 2e^{-rn}\exp\left(-ie^{-rn}\right)+o\left(\frac{1}{n}\right),\ as\ n\to\infty \tag{4.1.19}$$

Because $np \leq \lceil np \rceil < np+1$, $\frac{\lceil np \rceil}{np} \to 1$ as $n \to \infty$. From (4.1.19), we have

$$\lim_{n\to\infty}\sum_{i=\lceil np\rceil}^{n-1} S_1(i)$$
$$=\lim_{n\to\infty}\sum_{i=np}^{n-1} S_1(i)$$
$$=2\lim_{n\to\infty}\left[e^{-rn}\times\sum_{i=np}^{n-1}\left(\exp\left(-e^{-rn}\right)\right)^i\right]$$
$$=2\lim_{n\to\infty}\frac{e^{-rn}}{1-\exp\left(-e^{-rn}\right)}\times\lim_{n\to\infty}\left[\exp\left(-pne^{-rn}\right)\left(1-\exp\left(-(1-p)ne^{-rn}\right)\right)\right]$$

By Lemma 84, we have $ne^{-rn} \to e^{-c}$ and $e^{-rn} \to 0$. Further using the fact $e^{-rn} \to 0$ and Lemma 85, the first term in the last equation can be shown to approach 1. Thus, we have

$$\lim_{n\to\infty}\sum_{i=\lceil np\rceil}^{n-1} S_1(i) = 2e^{-pe^{-c}}\left(1-e^{-(1-p)e^{-c}}\right) \tag{4.1.20}$$

Secondly, using the same approach as that used in analyzing $S_1(i)$, we can obtain the second sub-sum $S_2(i)$:

$$S_2(i)$$
$$=(n-i+1)\sum_{j\in J_1'}\binom{i-1}{j}(-1)^j(1-(j+2)r)^n+o\left(\frac{1}{n}\right)$$
$$\sim (n-i+1)e^{-2rn}e^{-(i-1)e^{-rn}}+o\left(\frac{1}{n}\right),\ as\ n\to\infty \tag{4.1.21}$$

where J_1' is defined as

$$J_1' \triangleq \left\{ j : j \in \mathbb{N} \cup \{0\},\ 0 \le j \le \lfloor \frac{1}{r} \rfloor - 2,\ (j+2)r \to 0 \right\}$$

Hence, in the same way as that used in deriving $\lim_{n\to\infty} \sum_{i=\lceil np \rceil}^{n-1} S_1(i)$, using (4.1.21), Lemma 85 and Lemma 86, we have

$$\begin{aligned}
&\lim_{n\to\infty} \sum_{i=\lceil np \rceil}^{n-1} S_2(i) \\
&= \lim_{n\to\infty} \sum_{i=np}^{n-1} S_2(i) \\
&= \lim_{n\to\infty} \sum_{i=np}^{n-1} \left((n-i+1)e^{-2rn} e^{-(i-1)e^{-rn}} + o\left(\frac{1}{n}\right) \right) \\
&= e^{-c} e^{-pe^{-c}} \left(1 - e^{-(1-p)e^{-c}}\right) - pe^{-c} e^{-pe^{-c}} \\
&\quad + e^{-c} e^{-e^{-c}} - e^{-pe^{-c}} \left(1 - e^{-(1-p)e^{-c}}\right)
\end{aligned} \tag{4.1.22}$$

Thirdly, the second sum of (4.1.1) can also be obtained in the same way, i.e.,

$$\begin{aligned}
&\sum_{j=0}^{\min\{n-1, \lfloor \frac{1}{r} \rfloor\}} \binom{n-1}{j} (-1)^j (1-jr)^n \\
&\sim \exp\left(-(n-1)e^{-rn}\right) \sim e^{-e^{-c}}, \quad as\ n \to \infty
\end{aligned} \tag{4.1.23}$$

From (4.1.1) and the definitions of $S_1(i)$ and $S_2(i)$, we have

$$\begin{aligned}
\lim_{n\to\infty} P(n, r, p) &= \lim_{n\to\infty} \sum_{i=\lceil np \rceil}^{n-1} \left(S_1(i) + S_2(i)\right) \\
&\quad + \lim_{n\to\infty} \sum_{j=0}^{\min\{n-1, \lfloor \frac{1}{r} \rfloor\}} \binom{n-1}{j} (-1)^j (1-jr)^n
\end{aligned} \tag{4.1.24}$$

Substituting (4.1.20), (4.1.22), and (4.1.23) into (4.1.24), we can readily obtain (4.1.14).

Next, we shall prove that asymptotically almost surely there are only components whose orders are $\Theta(n)$ when $r = \frac{\log n + c + o(1)}{n}$.

For components whose orders are *asymptotically infinite and* are $o(n)$ (e.g., $\log n$), in a similar way as that used in the above analysis, we can obtain that the probability of having such components tends to zero as $n \to \infty$.

For components of finite order, let z be any fixed positive integer. In the same way as described in the proof of Theorem 80, the probability that $G_1(\mathcal{X}_n, r)$ has a component containing exactly z vertices, denoted as $P(n, r, z)$, can be calculated as

$$P(n,r,z) = 2\Pr\left(\mathcal{F}_z^2\right) + \Pr\left(\mathcal{F}_z^1\right)$$

Hence, by Lemma 82, Lemma 83, and Lemma 84, we have

$$\begin{aligned} P(n,r,z) &\sim 2\sum_{j=0}^{z}\binom{z}{j}(-1)^j e^{-(j+1)rn} + n\sum_{j=0}^{z-1}\binom{z-1}{j}(-1)^j e^{-(j+2)rn} \\ &= 2e^{-rn}\left(1-e^{-rn}\right)^z + ne^{-2rn}\left(1-e^{-rn}\right)^{z-1} \\ &= o(1), \quad as \quad n \to \infty \end{aligned} \tag{4.1.25}$$

Since z is an arbitrary but fixed positive integer, (4.1.25) implies that as $n \to \infty$, almost surely there is no finite component.

For isolated vertices, let I denote the number of isolated vertices in $G_1(\mathcal{X}_n, r)$, then the expected value of I can be obtained as

$$E(I) \sim n(1-2r)^{n-1} \sim ne^{-2r(n-1)} = o(1), \quad as \quad n \to \infty$$

Using Markov's inequality and noting that I can only be a nonnegative integer, we have

$$\Pr(I \geq 1) \leq E(I) = o(1), \quad as \quad n \to \infty, \tag{4.1.26}$$

Hence, as $n \to \infty$, almost surely no isolated vertex exists. □

4.1.2.2 When $n(1-2r)^n \to 0$ and $n(1-r)^n \to \infty$ as $n \to \infty$

Following the same procedure as that leading to Theorem 87, the following result on $P(n, r, p)$ when $n(1-2r)^n \to 0$ and $n(1-r)^n \to \infty$ as $n \to \infty$ can be established:

Theorem 88 *Adopt the same assumptions as in Theorem 80. If* $n(1-2r)^n \to 0$ *and* $n(1-r)^n \to \infty$ *as* $n \to \infty$, *then* $\lim_{n\to\infty} P(n, r, p) = 0$. *Furthermore, there is almost surely no isolated vertex or finite component as* $n \to \infty$.

The proof is omitted.

4.1.2.3 When $n(1-2r)^n \to e^{-c}$ and $n(1-r)^n \to \infty$ as $n \to \infty$

When $n(1-2r)^n \to e^{-c}$ and $n(1-r)^n \to \infty$ as $n \to \infty$, the following theorem on components of different sizes can be established:

Theorem 89 *Adopt the same assumptions as in Theorem 1. If* $n(1-2r)^n \to e^{-c}$ *as* $n \to \infty$*, or equivalently if* $r = \frac{\log n + c + o(1)}{2n}$*, then* $\lim_{n\to\infty} P(n,r,p) = 0$*. Furthermore, there is almost surely a nonvanishing probability of having isolated vertices and finite components.*

Proof Similar to the proof of Theorem 87, it can be obtained that $P(n,r,p) \to 0$ as $n \to \infty$.

For isolated vertices, let I denote the number of isolated vertices in $G_1(\mathcal{X}_n, r)$. Then

$$\begin{aligned}
\Pr(I > 0) &\geq \Pr(I = 1) \\
&\sim \binom{n}{1} \times (1-2r)^{n-1} \times \left(1 - (1-2r)^{n-1}\right)^{n-1} \\
&\sim ne^{-2rn} \times \exp(-ne^{-2rn}) \\
&\sim e^{-c} \exp(-e^{-c}), \quad as \quad n \to \infty
\end{aligned}$$

For finite components, let z be any fixed positive integer. In the same way as shown in the proof of Theorem 80, we have

$$\begin{aligned}
P(n,r,z) &\sim 2e^{-rn}\left(1-e^{-rn}\right)^z + ne^{-2rn}\left(1-e^{-rn}\right)^{z-1} \\
&\sim 2 \times 0 \times 1 + e^{-c} \times 1 = e^{-c}, \quad as \quad n \to \infty
\end{aligned}$$

The result follows. □

Remark 90 The above analysis reveals an interesting finding that is different from higher dimensional networks (e.g., $d = 2$ for d-dimensional networks). For $d \geq 2$, if $nr^d \to \infty$ as $n \to \infty$, almost surely the network only consists of isolated nodes and a unique giant component as $n \to \infty$. In addition, when the last isolated node vanishes, the network becomes connected almost surely. However, for $d = 1$, there may be multiple giant components (Theorem 87); and when the last isolated vertex vanishes, the network may still be disconnected (Theorem 88).

4.2 Securing a Giant Component with the Unit Disk Connection Model

For two-dimensional wireless networks, it is difficult to obtain an analytical formula comparable to the one-dimensional case for finite networks. In this section, we shall derive an asymptotic analytical upper bound on the minimum transmission range at which the probability of having a giant component of order above qn tends to one as $n \to \infty$, where q is any fixed real number in $(0, 1)$. In what follows, let r_q denote this minimum transmission range.

We shall use Poissonization and de-Poissonization techniques to derive our main results for two-dimensional networks. Poissonization and de-Poissonization are key techniques in the analysis of random geometric graphs [165] that allow one to extend results obtained assuming Poisson node distribution to uniform node distribution and the converse. Let $\{X_1, X_2, X_3, \ldots\}$ be a set of points randomly, independently, and uniformly distributed on a unit square $[-\frac{1}{2}, \frac{1}{2}]^2$ in $\Re^2$. Given $\lambda > 0$, let N_λ be a Poisson random variable with mean λ, independent of $\{X_1, X_2, X_3, \ldots\}$, and let

$$\mathcal{P}_{\lambda,1} \triangleq \{X_1, X_2, \ldots, X_{N_\lambda}\} \tag{4.2.1}$$

Then, $\mathcal{P}_{\lambda,1}$ is a Poisson point process on the unit square where the subscript 1 emphasizes that these points are on a unit square. Consider a graph with a vertex set $\mathcal{P}_{\lambda,1}$ and any two vertices are directly connected if and only if their Euclidean distance is smaller than or equal to r, i.e., assuming a unit disk connection model with a transmission range r. Denote the two-dimensional graph by $G_2(\mathcal{P}_{\lambda,1}, r)$. Furthermore, let

$$\mathcal{X}_n \triangleq \{X_1, X_2, \ldots, X_n\}$$

In this section, we shall start by presenting results on a graph with vertex set $\mathcal{P}_{\lambda,1}$, and on that basis derive results on a graph with vertex set $\mathcal{X}_n$.

We shall frequently use the following bound.

Lemma 91 ([92]) *For any $x \in [0, 1]$, $(1 - x) \leq e^{-x}$.*

Our main result for the upper bound on the minimum transmission range r_q is given in the following theorem.

Theorem 92 *Consider $G_2(\mathcal{X}_n, r)$ in $\Re^2$. Let q be any fixed real number within $(0, 1)$. Let c be any fixed real number. Let $f(n)$ be a function of n satisfying*

$$f(n) > 0, \quad \lim_{n\to\infty} f(n) = \infty \text{ and } f(n) = o(\log n) \tag{4.2.2}$$

If $\pi r^2 = \frac{f(n)+c}{n}$, *then*

$$\lim_{n\to\infty} \Pr\left(L\left(G_2(\mathcal{X}_n, r)\right) \geq qn\right) = 1 \tag{4.2.3}$$

where $L\left(G_2(\mathcal{X}_n, r)\right)$ *denotes the order of the largest component in* $G_2(\mathcal{X}_n, r)$.

The rest of this section is dedicated to the proof of Theorem 92.

Remark 93 At the first glance, the above result appears abnormal as it suggests the probability of having a giant component of order qn as $n \to \infty$ is independent of q. Here we offer the following intuitive explanation for the result. It is well known that the width of the phase transition region of the transmission range from an almost disconnected network to an almost connected network approaches zero as $n \to \infty$ ([97], see also Chap. 6 for a discussion on the phase transition region). This means at large n, the probability of having a connected network as a function of r is almost like a step function such that at a certain value of r (termed the critical transmission range), a small variation in r causes a large change in the probability. The above result indicates that the same phase transition phenomenon may also be observed for the probability of having a giant component. Note that the average node degree of $G_2(\mathcal{X}_n, r)$ is equal to $f(n) + c$. Consider a graph with nodes Poissonly distributed in $\Re^2$ with density $\frac{f(n)+c}{\pi}$ and assuming a unit disk connection model with a transmission range 1, denoting the graph by $G_2(\mathcal{P}_{\lambda,\Re^2}, 1)$ where $\lambda = \frac{f(n)+c}{\pi}$. A refined set of conditions of $f(n)+c$ in (4.2.2) that distinguishes the different values of q can be obtained by utilizing the connection between $G_2(\mathcal{X}_n, r)$ and $G_2(\mathcal{P}_{\lambda,\Re^2}, 1)$ through coupling and scaling. Specifically, let λ_c be the critical density of a network with nodes Poissonly distributed with density λ in $\Re^2$ and assuming a unit disk connection model with a transmission range 1, and let $p_\infty(\lambda)$ be the corresponding percolation probability. It has been shown [165, Chapter 10] that when $\lambda < \lambda_c$, as $s \to \infty$ there is no giant component in $G_2(\mathcal{P}_{\lambda,s^2}, 1)$ where $\mathcal{P}_{\lambda,s}$ is the restriction of $\mathcal{P}_{\lambda,\Re^2}$ to a square of side s; when $\lambda \geq \lambda_c$, the order of the largest component in $G_2(\mathcal{P}_{\lambda,s^2}, 1)$ converges in probability to $\lambda s^2 p_\infty(\lambda)$ as $s \to \infty$ and that giant component is unique.

In order to prove Theorem 92, we make use of some results from random geometric graph theory [165]. Specifically, consider a graph where nodes are distributed following a homogeneous Poisson point process of density λ on a square of side s and two nodes are connected directly if and only if their Euclidean distance is at most 1. Denote the graph by $G_2(\mathcal{P}_{\lambda,s^2}, 1)$. Let $L\left(G_2(\mathcal{P}_{\lambda,s^2}, 1)\right)$ be the order of the largest component in $G_2(\mathcal{P}_{\lambda,s^2}, 1)$ and let $p_\infty(\lambda)$ be the percolation probability of $G_2(\mathcal{P}_{\lambda,\Re^2}, 1)$. It has been shown [165, Theorem 10.9] that as $s \to \infty$,

$$s^{-2} L\left(G_2(\mathcal{P}_{\lambda,s^2}, 1)\right) \xrightarrow{p} \lambda p_\infty(\lambda) \tag{4.2.4}$$

where $\xrightarrow{p}$ denotes convergence in probability. When $\lambda < \lambda_c$, $p_\infty(\lambda) = 0$. Further note that $p_\infty(\lambda) \to 1$ as $\lambda \to \infty$ [165, Proposition 9.21]. Using these results, the following lemma can be established:

Lemma 94 *Consider $G_2(\mathcal{P}_{m(n),1}, r)$, where $m(n) = \lfloor n - n^{\frac{3}{4}} \rfloor$. Let q be any fixed real number within $(0, 1)$. Let c be any fixed real number. Let $f(n)$ be a function of n satisfying (4.2.2). If $\pi r^2 = \frac{f(n)+c}{n}$, then as $n \to \infty$*

$$\lim_{n\to\infty} \Pr\left(L(G_2(\mathcal{P}_{m(n),1}, r)) \geq qn\right) = 1 \tag{4.2.5}$$

Proof Using the scaling technique, a coupling between the two graphs $G_2\left(\mathcal{P}_{m(n),1}, r\right)$ and $G_2\left(\mathcal{P}_{m(n)r^2, \frac{1}{r^2}}, 1\right)$ can be readily established. Further noting that $m(n)r^2 = \frac{\lfloor n - n^{\frac{3}{4}} \rfloor}{\pi n}(f(n) + c) \to \infty$ as $n \to \infty$ and $\frac{1}{r^2} = \frac{\pi n}{f(n)+c} \to \infty$ as $n \to \infty$, the result follows. □

Now, we prove Theorem 92 using the de-Poissonization technique and Lemma 94.

Let $m(n) = \lfloor n - n^{\frac{3}{4}} \rfloor$. Define $Y(\mathcal{P}_{m(n),1}, r)$ and $Y(\mathcal{X}_n, r)$ as

$$Y(\mathcal{P}_{m(n),1}, r) \triangleq \Pr\left(L(G_2(\mathcal{P}_{m(n)}, r)) < qn\right)$$
$$Y(\mathcal{X}_n, r) \triangleq \Pr\left(L(G_2(\mathcal{X}_n, r)) < qn\right)$$

Because $m(n) = \lfloor n - n^{\frac{3}{4}} \rfloor$, we have $Y(\mathcal{P}_{m(n),1}, r) \to 0$ as $n \to \infty$ by Lemma 94.

It can be further shown that

$$\begin{aligned} Y(\mathcal{P}_{m(n),1}, r) &= \sum_{j=0}^{\infty} \frac{(m(n))^j}{j!} e^{-m(n)} Y(\mathcal{X}_j, r) \\ &\geq \sum_{\lfloor n-2n^{\frac{3}{4}} \rfloor \leq j \leq n} \frac{(m(n))^j}{j!} e^{-m(n)} Y(\mathcal{X}_j, r) \end{aligned} \tag{4.2.6}$$

Let $\mathcal{E}(\mathcal{X}_n, \mathcal{X}_j)$ denote the event that in $G_2(\mathcal{X}_n, r)$, all nodes in $\mathcal{X}_n \setminus \mathcal{X}_j$ are not directly connected to any node in $\mathcal{X}_j$. Then, for a fixed range r and any integer j satisfying $0 < j < n$, using Lemma 91, we have

$$\begin{aligned} Y(\mathcal{X}_n, r) &\leq \Pr\left(\mathcal{E}(\mathcal{X}_n, \mathcal{X}_j)\right) + Y(\mathcal{X}_j, r) \\ &\leq \left[\left(1 - \frac{1}{4}\pi r^2\right)^j\right]^{n-j} + Y(\mathcal{X}_j, r) \\ &\leq e^{-j\frac{1}{4}\pi r^2(n-j)} + Y(\mathcal{X}_j, r) \end{aligned} \tag{4.2.7}$$

Substituting (4.2.7) into (4.2.6), it can be obtained that

$$
\begin{aligned}
&Y(\mathcal{P}_{m(n)}, r) \\
&\geq \sum_{\lfloor n-2n^{\frac{3}{4}} \rfloor \leq j \leq n} \frac{(m(n))^j}{j!} e^{-m(n)} \left[Y(\mathcal{X}_n, r) - e^{-j\frac{1}{4}\pi r^2 (n-j)} \right] \\
&= Y(\mathcal{X}_n, r) \sum_{\lfloor n-2n^{\frac{3}{4}} \rfloor \leq j \leq n} \frac{(m(n))^j}{j!} e^{-m(n)} - \sum_{\lfloor n-2n^{\frac{3}{4}} \rfloor \leq j \leq n} \frac{(m(n))^j}{j!} e^{-m(n)} e^{-j\frac{1}{4}\pi r^2 (n-j)} \\
&\geq Y(\mathcal{X}_n, r) \sum_{\lfloor n-2n^{\frac{3}{4}} \rfloor \leq j \leq n} \frac{(m(n))^j}{j!} e^{-m(n)} - e^{\frac{1}{4}\pi r^2 \lfloor n-2n^{\frac{3}{4}} \rfloor \left(n - \lfloor n-2n^{\frac{3}{4}} \rfloor \right)} \qquad (4.2.8)
\end{aligned}
$$

Using Chebyshev's inequality,

$$
1 - \sum_{\lfloor n-2n^{\frac{3}{4}} \rfloor \leq j \leq n} \frac{(m(n))^j}{j!} e^{-m(n)} \leq n^{-\frac{1}{2}}
$$

It follows from (4.2.8) and Lemma 94 that $\lim_{n\to\infty} \Pr\left(L\left(G_2(\mathcal{X}_n, r)\right) \geq qn\right) = 1$. This completes the proof of Theorem 92.

Remark 95 Let r_1 denote the minimum transmission range above which a network is connected with probability one as $n \to \infty$. Using Theorem 79, $\sqrt{\frac{\log n + c'}{n\pi}}$ is a lower bound of r_1 where c' is any fixed real number. By Theorem 92, $\sqrt{\frac{f(n)+c}{n\pi}}$ is an upper bound of r_q. Hence, we have

$$
\lim_{n\to\infty} \frac{r_q}{r_1} \leq \lim_{n\to\infty} \frac{\sqrt{\frac{f(n)+c}{n\pi}}}{\sqrt{\frac{\log n + c'}{n\pi}}} = \lim_{n\to\infty} \sqrt{\frac{f(n) + c}{\log n + c'}} = 0.
$$

The implication of the above result is that when $n \to \infty$, the transmission range required for having a giant component is vanishingly small compared to the transmission range for a connected network. Therefore, in a large network, a significant energy saving can be achieved by requiring most nodes, instead of all nodes, to be connected. Furthermore, in a network where almost (but not) all nodes are connected, a large leap in the transmission range may be required to connect the remaining few nodes and the transmission range required for a large network to be connected is dominated by these few nodes, i.e., rare events.

4.3 Extension into More General Connection Models

In the last section, assuming the unit disk connection model, we established the result that in a two-dimensional network with a total of n nodes distributed on a unit square, as $n \to \infty$, the transmission range required for having a giant component with at least q percentage of nodes is vanishingly small compared with that required for having a connected network. This result established assuming the unit disk connection model can be extended into more general connection models.

Let us consider a random connection function g and some value $0 < p < 1$. Define g_p by $g_p(\boldsymbol{x}) = pg\left(\sqrt{p}\boldsymbol{x}\right)$ and g_p is a scaled version of g in the sense that the probabilities are reduced by a factor of p but the function is spatially stretched such that $\int_{\Re^2} g_p(\boldsymbol{x})\, d\boldsymbol{x} = \int_{\Re^2} g(\boldsymbol{x})\, d\boldsymbol{x}$. Let $\mathcal{G}\left(\mathcal{X}_\lambda, g, \Re^2\right)$ be a network with nodes Poissonly distributed on $\Re^2$ with density $\boldsymbol{\lambda}$ and nodes are connected following the random connection model with a connection function g. Denote by $\mathcal{G}\left(\mathcal{X}_\lambda, g_p, \Re^2\right)$ the same network as $\mathcal{G}\left(\mathcal{X}_\lambda, g, \Re^2\right)$ except that nodes are connected following the random connection model with a connection function g_p. Obviously, the expected number of connection per node, i.e., the average node degree, is the same in $\mathcal{G}\left(\mathcal{X}_\lambda, g, \Re^2\right)$ and $\mathcal{G}\left(\mathcal{X}_\lambda, g_p, \Re^2\right)$. Let $\lambda_c(g)$ be the critical density for $\mathcal{G}\left(\mathcal{X}_\lambda, g, \Re^2\right)$ to percolate and let $\lambda_c\left(g_p\right)$ be the critical density for $\mathcal{G}\left(\mathcal{X}_\lambda, g_p, \Re^2\right)$. The following theorem was established in [71] using the coupling technique and the fact that site percolation is easier to achieve than bond percolation [71, Theorem 2.2.7]:

Theorem 96 ([71, Theorem 2.5.2]) *For a random connection model with a connection function g and $0 < p < 1$, we have*

$$\lambda_c(g) \geq \lambda_c\left(g_p\right) \tag{4.3.1}$$

Broadly speaking, Theorem 96 implies that when stretching the connection function while maintaining the average node degree constant, or equivalently replacing some short-distance links by the same number of long-distance links, it becomes easier for the network to become connected or form a giant component. In this sense, the unit disk connection model constitutes the worst case in the connection model for forming a connected network or a giant component.

Let g_u be a connection function corresponding to a unit disk connection model such that $g_u(\boldsymbol{x}) = 1$ if and only if $\|\boldsymbol{x}\| \leq r$ and $g_u(\boldsymbol{x}) = 0$ otherwise. Using the same approach as that in the proof of Theorem 96 and the de-Poissonization technique used in Sect. 4.2, it is straightforward to conclude that as $n \to \infty$, if a network with n nodes on a unit square and assuming the connection function g_u has a giant component with q percentage of nodes asymptotically almost surely, the same network with connection function $g_{u,p} = pg_u\left(\sqrt{p}\boldsymbol{x}\right)$ also has a giant component with q percentage of nodes asymptotically almost surely. Further using the results established in Chap. 3 on the sufficient and necessary condition for a network assuming a general random connection model to be asymptotically almost surely connected, the conclusion readily follows that the results established in last section assuming the unit disk connection model can be extended to a more general connection model with connection function $g_{u,p} = pg_u(\sqrt{p}x)$ where $0 < p < 1$. Indeed, using the same methodology, the results can be further extended to a general random connection model satisfying the three conditions of nonincreasing monotonicity, integral boundedness, and rotational invariance in (1.1.8), (1.1.9) and (1.1.10) only.

4.4 Notes and Further Readings

In this chapter, we investigated the order of the giant component in large wireless networks. In one-dimensional networks, we derived a closed-form formula for calculating the probability $P(n, r, p)$ that a network has a giant component of order above pn with any fixed $0.5 < p \leq 1$. We also studied the asymptotic behavior of one-dimensional networks as $n \to \infty$. Interesting results were found on the asymptotic behavior of one-dimensional networks having a giant component, which is distinctly different from its two-dimensional counterparts. In two-dimensional networks, we derived an asymptotic analytical upper bound on the minimum transmission range r_q required for having a giant component of order above qn, where q is any fixed real number in $(0, 1)$. We showed that the ratio of r_q to r_1, the transmission range required for a connected network, approaches 0 as $n \to \infty$. This indicates that a significant energy saving may be achieved if we only require a giant component rather than a connected network, especially for a network with a large number of nodes.

The concept of the giant component has been extensively investigated in the literature for *Bernoulli random graphs* [27], and an analytical formula relating the giant component size, where the giant component size is the ratio of the number of nodes in the giant component to the total number of nodes, and the average node degree has been found [27, 150]. However, the Bernoulli random graph is not suitable for modeling wireless networks (see discussions in Sect. 1.1). Hence, it is inappropriate to apply the results on the giant component size obtained from Bernoulli random graphs directly to wireless networks.

In [100], Hekmat et al. investigated the giant component size in a log-normal shadowing environment, where a total of n nodes are randomly and uniformly distributed on a square and nodes are connected following the log-normal connection model. Based on the analytical results obtained from Bernoulli random graphs, the authors proposed an empirical formula relating the giant component size and the average node degree in random geometric graphs. In [156], Németh et al. investigated the giant component size by using a fractal propagation model where the probability of having a link between two nodes is determined by their Euclidean distance and two nonnegative constants. They found that the giant component size can be characterized by a single parameter, viz., the average node degree. However, both papers investigated the giant component size empirically.

In [170], Raghavan et al. investigated the phase transition behaviors for the emergence of a giant component in large wireless networks using the same network model as that considered in this chapter. The authors proposed an empirical formula for the critical transmission range at which the network has a giant component with a high probability, and they showed that the critical range is approximately inversely proportional to $\sqrt{n}$.

In [31], Bradonjić et al. studied the giant component using a network model based on a *geographical threshold graph* which is almost the same as the random geometric graph except that the existence of a link between any two nodes is determined not only by their Euclidean distance but also by the node weights assigned to them. The authors derived the conditions for the absence and existence of a giant component.

Chapter 5
Critical Density for Percolation

Abstract In this chapter we investigate the critical node density required to ensure that an arbitrary node in a large wireless network is connected (via multi-hop paths) to infinitely many other nodes with a positive probability, known as the percolation probability. A network is said to percolate if there exists a component of infinite order in the network. Specifically, assuming the infinite network model and the random connection model, we first obtain an upper and a lower bounds for the critical density. Then, we compare the bounds with other existing bounds in the literature under the unit disk connection model and the log-normal connection model, which are special cases of the random connection model. Percolation is an important subject in the study of connectivity of large random networks.

In this chapter we study the critical node density required to ensure that an arbitrary node in a large wireless network is connected (via multi-hop paths) to infinitely many other nodes with a positive probability. This probability is known as the *percolation probability*, which has been formally defined in Sect. 1.3.1. Specifically, we consider a network where nodes are distributed in $\Re^d$ ($d = 2, 3$) following a homogeneous Poisson point process with known and finite density λ and nodes are directly connected following the random connection model with a connection function g satisfying the three conditions of rotational and translational invariance, nonincreasing monotonicity and integral boundedness. Denote the network by $\mathcal{G}\left(\mathcal{X}_\lambda, g, \Re^d\right)$. We first obtain analytically an upper and a lower bounds for the critical density. Then, we compare the bounds with other existing bounds in the literature under the unit disk connection model and the log-normal connection model, respectively.

Denote by $\mathcal{X}_{\lambda,\boldsymbol{x}_0}$ a homogeneous Poisson point process $\mathcal{X}_\lambda$ on the d-dimensional Euclidean space $\Re^d$ with an additional node at $\boldsymbol{x}_0 \in \mathbb{R}^d$. Denote by $\mathcal{G}\left(\mathcal{X}_{\lambda,\boldsymbol{x}_0}, g, \Re^d\right)$ the Poisson random network with node set $\mathcal{X}_{\lambda,\boldsymbol{x}_0}$. Let $\mathcal{W}$ be the set of nodes in $\mathcal{G}\left(\mathcal{X}_{\lambda,\boldsymbol{x}_0}, g, \Re^d\right)$ connected (by multi-hop paths) to the node at $\boldsymbol{x}_0$. Denote by $|\mathcal{W}|$ the number of nodes in $\mathcal{W}$. Then the *percolation probability* is defined by

$$\theta(\lambda) = \Pr\left(|\mathcal{W}| = \infty\right) \tag{5.0.1}$$

which is the probability that $\mathcal{W}$ contains an infinite number of nodes.

G. Mao, *Connectivity of Communication Networks*,
DOI 10.1007/978-3-319-52989-9_5

The fact that the location of the chosen node in the definition of $\mathcal{W}$ is specified to be at $\boldsymbol{x}_0 \in \mathbb{R}^d$ is of no importance: due to the stationarity property of a homogeneous Poisson point process, the node can be anywhere in $\Re^d$. Also as pointed out by the Palm theory [165] on the other hand, assuming a node at $\boldsymbol{x}_0$ does not prevent the distribution of the rest of the nodes to be maintained the same as $\mathcal{X}_\lambda$. Evidently from the definition of the percolation probability, $\theta(\lambda) > 0$ means that an arbitrarily chosen node is connected to an infinite number of nodes with a positive probability. It can be shown that $\theta(\lambda)$ displays the following phase transition phenomenon for $d = 2, 3$.

Theorem 97 ([143, Theorem 6.1]) *For $\mathcal{G}\left(\mathcal{X}_{\lambda,\boldsymbol{x}_0}, g, \Re^d\right)$ with $d = 2, 3$, there exists a critical density $0 < \lambda_c < \infty$ such that $\theta(\lambda) = 0$ for $\lambda < \lambda_c$ and $\theta(\lambda) > 0$ for $\lambda > \lambda_c$.*

Theorem 97 implies that for $\mathcal{G}\left(\mathcal{X}_{\lambda,\boldsymbol{x}_0}, g, \Re^d\right)$ there exists a critical density above which an arbitrarily chosen node is connected to an infinite number of other nodes via multi-hop paths with a positive probability, and below which the node is almost surely connected to finite number of other nodes only.

The exact value of the critical density is however difficult to obtain [60, 94, 118, 119]. In this chapter we are motivated to obtain an upper and a lower bound for the critical density in $\mathcal{G}\left(\mathcal{X}_{\lambda,\boldsymbol{x}_0}, g, \Re^d\right)$. The lower bound for the critical density is obtained using a Galton-Watson branching process ([71], see also Sect. 1.3.2) and the upper bound is obtained by relating the problem to that of site percolation on a lattice [29]. We then consider two special cases of the random connection model, viz., the unit disk connection model and the log-normal connection model, and obtain specific bounds under both models.

Recall that in the unit disk connection model, two nodes located at $\boldsymbol{x}$ and $\boldsymbol{y}$, respectively, are directly connected if and only if their Euclidean distance is less than or equal to the transmission range r. The associated connection function for a unit disk connection model can be written as

$$g\left(\boldsymbol{x}-\boldsymbol{y}\right) = \begin{cases} 1 & \text{if } \|\boldsymbol{x}-\boldsymbol{y}\| \leq r, \\ 0 & \text{otherwise.} \end{cases} \tag{5.0.2}$$

In the log-normal connection model, the two nodes are directly connected with probability

$$g\left(\boldsymbol{x}-\boldsymbol{y}\right) = Q\left(\frac{10\alpha}{\sigma}\log_{10}\frac{\|\boldsymbol{x}-\boldsymbol{y}\|}{r}\right) \tag{5.0.3}$$

where $Q(y) = \frac{1}{\sqrt{2\pi}}\int_y^\infty \exp(-\frac{z^2}{2})dz$ is the tail probability of the standard normal distribution, α is the path loss exponent, σ^2 is the shadowing variance, r is the transmission range ignoring shadowing effect (i.e., $\sigma = 0$). The log-normal connection model reduces to the unit disk connection model when $\sigma = 0$.

5.1 A Lower Bound for the Critical Density

In this section, we derive a lower bound of the critical density λ_c in $\mathcal{G}\left(\mathcal{X}_\lambda, g, \Re^d\right)$ $(d = 2, 3)$ using a Galton-Watson branching process [71].

The following lemma is used in obtaining the lower bound.

Lemma 98 *Consider* $\mathcal{G}\left(\mathcal{X}_{\lambda,\boldsymbol{x}_0}, g, \Re^d\right)$ *for* $d = 2, 3$ *and denote by* X_0 *the node at* $\boldsymbol{x}_0$. *A node* $Y \in \mathcal{X}_{\lambda,\boldsymbol{x}_0}$ *is called a k-hop node if the length of the shortest path between Y and* X_0, *measured by the number of hops, is k. Let* N_k *be the total number of k-hop nodes. Then for all k,* $E[N_k]$ *is finite, with*

$$E[N_1] = \int_{\Re^d} \lambda g(\boldsymbol{y} - \boldsymbol{x}_0) d\boldsymbol{y} = \int_{\Re^d} \int_{\Re^d} h_1(\boldsymbol{y}, \boldsymbol{z}) d\boldsymbol{z} d\boldsymbol{y} \tag{5.1.1}$$

where

$$h_1(\boldsymbol{y}, \boldsymbol{z}) = \lambda g(\boldsymbol{y} - \boldsymbol{z}) \phi(\boldsymbol{z} - \boldsymbol{x}_0) \tag{5.1.2}$$

$\phi(.)$ *is the Dirac delta function; and*

$$E[N_k] \leq \int_{\Re^d} \int_{\Re^d} h_k(\boldsymbol{y}, \boldsymbol{z}) d\boldsymbol{z} d\boldsymbol{y} \tag{5.1.3}$$

for $k \geq 2$ *where*

$$h_k(\boldsymbol{y}, \boldsymbol{z}) = \int_{\Re^d} \lambda g(\boldsymbol{y} - \boldsymbol{z}) \left(1 - g(\boldsymbol{y} - \boldsymbol{w})\right) h_{k-1}(\boldsymbol{z}, \boldsymbol{w}) d\boldsymbol{w} \tag{5.1.4}$$

Proof This proof consists of two parts. First we derive (5.1.1) and (5.1.3), and then we prove that $E[N_k]$ is finite. The derivation of (5.1.1) and (5.1.3) is based on the use of the Galton-Watson branching process, which have been introduced earlier in Sect. 1.3.2. Particularly, the k-hop nodes can be considered as the members of the k-th generation in the Galton-Watson branching process, with node X_0 being considered as the root of the branching process. The distribution of the k-hop nodes is then obtained by evaluating the impact from the previous two hops, i.e., $(k-1)$-hop nodes and $(k-2)$-hop nodes. Using this approach, the expected number of 1-hop nodes and an upper bound of the expected number of k-hop nodes are then derived. The detailed derivations are outlined in the next several paragraphs.

Imagine we partition the $\Re^d$ space $(d = 2, 3)$ into small and nonoverlapping d-cubes of side length Δ. Assume that one of the d-cubes is centered at $\boldsymbol{x}_0$. Then, we have a collection of d-cubes centered at $\mathbb{D}^d = \left\{\boldsymbol{x}_0 + (\boldsymbol{v} \cdot \Delta) : \boldsymbol{v} \in \mathbb{Z}^d\right\}$ where $\mathbb{Z}^d$ is the integer set in the d-dimensional space. Denote by $B_{\boldsymbol{x}}$ the d-cube centered at $\boldsymbol{x}$. The Lebesgue measure of $B_{\boldsymbol{x}}$ is $\delta\boldsymbol{x} \triangleq |B_{\boldsymbol{x}}| = \Delta^d$. Since nodes are Poissonly distributed in $\Re^d$, for a sufficiently small Δ, the probability that there exists exactly one node within $B_{\boldsymbol{x}}$ is $p_1(B_{\boldsymbol{x}}) = \lambda\delta\boldsymbol{x} + o(\delta\boldsymbol{x})$ where $o(\delta\boldsymbol{x})$ denotes a quantity which, for small $\delta\boldsymbol{x}$, is of lower order than $\delta\boldsymbol{x}$, i.e., $\lim_{\delta x \to 0} \frac{o(\delta x)}{\delta x} = 0$. The probability that there are more than one nodes in $B_{\boldsymbol{x}}$ is $o(\delta\boldsymbol{x})$. Let $I_{\boldsymbol{x}}^k$ be the indicator random variable of the event that there exists exactly one node within the d-cube $B_{\boldsymbol{x}}$ *and* the node is a k-hop node. Using the monotone convergence theorem, we have that

$$E[N_k] = \lim_{\Delta \to 0} \sum_{\boldsymbol{x} \in \mathbb{D}^d \backslash \{\boldsymbol{x}_0\}} E\left[I_{\boldsymbol{x}}^k\right] = \lim_{\Delta \to 0} \sum_{\boldsymbol{x} \in \mathbb{D}^d \backslash \{\boldsymbol{x}_0\}} \Pr\left(I_{\boldsymbol{x}}^k = 1\right) \tag{5.1.5}$$

and the limit exists. Similarly, let $J_{\boldsymbol{x}}$ be the indicator random variable of the event that there exists exactly one node within $B_{\boldsymbol{x}}$. Without loss of generality, we assume that when $J_{\boldsymbol{x}} = 1$ or $I_{\boldsymbol{x}}^k = 1$, the node within $B_{\boldsymbol{x}}$ is located at $\boldsymbol{x}$. The difference between the actual location of the node within $B_{\boldsymbol{x}}$ and $\boldsymbol{x}$ becomes negligibly small as $\Delta \to 0$. Let $H_{\boldsymbol{x}\boldsymbol{y}}$ be the indicator random variable of the event that a node at $\boldsymbol{x}$ and another node at $\boldsymbol{y}$, if exist, are directly connected.

The probability that a 1-hop node exists in $B_{\boldsymbol{y}}$ is

$$\begin{aligned}\Pr\left(I_{\boldsymbol{y}}^1 = 1\right) =& \Pr\left(J_{\boldsymbol{y}} = 1, H_{\boldsymbol{y}\boldsymbol{x}_0} = 1\right)\\ =& g(\boldsymbol{y} - \boldsymbol{x}_0)\left[\lambda\delta\boldsymbol{y} + o(\delta\boldsymbol{y})\right]\end{aligned} \tag{5.1.6}$$

Using (5.1.5) and (5.1.6), we obtain

$$E[N_1] = \lim_{\Delta \to 0} \sum_{\boldsymbol{y} \in \mathbb{D}^d \backslash \{\boldsymbol{x}_0\}} \Pr\left(I_{\boldsymbol{y}}^1 = 1\right) = \int_{\Re^d} \lambda g(\boldsymbol{y} - \boldsymbol{x}_0) d\boldsymbol{y} \tag{5.1.7}$$

which is (5.1.1).

A node is a 2-hop node if it is directly connected to *at least* one of the 1-hop nodes but not directly connected to the node at $\boldsymbol{x}_0$. By applying the union bound and with some arithmetic steps, we obtain

$$\begin{aligned}
&E[N_2]\\
&= \lim_{\Delta \to 0} \sum_{\boldsymbol{y} \in \mathbb{D}^d \backslash \{\boldsymbol{x}_0\}} \Pr\left(I_{\boldsymbol{y}}^2 = 1\right)\\
&\le \lim_{\Delta \to 0} \sum_{\boldsymbol{y} \in \mathbb{D}^d \backslash \{\boldsymbol{x}_0\}} \sum_{\boldsymbol{z} \in \mathbb{D}^d \backslash \{\boldsymbol{y}, \boldsymbol{x}_0\}} \left(\Pr\left(I_{\boldsymbol{z}}^1 = 1\right) \Pr\left(J_{\boldsymbol{y}} = 1, H_{\boldsymbol{y}\boldsymbol{z}} = 1, H_{\boldsymbol{y}\boldsymbol{x}_0} = 0 \middle| I_{\boldsymbol{z}}^1 = 1\right)\right)\\
&= \lim_{\Delta \to 0} \sum_{\boldsymbol{y} \in \mathbb{D}^d \backslash \{\boldsymbol{x}_0\}} \sum_{\boldsymbol{z} \in \mathbb{D}^d \backslash \{\boldsymbol{y}, \boldsymbol{x}_0\}} \left[g(\boldsymbol{z} - \boldsymbol{x}_0)\left(\lambda\delta\boldsymbol{z} + o(\delta\boldsymbol{z})\right) g(\boldsymbol{y} - \boldsymbol{z})\left(1 - g(\boldsymbol{y} - \boldsymbol{x}_0)\right)\left(\lambda\delta\boldsymbol{y} + o(\delta\boldsymbol{y})\right)\right]\\
&= \int_{\Re^d}\int_{\Re^d} \lambda^2 g(\boldsymbol{y} - \boldsymbol{z})\left[1 - g(\boldsymbol{y} - \boldsymbol{x}_0)\right] g(\boldsymbol{z} - \boldsymbol{x}_0) d\boldsymbol{z} d\boldsymbol{y}
\end{aligned} \tag{5.1.8}$$

where the first step in the above equations results from the following analysis. Note that if a node Y is a 2-hop node, then it is not directly connected to the node at $\boldsymbol{x}_0$ but directly connected to at least one 1-hop node. Hence for any positive integer n_1, we have

$$
\begin{aligned}
&\Pr\left(I_y^2 = 1, \bigcap_{i=1}^{n_1}\{I_{z_i}^1 = 1\} \middle| N_1 = n_1\right) \\
&= \Pr\left(J_y = 1, H_{y,x_0} = 0, \bigcup_{i=1}^{n_1}\{H_{y,z_i} = 1\}, \bigcap_{i=1}^{n_1}\{I_{z_i}^1 = 1\} \middle| N_1 = n_1\right) \\
&\le \sum_{j=1}^{n_1}\Pr\left(J_y = 1, H_{y,x_0} = 0, H_{y,z_j} = 1, \bigcap_{i=1}^{n_1}\{I_{z_i}^1 = 1\} \middle| N_1 = n_1\right) \qquad (5.1.9)
\end{aligned}
$$

where the last step results due to the union bound. The conditional probability $\Pr\left(I_y^2 = 1 \middle| N_1 = n_1\right)$ is then obtained as follows, where (5.1.9) is used in the second step below and the third step results by moving the summation on j and z_j to the outermost:

$$
\begin{aligned}
&\Pr\left(I_y^2 = 1 \middle| N_1 = n_1\right) \\
&= \frac{1}{n_1!}\sum_{\substack{z_1,\cdots,z_{n_1}\in\mathbb{D}^d\setminus\{y,x_0\}\\ z_m\neq z_n \text{ for } m\neq n}}\Pr\left(I_y^2 = 1, \bigcap_{i=1}^{n_1}\{I_{z_i}^1 = 1\} \middle| N_1 = n_1\right) \\
&\le \frac{1}{n_1!}\sum_{\substack{z_1,\cdots,z_{n_1}\in\mathbb{D}^d\setminus\{y,x_0\}\\ z_m\neq z_n \text{ for } m\neq n}}\sum_{j=1}^{n_1}\Pr\Bigg(J_y = 1, H_{y,x_0} = 0, \\
&\qquad H_{y,z_j} = 1, \bigcap_{i=1}^{n_1} I_{z_i}^1 = 1 \middle| N_1 = n_1\Bigg) \\
&= \frac{1}{n_1!}\sum_{j=1}^{n_1}\sum_{z_j\in\mathbb{D}^d\setminus\{y,x_0\}}\Bigg(\sum_{\substack{z_1,\cdots,z_{j-1},z_{j+1},\cdots,z_{n_1}\in\mathbb{D}^d\setminus\{y,x_0,z_j\}\\ z_m\neq z_n \text{ for } m\neq n}}\Pr\Bigg(J_y = 1, \\
&\qquad H_{y,x_0} = 0, H_{y,z_j} = 1, \bigcap_{i=1}^{n_1} I_{z_i}^1 = 1 \middle| N_1 = n_1\Bigg)\Bigg) \\
&= \frac{1}{n_1!}\sum_{j=1}^{n_1}\sum_{z_j\in\mathbb{D}^d\setminus\{y,x_0\}}\left((n_1-1)!\Pr\left(J_y = 1, H_{y,x_0} = 0, H_{y,z_j} = 1, I_{z_j}^1 = 1 \middle| N_1 = n_1\right)\right) \\
&= \sum_{z\in\mathbb{D}^d\setminus\{y,x_0\}}\left(\Pr\left(J_y = 1, H_{y,x_0} = 0, H_{y,z} = 1 \middle| I_z^1 = 1, N_1 = n_1\right)\Pr\left(I_z^1 = 1 \middle| N_1 = n_1\right)\right)
\end{aligned}
\qquad (5.1.10)
$$

Since

$$\Pr\left(I_z^1 = 1 \middle| N_1 = n_1\right) = \frac{\Pr\left(I_z^1 = 1, N_1 = n_1\right)}{\Pr\left(N_1 = n_1\right)} \tag{5.1.11}$$

and

$$\sum_{n_1=1}^{\infty} \Pr\left(I_z^1 = 1, N_1 = n_1\right) = \Pr\left(I_z^1 = 1\right), \tag{5.1.12}$$

using (5.1.10), (5.1.11), and (5.1.12), we obtain

$$\begin{aligned}
&\Pr\left(I_y^2 = 1\right) \\
&= \sum_{n_1=1}^{\infty} \left(\Pr\left(I_y^2 = 1 \middle| N_1 = n_1\right) \Pr\left(N_1 = n_1\right)\right) \\
&\leq \sum_{z \in \mathbb{D}^d \setminus \{y, x_0\}} \left(\Pr\left(J_y = 1, H_{y,x_0} = 0, H_{y,z} = 1 \middle| I_z^1 = 1\right) \Pr\left(I_z^1 = 1\right)\right)
\end{aligned} \tag{5.1.13}$$

Hence we obtain the result in the first step of (5.1.8).

This proves (5.1.3) for $k = 2$.

From (5.1.2), $h_1(\boldsymbol{y}, \boldsymbol{z})$ has the meaning of being the probability that the node at $\boldsymbol{y}$ is directly connected to the node at $\boldsymbol{x}_0$. That is, the node at $\boldsymbol{y}$ is a 1-hop node. Applying (5.1.2) into (5.1.4), it can be shown that $h_2(\boldsymbol{y}, \boldsymbol{z})$ is the probability that the node at $\boldsymbol{y}$ is directly connected to the 1-hop node at $\boldsymbol{z}$ but not directly connected to the node at $\boldsymbol{x}_0$. That is, the node at $\boldsymbol{y}$ is a 2-hop node and the node at z is a 1-hop node. For $k = 3$, we have that $h_3(\boldsymbol{y}, \boldsymbol{z})$ is the probability that the node at $\boldsymbol{y}$ is directly connected to the 2-hop node at $\boldsymbol{z}$ but not directly connected to the 1-hop node at $\boldsymbol{w}$. That is, the node at $\boldsymbol{y}$ is a "potential" 3-hop node as we do not specify whether or not the node at $\boldsymbol{y}$ is directly connected to the node at $\boldsymbol{x}_0$. Therefore, $h_3(\boldsymbol{y}, \boldsymbol{z})$ is an upper bound on the probability that there exists a 3-hop node at $\boldsymbol{y}$ *and* a 2-hop node at z *and* the two nodes are directly connected. By recursion, $h_k(\boldsymbol{y}, \boldsymbol{z})$ forms an upper bound on the probability that there exists a k-hop node at $\boldsymbol{y}$ *and* a $(k-1)$-hop node at $\boldsymbol{z}$ *and* the two nodes are directly connected. Note that a node is a k-hop node if and only if it is directly connected to at least one $(k-1)$-hop node *and* not directly connected to any of the i-hop nodes where $i < k - 1$. Therefore, using the union bound (with some arithmetic steps omitted) and only considering the $(k-1)$-hop nodes and $(k-2)$-hop nodes, we obtain

$$E\left[N_k\right] = \lim_{\Delta \to 0} \sum_{y \in \mathbb{D}^d \setminus \{x_0\}} \Pr\left(I_y^k = 1\right)$$

$$\leq \lim_{\Delta\to 0} \sum_{\mathbf{y}\in\mathbb{D}^d\setminus\{\mathbf{x_0}\}} \sum_{\substack{z,w\in\mathbb{D}^d\setminus\{y,x_0\}\\ z\neq w}} \Big(\Pr\Big(J_y{=}1, H_{y,z}{=}1, H_{y,w}{=}0 \Big| I_z^{k-1}=1, I_w^{k-2}=1, H_{z,w}=1\Big)$$

$$\Pr\Big(I_z^{k-1}=1, I_w^{k-2}=1, H_{z,w}=1\Big)\Big) \tag{5.1.14}$$

$$\leq \lim_{\Delta\to 0} \sum_{\mathbf{y}\in\mathbb{D}^d\setminus\{\mathbf{x_0}\}} \sum_{\substack{z,w\in\mathbb{D}^d\setminus\{y,x_0\}\\ z\neq w}} \left[g(y-z)\left(1-g(y-w)\right)\left(\lambda\delta y+o(\delta y)\right)h_{k-1}(z,w)\right]\delta w\delta z$$

$$= \int_{\mathbb{R}^d}\int_{\mathbb{R}^d} h_k(y,z)dzdy \tag{5.1.15}$$

where inequality (5.1.14) results from the following analysis. An upper bound for the probability that a k-hop node exists within B_y conditioned on the event that $\{N_{k-2}=n_{k-2}\}$ can be obtained as follows.

$$\Pr\left(I_y^k=1 \middle| N_{k-2}=n_{k-2}\right) \tag{5.1.16}$$

$$= \frac{1}{n_{k-2}!} \sum_{\substack{w_1,\cdots,w_{n_{k-2}}\in\mathbb{D}^d\setminus\{y,x_0\}\\ \mathbf{w_m}\neq\mathbf{w_n}\text{ for } m\neq n}} \Pr\left(I_{\mathbf{y}}^k=1, \bigcap_{i=1}^{n_{k-2}} I_{\mathbf{w_i}}^{k-2}=1 \middle| N_{k-2}=n_{k-2}\right)$$

By generalizing the derivation of (5.1.13), it can be shown that

$$\Pr\left(I_y^k=1, \bigcap_{i=1}^{n_{k-2}} I_{w_i}^{k-2}=1 \middle| N_{k-2}=n_{k-2}\right)$$

$$\leq \sum_{z\in\mathbb{D}^d\setminus\{y,x_0\}} \Pr\left(J_y=1, \bigcap_{i=1}^{n_{k-2}} H_{y,w_i}=0, H_{y,z}=1, \bigcup_{i=1}^{n_{k-2}} H_{z,w_i}=1, I_z^{k-1}=1, \bigcap_{i=1}^{n_{k-2}} I_{w_i}^{k-2}=1 \middle| N_{k-2}=n_{k-2}\right)$$

$$\leq \sum_{j=1}^{n_{k-2}} \sum_{z\in\mathbb{D}^d\setminus\{y,x_0\}} \Pr\left(J_y=1, H_{y,w_j}=0, H_{y,z}=1, H_{z,\mathbf{w_j}}=1, I_z^{k-1}=1, \bigcap_{i=1}^{n_{k-2}} I_{w_i}^{k-2}=1 \middle| N_{k-2}=n_{k-2}\right)$$

$$= \sum_{j=1}^{n_{k-2}} \sum_{z\in\mathbb{D}^d\setminus\{y,x_0\}} \Big(\Pr\Big(J_y=1, H_{y,w_j}=0, H_{y,z}=1 \Big| H_{z,w_j}=1, I_z^{k-1}=1, I_{\mathbf{w_j}}^{k-2}=1, N_{k-2}=n_{k-2}\Big)$$

$$\times \Pr\left(H_{z,w_j}=1, I_z^{k-1}=1, \bigcap_{i=1}^{n_{k-2}} I_{w_i}^{k-2}=1 \middle| N_{k-2}=n_{k-2}\right)\Big) \tag{5.1.17}$$

Note that for any integer $1\leq j\leq n_{k-2}$,

$$\sum_{\substack{w_1,\cdots,w_j,w_{j+1},\cdots,w_{n_{k-2}}\in\mathbb{D}^d\setminus\{y,x_0,z,w_j\}\\ w_m\neq w_n\text{ for } m\neq n}} \Pr\left(H_{z,w_j}=1, I_z^{k-1}=1, \bigcap_{i=1}^{n_{k-2}} I_{\mathbf{w_i}}^{k-2}=1 \middle| N_{k-2}=n_{k-2}\right)$$

$$= (n_{k-2}-1)!\Pr\left(H_{z,w_j}=1, I_z^{k-1}=1, I_{w_j}^{k-2}=1 \middle| N_{k-2}=n_{k-2}\right) \tag{5.1.18}$$

Substituting (5.1.17) and (5.1.18) into (5.1.16),

$$\begin{aligned}
&\Pr\left(I_y^k = 1 \middle| N_{k-2} = n_{k-2}\right)\\
&\le \frac{1}{n_{k-2}!}\sum_{j=1}^{n_{k-2}} \sum_{\substack{z,w_j\in\mathbb{D}^d\setminus\{y,x_0\}\\ z\neq w_j}} \left(\Pr\left(J_{0y} = 1, H_{y,w_j} = 0, H_{y,z} = 1 \middle| H_{z,w_j} = 1,\right.\right.\\
&\left. I_z^{k-1} = 1, I_{w_j}^{k-2} = 1, N_{k-2} = n_{k-2}\right)\\
&\quad \left.\times (n_{k-2}-1)! \Pr\left(H_{z,w_j} = 1, I_z^{k-1} = 1, I_{w_j}^{k-2} = 1 \middle| N_{k-2} = n_{k-2}\right)\right)\\
&= \sum_{\substack{z,w\in\mathbb{D}^d\setminus\{y,x_0\}\\ z\neq w}} \left(\Pr\left(J_y = 1, H_{y,w} = 0, H_{y,z} = 1 \middle| H_{z,w} = 1, I_z^{k-1} = 1, I_w^{k-2} = 1, N_{k-2} = n_{k-2}\right)\right.\\
&\quad \left.\times \Pr\left(H_{z,w} = 1, I_z^{k-1} = 1, I_w^{k-2} = 1 \middle| N_{k-2} = n_{k-2}\right)\right)
\end{aligned} \tag{5.1.19}$$

Since

$$\begin{aligned}
&\Pr\left(H_{z,w} = 1, I_z^{k-1} = 1, I_w^{k-2} = 1 \middle| N_{k-2} = n_{k-2}\right)\\
&\quad = \frac{\Pr\left(H_{z,w} = 1, I_z^{k-1} = 1, I_w^{k-2} = 1, N_{k-2} = n_{k-2}\right)}{\Pr\left(N_{k-2} = n_{k-2}\right)}
\end{aligned} \tag{5.1.20}$$

and

$$\begin{aligned}
&\sum_{n_{k-2}=1}^{\infty} \Pr\left(H_{z,w} = 1, I_z^{k-1} = 1, I_w^{k-2} = 1, N_{k-2} = n_{k-2}\right)\\
&\quad = \Pr\left(H_{z,w} = 1, I_z^{k-1} = 1, I_w^{k-2} = 1\right)
\end{aligned} \tag{5.1.21}$$

using (5.1.19), (5.1.20) and (5.1.21), we obtain

$$\begin{aligned}
&\Pr\left(I_y^k = 1\right)\\
&= \sum_{n_{k-2}=1}^{\infty} \left(\Pr\left(I_y^k = 1 \middle| N_{k-2} = n_{k-2}\right) \Pr\left(N_{k-2} = n_{k-2}\right)\right)\\
&\le \sum_{\substack{z,w\in\mathbb{D}^d\setminus\{y,x_0\}\\ z\neq w}} \left(\Pr\left(J_y = 1, H_{y,w} = 0, H_{y,z} = 1 \middle| H_{z,w} = 1, I_z^{k-1} = 1, I_w^{k-2} = 1\right)\right.\\
&\quad \left.\times \Pr\left(H_{z,w} = 1, I_z^{k-1} = 1, I_w^{k-2} = 1\right)\right)
\end{aligned} \tag{5.1.22}$$

Hence we obtain inequality (5.1.14).

Now (5.1.3) is proved for general k.

Next, we prove that $E[N_k]$ is finite by showing that the integral $\int_{\Re^d}\int_{\Re^d} h_k(\mathbf{y}, z)\, dzdy$ at the right-hand side of (5.1.3) is finite using mathematical induction. For $k = 1$, we have

$$E[N_1] = \int_{\Re^d}\int_{\Re^d} h_1(\mathbf{y},z)dzd\mathbf{y} = \int_{\Re^d} \lambda g(\mathbf{y}-\mathbf{x}_0)d\mathbf{y} < \infty$$

due to the integral boundedness of the connection function g. Supposing $\int_{\Re^d}\int_{\Re^d} h_{k-1}(\mathbf{y},z)dzd\mathbf{y} < \infty$ and then using (5.1.3), it can be shown that

$$\begin{aligned}
&E[N_k] \\
&\le \int_{\Re^d}\int_{\Re^d} h_k(\mathbf{y},z)dzd\mathbf{y} \\
&= \int_{\Re^d}\int_{\Re^d}\int_{\Re^d} \lambda g(\mathbf{y}-z)\left[1-g(\mathbf{y}-\mathbf{w})\right] h_{k-1}(z,\mathbf{w})d\mathbf{w}dzd\mathbf{y} \\
&\le \int_{\Re^d}\int_{\Re^d}\int_{\Re^d} \lambda g(\mathbf{y}-z)h_{k-1}(z,\mathbf{w})d\mathbf{w}dzd\mathbf{y} \\
&= \int_{\Re^d}\left[\int_{\Re^d} \lambda g(\mathbf{y}-z)d\mathbf{y}\right]\left[\int_{\Re^d} h_{k-1}(z,\mathbf{w})d\mathbf{w}\right]dz \\
&= \left[\int_{\Re^d} \lambda g(\mathbf{y})d\mathbf{y}\right] \times \int_{\Re^d}\int_{\Re^d} h_{k-1}(z,\mathbf{w})d\mathbf{w}dz \\
&< \infty
\end{aligned}$$

This completes the proof of the lemma. □

On the basis of Lemma 98, the following theorem on a lower bound for λ_c can be obtained.

Theorem 99 *For $\mathcal{G}\left(\mathcal{X}_{\lambda,\mathbf{x}_0}, g, \Re^d\right)$ with $d = 2, 3$ the critical density λ_c is lower bounded by*

$$\lambda_c \ge \sup_{m\in\mathbb{Z}^+}\left\{\sqrt[m]{1/f_{\sup}(m)}\right\} \tag{5.1.23}$$

where

$$f_{\sup}(m) = \sup_{\mathbf{y}_{m+1}\,\mathbf{y}_{m+2}\in\Re^d}\left\{\int_{\Re^d}\cdots\int_{\Re^d}\left[\prod_{i=1}^{m}\left[g(\mathbf{y}_i-\mathbf{y}_{i+1})\left(1-g(\mathbf{y}_i-\mathbf{y}_{i+2})\right)\right]d\mathbf{y}_1\cdots d\mathbf{y}_m\right.\right\} \tag{5.1.24}$$

Proof From (5.1.4), we have

$$\begin{aligned}
&h_k(\mathbf{y}_1,\mathbf{y}_2) \\
&= \int_{\Re^d} \lambda g(\mathbf{y}_1-\mathbf{y}_2)[1-g(\mathbf{y}_1-\mathbf{y}_3)]h_{k-1}(\mathbf{y}_2,\mathbf{y}_3)d\mathbf{y}_3 \\
&= \int_{\Re^d}\cdots\int_{\Re^d}\prod_{i=1}^{m}\left[\lambda g(\mathbf{y}_i-\mathbf{y}_{i+1})\left(1-g(\mathbf{y}_i-\mathbf{y}_{i+2})\right)\right]h_{k-m}(\mathbf{y}_{m+1},\mathbf{y}_{m+1})d\mathbf{y}_3\cdots d\mathbf{y}_{m+2}
\end{aligned} \tag{5.1.25}$$

for any integer $1 \leq m < k$. Using (5.1.24) and (5.1.25), it can be shown that

$$\int_{\Re^d}\int_{\Re^d} h_k(\mathbf{y},\mathbf{z})d\mathbf{y}d\mathbf{z} \leq \lambda^m f_{\sup}(m)\int_{\Re^d}\int_{\Re^d} h_{k-m}(\mathbf{y},\mathbf{z})d\mathbf{y}d\mathbf{z} \tag{5.1.26}$$

for any integer $1 \leq m < k$. Applying (5.1.26) recursively into itself and using (5.1.3), we obtain

$$E[N_k] \leq \left[\lambda^m f_{\sup}(m)\right]^{\lfloor \frac{k}{m} \rfloor}\int_{\Re^d}\int_{\Re^d} h_i(\mathbf{y},\mathbf{z})d\mathbf{y}d\mathbf{z} \tag{5.1.27}$$

where $i = k - \left\lfloor \frac{k}{m} \right\rfloor m$ and $\lfloor x \rfloor$ is the largest integer smaller than or equal to x. Since $\int_{\Re^d}\int_{\Re^d} h_i(\mathbf{y},\mathbf{z})d\mathbf{y}d\mathbf{z} < \infty$ for any positive integer i (refer to the proof of Lemma 98), and $E[|\mathcal{W}|] = \sum_{k=1}^{\infty} E[N_k]$, (5.1.27) implies that $E[|\mathcal{W}|]$ is finite if $\lambda^m f_{\sup}(m) < 1$. Given the fact that $E[|\mathcal{W}|]$ is finite implies $\theta(\lambda) = 0$ (see [143, Theorem 6.1] or Sect. 1.3.1), we have $\lambda_c \geq \sqrt[m]{1/f_{\sup}(m)}$ for all positive integer m. The result follows. □

Alternatively, Theorem 99 can be rewritten as follows.

Theorem 100 *For $\mathcal{G}\left(\mathcal{X}_{\lambda,\mathbf{x}_0}, g, \Re^d\right)$ with $d = 2, 3$ the critical density λ_c is lower bounded by*

$$\lambda_c \geq \sup_{m\in\mathbb{Z}^+}\left\{\sqrt[m]{1/f_{\sup}(m)}\right\} = \lim_{m\to\infty}\left\{\sqrt[m]{1/f_{\sup}(m)}\right\} \tag{5.1.28}$$

where $f_{\sup}(m)$ is given in (5.1.24).

The proof of Theorem 100 relies on the following lemma from [107].

Lemma 101 ([107, Lemma 2.1]) *Let $(a_n)_{n\in\mathbb{N}}$ be a sequence of elements of $\mathbb{R}^+ \cup \{\infty\}$ such that*

$$a_{n+m} \leq a_n a_m \qquad \textit{for all } n, m \in \mathbb{N}$$

If $a_1 < \infty$, then $a_n < \infty$ for all $n \in \mathbb{N}$, the sequence $\left(a_n^{1/n}\right)_{n\in\mathbb{N}}$ is convergent and

$$\lim_{n\to\infty} a_n^{1/n} \leq a_m^{1/m} \quad \textit{for each } m \in \mathbb{N}$$

On the basis of Lemma 101, we can prove Theorem 100 as follows.

Proof From (5.1.24), we have for all $n, m \in \mathbb{Z}^+$,

$$
\begin{aligned}
& f_{\sup}(m+n) \\
&= \sup_{y_{m+n+1} y_{m+n+2} \in \Re^d} \left\{ \int_{\Re^d} \cdots \int_{\Re^d} \prod_{i=1}^{m+n} g(y_i - y_{i+1})\,(1 - g(y_i - y_{i+2}))\, dy_1 \cdots dy_{m+n} \right\} \\
&= \sup_{y_{m+n+1} y_{m+n+2} \in \Re^d} \left\{ \int_{\Re^d} \cdots \int_{\Re^d} \left[\int_{\Re^d} \cdots \int_{\Re^d} \prod_{i=1}^{m} g(y_i - y_{i+1})\,(1 - g(y_i - y_{i+2}))\, dy_1 \cdots dy_m \right] \right. \\
&\qquad \left. \prod_{i=m+1}^{m+n} g(y_i - y_{i+1})\,(1 - g(y_i - y_{i+2}))\, dy_{m+1} \cdots dy_{m+n} \right\} \\
&\le \sup_{y_{m+1} y_{m+2} \in \Re^d} \left\{ \int_{\Re^d} \cdots \int_{\Re^d} \prod_{i=1}^{m} g(y_i - y_{i+1})\,(1 - g(y_i - y_{i+2}))\, dy_1 \cdots dy_m \right\} \\
&\quad \times \sup_{y_{m+n+1} y_{m+n+2} \in \Re^d} \left\{ \int_{\Re^d} \cdots \int_{\Re^d} \prod_{i=m+1}^{m+n} g(y_i - y_{i+1})\,(1 - g(y_i - y_{i+2}))\, dy_{m+1} \cdots dy_{m+n} \right\} && (5.1.29) \\
&= f_{\sup}(m) f_{\sup}(n) && (5.1.30)
\end{aligned}
$$

where (5.1.29) follows from the monotonicity of the integral. Furthermore,

$$
\begin{aligned}
f_{\sup}(1) &= \sup_{y_2, y_3 \in \Re^d} \left\{ \int_{\Re^d} g(y_1 - y_2)\,(1 - g(y_1 - y_3))\, dy_1 \right\} \\
&= \sup_{y_3 \in \Re^d} \left\{ \int_{\Re^d} g(y_1)\,(1 - g(y_1 - y_3))\, dy_1 \right\} \\
&\le \int_{\Re^d} g(y) dy < \infty && (5.1.31)
\end{aligned}
$$

Equations (5.1.30), (5.1.31) and Lemma 101 together show that the sequence $\left(\sqrt[m]{f_{\sup}(m)}\right)_{m \in \mathbb{Z}^+}$ is convergent and

$$
\begin{aligned}
& \lim_{m \to \infty} \sqrt[m]{f_{\sup}(m)} \le \sqrt[n]{f_{\sup}(n)} \quad \text{for each } n \in \mathbb{Z}^+ \\
& \Leftrightarrow \lim_{m \to \infty} \sqrt[m]{f_{\sup}(m)} = \inf_{n \in \mathbb{Z}^+} \left\{ \sqrt[n]{f_{\sup}(n)} \right\} && (5.1.32)
\end{aligned}
$$

With (5.1.32) and Theorem 99, the result follows. □

Remark 102 Although we consider $\mathcal{G}\left(\mathcal{X}_{\lambda,x_0}, g, \Re^d\right)$ with $d = 2, 3$ in Lemma 98, Theorems 99 and 100, it can be seen from their proofs that they are applicable to an arbitrary integer d. However this is not the case for Theorems 103 and 105 in Sect. 5.2.

5.1.1 Application of the Lower Bound on the Critical Density to the Unit Disk Connection Model and the Log-Normal Connection Model

In this subsection, we compare our lower bound with existing results in the literature obtained assuming the unit disk connection model and the log-normal connection model, which are two special cases of the random connection model.

To obtain the lower bound under the unit disk connection model, we apply (5.0.2) into (5.1.24) and obtain

$$\begin{aligned} f_{\text{sup}}(1) &= \sup_{y_2, y_3 \in \Re^d} \left\{ \int_{\Re^d} g(y_1 - y_2)\left(1 - g(y_1 - y_3)\right) dy_1 \right\} \\ &= \int_{\|y_1\| \le r} dy_1 = V_d r^d \end{aligned} \tag{5.1.33}$$

where $V_d = \frac{\pi^{d/2}}{\Gamma(\frac{d}{2}+1)}$ is the volume of $(d-1)$-dimensional sphere with unit radius and $\Gamma(.)$ is the gamma function. Similarly,

$$\begin{aligned} f_{\text{sup}}(2) &= \sup_{y_3, y_4 \in \Re^d} \left\{ \left\{ \int_{\Re^d} \int_{\Re^d} g(y_1 - y_2)\left(1 - g(y_1 - y_3)\right) g(y_2 - y_3)\left(1 - g(y_2 - y_4)\right) dy_1 dy_2 \right\} \right\} \\ &= \int_{\|y_2\| \le r} \int_{\|y_1 - y_2\| \le r, \|y_1\| > r} dy_1 dy_2 \end{aligned} \tag{5.1.34}$$

For $d = 2$, (5.1.34) can be simplified to

$$f_{\text{sup}}(2) = \int_0^r \left(\int_{r-u}^r 2v \arccos\left(\frac{r^2 - u^2 - v^2}{2uv}\right) dv \right) 2\pi u du \tag{5.1.35}$$

which is obtained using elementary geometric calculations for finding an arc length, as illustrated in Fig. 5.1.

For $d = 3$, (5.1.34) can be simplified to

$$f_{\text{sup}}(2) = \int_0^r \left[\int_{r-u}^r 2\pi v^2 \left(1 - \frac{r^2 - u^2 - v^2}{2uv}\right) dv \right] 4\pi u^2 du \tag{5.1.36}$$

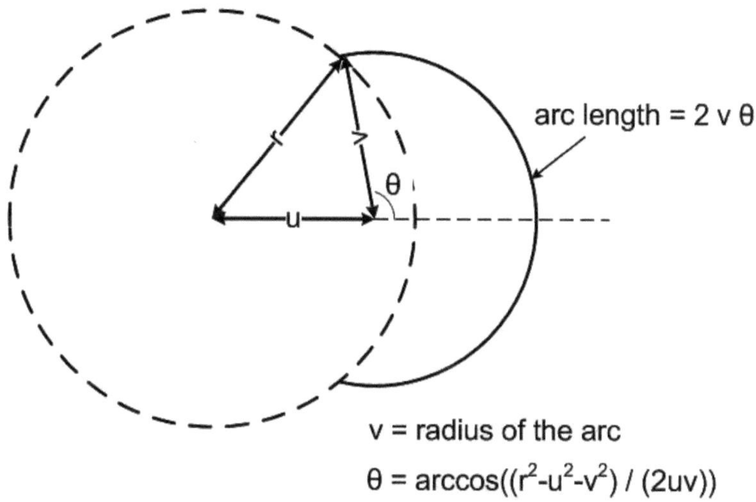

Fig. 5.1 An illustration of arc length calculation related to $f_{\text{sup}}(2)$ in $\Re^2$ under the unit disk connection model

which follows from elementary geometric calculations for finding the curved surface area of a spherical cap [130], as illustrated in Fig. 5.2.

Using a similar approach as that in (5.1.35) and (5.1.36), we can extend and generalize the calculation of $f_{\text{sup}}(m)$ under the unit disk connection model for $m \geq 3$, i.e., considering $\|\boldsymbol{y}_i - \boldsymbol{y}_{i+1}\| \leq r$ and $\|\boldsymbol{y}_i - \boldsymbol{y}_{i+2}\| > r$ (for all $1 \leq i \leq m-2$) for the integrations in (5.1.24). Then, we obtain a lower bound for λ_c using Theorem 99. The result is shown with other existing results in the literature in Figs. 5.3 and 5.4.

Figure 5.3 shows that our lower bound in $\Re^2$ is as tight as the lower bound obtained by Meester and Roy [143]. The lower bounds reported by Kong and Yeh [118], and Gu and Hong [88] are tighter than our lower bound, however their results are valid for the unit disk connection model only. In this chapter we consider the random connection model which is applicable to broader class of connection models, including the unit disk connection model. On the other hand, Fig. 5.4 also shows that our lower bound in $\Re^3$ is not as tight as the bound obtained by Kong and Yeh [118] (though it is within 6%). The tightness of our lower bound largely depends on how the distribution of the k-hop nodes is obtained. When we obtain the distribution of the k-hop nodes through the technique in Lemma 98, we only considered the impact of the previous two hops, i.e., $(k-1)$-hop nodes and $(k-2)$-hop nodes, on the distribution of the k-hop nodes. The impact of nodes three or more hops away is not taken into account. The tightness of our lower bound is therefore sacrificed for simplicity. A tighter lower bound can be obtained if we take into account the impact of nodes three or more hops away. However it will involve more intricate analysis.

To obtain the lower bound under the log-normal connection model, we first apply (5.0.3) into (5.1.24). Then, the lower bound under the log-normal connection

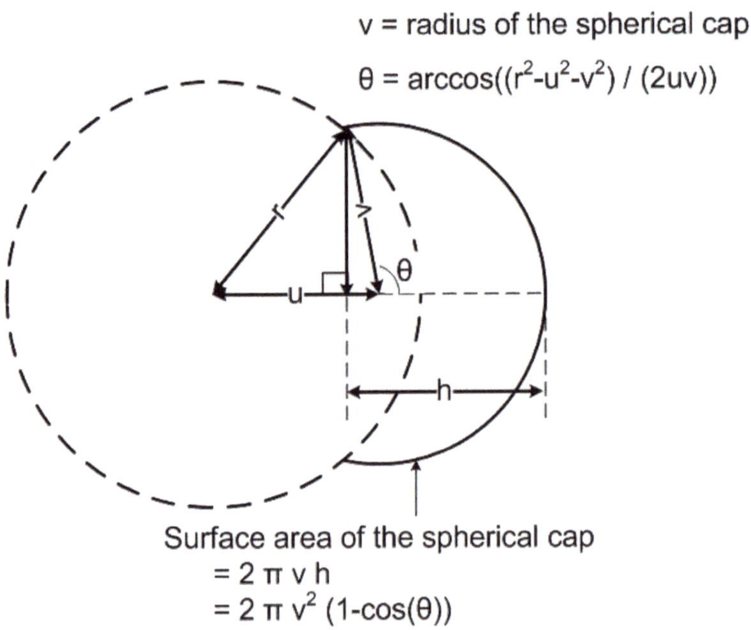

Fig. 5.2 An illustration of the surface area calculation of a spherical cap related to $f_{\text{sup}}(2)$ in $\Re^3$ under the unit disk connection model (two-dimensional view)

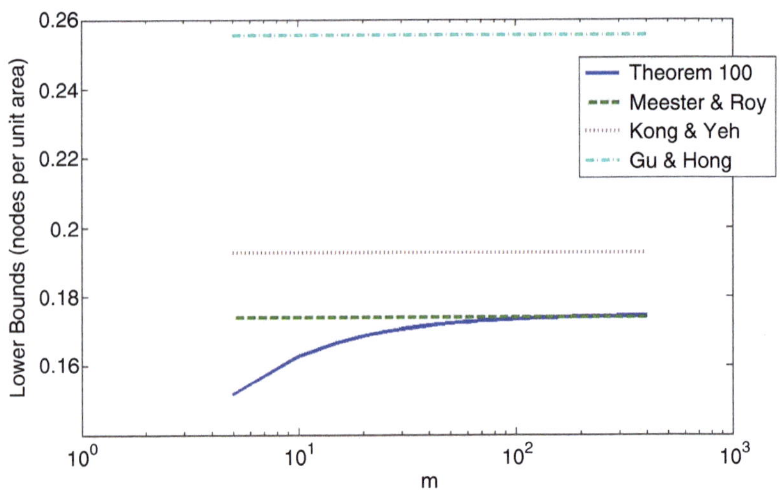

Fig. 5.3 Lower bounds for λ_c in $\Re^2$ under the unit disk connection model with $r = 2$

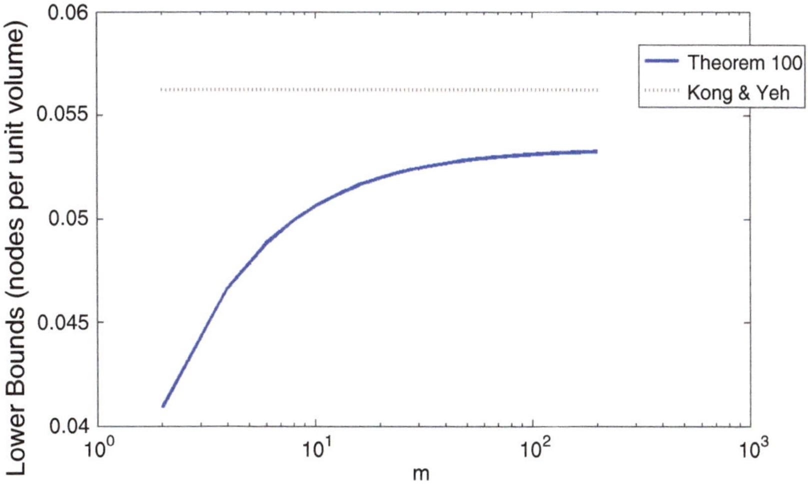

Fig. 5.4 Lower bounds for λ_c in $\Re^3$ under the unit disk connection model with $r = 2$

model can be calculated by using a similar approach as that used in obtaining the lower bound under the unit disk connection model, which involves converting (5.1.24) from Cartesian coordinate system to Polar coordinate system in $\Re^2$ or Spherical coordinate system in $\Re^3$. Due to its complexity, the end form of the equation is omitted here. Instead, the results are plotted in Figs. 5.5 and 5.6 but with no comparison to bounds from another method, since we know of none. Note that in order to have a fairer comparison between lower bounds under the log-normal model with different shadowing variance, the results in Figs. 5.5 and 5.6 have been normalized so that the average node degree is preserved while changing the shadowing variance, similarly for Figs. 5.10 and 5.11 in the next section.

5.2 An Upper Bound for the Critical Density

In this section, we derive an upper bound of the critical density λ_c in $\mathcal{G}\left(\mathcal{X}_\lambda, g, \Re^d\right)$ $(d = 2, 3)$ by coupling the problem to that of site percolation on a lattice [29]. First, we consider the problem in $\Re^2$ as follows. Let us partition the $\Re^2$ plane into nonoverlapping hexagons, where the Euclidean distance between the centers of two neighboring hexagons is a constant $a > 0$. We further partition each hexagon into six nonoverlapping equilateral triangles. As shown in Fig. 5.7, the hexagon labeled H_2 is partitioned into six triangles.

Consider a hexagon, e.g., H_2 in Fig. 5.7, and an equilateral triangle, e.g., $T_{2,2}$, in the hexagon, there is exactly one hexagon side that is located directly opposite to the triangle. Centered at the middle point of that hexagon side, i.e., $M_{2,2}$, we draw a circle with a radius a and obtain its intersectional area with the triangle, i.e., $S_{2,2}$.

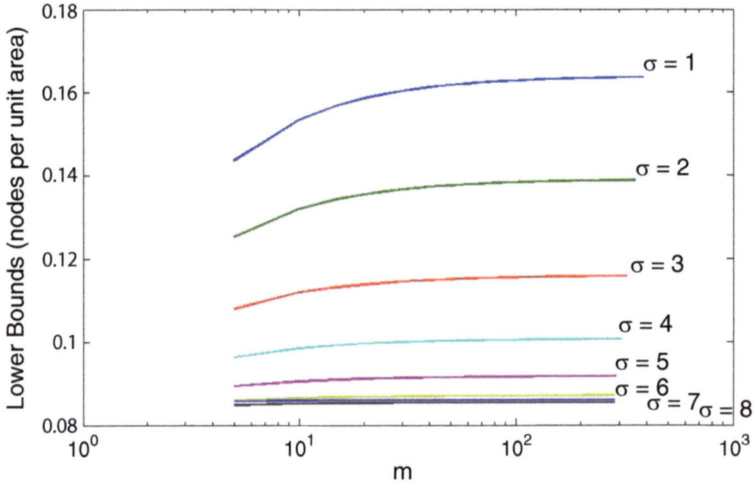

Fig. 5.5 Our lower bounds for λ_c in $\Re^2$ under the log-normal model with different shadowing variances

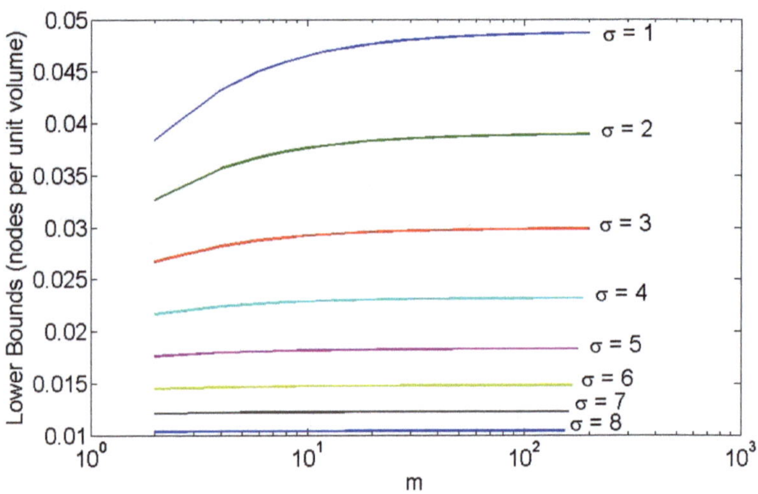

Fig. 5.6 Our lower bounds for λ_c in $\Re^3$ under the log-normal model with different shadowing variances

Repeat the action for the other five equilateral triangles in the hexagon. Merging the six intersectional areas, we obtain a flower-shaped cell within the hexagon, i.e., A_2.

Next, we want to obtain a lower bound on the probability that any two nodes inside two neighboring flower-shaped cells are directly connected. Consider two neighboring flower-shaped cells, e.g., A_4 and A_5 in Fig. 5.7, and two nodes Y and Z (one in each cell). Among the six intersectional areas in A_4, consider that node Y is located in the intersectional area furthest to A_5, i.e., $S_{4,3}$ in A_4. Denote by b

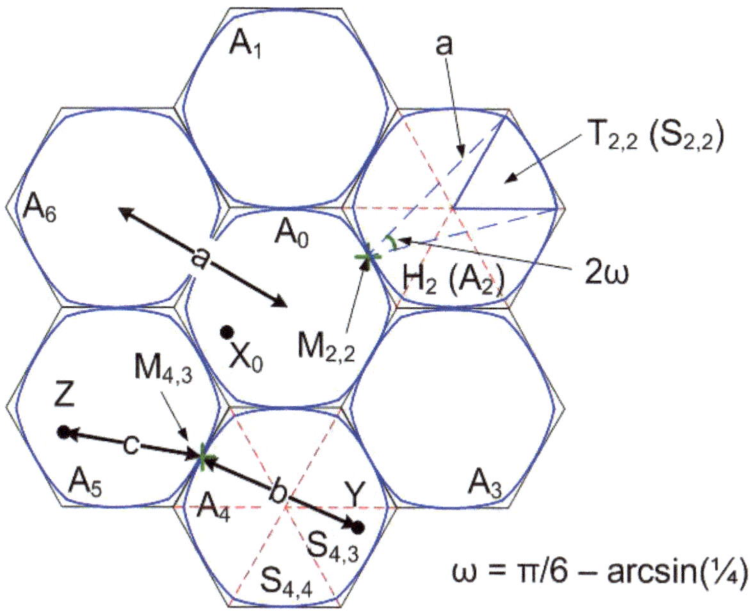

Fig. 5.7 An illustration of the hexagons and the flower-shaped cell in each hexagon. The partition of each hexagon into six nonoverlapping equilateral triangles and construction of an intersectional area inside each triangle are illustrated in hexagon H_2. An example of how the distance b is defined for a node is shown in flower-shaped cell A_4

the Euclidean distance between the node and the middle point, i.e., $M_{4,3}$, of the hexagon side shared by A_4 and A_5. Then, we have $b \in [\frac{a}{2}, a]$. Consider that node Z is located anywhere within A_5 and denote by c the Euclidean distance between Z and the middle point $M_{4,3}$. Due to the triangular inequality, the distance between Y and Z is less than or equal to $b + c$. Since $c \leq a$, it can be easily shown that the distance between Y and a node in A_5 is at most $b + a$. Let $f_a(b)$ be the probability density function of the above defined distance b. Using elementary geometric calculations,

$$f_a(b) = \frac{12b}{|A(a)|}\left(\frac{\pi}{6} - \arcsin(\frac{a}{4b})\right) \tag{5.2.1}$$

where $|A(a)|$ is the area of a flower-shaped cell and is given by

$$|A(a)| = 6a^2\left(\omega - \frac{1}{2}\sin(\omega)\right) \tag{5.2.2}$$

with $\omega = \frac{\pi}{6} - \arcsin(\frac{1}{4})$ (see also Fig. 5.7). Due to the nonincreasing monotonicity of g, the probability that Y is directly connected to any node in A_0 is lower bounded by $g\left((b + a)\,\boldsymbol{u}\right)$ where $\boldsymbol{u}$ is a vector of unit size. Due to the rotational invariance property of g, the direction of $\boldsymbol{u}$ does not affect $g\left((b + a)\,\boldsymbol{u}\right)$. Therefore, the probability that any two nodes inside two neighboring flower-shaped cells are directly connected, denoted by $\bar{p}(a)$, satisfies the following condition:

$$\bar{p}(a) \geq \int_{a/2}^{a} g\left((b+a)\,\boldsymbol{u}\right) f_a(b) db \triangleq \hat{p}(a) \tag{5.2.3}$$

Now, starting with the flower-shaped cell A_0 with a node X_0 located anywhere within it, we examine the connections between X_0 and the nodes in the six neighboring flower-shaped cells, $A_1, A_2, \cdots, A_6$ (see Fig. 5.7). We say A_i, $1 \leq i \leq 6$, is "*occupied*" if and only if there exists at least one node in A_i *and* X_0 is directly connected to at least one of the nodes in A_i. The probability that A_i (with size $|A(a)|$) is "occupied," denoted by p, is then

$$\begin{aligned} p >& \sum_{m=1}^{\infty} \left[\frac{(\lambda\,|A(a)|)^m \exp(-\lambda\,|A(a)|)}{m!} \times [1-(1-\hat{p}(a))^m] \right] \\ =& \left(1-e^{-\lambda|A(a)|}\right) - e^{-\lambda|A(a)|\hat{p}(a)} \left[1-e^{-\lambda|A(a)|(1-\hat{p}(a))}\right] \\ =& 1-e^{-\lambda|A(a)|\hat{p}(a)} \end{aligned} \tag{5.2.4}$$

Note that the events that neighboring flower-shaped cells are "occupied" are independent. Next, for each A_i marked as "occupied," we focus on a node X_i in A_i which is directly connected to X_0 and examine the direct connections from X_i to other nodes in the neighboring flower-shaped cells that have not been considered before. The process continues in a similar way until every flower-shaped cell is marked to be either "occupied" or "empty."

Following from the above marking process, it can be seen that $\Pr(|\mathcal{W}| = \infty) > 0$ *if* the probability that there are infinite number of flower-shaped cells, which are marked as "occupied," is positive. We say that the flower-shaped cells percolate if there are infinite number of flower-shaped cells marked as "occupied". Imagine we replace each flower-shaped cell by a vertex and draw an edge between two neighboring vertices. Then, the vertices form an equilateral triangular lattice. If the vertices inherit the status of respective flower-shaped cells, i.e., a vertex is marked as "occupied" if and only if the corresponding flower-shaped cell is marked as "occupied," then the site percolation on the accompanying equilateral triangular lattice implies the percolation of the flower-shaped cells, and the converse. Hence, the flower-shaped cells percolates if $p > 0.5$ [29, p. 132]. That is, $\Pr(|\mathcal{W}| = \infty) > 0$ if

$$1-e^{-\lambda|A(a)|\hat{p}(a)} > 0.5 \Leftrightarrow \lambda > \frac{\log_e(2)}{|A(a)|\,\hat{p}(a)} \tag{5.2.5}$$

Indeed, $\Pr(|\mathcal{W}| = \infty) > 0$ if (5.2.5) holds *for any* value of a. The above analysis can be summarized into the following theorem.

Theorem 103 *For $\mathcal{G}\left(\mathcal{X}_\lambda, g, \Re^2\right)$ the critical density λ_c is upper bounded by*

$$\lambda_c \leq \inf_{a \in \mathbb{R}+} \left\{ \frac{\log_e(2)}{12 \int_{a/2}^{a} b\bar{g}(b+a)\left(\frac{\pi}{6} - \arcsin(\frac{a}{4b})\right) db} \right\} \tag{5.2.6}$$

Remark 104 In Theorem 103, we obtained an upper bound for λ_c using the probability density function of b which is given in (5.2.1). Note that (5.2.1) is obtained by considering a node located in just one of the six intersectional areas in a flower-shaped cell. Particularly, the chosen intersectional area is the one which will maximize the distance b. An improved upper bound can be obtained by considering the situation that the node can be located in any intersectional area within the flower-shaped cell. However, due to the difficulty in computing the probability density function of b, no closed form can be obtained for the upper bound. Therefore, we continue to use the existing approach to obtain an upper bound in $\mathbb{R}^2$ and later in $\mathbb{R}^3$. In Fig. 5.10, the improved upper bound is plotted numerically.

To extend the above analysis for $\Re^2$ to $\Re^3$, we first partition the $\Re^3$ space into nonoverlapping rhombic dodecahedra. The logic behind the transformation from hexagons in $\Re^2$ to rhombic dodecahedra in $\Re^3$ is related to the close-packing of equal-radius $(d-1)$-spheres in $\Re^d$. In $\Re^2$, the hexagon is the shape of the Voronoi cells constructed from the equilateral triangular lattice [9], which delivers the densest arrangement of equal-radius disks in $\Re^2$ [46, p. 6]. In $\Re^3$, the rhombic dodecahedron is the shape of the Voronoi cells constructed from the face-centered cubic lattice [46, p. 34], which delivers one of the densest possible arrangements of equal-radius spheres in $\Re^3$ [9]. Refer to Figs. 5.8 and 5.9 for a visualization of the face-centered cubic packing of equal-radius spheres and rhombic dodecahedron and the shape of the associated Voronoi cells.

Using a similar approach in $\Re^2$, we then construct a three-dimensional flower-shaped cell within each rhombic dodecahedron so that the maximum Euclidean distance between any pair of nodes within two neighboring flower-shaped cells is $2a$, where a is the distance between the centers of two neighboring rhombic dodecahedra. Based on the constructed flower-shaped cells, we can rewrite (5.2.1) into

Fig. 5.8 Face-centered cubic packing of equal spheres

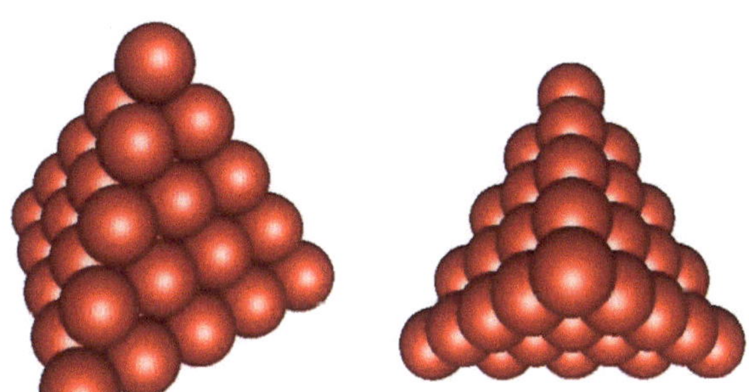

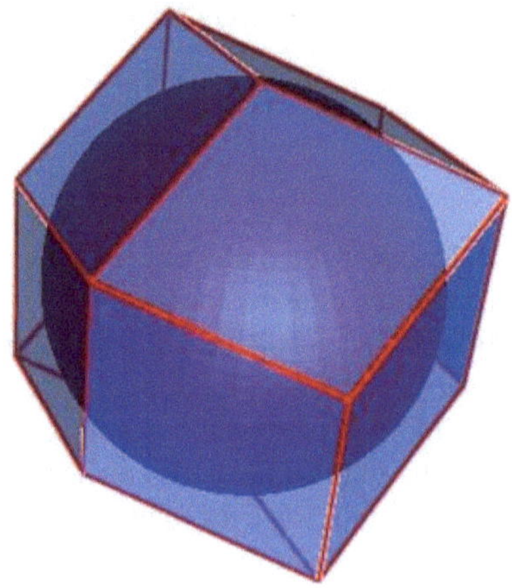

Fig. 5.9 Rhombic dodecahedron as the shape of the Voronoi cells

$$f_a(b) = \frac{48b^2}{|V(a)|}\int_{\theta_{\min}}^{\frac{\pi}{2}} \sin(\theta)\left(\frac{\pi}{4} - \arcsin(\frac{a/2+\sqrt{2}b\cos(\theta)}{\sqrt{2}b\sin(\theta)})\right)d\theta \tag{5.2.7}$$

where $\theta_{\min} = \frac{\pi}{2} - \arcsin(\frac{1}{\sqrt{3}}) + \arcsin(\frac{a}{2\sqrt{3}b})$ and $|V(a)|$ is the volume of a three-dimensional flower-shaped cell. Substituting (5.2.7) into (5.2.3), and bearing in mind that the site percolation on the face-centered cubic lattice (where the critical probability is approximately 0.199 [134]) implies the site percolation of the flower-shaped cells, we obtain the follow theorem.

Theorem 105 *For $\mathcal{G}\left(\mathcal{X}_\lambda, g, \Re^3\right)$ the critical density λ_c is upper bounded by*

$$\lambda_c \le \inf_{a\in\mathbb{R}+}\left\{-\frac{\log_e(0.801)}{\int_{\frac{a}{2}}^{a} g\left((b+a)\boldsymbol{u}\right)\bar{f}_a(b)db}\right\} \tag{5.2.8}$$

where

$$\bar{f}_a(b) = 48b^2\int_{\theta_{\min}}^{\pi/2} \sin(\theta)\left(\frac{\pi}{4} - \arcsin\left(\frac{a/2+\sqrt{2}b\cos(\theta)}{\sqrt{2}b\sin(\theta)}\right)\right)d\theta,$$

and $\theta_{\min} = \frac{\pi}{2} - \arcsin(\frac{1}{\sqrt{3}}) + \arcsin(\frac{a}{2\sqrt{3}b})$.

5.2.1 *Application of the Upper Bound on the Critical Density to the Unit Disk Connection Model and the Log-Normal Connection Model*

To evaluate the tightness of our upper bound, we compare our result with existing upper bounds in the literature obtained under the unit disk connection model [88, 143, 167] and the log-normal connection model [131].

To obtain the upper bound in $\Re^2$ under the unit disk connection model, first let $\lambda_{upper}^{(r)}$ be the upper bound for λ_c under the unit disk connection model with a transmission range r. Applying (5.0.2) into (5.2.6), it can be shown that the infimum is achieved at $a = \frac{1}{2}r$. That is,

$$\lambda_{upper}^{(r)} = \frac{\log_e(2)}{|A(r/2)|} = \frac{\log_e(2)}{\frac{3}{2}r^2\left(\omega - \frac{1}{2}\sin(\omega)\right)} \tag{5.2.9}$$

with $\omega = \frac{\pi}{6} - \arcsin(\frac{1}{4})$. Then (5.2.6) reduces to the upper bound obtained in [88, 143]. As a specific example with $r = 2$, we have $\lambda_{upper}^{(2)} \approx 0.843$ and it is shown in Fig. 5.10 (as the log-normal connection model with $\sigma = 0$). On the other hand, it means that our upper bound is not as tight as the upper bound obtained by Philips et al. [167] (0.8376 for $r = 2$). In our approach to obtaining the upper bound, the constructed flower-shaped cells must lie inside the corresponding hexagons. Then, we consider the nodes that fall inside each flower-shaped cell. In contrast, Philips

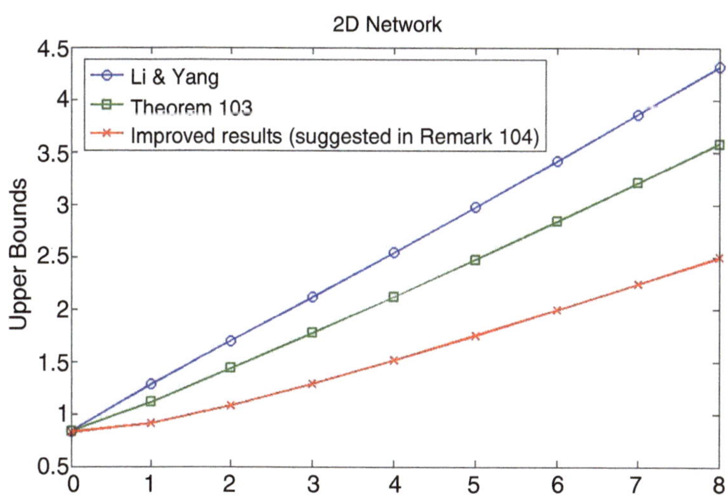

Fig. 5.10 Upper bounds for λ_c in $\Re^2$ under the log-normal connection model with different shadowing variances and $r = 2$, $\alpha = 2$. Note that the log-normal connection model reduces to the unit disk connection model when $\sigma = 0$

et al. [167] allow some portion of each flower-shaped cell to exceed the boundary of the corresponding hexagon. Then they consider those nodes that fall inside both flower-shaped cell *and* the corresponding hexagon. The ratio of flower-shaped cell size to hexagon size is adjusted to obtain the tightest upper bound. The approach used in [167] can be adapted into our analysis to achieve a tighter upper bound. However, it will involve more intricate analysis.

The upper bound in $\Re^2$ under the log-normal connection model can be calculated by substituting (5.0.3) into (5.2.6). The end form of the equation is omitted due to its complexity. To compare our result with other upper bounds in the literature [131], first let

$$\lambda_{upper}(g) = \inf_{a\in\mathbb{R}+}\left\{\frac{\log_e(2)}{12\int_{a/2}^{a} b\bar{g}(b+a)\left(\frac{\pi}{6}-\arcsin(\frac{a}{4b})\right)db}\right\} \tag{5.2.10}$$

be the upper bound for λ_c given by (5.2.6). From (5.2.10),

$$\begin{aligned}\lambda_{upper}(g) &\le \inf_{a\in\mathbb{R}+}\left\{\frac{\log_e(2)}{|A(a)|\,g\,(2a\boldsymbol{u})}\right\}\\ &= \inf_{a\in\mathbb{R}+}\left\{\frac{\log_e(2)}{|A(a/2)|\,g\,(a\boldsymbol{u})}\right\} = \inf_{a\in\mathbb{R}+}\left\{\frac{\lambda_{upper}^{(1)}}{a^2 g\,(a\boldsymbol{u})}\right\}\end{aligned} \tag{5.2.11}$$

where $\lambda_{upper}^{(1)}$ is given by (5.2.9) with $r = 1$. Note that the right-hand side of (5.2.11) was reported by Li and Yang [131] but only for the log-normal connection model. The derivation of (5.2.11) shows that Theorem 103 provides a tighter upper bound than that in [131] (see Fig. 5.10 for an illustration).

To the best of our knowledge, no upper bound has been obtained in $\Re^3$ under either the unit disk connection model or the log-normal connection model. An illustration of the upper bound in Theorem 105 is shown in Fig. 5.11.

5.3 Notes and Further Reading

In this chapter we investigated the analytical bounds for the critical density in large wireless networks where nodes are Poissonly distributed in $\Re^d$ ($d = 2, 3$) and connections are established following the random connection model. We obtained a lower bound for the critical density using a Galton-Watson branching process and an upper bound by coupling the problem to that of site percolation on a lattice. For both the upper and the lower bound, results for the unit disk connection model and the log-normal connection model are provided as special cases of the random connection model.

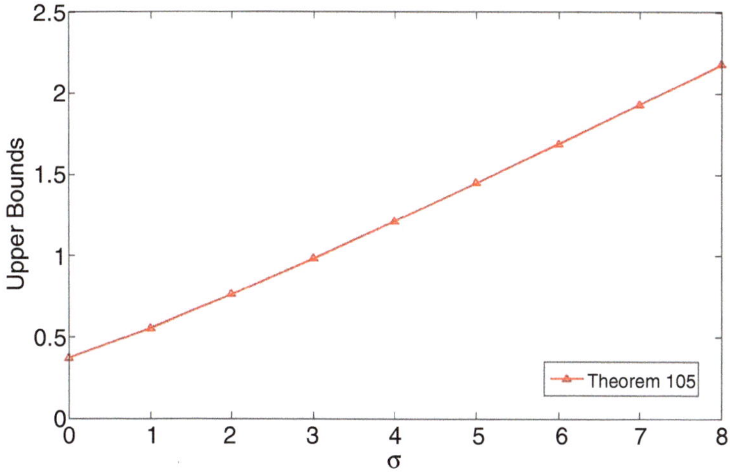

Fig. 5.11 Upper bounds for λ_c in $\Re^3$ under the log-normal connection model with different shadowing variances and $r = 2$, $\alpha = 2$. Note that the log-normal connection model reduces to the unit disk connection model when $\sigma = 0$

Some bounds for the critical density have been given in the literature and mostly for the unit disk connection model [88, 118, 143, 167] in $\Re^2$. Since these results in $\Re^2$ were obtained with different transmission ranges, we rescale the results using the scaling technique to a common transmission range of 2, and report them as follows: a well-known set of analytical lower and upper bounds for the critical density was given by Meester and Roy [143], i.e., the critical density should lie between 0.174 and 0.843. Philips et al. [167] obtained the same lower bound as that in [143] but a slightly tighter upper bound (0.8376). On the other hand, Gu and Hong [88] reported the same upper bound as in [143] but a tighter lower bound (0.2553). Kong and Yeh [118] obtained another lower bound (0.1925) for the critical density. No upper bound was obtained in [118]. Note that a lower bound in $\Re^2$ under the unit disk model is also reported in [118].

Limited work was reported under connection models other than the unit disk connection model. In [119], Kong and Yeh extended the unit disk connection model to a unit disk connection model with unreliable links. In the new model, two nodes are no longer directly connected with probability 1 but with some lesser probability, provided their Euclidean distance is within the transmission range. The lower bound obtained is comparable with their earlier result obtained under the unit disk connection model [118]. In $\Re^2$ and under the log-normal connection model, Li and Yang [131] obtained analytically an upper bound for the critical density. Note that the connection models considered in [119] and [131] are special cases of the random connection model considered in this chapter.

Chapter 6
Phase Transitions in Large Networks

Abstract In this chapter, we study the phase transition behavior of k-connectivity ($k = 1, 2, \ldots$) in large wireless networks where a total of n nodes are randomly and independently distributed following a uniform distribution in a unit cube $[0, 1]^d$ ($d = 1, 2, 3$), and each node has an identical transmission range $r(n)$. It has been shown that the phase transition of k-connectivity becomes sharper as the total number of nodes n increases. In this chapter, we investigate how fast such phase transition happens, and derive a general formula for the phase transition width of k-connectivity for large enough n and for any fixed positive integer k in d-dimensional space by resorting to a Poisson approximation for the node placement. The results in this chapter are important for understanding the phase transition phenomenon in network connectivity.

In this chapter, we study the phase transition phenomenon in connectivity of large wireless networks.

We illustrate the concept of phase transition first using an example of rainfall. Before the rain starts, the ground is completely dry. When a raindrop falls, the ground soaks up the water and a wet patch is formed around the point where the raindrop falls. When the first raindrops fall, we see small wet regions inside a large dry ground. As the rain continues, more raindrops fall and the wet regions grow. This process continues until at a certain point so many raindrops have reached the ground to the extent that the ground suddenly changes from dry ground with a few wet regions into wet ground with a few dry islands. Similarly, consider a second example of an infinite plane covered by disks. The infinite plane is initially empty. Then, we throw in disks where the centers of the disks are Poisson point process with a given density and the radius of the disks are a constant resembling that of the unit disk connection model. A point in the infinite plane is *occupied* if it is inside one of the disks, otherwise it is empty. When the density of the disks is low, the plane is largely empty with a few disconnected occupied regions. As the density increases, more and more regions of the plane become occupied until when the density reaches a certain threshold, the plane suddenly transforms from an empty plane with a few occupied regions into an occupied plane with a few empty regions. Informally, a *phase transition* is defined as a phenomenon where a small change in the local parameters of a system results in an abrupt change in the macroscopic

G. Mao, *Connectivity of Communication Networks*,
DOI 10.1007/978-3-319-52989-9_6

behavior of the system. Phase transitions are common in nature and used today in many technologies, for example, the transition of water from gas to liquid and then to ice as the temperature drops.

In an infinite network, such phase transition can occur abruptly while in a finite network, the phase transition is more gradual. Obviously, one also expects the behavior that can be observed at the infinite scale to be a good indication of what happens when we consider finite models that grow larger and larger in size. We shall see that this is indeed the case when considering scaling properties of finite networks.

In this chapter, we are particularly interested in studying the phase transition that occurs when a large wireless network transits from an almost disconnected network to an almost connected network, to be defined more precisely shortly later.

Next we recall some definitions and concepts on network connectivity that will be used later in the chapter. A network is said to be *k-vertex connected* (*k-connected* for simplicity) if and only if for any pair of two nodes there exist at least k mutually independent paths connecting them, i.e., these paths do not share a common node except for the beginning and the end of the path (refer to Fig. 6.1 for an illustration). Equivalently, a connected network is k-connected if and only if there is no set of $k-1$ nodes whose removal will make the network trivial or disconnected. In other words, a k-connected network is able to sustain the failure of $k-1$ nodes. A network is said to be *k-edge connected* if and only if for any pair of two nodes there exist at least k mutually independent edge disjoint paths connecting them. Throughout this chapter, we use the term k-connectivity as shorthand for k-vertex connectivity.

Assuming a unit disk connection model and for a finite network of n nodes randomly and uniformly distributed in a finite region, it has been shown that there exists a threshold of the transmission range above which the network is k-connected with a high probability; and there also exists a threshold of transmission range below which the network is k-connected with a low probability, i.e., the network is more likely not k-connected. The *difference* between the two thresholds defines the so-called *phase transition width*. Intuitively, the phase transition width gives an indication on how easy or difficult it is to transform a network that is not k-connected into a k-connected network. It has been shown that the phase transition width becomes sharper as the total number of nodes n increases [19, 97, 121, 122].

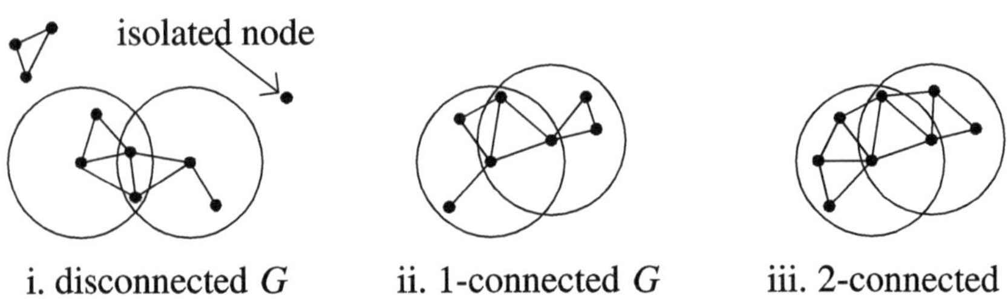

Fig. 6.1 An illustration of a k-connected network. Note that 1-connectivity has the same meaning as connectivity

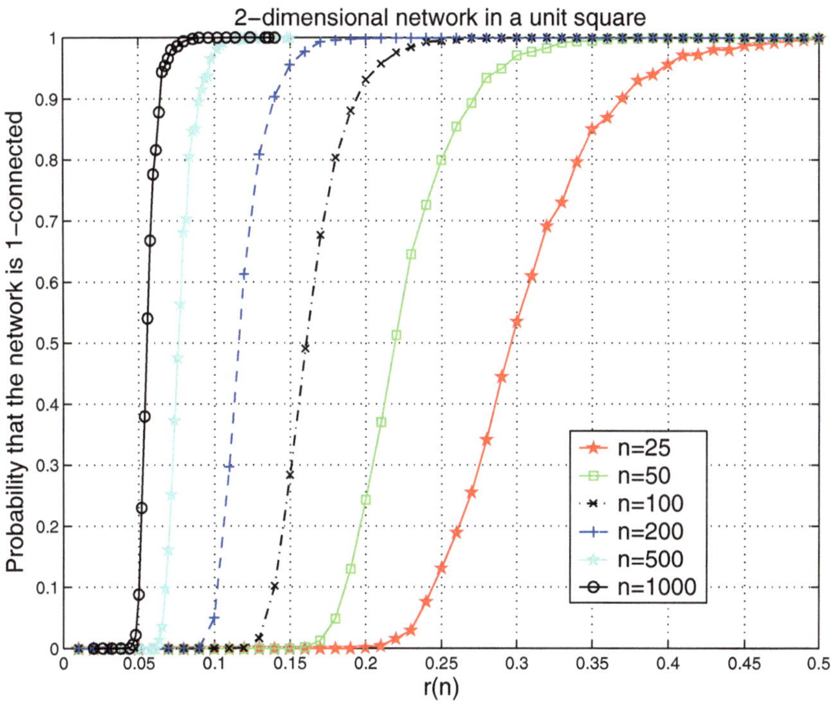

Fig. 6.2 An illustration of the phase transition behavior of 1-connectivity in 2-dimensional networks through simulations. Consider a network with a total of n nodes randomly and uniformly distributed on a unit square and a pair of nodes in the network are directly connected following a unit disk connection model with a transmission range of $r(n)$. The probability that the network is 1-connected transits from nearly zero to nearly one over a small range of values of the transmission range $r(n)$, and the transition becomes steeper as the number of nodes n increases

When n approaches infinity, the phase transition width approaches zero and the two aforementioned thresholds will converge to the same value. As an example, Fig. 6.2 shows the phase transition behavior of 1-connectivity in two-dimensional networks through simulations. As shown in the figure, when the total number of nodes n is large (e.g., $n = 1000$), it takes a small increase in the transmission range to transform a disconnected network into a connected network. A good understanding of such a phase transition phenomenon is of practical importance for the design of large wireless networks and is also of theoretical significance in the study of network connectivity. For higher dimensional networks (i.e., $d = 2, 3$), all the analytical results hitherto derived on the critical transmission range, at which the network is connected or k-connected, are valid for large networks only [92, 173, 185]. For an infinite network with an infinite number of nodes, the phase transition from not being k-connected to being k-connected occurs at a precise transmission range, termed the *critical transmission range*, and the phase transition width is zero. However, in practice, the total number of nodes in the network is *finite*.

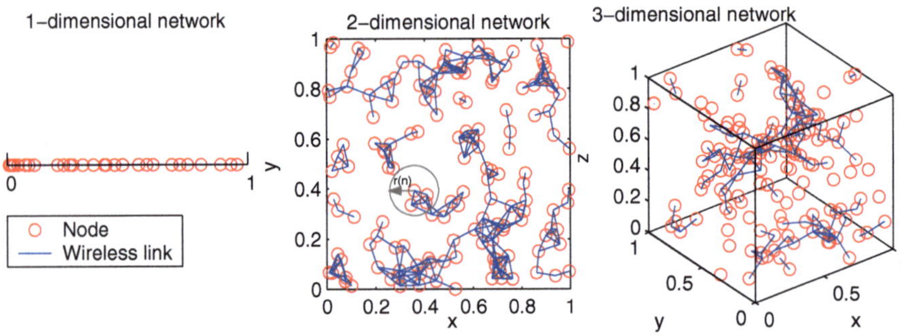

Fig. 6.3 An illustration of a network in d-dimensional space ($d = 1, 2, 3$). A total of n nodes are randomly and independently distributed in a unit cube $[0, 1]^d$ following a uniform distribution. Each node has an identical transmission range $r(n)$, and any two nodes can communicate with each other if and only if their Euclidean distance is at most $r(n)$

So for large but not infinite networks it is crucial to study not just the critical transmission range itself, but the width of the phase transition region, since the phase transition from not being k-connected to being k-connected no longer occurs abruptly.

The central aim of this chapter is to investigate analytically how quickly the phase transition of k-connectivity happens in a large wireless network. Specifically, we consider that the network with a total of n nodes randomly and independently distributed in a unit cube $[0, 1]^d$ ($d = 1, 2, 3$) following a uniform distribution and connections are established following the unit disk connection model with a transmission range $r(n)$. Denote the network by $G(n, r(n))$. As an example, Fig. 6.3 gives an illustration of such network model for different values of d. In addition, the results obtained using this model can also be extended to other network models, e.g., nodes distributed following a Poisson point process. We refer readers to the analysis in Chap. 4 for relevant techniques on converting the results obtained assuming a uniform node distribution to that assuming a Poisson node distribution and the converse. To quantify how fast the phase transition of k-connectivity occurs, we use a metric called the *phase transition width*, which is also referred to as the *threshold width* in the literature. Let $P_k(n, r(n))$ denote the probability that an instance of a randomly generated network $G(n, r(n))$ is k-connected. The mapping $r(n) \to P_k(n, r(n))$ is strictly monotonically increasing with $0 < P_k(n, r(n)) < 1$ in some finite interval of $r(n)$, and $P_k(n, r(n)) = 0$ or 1 outside the interval [97].

A *monotone graph property* is usually defined as follows:

Definition 106 A graph property Λ is increasing if and only if

$$G \in \Lambda \Rightarrow (\forall G')[(V(G') = V(G) \text{ and } E(G) \subseteq E(G')) \Rightarrow G' \in \Lambda]$$

where $V(G)$ and $E(G)$ denote, respectively, the vertex set and the edge set of G. A graph property Λ is said to be monotone if either Λ or its complement Λ^c is increasing.

The *minimum node degree* of a graph G is defined as

$$deg_{min}(G) \triangleq \min_{u \in V(G)} \{deg(u)\}$$

where $deg(u)$ is the degree of a node $u \in V(G)$, and the *average node degree* of a graph G is

$$C(G) \triangleq \frac{1}{n} \sum_{u \in V(G)} deg(u)$$

Let α denote a positive real number between 0 and 1. Then, define

$$r_k(n, \alpha) \triangleq \inf(r > 0 : P_k(n, r) \geq \alpha), \ \alpha \in (0, 1) \tag{6.0.1}$$

By the definition of $r_k(n, \alpha)$ and the strict monotonicity of $r(n) \to P_k(n, r(n))$, it readily follows that $P_k(n, r_k(n, \alpha)) = \alpha$. The *phase transition width* over the probability interval $[\alpha, 1 - \alpha]$ of k-connectivity is then defined as

$$\delta_k(n, \alpha) \triangleq r_k(n, 1 - \alpha) - r_k(n, \alpha), \ \alpha \in (0, \frac{1}{2}) \tag{6.0.2}$$

In addition to the above definition of $\delta_k(n, \alpha)$, we shall define the phase transition width $\delta'_k(n, \alpha)$ measured using the average node degree as follows. The reason for defining the phase transition width in terms of the average node degree is that in random geometric graphs, connectivity is often directly related to the average node degree, which combines the effects of the node density and the transmission range. Let $C(n)$ denote the average node degree of $G(n, r(n))$, and let $P_k(n, C(n))$ denote the probability that $G(n, r(n))$ is k-connected. Define

$$C_k(n, \alpha) \triangleq \inf(C > 0 : P_k(n, C) \geq \alpha), \ \alpha \in (0, 1) \tag{6.0.3}$$

The new phase transition width $\delta'_k(n, \alpha)$ of k-connectivity in terms of the average node degree is

$$\delta'_k(n, \alpha) \triangleq C_k(n, 1 - \alpha) - C_k(n, \alpha), \ \alpha \in (0, \frac{1}{2}) \tag{6.0.4}$$

Henceforth, unless otherwise indicated, the short term phase transition width will be used with α being simply understood. Note that the definition of the phase transition width given by (6.0.2) is closely related to the so-called *finite size scaling* in the physics of percolation and related phenomenon [37, 120].

In this chapter, we derive a general analytical formula for the phase transition width $\delta_k(n,\alpha)$ of k-connectivity for large n and for any fixed positive integer $k > 0$ in d-dimensional space ($d = 1, 2, 3$). Based on the result, we then compare the phase transition width for different values of d by fixing k, n and α, which shows that the phase transition width $\delta_k(n,\alpha)$ is larger for higher dimensional networks than that for lower dimensional networks. Similarly we compare the phase transition width for different values of k by fixing d, n and α, which shows that for large n the phase transition width of k-connectivity is approximately the same as the phase transition width of $(k+1)$-connectivity. Alternatively, one may also investigate the phase transition width in terms of the average node degree. This chapter also provides an analytical result for the phase transition width measured in the average node degree. Surprisingly, the newly defined phase transition width turns out to be independent of n and k for large n. We also conduct corresponding simulations to verify our theoretical analysis for large networks. The simulation results show that to achieve a good accuracy, n should be larger than 200 when $k = 1$ and $d = 1$; and n should be larger than 600 when $k \leq 3$ and $d = 2, 3$.

6.1 Phase Transition Width for Network with Different Orders of Connectivity

In this section, we present the main results on the phase transition width $\delta_k(n,\alpha)$ of k-connectivity ($k > 0$) in d-dimensional space ($d = 1, 2, 3$). In our analysis, we ignore the boundary effect for $d = 2, 3$. We refer interested readers to Chaps. 2 and 3 for techniques on handling the boundary effect and for discussions on the impact of boundary effect on network connectivity.

Our main result in this section on $r_k(n,\alpha)$ for large n is given in the following theorem:

Theorem 107 *For a random geometric graph $G(n, r(n))$ in $\Re^d$ ($d = 1, 2, 3$), a fixed positive integer $k > 0$ and a positive real number $\alpha \in (0, 1)$, let $r_k(n,\alpha)$ denote the smallest transmission range at which the probability that $G(n, r(n))$ is k-connected is at least α. Ignore the boundary effect except when $d = 1$. Then, for large n, $r_k(n,\alpha)$ is given by*

$$r_k(n,\alpha) = \left(\frac{F(n,k)}{\pi_d n}\right)^{\frac{1}{d}} - \frac{\log\left(\log\left(\frac{1}{\alpha}\right)\right)(1+o(1))}{d(\pi_d n)^{\frac{1}{d}}\,(F(n,k))^{\frac{d-1}{d}}} \tag{6.1.1}$$

where

$$F(n,k) \triangleq \log n + (k-1)\log(\log n) - \log(k-1)! \tag{6.1.2}$$

and π_d ($d = 1, 2, 3$) is defined as $\pi_1 = 1$, $\pi_2 = \pi$ and $\pi_3 = \frac{4}{3}\pi$.

In the rest of this section, we shall prove Theorem 107. We first present the following Lemma which will be used later in the proof of Theorem 107.

Lemma 108 *Consider a random geometric graph $G(n, r(n))$ in $\Re^d$ ($d = 2, 3$), a fixed positive integer $k > 0$ and a real number $\omega \in \Re$. Let $\xi(k, n, r(n))$ be the expected number of nodes with degree k. If the boundary effect is ignored, and $r(n)$ is given by*

$$r(n) = r_n(\omega) = \left(\frac{F(n,k) + \omega}{\pi_d n}\right)^{\frac{1}{d}} \tag{6.1.3}$$

where $F(n, k)$ is as defined in (6.1.2), the following holds:

$$\xi(k-1, n, r_n(\omega)) = e^{-\omega},\ k = 1 \tag{6.1.4}$$

$$\lim_{n\to\infty} \xi(k-1, n, r_n(\omega)) = e^{-\omega},\ k > 1 \tag{6.1.5}$$

Proof As shown in [19, 23, 96, 165, 173] (see also Sect. 1.2), for a set of n nodes, where each node is randomly, independently, and uniformly placed in a finite region of volume V in $\Re^d$, the limiting case obtained by letting $n \to \infty$ and $V \to \infty$ while keeping n/V constant can be regarded as defining a homogeneous Poisson point process of density $\rho = n/V$. For large n and large V, i.e., $n \gg 1$ and $V \gg \pi_d r^d(n)$, a homogeneous Poisson point process of density $\rho = n/V$ is a close approximation for the corresponding uniform distribution. Furthermore, using the scaling and coupling technique, any realization of $G(n, r(n))$ in a *unit* cube in $\Re^d$ coincides with another realization of $G(n, \sqrt[d]{V}r(n))$ placed in a cube of volume V in $\Re^d$. Hence, throughout this chapter, we focus on $G(n, r(n))$ distributed in a unit cube in $\Re^d$, and we assume $n \gg 1$ and $1 \gg \pi_d r^d(n)$ so that a homogeneous Poisson point process of density $\rho = n/1 = n$ can be used to approximate the uniform distribution for the node spatial distribution [164, 165]. For a more formal treatment of the conversion between the results obtained assuming a uniform distribution to that assuming a Poisson distribution and the converse, we refer readers to the Poissonization and de-Poissonization techniques that have been used in Chap. 4 and the discussions in Chap. 1.

Using the Poisson approximation and ignoring the boundary effect, the probability that a particular node i ($i = 1, 2, \ldots, n$) has m ($m \geq 0$) neighbors, denoted as $p_i(m)$, is given by

$$p_i(m) = \frac{\left(n\pi_d r^d(n)\right)^m}{m!} e^{-n\pi_d r^d(n)},\quad d = 2, 3 \tag{6.1.6}$$

Therefore, by using the Palm Theorem [165] which captures a form of spatial ergodicity property relating the probabilities that a given node has a certain degree with the expected number of nodes in a network that have a certain degree, and ignoring the boundary effect, the expected number of nodes with degree $(k - 1)$,

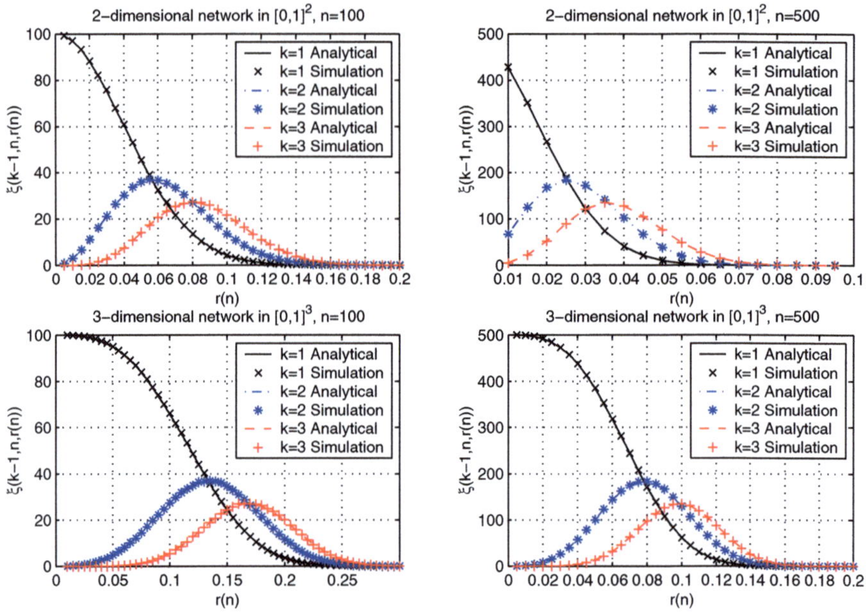

Fig. 6.4 An illustration of the expected number of nodes with degree k ($k = 1, 2, 3$). For the analytical results, the boundary effect is ignored; for the simulation results, the boundary effect is eliminated by using the toroidal distance metric

denoted by $\xi(k-1, n, r(n))$, is given by

$$\xi\,(k-1, n, r(n)) = n \cdot p_i(k-1) = n \cdot \frac{(n\pi_d r^d(n))^{k-1}}{(k-1)!} e^{-n\pi_d r^d(n)} \tag{6.1.7}$$

We also conduct simulations to establish the accuracy of the analytical formula given by (6.1.7). The analytical formula given by (6.1.7) is derived by approximating the uniform point process with a Poisson point process. In Fig. 6.4, a comparison between the analytical results obtained using the Poisson point process approximation and the simulation results where nodes are distributed following a uniform point process verifies the accuracy of the Poisson approximation.

For $k = 1$, substituting (6.1.3) into (6.1.7), we have

$$\begin{aligned}
\xi(0, n, r_n(\omega)) &= n \times \frac{(n\pi_d r^d(n))^0}{0!} e^{-n\pi_d r^d(n)} \\
&= n \times \exp\left(-n \cdot \pi_d \cdot \frac{\log n + \omega}{n\pi_d}\right) \\
&= e^{-\omega}
\end{aligned}$$

which proves (6.1.4).

For $k > 1$, substituting (6.1.3) into (6.1.7), we have

$$\begin{aligned}
&\lim_{n\to\infty} \xi(k-1, n, r_n(\omega)) \\
&= \lim_{n\to\infty} \left(n \cdot \frac{(F(n,k)+\omega)^{k-1}}{(k-1)!} \cdot \frac{1}{n} \cdot \frac{(k-1)!}{(\log n)^{k-1}} \cdot e^{-\omega} \right) \\
&= e^{-\omega} \lim_{n\to\infty} \left(1 + \frac{(k-1)\log(\log n)}{\log n} - \frac{\log(k-1)!}{\log n} + \frac{\omega}{\log n} \right)^{k-1} \\
&= e^{-\omega}
\end{aligned}$$

which proves (6.1.5). □

A more general form of Lemma 108 considering the random connection model has also been given in Chap. 2.

Now, we are able to prove Theorem 107. We first prove the result for $d = 2, 3$ based on Theorem 1.1 and Theorem 2.2 in [164] and Lemma 108. Since the two theorems in [164] are not valid for $d = 1$, we shall then prove the result for $d = 1$ separately based on Theorem 15 given in [79].

6.1.1 Case When $d = 2,\ 3$

For $d = 2,\ 3$, we first introduce two theorems by Penrose [164] that are important for our proof. Let $\gamma_k(n)$ (respectively, $\tau_k(n)$) denote the minimum transmission range at which a random geometric graph $G(n, r(n))$ is k-connected (respectively, has minimum node degree k). The following two theorems have been established in [164]:

Proposition 109 *Consider a random geometric graph $G(n, r(n))$ in $\Re^d$ ($d \geq 2$). Given any integer $k > 0$,*

$$\lim_{n\to\infty} \Pr(\gamma_k(n) = \tau_k(n)) = 1 \tag{6.1.8}$$

Theorem 110 *Consider a random geometric graph $G(n, r(n))$ in $\Re^d$ ($d \geq 1$). Let $\omega \in \Re$. Given any integer $k > 0$ and $(r_n)_{n\geq 1}$ satisfying the following condition*

$$\lim_{n\to\infty} \xi(k-1, n, r_n) = e^{-\omega} \tag{6.1.9}$$

then it follows that

$$\lim_{n\to\infty} \Pr(\tau_k(n) \leq r_n) = \exp(-e^{-\omega}) \tag{6.1.10}$$

Remark 111 Theorem 109 implies that for very large value of n, k-connectivity is predicted by the minimum node degree. Theorem 110 shows that there is a relation between the proportion of nodes with degree $(k-1)$ and the probability that the network has a minimum node degree k. It is also important to notice that Theorem 109 is valid for $d \geq 2$ and Theorem 110 is valid for $d \geq 1$.

For any $\omega \in \Re$ and any positive integer $k > 0$, Theorem 109 and Theorem 110 immediately yield

$$\begin{aligned}
&\lim_{n\to\infty} \Pr\left(\gamma_k(n) \leq r_n(\omega)\right) \\
&\quad = \lim_{n\to\infty} \Pr\left(\tau_k(n) \leq r_n(\omega)\right) \\
&\quad = \exp(-e^{-\omega})
\end{aligned} \tag{6.1.11}$$

Hence, (6.1.11) and Lemma 108 yield

$$\lim_{n\to\infty} P_k\left(n, r_n(\omega)\right) = \exp(-e^{-\omega}) \tag{6.1.12}$$

which plays a key role in the proof of Theorem 107.

For each $x \in \Re$, define the $[0, 1]$-valued sequence $\{\sigma_n(x), n = 1, 2, 3 \ldots\}$ by

$$\sigma_n(x) = \min\left(1, \left(\frac{F(n,k)+x}{\pi_d n}\right)_+^{\frac{1}{d}}\right), \quad n = 1, 2, \ldots \tag{6.1.13}$$

Because for any fixed integer $k > 0$ and $x \in \Re$, $\frac{F(n,k)}{n} \to 0$ as $n \to \infty$, there exists a finite integer $N(k, x)$ such that

$$0 < \left(\frac{F(n,k)+x}{\pi_d n}\right)^{\frac{1}{d}} < 1, \quad \forall n > N(k,x)$$

Hence, we have

$$\sigma_n(x) = \left(\frac{F(n,k)+x}{\pi_d n}\right)^{\frac{1}{d}}, \quad \forall n > N(k,x) \tag{6.1.14}$$

Therefore, from Lemma 108 and (6.1.12), we have

$$\lim_{n\to\infty} P_k(n, \sigma_n(x)) = \exp(-e^{-x}). \tag{6.1.15}$$

Now fixing x in $\Re$, from (6.1.15), we can obtain that for each $\varepsilon > 0$, there exists a finite integer $N(\varepsilon, k, x)$ such that

$$|P_k(n, \sigma_n(x)) - \exp(-e^{-x})| < \varepsilon, \quad \forall n > N(\varepsilon, k, x) \tag{6.1.16}$$

It can be easily established that the mapping $\Re \to \Re^+ : x \to \exp(-e^{-x})$ is strictly monotonically increasing and continuous with $\lim_{x\to-\infty}\exp(-e^{-x}) = 0$ and $\lim_{x\to\infty}\exp(-e^{-x}) = 1$. Therefore, for each $\alpha \in (0,1)$, there exists a unique value of x in $\Re$, denoted as x_α, such that $\exp(-e^{-x_\alpha}) = \alpha$. In fact, from the equality $\exp(-e^{-x_\alpha}) = \alpha$, we have

$$x_\alpha = -\log(-\log\alpha) \tag{6.1.17}$$

Hence, fixing x in $\Re$ is equivalent to fixing α in $(0,1)$. Now fix α in the interval $(0,1)$, and let ε be sufficiently small such that $0 < 2\varepsilon < \alpha$ and $\alpha + 2\varepsilon < 1$. Then, applying (6.1.16) with $x = x_{\alpha+\varepsilon}$ and $x = x_{\alpha-\varepsilon}$, respectively, we have

$$|P_k(n, \sigma_n(x_{\alpha+\varepsilon})) - \exp(-e^{-x_{\alpha+\varepsilon}})| < \varepsilon, \ \forall n > N(\varepsilon, k, x_{\alpha+\varepsilon}) \tag{6.1.18}$$

and

$$|P_k(n, \sigma_n(x_{\alpha-\varepsilon})) - \exp(-e^{-x_{\alpha-\varepsilon}})| < \varepsilon, \ \forall n > N(\varepsilon, k, x_{\alpha-\varepsilon}) \tag{6.1.19}$$

We always assume that n is sufficiently large when necessary. In the rest of this subsection, we assume that $n > N(\varepsilon, k, \alpha)$ with

$$N(\varepsilon, k, \alpha) = \max\{N(k, x_\alpha),\ N(\varepsilon, k, x_{\alpha+\varepsilon}),\ N(\varepsilon, k, x_{\alpha-\varepsilon})\}$$

where $N(k, x_\alpha)$ represents the finite integer above which (6.1.14) holds.

Since $\exp(-e^{-x_{\alpha\pm\varepsilon}}) = \alpha \pm \varepsilon$, it can be readily obtained from (6.1.18) and (6.1.19) that

$$\alpha < P_k(n, \sigma_n(x_{\alpha+\varepsilon})) < \alpha + 2\varepsilon$$

and

$$\alpha - 2\varepsilon < P_k(n, \sigma_n(x_{\alpha-\varepsilon})) < \alpha$$

According to the definition of $r_k(n,\alpha)$, we have $P_k(n, r_k(n,\alpha)) = \alpha$. Hence, from the last two inequalities, it follows that

$$P_k(n, \sigma_n(x_{\alpha-\varepsilon})) < P_k(n, r_k(n,\alpha)) < P_k(n, \sigma_n(x_{\alpha+\varepsilon}))$$

Because of the strict monotonicity of the mapping $r(n) \to P_k(n, r(n))$, we have

$$\sigma_n(x_{\alpha-\varepsilon}) < r_k(n,\alpha) < \sigma_n(x_{\alpha+\varepsilon}) \tag{6.1.20}$$

Define $\eta(n,\alpha) \triangleq r_k(n,\alpha) - \sigma_n(x_\alpha)$, then it can be obtained from (6.1.20) that

$$\sigma_n(x_{\alpha-\varepsilon}) - \sigma_n(x_\alpha) < \eta(n,\alpha) < \sigma_n(x_{\alpha+\varepsilon}) - \sigma_n(x_\alpha) \tag{6.1.21}$$

For any fixed $k > 0$ and $x \in \Re$, it is true that

$$\begin{aligned}\lim_{n\to\infty} \frac{x}{F(n,k)} &= \lim_{n\to\infty} \frac{x}{\log n + (k-1)\log(\log n) - \log(k-1)!} \\ &= 0\end{aligned}$$

Hence, from (6.1.14), we have

$$\begin{aligned}\sigma_n(x) &= \left(\frac{F(n,k)+x}{\pi_d n}\right)^{\frac{1}{d}} \\ &= \left(\frac{F(n,k)}{\pi_d n}\right)^{\frac{1}{d}} \left(1 + \frac{x}{F(n,k)}\right)^{\frac{1}{d}} \\ &= \left(\frac{F(n,k)}{\pi_d n}\right)^{\frac{1}{d}} \left(1 + \frac{1}{d}\frac{x(1+o(1))}{F(n,k)}\right), \text{ as } n \to \infty\end{aligned}$$

Therefore, for any fixed $k > 0$ and $\alpha \in (0, 1)$, we have

$$\begin{aligned}&\sigma_n(x_{\alpha\pm\varepsilon}) - \sigma_n(x_\alpha) \\ &= \left(\frac{F(n,k)}{\pi_d n}\right)^{\frac{1}{d}} \left(1 + \frac{1}{d}\frac{x_{\alpha\pm\varepsilon}(1+o(1))}{F(n,k)}\right) \\ &\quad - \left(\frac{F(n,k)}{\pi_d n}\right)^{\frac{1}{d}} \left(1 + \frac{1}{d}\frac{x_\alpha(1+o(1))}{F(n,k)}\right) \\ &= \left(\frac{F(n,k)}{\pi_d n}\right)^{\frac{1}{d}} \frac{x_{\alpha\pm\varepsilon} - x_\alpha}{F(n,k)d}(1+o(1)), \text{ as } n \to \infty \end{aligned} \tag{6.1.22}$$

Because (6.1.21) holds for all $n > N(\varepsilon, k, \alpha)$, it must be valid when $n \to \infty$ as well. And as $n \to \infty$, the small order part $o(1)$ in (6.1.22) goes to zero. Hence, from (6.1.21) and (6.1.22), we have

$$\begin{aligned}x_{\alpha-\varepsilon} - x_\alpha &\le \liminf_{n\to\infty} \left(\frac{F(n,k)d(1+o(1))}{\left(\frac{F(n,k)}{\pi_d n}\right)^{\frac{1}{d}}}\eta(n,\alpha)\right) \\ &= \liminf_{n\to\infty} \left(\frac{F(n,k)d}{\left(\frac{F(n,k)}{\pi_d n}\right)^{\frac{1}{d}}}\eta(n,\alpha)\right)\end{aligned}$$

and

$$x_{\alpha+\varepsilon} - x_\alpha \geq \limsup_{n\to\infty} \left(\frac{F(n,k)d(1+o(1))}{\left(\frac{F(n,k)}{\pi_d n}\right)^{\frac{1}{d}}} \eta(n,\alpha) \right)$$

$$= \limsup_{n\to\infty} \left(\frac{F(n,k)d}{\left(\frac{F(n,k)}{\pi_d n}\right)^{\frac{1}{d}}} \eta(n,\alpha) \right)$$

Because ε can be chosen to be arbitrarily small, and as stated earlier $x_\alpha = -\log(-\log\alpha)$ is a continuous and strictly monotonically increasing function of α for $\alpha \in (0, 1)$, it can be shown that

$$\lim_{\varepsilon\downarrow 0}(x_{\alpha-\varepsilon} - x_\alpha) = \lim_{\varepsilon\downarrow 0}(x_{\alpha+\varepsilon} - x_\alpha) = 0$$

Hence, we have

$$\liminf_{n\to\infty} \left(\frac{F(n,k)d}{\left(\frac{F(n,k)}{\pi_d n}\right)^{\frac{1}{d}}} \eta(n,\alpha) \right)$$

$$= \limsup_{n\to\infty} \left(\frac{F(n,k)d}{\left(\frac{F(n,k)}{\pi_d n}\right)^{\frac{1}{d}}} \eta(n,\alpha) \right)$$

$$= 0$$

Thus, given that

$$\lim_{n\to\infty} \left(\frac{F(n,k)d}{\left(\frac{F(n,k)}{\pi_d n}\right)^{\frac{1}{d}}} \eta(n,\alpha) \right) = 0$$

holds, it must be true that

$$\eta(n,\alpha) = o\left(\frac{\left(\frac{F(n,k)}{\pi_d n}\right)^{\frac{1}{d}}}{F(n,k)d} \right) = o\left(\frac{1}{d\,(\pi_d n)^{\frac{1}{d}}\,(F(n,k))^{\frac{d-1}{d}}} \right) \tag{6.1.23}$$

Hence, we have

$$
\begin{aligned}
r_k(n,\alpha) &= \sigma_n(x_\alpha) + \eta(n,\alpha)\\
&= \left(\frac{F(n,k)+x_\alpha}{\pi_d n}\right)^{\frac{1}{d}} + o\left(\frac{1}{d\,(\pi_d n)^{\frac{1}{d}}\,(F(n,k))^{\frac{d-1}{d}}}\right)\\
&= \left(\frac{F(n,k)}{\pi_d n}\right)^{\frac{1}{d}}\left(1+\frac{x_\alpha(1+o(1))}{F(n,k)d}\right) + o\left(\frac{1}{d\,(\pi_d n)^{\frac{1}{d}}\,(F(n,k))^{\frac{d-1}{d}}}\right)\\
&= \left(\frac{F(n,k)}{\pi_d n}\right)^{\frac{1}{d}} - \frac{\log\left(\log\left(\frac{1}{\alpha}\right)\right)(1+o(1))}{d(\pi_d n)^{\frac{1}{d}}\,(F(n,k))^{\frac{d-1}{d}}}
\end{aligned}
$$

The proof of Theorem 107 for $d = 2, 3$ is complete.

6.1.2 Case When $d = 1$

For $d = 1$, when $k = 1$, (6.1.1) readily becomes the result already established by Han et al. (see Eq. 4 in [97]), therefore, the result given in Theorem 107 is true for $d = 1$ and $k = 1$.

When $k > 1$, we shall prove the result based on Theorem 15 given in [79]. It is noted in [79, Theorem 15] that given a random geometric graph $G(n, r(n))$ in 1-dimensional space and $\omega \in \Re$, for any positive integer $k > 1$, if $r(n)$ is given by

$$
r(n) = \frac{1}{n}(\log n + (k-1)\log(\log n) - \log(k-1)! + \omega)
$$

then

$$
\lim_{n\to\infty} P_k(n, r(n)) = \exp(-e^{-\omega})
$$

Based on the critical transmission range given in [79] and using the same technique as that described in Sect. 6.1.1, we can obtain

$$
r_k(n,\alpha) = \frac{F(n,k)}{n} - \frac{\log\left(\log\left(\frac{1}{\alpha}\right)\right)(1+o(1))}{n},\ d = 1
$$

which agrees with the result given in Theorem 107 for $d = 1$ and $k > 1$. The proof of Theorem 107 for $d = 1$ is thus complete. It is important to notice that the boundary effect affects neither the derivation of Han et al.'s result in [97] nor the derivation of Theorem 15 in [79]. Hence, our result in this chapter is not affected by the boundary effect when $d = 1$.

Combining Sects. 6.1.1 and 6.1.2, we have finally proved Theorem 107.

Remark 112 Equation (6.1.1) given in Theorem 107 confirms Theorem 11 in [8]. It was noted in Theorem 11 of [8] that given a 2-connected network in $\Re^2$, if we double the transmission range, then the resulting network becomes globally rigid, where global rigidity is a graph property that has been widely investigated for network localization and for control [141]. Equation (6.1.1) yields $2r_2(n,\alpha) > r_6(n,\alpha)$ for all large n. Given that 6-connectivity is a sufficient condition for global rigidity in $\Re^2$ [108], we can obtain that the network with a transmission range $2r_2(n,\alpha)$ is globally rigid with high probability. In addition, the transmission range given by (6.1.1) has a similar asymptotic behavior compared with the one given in [181] as n goes to infinity.

Using Theorem 107, the desired result on the phase transition width $\delta_k(n,\alpha)$ for large n is given in the following corollary, which forms another main result of this subsection.

Corollary 113 *Consider a random geometric graph $G(n, r(n))$ in $\Re^d$ $(d = 1, 2, 3)$, a fixed positive integer $k > 0$ and a positive real number $\alpha \in (0, \frac{1}{2})$. Ignore the boundary effect except when $d = 1$. Then, for large n, the phase transition width of k-connectivity $\delta_k(n,\alpha)$ is given by*

$$\delta_k(n,\alpha) = \frac{\log\left(\frac{\log\alpha}{\log(1-\alpha)}\right)}{d\,(\pi_d n)^{\frac{1}{d}}\,(F(n,k))^{\frac{d-1}{d}}}(1+o(1)) \tag{6.1.24}$$

Proof Based on Theorem 107, for any fixed positive integer $k > 0$ and $d \in \{1, 2, 3\}$, the phase transition width $\delta_k(n,\alpha)$ for large n given in Corollary 113 can be readily derived as

$$\begin{aligned}
\delta_k(n,\alpha) &= r_k(n, 1-\alpha) - r_k(n,\alpha)\\
&= \left(\frac{F(n,k)}{\pi_d n}\right)^{\frac{1}{d}} - \frac{\log\left(\log\left(\frac{1}{1-\alpha}\right)\right)(1+o(1))}{d(\pi_d n)^{\frac{1}{d}}\,(F(n,k))^{\frac{d-1}{d}}}\\
&\quad - \left(\frac{F(n,k)}{\pi_d n}\right)^{\frac{1}{d}} + \frac{\log\left(\log\left(\frac{1}{\alpha}\right)\right)(1+o(1))}{d(\pi_d n)^{\frac{1}{d}}\,(F(n,k))^{\frac{d-1}{d}}}\\
&= \frac{\log\left(\frac{\log\alpha}{\log(1-\alpha)}\right)}{d(\pi_d n)^{\frac{1}{d}}\,(F(n,k))^{\frac{d-1}{d}}}(1+o(1))
\end{aligned}$$

The proof of Corollary 113 is complete. □

Remark 114 Comparing (6.4.2) on the phase transition width, which was first obtained in [97] and shown later in the chapter, and (6.1.24), we can see that when $k = 1$ and $d = 1, 2$, (6.1.24) readily reduces to (6.4.2) appearing in [97]. Furthermore, although we have derived the phase transition width of k-connectivity given by (6.1.24), this result holds only when n is sufficiently large. When n is a small number (especially when n is comparable with $k!$), this result does not hold any more. This observation is manifested in our proof of Theorem 107 and our later simulation results, which show significant discrepancy between analytical results and simulation results for small n. Also, the simulation results show that n should be larger than 200 when $d = 1$ and $k = 1$; and n should be larger than 600 if $k \leq 3$ and $d = 2,\ 3$.

6.2 A Discussion on Properties of the Phase Transition Width of a k-Connected Network

In this section, we continue to discuss properties of the phase transition width of a k-connected network using the results established in the previous section.

Based on Corollary 113, three further corollaries can be established on the properties of the phase transition width.

The following corollary compares the phase transition width for a k-connected network for different values of k.

Corollary 115 *Consider a random geometric graph $G(n, r(n))$ in $\Re^d$ $(d = 1, 2, 3)$, a fixed positive integer $k > 0$ and a positive real number $\alpha \in (0, \frac{1}{2})$. Ignore the boundary effect except when $d = 1$. Then, the phase transition width of k-connectivity $\delta_k(n, \alpha)$ and the phase transition width of $(k+1)$-connectivity $\delta_{k+1}(n, \alpha)$ satisfy*

$$\lim_{n\to\infty} \frac{\delta_{k+1}(n,\alpha)}{\delta_k(n,\alpha)} = 1 \tag{6.2.1}$$

Proof Based on Corollary 113, for any fixed positive integer $k > 0$ and $d \in \{1, 2, 3\}$, we have

$$\begin{aligned}
&\lim_{n\to\infty} \frac{\delta_{k+1}(n,\alpha)}{\delta_k(n,\alpha)} \\
&= \lim_{n\to\infty} \left(\frac{d(\pi_d n)^{\frac{1}{d}} (F(n,k))^{\frac{d-1}{d}} (1 + o(1))}{d(\pi_d n)^{\frac{1}{d}} (F(n,k+1))^{\frac{d-1}{d}} (1 + o(1))} \right) \\
&= \lim_{n\to\infty} \left(\frac{\log n + (k-1)\log(\log n) - \log(k-1)!}{\log n + k\log(\log n) - \log k!} \right)^{\frac{d-1}{d}} \\
&= \lim_{n\to\infty} \left(\frac{1 + \frac{(k-1)\log(\log n)}{\log n} - \frac{\log(k-1)!}{\log n}}{1 + \frac{k\log(\log n)}{\log n} - \frac{\log k!}{\log n}} \right)^{\frac{d-1}{d}} \\
&= 1.
\end{aligned}$$

The proof of Corollary 115 is complete. □

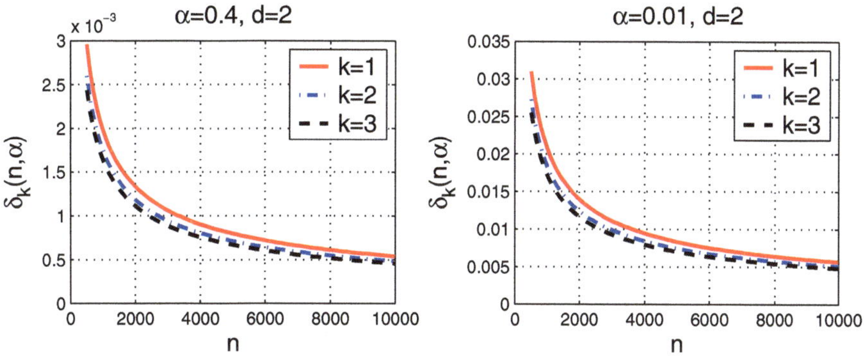

Fig. 6.5 Analytical results for the phase transition width of k-connectivity ($k = 1, 2, 3$) in two-dimensional space. The value of α is fixed in each scenario

As an example, Fig. 6.5 shows the analytical results for the phase transition width of k-connectivity ($k = 1, 2, 3$) in two-dimensional space, which are calculated from (6.1.24) (with $d = 2$ and $k = 1, 2, 3$). Parameter α is set to two typical values, i.e., 0.4 (close to 0.5) and 0.01 (close to 0). Parameter n is set to be a large value (i.e., $10,000$) so that the way in which $\delta_k(n,\alpha)$ varies with n can be observed. Note that in the calculation, we omit the small order part, i.e., $o(1)$ in $(1 + o(1))$ which becomes vanishingly small as n becomes large.

Remark 116 Corollary 115 means that for large enough n, $\delta_{k+1}(n,\alpha) \approx \delta_k(n,\alpha)$. In other words, the increase in the transmission range for making the probability that the network is k-connected increase from almost zero to almost one is approximately the same as that required for making the probability that the network is $(k+1)$-connected increase from almost zero to almost one.

Corollary 117 further compares the phase transition width for a k-connected network in space of different dimensions.

Corollary 117 *Consider a random geometric graph $G(n, r(n))$ in $\Re^d$ ($d = 1, 2, 3$), a fixed positive integer $k > 0$ and a positive real number $\alpha \in (0, \frac{1}{2})$. Ignore the boundary effect except when $d = 1$. Then, for large n, the phase transition width of k-connectivity $\delta_k(n,\alpha)$ in $(j+1)$-dimensional space is larger than that in j-dimensional space, where $j = 1, 2$.*

Proof Let $\delta_k^{(d)}(n,\alpha)$ denote the phase transition width of k-connectivity in d-dimensional space as we want to emphasize the dependence of $\delta_k(n,\alpha)$ on d. Let j be either 1 or 2. Then, based on Corollary 113, we have

$$\frac{\delta_k^{(j+1)}(n,\alpha)}{\delta_k^{(j)}(n,\alpha)}$$

$$= \frac{\dfrac{\log\left(\frac{\log\alpha}{\log(1-\alpha)}\right)(1+o(1))}{(j+1)(\pi_{j+1}n)^{\frac{1}{j+1}}(F(n,k))^{\frac{j}{j+1}}}}{\dfrac{\log\left(\frac{\log\alpha}{\log(1-\alpha)}\right)(1+o(1))}{j(\pi_j n)^{\frac{1}{j}}(F(n,k))^{\frac{j-1}{j}}}}$$

$$= \frac{j}{j+1}\frac{(\pi_j)^{\frac{1}{j}}}{(\pi_{j+1})^{\frac{1}{j+1}}}\left(\frac{n}{F(n,k)}\right)^{\frac{1}{j(j+1)}}(1+o(1)) \tag{6.2.2}$$

For any fixed positive integer $k > 0$ and j (either $j = 1$ or $j = 2$), there exists a finite integer $N(k,j)$ such that

$$\frac{j}{j+1}\frac{(\pi_j)^{\frac{1}{j}}}{(\pi_{j+1})^{\frac{1}{j+1}}}\left(\frac{n}{F(n,k)}\right)^{\frac{1}{j(j+1)}}(1+o(1)) > 1$$

for all $n > N(k,j)$. Hence, from (6.2.2), we have $\delta_k^{(j+1)}(n,\alpha) > \delta_k^{(j)}(n,\alpha)$ $(j = 1,2)$ for large n. The proof of Corollary 117 is complete. □

As an example, Fig. 6.6 shows the analytical results for the phase transition width of 1-connectivity in d-dimensional space ($d = 1,2,3$), which are calculated using (6.1.24) ($k = 1$ and $d = 1,2,3$). Other settings are the same as those in Fig. 6.5. In the calculation, we still omit the small order part $o(1)$ in $(1 + o(1))$. We can see that $\delta_k^{(3)}(n,\alpha) > \delta_k^{(2)}(n,\alpha) > \delta_k^{(1)}(n,\alpha)$ when n is larger than a certain threshold.

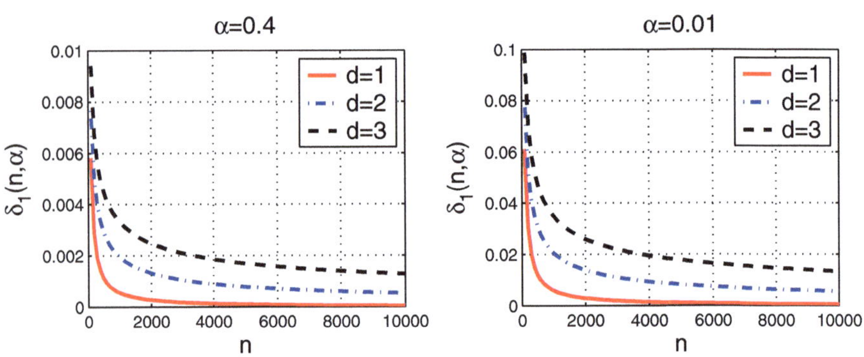

Fig. 6.6 Analytical results for the phase transition width of 1-connectivity in d-dimensional space ($d = 1,2,3$). The value of α is fixed in each scenario

Remark 118 Corollary 117 is an easy consequence of the main result in Corollary 113. It indicates that in a higher dimensional network, more transmission power/range is needed in order to make the probability that the network is k-connected increase from almost zero to almost one.

Corollary 119 allows us to separately evaluate the impact of the four key parameters α, n, k, and d on the phase transition width.

Corollary 119 *Consider a random geometric graph $G(n, r(n))$ in $\Re^d$ $(d = 1, 2, 3)$, a fixed positive integer $k > 0$, and a positive real number $\alpha \in (0, \frac{1}{2})$. Ignore the boundary effect except when $d = 1$. Then, for large n, the variation of the phase transition width of k-connectivity $\delta_k(n, \alpha)$ with α is separable from the variation with n, k, and d in the sense that for some functions T and Y, there holds $\delta_k(n, \alpha) = T(\alpha)Y(n, k, d)$.*

Proof Let $\delta_k^{(d)}(n, \alpha)$ denote the phase transition width of k-connectivity in d-dimensional space as we want to emphasize the dependence of $\delta_k(n, \alpha)$ on d. We assume that n is large enough such that the small order part $o(1)$ in (6.1.24) can be ignored. Then, for any fixed n, k, and d, we have

$$\begin{aligned}\delta_k^{(d)}(n, \alpha) &\approx \frac{\log\left(\frac{\log\alpha}{\log(1-\alpha)}\right)}{d\,(\pi_d n)^{\frac{1}{d}}\,(F(n,k))^{\frac{d-1}{d}}} \\ &= \log\left(\frac{\log\alpha}{\log(1-\alpha)}\right) \times \frac{1}{d\,(\pi_d n)^{\frac{1}{d}}\,(F(n,k))^{\frac{d-1}{d}}} \\ &= T(\alpha) \times Y(n, k, d)\end{aligned}$$

It is obvious that for any fixed n, k, and d, the term $Y(n, k, d)$ is fixed. In addition, the term $Y(n, k, d)$ is independent of α. Therefore, the value of $Y(n, k, d)$ does not change as α varies, given that n, k, and d are fixed. If we know $\delta_k^{(d)}(n, \alpha)$ for any $\alpha \in (0, \frac{1}{2})$, we can derive the value of $Y(n, k, d)$, which leads us to derive $\delta_k^{(d)}(n, \alpha)$ for any other $\alpha \in (0, \frac{1}{2})$ with n, k and d fixed.

The proof of Corollary 119 is complete. □

As an example, Fig. 6.7 shows the dependence of the phase transition width of k-connectivity on α in d-dimensional space $(d = 1, 2, 3)$. These analytical results are calculated from (6.1.24), and the small order part $o(1)$ in $(1 + o(1))$ is omitted. We can see that the variation of $\delta_k^{(d)}(n, \alpha)$ with α has the same functional dependence irrespective of n, k, and d, saving for a scaling constant defined by these latter variables.

Remark 120 Corollary 119 means that if we learn the phase transition width for any $\alpha \in (0, \frac{1}{2})$, it is easy to obtain it for any other $\alpha \in (0, \frac{1}{2})$ given that n, k, and d are fixed.

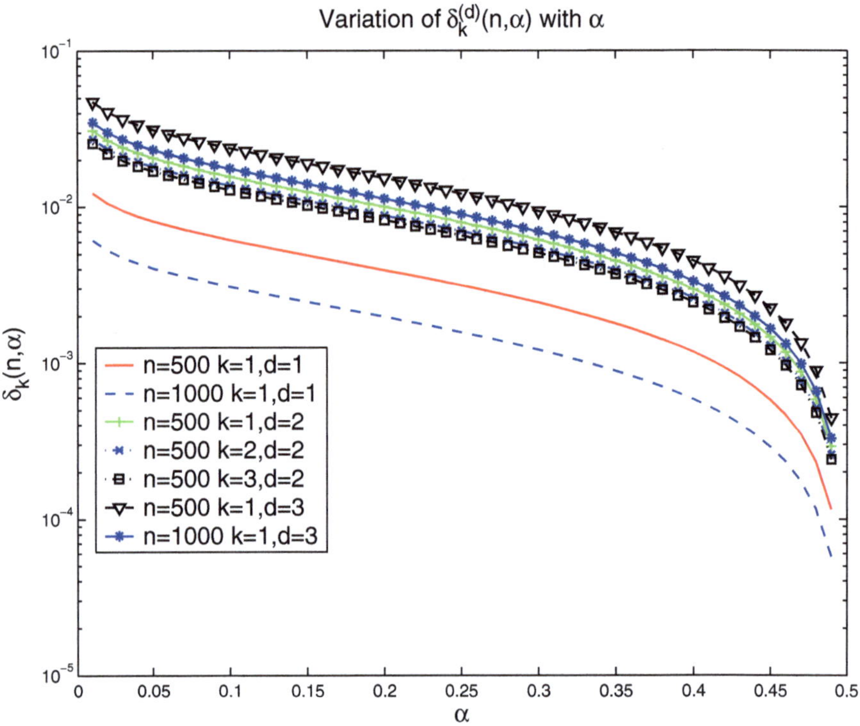

Fig. 6.7 Analytical results for the phase transition width of k-connectivity *versus* α in d-dimensional space $d = 1, 2, 3$

In addition to the results for $\delta_k(n, \alpha)$ in terms of the transmission range, using the same technique for proving Theorem 107 and Corollary 113, and the relation between the average node degree and the transmission range, we can obtain the following theorem for $\delta'_k(n, \alpha)$ in terms of the average node degree:

Theorem 121 *Consider a random geometric graph $G(n, r(n))$ in $\Re^d$ $(d = 1, 2, 3)$, a fixed positive integer $k > 0$, and a positive real number $\alpha \in (0, \frac{1}{2})$. Ignore the boundary effect except when $d = 1$. Then, for large n, the phase transition width of k-connectivity $\delta'_k(n, \alpha)$ in terms of the average node degree is given by*

$$\delta'_k(n, \alpha) = \begin{cases} 2\log\left(\frac{\log\alpha}{\log(1-\alpha)}\right) + o(1), & d = 1; \\ \log\left(\frac{\log\alpha}{\log(1-\alpha)}\right) + o(1), & d \geq 2 \end{cases} \tag{6.2.3}$$

Proof For $d = 2, 3$, ignoring the boundary effect, the average node degree is $n\pi_d r^d(n)$. In the same way as shown in the proof of Theorem 107 and Corollary 113, we can readily obtain the result given in (6.2.3) for $d = 2, 3$.

For $d = 1$, the average node degree is $2nr(n)$. Again, using the same method as shown in the proof of Theorem 107 and Corollary 113, the result follows. □

Theorem 121 indicates that for large enough n, the phase transition width $\delta'_k(n, \alpha)$ measured in the average node degree is only determined by α and is independent of n and k. In addition, the phase transition width $\delta'_k(n, \alpha)$ for $d = 2$ and $d = 3$ are the same and are half of that for $d = 1$.

6.3 Simulation Studies of the Phase Transition Width

We have theoretically investigated the phase transition width of k-connectivity $\delta_k(n, \alpha)$ for any positive integer k and for large n in d-dimensional space ($d = 1, 2, 3$) in the previous sections. As emphasized in our analysis, the results established on the phase transition width only apply to large wireless networks. The theoretical results will become increasingly accurate as the number of node n in the network increases. In this section, we conduct simulations to investigate empirically how large a network needs to be in order for the theoretical results to be considered reasonably accurate and the errors between the theoretical results and those obtained from simulations for a network of finite size.

In the simulations, we consider that a total of n nodes are randomly and independently distributed in a unit cube $[0, 1]^d$ ($d = 1, 2, 3$) according to a uniform distribution. All nodes have the same transmission range $r(n)$. Simulations become very computationally intensive and time consuming for $k > 3$ and large values of n. Therefore, we limited k to 3 and n to 1500 in the simulations.

Computing the Phase Transition Width $\delta_k(n, \alpha)$

Here we give a brief description of the determination of the phase transition width of k-connectivity $\delta_k(n, \alpha)$ in simulations. The following are the main steps:

(1) For any given n, distribute n nodes randomly and independently in a unit cube $[0, 1]^d$ ($d = 1, 2, 3$) following a uniform distribution. Then, we obtain an instance of a random network with n nodes. For this network and each $k \in \{1, 2, 3\}$, find the corresponding minimum transmission range $r_k(i)$ which makes the network k-connected.
(2) Repeat step 1) for a large number of times N (e.g., $N = 10,000$), then we obtain a set of N random network instances and three sets of N corresponding minimum transmission ranges $\{r_k(i), i = 1, 2, 3, \ldots, N\}$ for $k = 1$, 2 and 3, respectively, where each set contains N transmission range values.
(3) Record each set of the N transmission ranges (i.e., $\{r_k(i), i = 1, 2, 3, \ldots, N\}$ for $k = 1$ or 2 or 3 in an ascending order, such that $r'_k(i) \leq r'_k(i + 1)$ in the new ordered sets for all $i \in [1, N - 1]$.

(4) For each $k \in \{1, 2, 3\}$, let $j = \lceil N \times \alpha \rceil$ and $l = \lceil N \times (1 - \alpha) \rceil$. Then, the j-th (respectively, l-th) item $r'_k(j)$ (respectively, $r'_k(l)$) in the new ordered set is approximately the minimum transmission range at which the network is k-connected with probability α (respectively, $1 - \alpha$). Finally, the difference between these two transmission ranges $\tilde{\delta}_k(n, \alpha) = r'_k(l) - r'_k(j)$ is approximately the phase transition width over the probability interval $[\alpha, 1-\alpha]$ of k-connectivity. The larger N is, the more accurate the computed phase transition width is. However, a large N will cost a large amount of time.
(5) Repeat steps 1) to 4) for different values of n to obtain the phase transition width of k-connectivity for different values of n.

An important consideration in the simulation is to eliminate the boundary effect. Since the simulation is performed in a bounded area (e.g., unit cube $[0, 1]^d$), the average node degree of nodes located near borders of the network area is lower than that of the nodes located in the middle of the network area. The boundary effect in the simulation will make a fair comparison between the simulation results and the analytical results difficult because the analytical results are derived without considering the boundary effect and that in a large network, the impact of boundary effect will be small as demonstrated in our analysis in Chaps. 2 and 3.

Usually, there are two ways to eliminate the boundary effect. The first one is to divide the entire simulation area into two disjoint subareas: a boundary subarea Z_{out} with a width of at least $r(n)$, and an inner subarea Z_{in}. Figure 6.8 shows two examples for two-dimensional networks.

Only nodes located in the inner subarea Z_{in} are counted for computing statistics of the simulations. In fact, the nodes in the boundary subarea Z_{out} can be used as relay nodes for building links with the nodes located in the inner subarea Z_{in}, but they are not counted for the statistics. One of the disadvantages of this technique is

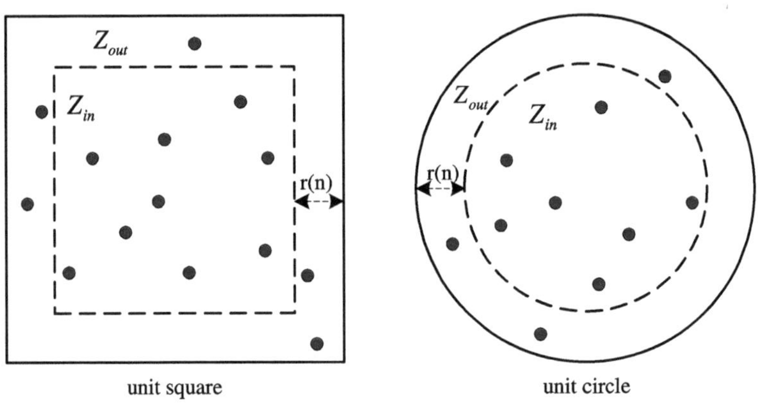

Fig. 6.8 An illustration of the first method for eliminating the boundary effect in a two-dimensional network area. Only nodes located in subarea Z_{in} are counted for computing statistics in the simulation

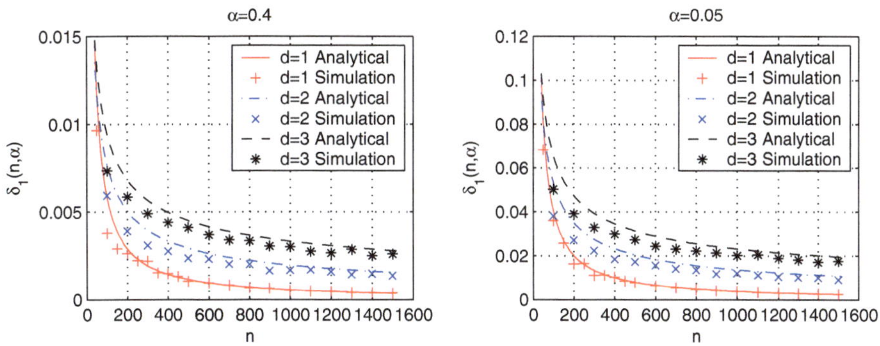

Fig. 6.9 Phase transition width of 1-connectivity versus the number of nodes n in d-dimensional space $d = 1, 2, 3$. The value of α is fixed in each scenario. For the analytical results, the boundary effect is ignored; for the simulation results, the boundary effect is eliminated by using the toroidal distance metric

that the number of nodes located in the inner subarea Z_{in} decreases as $r(n)$ increases when the size of the network area is fixed.

The second approach to eliminate the boundary effect is to use the *toroidal distance* metric introduced in Sect. 1.3.1.1.

Figure 6.9 shows the analytical results and the simulation results for the phase transition width of 1-connectivity $\delta_1(n, \alpha)$ in d-dimensional space $d = 1, 2, 3$. The value of n is varied between 100 and 1500, α is set to two typical values, i.e., 0.4 (close to 0.5) and 0.05 (close to 0). For the analytical results, the boundary effect is ignored; when calculating the analytical results by (6.1.24), the small order part is omitted, i.e., $o(1)$ in the term $(1 + o(1))$ is ignored. For the simulation results, the boundary effect is eliminated by using the toroidal distance metric. We can see that for each $d \in \{1, 2, 3\}$, the phase transition width of 1-connectivity decreases as n grows. It is obvious that there is significant discrepancy between the analytical results and the simulation results for small values of n (e.g., $n < 200$). This is because the small order part $o(1)$ in the analytical results is significant when n is small, especially when n is comparable with k! However, the small order part $o(1)$ goes to zero as n goes to infinity. So the discrepancy decreases as n increases. We can see that although there is significant discrepancy when n is not large enough, the simulation results exhibit the same decreasing trend as the analytical results. We can also see that the phase transition width of 1-connectivity is larger for $d = 3$ than that for $d = 2$, and similarly, the phase transition width is larger for $d = 2$ than that for $d = 1$, which validate Corollary 117. This means that in a higher dimensional network, more transmission power/range is needed in order to make the probability that the network is k-connected transit from almost zero to almost one.

Figure 6.10 shows the analytical results and the simulation results for the phase transition width of k-connectivity $\delta_k(n, \alpha)$ ($k = 1, 2, 3$) in two-dimensional space. Other settings are the same as those in Fig. 6.9. We can see that when $d = 2$, the phase transition width of k-connectivity decreases as n increases. The figure also

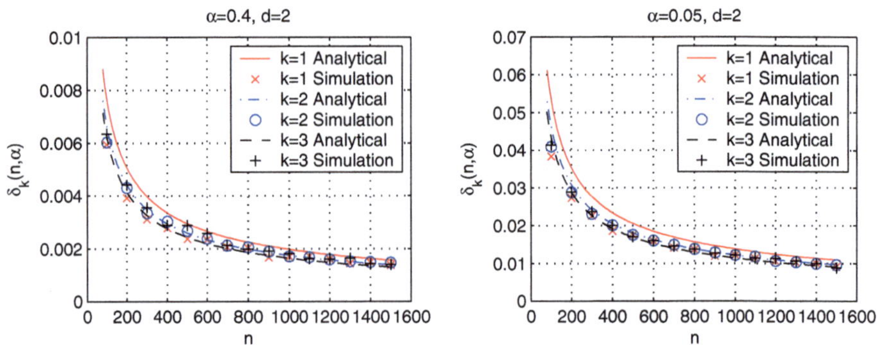

Fig. 6.10 Phase transition width of k-connectivity ($k = 1, 2, 3$) *versus* the number of nodes n in 2-dimensional space. The value of α is fixed in each scenario. For the analytical results, the boundary effect is ignored; for the simulation results, the boundary effect is eliminated by using the toroidal distance metric

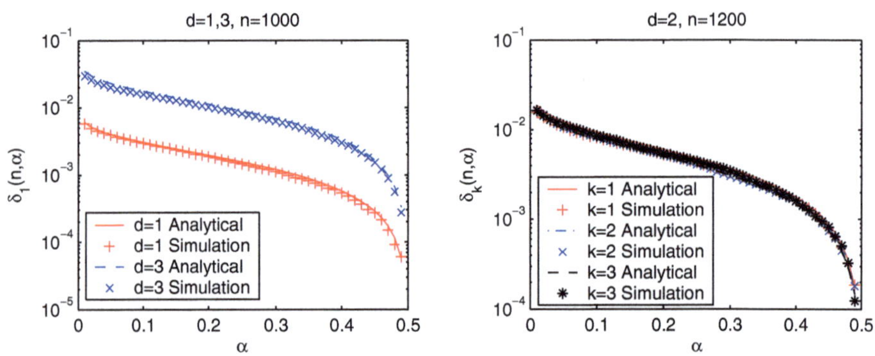

Fig. 6.11 Phase transition width of k-connectivity versus α in d-dimensional space with $d = 1, 2, 3$. The value of n is fixed in each scenario. For the analytical results, the boundary effect is ignored; for the simulation results, the boundary effect is eliminated by using the toroidal distance metric

indicates that the difference between $\delta_k(n, \alpha)$ and $\delta_{k+1}(n, \alpha)$ becomes smaller as n gets larger. It means that $\delta_k(n, \alpha) \approx \delta_{k+1}(n, \alpha)$ when n is large enough, which is consistent with Corollary 115. In other words, the transmission power/range required for making the probability that the network is k-connected increase from almost zero to almost one is approximately the same as the transmission power/range required for making the probability that the network is $(k + 1)$-connected increase from almost zero to almost one.

Figure 6.11 shows the dependence of the phase transition width of k-connectivity on α given that n, k, and d are fixed. We can see that the variation of $\delta_k(n, \alpha)$ with α has the same functional dependence irrespective of n, k, and d. This verifies Corollary 119. We can also see that for fixed n, k, and d, $\delta_k(n, \alpha)$ decreases as α increases. In addition, the discrepancy between the analytical results and the simulation results becomes significant when α is very small. The reason for this is that the small order part $o(1)$ in the analytical results becomes significant when α is very close to zero.

Simulation results also suggest that in order to achieve a good accuracy, n should be larger than 200 when $k = 1$ and $d = 1$; and n should be larger than 600 when $k \leq 3$ and $d = 2, 3$.

6.4 Notes and Further Readings

In this chapter, we investigated the phase transition behavior of k-connectivity with respect to the transmission range of nodes in large wireless networks, where a total of n nodes are randomly and uniformly distributed in a unit cube $[0, 1]^d$, and each node has a uniform transmission range $r(n)$. The phase transition behavior is associated with the transmission range, thus the transmission power of nodes. It is desirable to control the transmission range to be just above the right boundary of the phase transition region so that the network achieves k-connectivity with a high probability while minimizing the energy consumption. For large n, we derived a general analytical formula for calculating the phase transition width of k-connectivity for any fixed positive integer k in d-dimensional space ($d = 1, 2, 3$). We also derived an analytical formula for a modified version of the phase transition width in terms of the average node degree. We further conducted simulations to validate our theoretical analysis. Our results were derived for large enough n. Hence, they hold only when n is sufficiently large. When n is a small number (especially when n is comparable with $k!$), our results do not hold any more. Simulation results showed that n should be larger than 200 when $k = 1$ and $d = 1$; and n should be larger than 600 when $k \leq 3$ and $d = 2, 3$. These results are of practical value in the self configuration of large wireless networks, and provide us useful design principles for large wireless networks as well.

There has been extensive work on the phase transition phenomenon in the past several years. Extensive results have been obtained for Bernoulli random graphs [27]. Usually, a Bernoulli random graph is obtained by randomly distributing n vertices and connecting any pair of two vertices with probability $p(n)$, independently of all other pairs of vertices and the Euclidean distance between the two vertices. Friedgut et al. [72] proved that all monotone graph properties have a sharp threshold in the Bernoulli random graph, and the threshold width is $\delta(n, \varepsilon) = O(\log \varepsilon^{-1} / \log n)$. However, the techniques used for Bernoulli random graphs cannot be applied straightforwardly to random geometric graphs, because in random geometric graphs, the probability of existence of a link between two nodes is dependent on their Euclidean distance.

In [122], Krishnamachari et al. demonstrated the ubiquity of the phase transition phenomenon for monotone graph properties in Bernoulli random graphs and random geometric graphs, and presented some examples in wireless ad hoc networks where a phase transition phenomenon exists, such as connectivity, coordination and probabilistic flooding for route discovery. In [121], Krishnamachari et al. investigated three distributed configuration tasks in large wireless networks, i.e., partition into coordinating cliques, Hamiltonian cycle formation and conflict-free channel allocation. They showed that these tasks undergo phase transitions with respect to the transmission range, and argued that phase transition analysis is useful for quantifying the critical range of energy and bandwidth resources needed for a scalable performance of self-configuring wireless networks. In [173], Ravelomanana showed that the coverage property is subject to abrupt phase transition in three-dimensional wireless sensor networks. In [8], Aspnes et al. exhibited with simulation evidence the phase transition for localizability in wireless sensor networks. In [170], Raghavan et al. investigated phase transition behaviors for the emergence of a giant component in wireless sensor networks and obtained an expression for the critical radius at which the network has a giant component with high probability.

In [82], Goel et al. proved that all monotone properties in random geometric graph have a sharp threshold. Furthermore, the threshold width for random geometric graphs is much sharper than that for Bernoulli random graphs. They showed that for every monotone property, the threshold width $\delta(n, \varepsilon)$ is

$$\delta(n,\varepsilon) = \begin{cases} O(\log^{1/2}\frac{1}{\varepsilon}/n^{1/2}), & d=1 \\ O(\log^{3/4} n/n^{1/2}), & d=2 \\ O(\log^{1/d} n/n^{1/d}), & d\geq 3 \end{cases} \tag{6.4.1}$$

In [97], Han et al. found that while the results in [82] were derived for a general monotone property, they may be further sharpened for certain *specific* monotone graph properties such as connectivity. They were seeking to improve the results given by Goel for the property of connectivity in one and two-dimensional spaces, and derived the phase transition width for large n, i.e.,

$$\delta_1(n,\varepsilon) = \begin{cases} \frac{C(\varepsilon)}{n} + o(n^{-1}), & d=1 \\ \frac{C(\varepsilon)}{2}\sqrt{\frac{1}{\pi n \log n}}(1+o(1)), & d=2 \end{cases} \tag{6.4.2}$$

where $C(\varepsilon) = \log(\frac{\log \varepsilon}{\log(1-\varepsilon)})$. The results are much sharper than the results given in (6.4.1), which indicates that the results in (6.4.1) can be quite conservative for specific monotone properties.

Chapter 7
Connectivity of Large Wireless Networks in the Presence of Interference

Abstract In this chapter, we investigate connectivity of large wireless networks in the presence of interference. Different from previous chapters where connections are assumed to be independent, the presence of interference implies that signals transmitted at the same time will mutually interfere with each other. Hence, connections become mutually correlated. Specifically, consider the extended network model and the SINR connection model, we establish a sufficient condition and a necessary condition, i.e., an upper bound and a lower bound, on the transmission power required for a network adopting the carrier sense multiple access protocol to be asymptotically almost surely connected. The two bounds differ by a constant factor only. It is shown that the transmission power only needs to be increased by a constant factor to combat interference and maintain connectivity compared with that considering a unit disk connection model without interference.

In this chapter, we investigate connectivity of large wireless networks in the presence of interference. Due to the nature of wireless communications, signals transmitted at the same time will mutually interfere with each other. The SINR (signal to interference plus noise) connection model, which has been introduced in Sect. 1.1.5 and recited below for convenience, has been widely used to capture the impact of interference on network connectivity.

7.1 Connections in Carrier Sense Multiple Access (CSMA) Networks

Under the SINR connection model, the existence of a directional link between a pair of nodes is determined by the strength of the received signal from the desired transmitter, the interference caused by other concurrent transmissions and the background noise. Therefore, the existence of a connection between a pair of nodes is no longer independent of the existence of a connection between another distinct pair of nodes. This distinction sets the SINR connection model apart from the random connection model, the unit disk connection model, and the log-normal connection model discussed in the previous chapters.

G. Mao, *Connectivity of Communication Networks*,
DOI 10.1007/978-3-319-52989-9_7

Assume that all nodes transmit with the same power P and let $\boldsymbol{x}_k$, $k \in \Gamma$, be the location of node k, where Γ represents the set of indices of all nodes in the network. A node j can successfully receive the transmitted signal from a node i (i.e., node j *is directly connected to* node i) if the SINR at $\boldsymbol{x}_j$, denoted by $\text{SINR}\left(\boldsymbol{x}_i \to \boldsymbol{x}_j\right)$, is above a prescribed threshold β, i.e.,

$$\text{SINR}\left(\boldsymbol{x}_i \to \boldsymbol{x}_j\right) = \frac{P\ell\left(\boldsymbol{x}_i, \boldsymbol{x}_j\right)}{N_0 + \gamma \sum_{k \in \mathcal{T}_i} P\ell\left(\boldsymbol{x}_k, \boldsymbol{x}_j\right)} \geq \beta \tag{7.1.1}$$

where $\mathcal{T}_i \subseteq \Gamma$ denotes the subset of nodes that are transmitting at the same time as node i and N_0 is the background noise power. The function $\ell\left(\boldsymbol{x}_i, \boldsymbol{x}_j\right)$ is the power attenuation from $\boldsymbol{x}_i$ to $\boldsymbol{x}_j$. The coefficient $0 \leq \gamma \leq 1$ is the inverse of the processing gain of the system and it weighs the impact of interference. In a broadband system using CDMA (Code Division Multiple Access), γ depends on the orthogonality between codes used during concurrent transmissions and is smaller than 1; in a narrow-band system, γ is equal to 1 [57, 94].

Similarly, node i can receive from node j (i.e., node i is directly connected to node j) if and only if

$$\text{SINR}\left(\boldsymbol{x}_j \to \boldsymbol{x}_i\right) = \frac{P\ell\left(\boldsymbol{x}_j, \boldsymbol{x}_i\right)}{N_0 + \gamma \sum_{k \in \mathcal{T}_j} P\ell\left(\boldsymbol{x}_k, \boldsymbol{x}_i\right)} \geq \beta \tag{7.1.2}$$

Therefore, node i and node j are directly connected, i.e., a bidirectional or undirectional link exists between node i and node j, if and only if both (7.1.1) and (7.1.2) are satisfied.

A major difficulty in moving from the random connection model, including the unit disk connection model and the log-normal connection model as two important special cases, to the SINR connection model is that under the random connection model, connections are assumed to be independent, i.e., the event that a pair of nodes are directly connected and the event that another distinct pair of nodes are directly connected are independent. This independence assumption on connections is a critical assumption in the analysis of connectivity under the random connection model. In the SINR model however, due to the presence of interference, the existence of a direct connection between a pair of nodes depends on both the location and the activities of other nodes in the network.

In [57] Dousse et al. used the above SINR connection model to analyze the impact of interference on connectivity from the percolation perspective. They considered a network where all nodes are distributed in $\Re^2$ following a homogeneous Poisson point process with a constant density λ and an attenuation function ℓ with bounded support. By letting $\mathcal{T}_j = \Gamma / \{i,j\}$, i.e., all other nodes in the network transmit simultaneously with node i irrespective of their relative location to $\boldsymbol{x}_i$ and

$\boldsymbol{x}_j$, it is shown that there exists a very small positive constant γ' such that if $\gamma > \gamma'$ there is no infinite connected component in the network, i.e., the network does not *percolate*. Furthermore, when $\gamma < \gamma'$, there exists $0 < \lambda' < \infty$ such that percolation can occur when $\lambda > \lambda'$. An improved result by the same authors in [58] shows that under the more general conditions that $\lambda > \lambda_c$ and the attenuation function has unbounded support, percolation occurs when $\gamma < \gamma'$. Here, λ_c is the critical node density above which the network with $\gamma = 0$ (i.e., equivalent to *the unit disk connection model* without interference) percolates [71, p48]. The above results suggest that percolation under the SINR model can happen if and only if γ is sufficiently small. In their network setting, it is assumed that each node transmits randomly and independently, irrespective of any nearby transmitter. This corresponds to the ALOHA-type multiple access scheme [94]. In the ALOHA multiple access scheme, if a node has data to transmit, the node will transmit the data independently and irrespective of simultaneous transmissions from other nearby transmitters. If more than two nearby nodes transmit at the same time, a collision occurs and all transmitting nodes need to retransmit later. Since collisions can occur and data may have to be sent again, ALOHA can result in low utilization of the communications channel, especially when the density of transmitters is high. This ALOHA multiple access strategy has therefore become obsolete [74, 123].

The more advanced multiple access strategies, e.g., CSMA (Carrier Sense Multiple Access) and CSMA/CD (Carrier Sense Multiple Access with Collision Detection) [172], have become prevalent with widespread adoption. The general idea of CSMA schemes is that nearby nodes will not be scheduled to transmit simultaneously. Therefore, it is natural to expect that CSMA schemes could improve the performance of ALOHA schemes by alleviating interference, particularly under heavy traffic.

Note that in the ALOHA multiple access scheme, because each node transmits independently of other nodes, the subset of transmitters can be modeled by a Poisson point process if all nodes in the network are distributed following a homogeneous Poisson point process. In the CSMA multiple access scheme however, because a node's activity depends on the activities of nearby nodes, the subset of transmitters no longer follows a Poisson point process even if all nodes are distributed following a homogeneous Poisson point process.

In this chapter, we analyze connectivity of wireless CSMA networks under the SINR model. Specifically, we consider a network with n nodes uniformly, independently and identically distributed on a square $\left[-\frac{\sqrt{n}}{2}, \frac{\sqrt{n}}{2}\right]^2$, i.e., the so-called *extended network model*, where the network size scales with the network area while the node density is fixed. It is assumed that all nodes use the same transmission power P and there is always a packet at a node waiting to be transmitted. This later assumption allows us to focus on the network property without being disturbed by other factors, e.g., traffic distribution.

We consider that the attenuation function ℓ in (7.1.1) and (7.1.2) only depends on Euclidean distance, viz., $\ell : \Re^+ \to \Re^+$, and is a power-law function [57, 58]:

$$\ell(s) = s^{-\alpha} \tag{7.1.3}$$

where s represents the Euclidean distance between a pair of nodes and α is the path-loss exponent, which typically varies from 2 to 6 [172, p139]. We assume that $\alpha > 2$. The above assumptions on the attenuation function are widely used [57, 94, 95] and are also supported by measurement studies [172]. The results in this chapter can be readily extended to the situation where $\ell(s) = \min\{1, s^{-\alpha}\}$ at the expense of some lengthy but mostly straightforward discussions on the special case when $s^{-\alpha} < 1$. The path loss model $\ell(s) = \min\{1, s^{-\alpha}\}$ helps to avoid singularity or unreasonably large signal strength gains that may occur as s approaches 0. As commonly done in the connectivity analysis [57, 58, 71, 92, 165], the impact of small-scale fading is ignored and only bidirectional communication links are considered. Furthermore, since in dense wireless networks the background noise is typically negligibly small [94, 95], in the following analysis, we ignore the background noise N_0 in (7.1.1) and (7.1.2). The result can be readily extended to include the impact of background noise. In addition, we consider that all transmitters are using the same channel, i.e., $\gamma = 1$, which corresponds to a narrow-band system [57, 94, 95].

In CSMA networks, two nodes located at $\boldsymbol{x}_i$ and $\boldsymbol{x}_j$, respectively, are allowed to transmit at the same time if and only if they cannot detect each other's transmission, i.e., both $P\ell(\boldsymbol{x}_i, \boldsymbol{x}_j)$ and $P\ell(\boldsymbol{x}_j, \boldsymbol{x}_i)$ in (7.1.1) and (7.1.2) are under a certain detection threshold P_{th}. It then follows from (7.1.3) that the carrier-sensing range R_c, which determines the *minimum* Euclidean distance between two simultaneously active transmitters, is given by

$$R_c = (P/P_{\text{th}})^{1/\alpha} \tag{7.1.4}$$

Alternatively, one may consider a scenario where a node can only transmit when the aggregate interference is below the threshold P_{th}. This scenario forms a trivial extension of the situation considered in this chapter.

7.2 Sufficient Condition for Almost Surely Connected CSMA Networks

A major challenge in dealing with connectivity analysis under the SINR model is that under the SINR connection model, the existence of a direct connection between a pair of nodes depends on both the location and activities of other nodes in the network, i.e., connections are correlated. In this chapter, we resort to a de-coupling approach to handle the connection correlations. More specifically, we first establish an upper bound on the aggregate interference experienced by any receiver in CSMA networks. On that basis, we show that for an arbitrarily chosen SINR threshold, there exists a transmission range R_0 such that a pair of nodes are directly connected if their Euclidean distance is smaller than or equal to R_0. Given the above results, we can then use existing results on network connectivity obtained under the unit disk connection model to analyze network connectivity under the SINR connection model.

7.2.1 *An Upper Bound on Interference and the Associated Transmission Range*

A major result of this section is summarized in the following theorem, which provides an upper bound on the aggregate interference experienced by any receiver in CSMA networks.

Theorem 122 *Consider a CSMA network with nodes distributed arbitrarily on a finite area in* $\Re^2$ *where the carrier-sensing range is* R_c*, which is given by (7.1.4), and each node transmits at the same power* P*. Assume that the attenuation function* l *is a power-law function given in (7.1.3). Denote by* r_0 *the Euclidean distance between a receiver and its nearest transmitter in the network, which is also the intended transmitter for the receiver. When* $r_0 < R_c$*, the maximum interference experienced by the receiver is smaller than or equal to* $N(r_0) = N_1(r_0) + N_2$*, where*

$$N_1(r_0) = \frac{4P\left(\frac{5\sqrt{3}}{4}R_c - r_0\right)^{1-\alpha}\left(\frac{\sqrt{3}}{4}(3\alpha-1)R_c - r_0\right)}{R_c^2(\alpha-1)(\alpha-2)} + \frac{3P}{(R_c - r_0)^\alpha} + \frac{3P}{\left(\sqrt{3}R_c - r_0\right)^\alpha} + \frac{3P\left(\frac{3}{2}R_c - r_0\right)^{1-\alpha}}{(\alpha-1)R_c} \tag{7.2.1}$$

$$N_2 = \frac{3P}{R_c^\alpha} + \frac{3P(\frac{3}{2})^{1-\alpha}}{(\alpha-1)R_c^\alpha} + \frac{3P}{\left(\sqrt{3}R_c\right)^\alpha} + \frac{3P\left(\frac{5}{4}\right)^{1-\alpha}(3\alpha-1)}{(\alpha-1)(\alpha-2)\left(\sqrt{3}R_c\right)^\alpha} \tag{7.2.2}$$

Proof We first note that a network on a finite area, denoted by $A \subset \Re^2$, can always be obtained from a network on an infinite area $\Re^2$ with the same node density and distribution by removing these nodes outside A. Such removal process will also remove all transmitters outside A. Therefore, it follows that the interference experienced by a receiver in A is less than or equal to the interference experienced

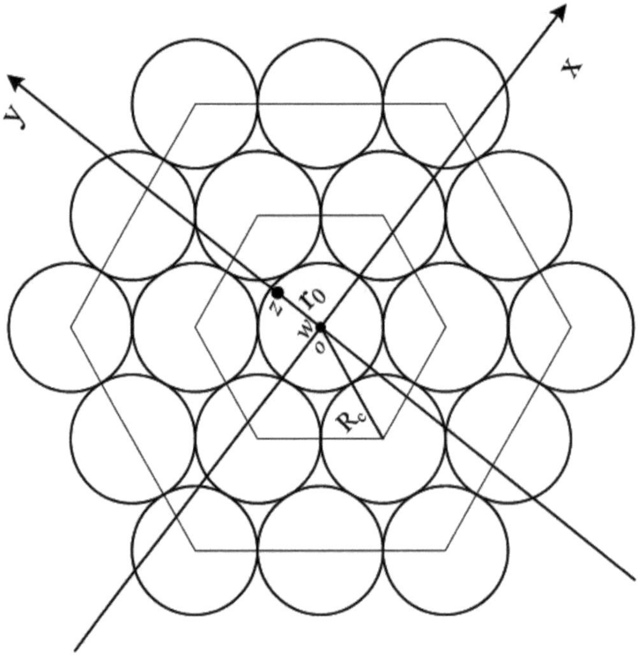

Fig. 7.1 An illustration of the densest equal-circle packing

by its counterpart in an infinite network in $\Re^2$. It then suffices to show that the interference experienced by a receiver in an infinite network in $\Re^2$ is bounded by $N(r_0)$.

Consider that an arbitrary receiver z is located at a Euclidean distance r_0 from its closest transmitter w, which is also the intended transmitter for z. We construct a coordinate system such that the origin of the coordinate system is at w and z is on the $+y$ axis, as shown in Fig. 7.1.

In a CSMA network, any two simultaneously active transmitters are separated by at least a Euclidean distance R_c. Draw a circle of radius $R_c/2$ centered at each simultaneously active transmitter. Then, the two circles centered at two closest transmitters cannot overlap except at single point. Therefore, the problem of determining the maximum interference can be transformed into one that determining the maximum number of equal-radius nonoverlapping circles that can be packed into $\Re^2$. In [46], the problem of the densest circle packing is studied. It was shown that for circles with equal radius, the densest circle packing, i.e., fitting the maximum number of nonoverlapping circles into $\Re^2$, is obtained by placing the circle centers at the vertices of a hexagonal lattice [46, p. 8], as shown in Fig. 7.1.

Partition the vertices of the hexagonal lattice into tiers of increasing distances from the origin. The six vertices of the first tier are within a Euclidean distance R_c to the origin. The $6m$ vertices in the mth tier are located at distances within $((m-1)R_c, mR_c]$ from the origin.

Let I_1 be the total interference caused by transmitters, hereinafter referred to as interferers in this subsection, above the x-axis at node z. Using the triangle inequality, it follows that $\|\boldsymbol{x}_i - \boldsymbol{z}\| \geq \|\boldsymbol{x}_i\| - r_0$ where $\boldsymbol{x}_i$ is the location of an interferer above the x-axis. Using some straightforward geometric analysis, it can be shown that among the $6m$ interferers in the mth group, half of them are located above the x-axis. Among these interferers in the mth group above the x-axis, three of them are at a Euclidean distance of exactly mR_c from the origin and the remaining $3(m-1)$ interferers are at Euclidean distances within $[\frac{\sqrt{3}}{2}mR_c,\ mR_c]$. It then follows from (7.1.3) that

$$I_1 \leq \sum_{m=1}^{\infty}\left(\frac{3\,(m-1)\,P}{\left(\frac{\sqrt{3}}{2}mR_c - r_0\right)^{\alpha}} + \frac{3P}{(mR_c - r_0)^{\alpha}}\right) \tag{7.2.3}$$

Consider the first summation in (7.2.3). Let $U_m,\ m = 3, \ldots, \infty$, be random variables uniformly, identically, independently distributed in $[m-1/2,\ m+1/2]$. It follows from the convexity of $\frac{3(m-1)P}{\left(\frac{\sqrt{3}}{2}mR_c - r_0\right)^{\alpha}}$ and Jensen's inequality (used in the second step) that

$$\begin{aligned}
&\sum_{m=3}^{\infty}\frac{3\,(m-1)\,P}{\left(\frac{\sqrt{3}}{2}mR_c - r_0\right)^{\alpha}} \\
&= \sum_{m=3}^{\infty}\frac{3\,(\mathrm{E}\,(U_m)-1)\,P}{\left(\frac{\sqrt{3}}{2}\mathrm{E}\,(U_m)\,R_c - r_0\right)^{\alpha}} \\
&\leq \sum_{m=3}^{\infty}\mathrm{E}\left(\frac{3\,(U_m-1)\,P}{\left(\frac{\sqrt{3}}{2}U_mR_c - r_0\right)^{\alpha}}\right) \\
&= \sum_{m=3}^{\infty}\int_{m-1/2}^{m+1/2}\frac{3\,(x-1)\,P}{\left(\frac{\sqrt{3}}{2}xR_c - r_0\right)^{\alpha}}dx \\
&= 3P\int_{5/2}^{\infty}(x-1)\left(\frac{\sqrt{3}}{2}xR_c - r_0\right)^{-\alpha}dx \\
&= \frac{4P\left(\frac{5\sqrt{3}}{4}R_c - r_0\right)^{1-\alpha}\left(\frac{\sqrt{3}}{4}\,(3\alpha-1)\,R_c - r_0\right)}{R_c^2\,(\alpha-1)\,(\alpha-2)}
\end{aligned} \tag{7.2.4, 7.2.5}$$

Likewise, it can also be shown that

$$\sum_{m=2}^{\infty}\frac{3P}{(mR_c - r_0)^{\alpha}} \leq \frac{3P\left(\frac{3}{2}R_c - r_0\right)^{1-\alpha}}{(\alpha-1)\,R_c} \tag{7.2.6}$$

As a result of (7.2.3), (7.2.5), (7.2.6), and (7.2.1), it follows that

$$I_1 \leq N_1(r_0) \tag{7.2.7}$$

Now we consider the total interference caused by interferers below the x-axis at node z, denoted by I_2. Let $\boldsymbol{x}_i$ be the location of an interferer below the x-axis, it follows from the triangle inequality that $\|\boldsymbol{x}_i - \boldsymbol{z}\| \geq \|\boldsymbol{x}_i\|$. Therefore,

$$I_2 \leq \sum_{m=1}^{\infty}\left(\frac{3P}{(mR_c)^{\alpha}} + \frac{3(m-1)P}{\left(\frac{\sqrt{3}}{2}mR_c\right)^{\alpha}}\right)$$

$$= \frac{3P}{R_c^{\alpha}} + \frac{3P(\frac{3}{2})^{1-\alpha}}{(\alpha-1)R_c^{\alpha}} + \frac{3P}{\left(\sqrt{3}R_c\right)^{\alpha}}$$

$$+ 3P\left(\frac{\sqrt{3}}{2}R_c\right)^{-\alpha}\sum_{m=3}^{\infty}\frac{(m-1)}{m^{\alpha}} \tag{7.2.8}$$

$$\leq \frac{3P}{R_c^{\alpha}} + \frac{3P(\frac{3}{2})^{1-\alpha}}{(\alpha-1)R_c^{\alpha}} + \frac{3P}{\left(\sqrt{3}R_c\right)^{\alpha}}$$

$$+ \frac{3P\left(\frac{5}{4}\right)^{1-\alpha}(3\alpha-1)}{(\alpha-1)(\alpha-2)\left(\sqrt{3}R_c\right)^{\alpha}} \tag{7.2.9}$$

Combining (7.2.7) and (7.2.9), Theorem 122 is proved. □

Remark 123 The upper bound in Theorem 122 is valid for any node distribution. In a sparse network or a network where nodes are placed in a coordinated or planned manner, by replacing the carrier-sensing range with the minimum separation distance between a pair of simultaneously active transmitters, Theorem 122 can be readily extended to be applicable for such situation.

Remark 124 The assumption that $r_0 < R_c$ is valid in most wireless systems which not only require the SINR from a transmitter to a receiver to be above a certain threshold but also require the received signal from the transmitter to be of sufficiently good quality. However Theorem 122 does not critically depend on the assumption. In the situation that $r_0 \geq R_c$, so long as there exists a positive integer c such that $r_0 < cR_c$, the upper bound in Theorem 122 can be revised to accommodate the situation by changing the range of the summation in (7.2.4) and (7.2.6) from $[3, \infty]$ and $[2, \infty]$ to $[c+2, \infty]$ and $[c+1, \infty]$ respectively and revising (7.2.7) accordingly.

The following result can be obtained as a ready consequence of Theorem 122.

Corollary 125 *Under the same settings as in Theorem 122, assume that the SINR threshold required for a successful transmission is β. There exists a transmission range $R_0 < R_c$ such that a pair of nodes are directly connected if their Euclidean distance is smaller than or equal to R_0. Furthermore, R_0 is given implicitly in the following equation*

$$\frac{PR_0^{-\alpha}}{N\left(R_0\right)} = \beta \tag{7.2.10}$$

Proof In Theorem 122, we have established that the interference experienced by a receiver z at a Euclidean distance r_0 from its transmitter w, denoted by $I\left(r_0\right)$, is upper bounded by $N\left(r_0\right)$. From (7.2.1) and (7.2.2), it can be shown that for $r_0 < R_c$, $N\left(r_0\right)$ is an increasing function of r_0. Note also that $Pr_0^{-\alpha}$ is a decreasing function of r_0. Therefore, using (7.2.10), the SINR of a receiver at $r_0 \le R_0$ from its transmitter, denoted by $\text{SINR}\left(r_0\right)$, satisfies

$$\text{SINR}\left(r_0\right) = \frac{Pr_0^{-\alpha}}{I\left(r_0\right)} \ge \frac{Pr_0^{-\alpha}}{N\left(r_0\right)} \ge \beta$$

i.e., the SINR at the receiver is greater than or equal to the threshold β.

By symmetry, when the packet transmission occurs in the opposite direction, i.e., from z to w, the interference generated by the set of nodes that are transmitting at the same time as z is also upper bounded by $N\left(r_0\right)$. Therefore, the SINR at w is also greater than or equal to β.

Finally, the existence of a (unique) solution to (7.2.10) can be proved by noting that $\frac{Pr_0^{-\alpha}}{N(r_0)} \to \infty$ as $r_0 \to 0$, $\frac{Pr_0^{-\alpha}}{N(r_0)} \to 0$ as $r_0 \to R_c^-$ and that $\frac{Pr_0^{-\alpha}}{N(r_0)}$ is a monotonically decreasing function of r_0. □

Corollary 125 relates the transmission range R_0 to the transmission power P. It allows the computation of R_0 given the transmission power P and the converse. A more convenient way to study the relation between P and R_0 is by noting that $P = P_{\text{th}}R_c^\alpha$ and considering R_0 as a function of R_c. Using (7.2.1) and (7.2.2), (7.2.10) can be rewritten as

$$\begin{aligned}\frac{1}{\beta} = & \frac{4\left(\frac{5\sqrt{3}}{4}\frac{R_c}{R_0} - 1\right)^{1-\alpha}\left(\frac{\sqrt{3}}{4}\left(3\alpha - 1\right)\frac{R_c}{R_0} - 1\right)}{\left(\frac{R_c}{R_0}\right)^2\left(\alpha - 1\right)\left(\alpha - 2\right)} \\ & + \frac{3}{\left(\frac{R_c}{R_0} - 1\right)^\alpha} + \frac{3}{\left(\sqrt{3}\frac{R_c}{R_0} - 1\right)^\alpha} + \frac{3\left(\frac{3R_c}{2R_0} - 1\right)^{1-\alpha}}{\left(\alpha - 1\right)\frac{R_c}{R_0}} \\ & + \frac{3}{\left(\frac{R_c}{R_0}\right)^\alpha} + \frac{3\left(\frac{3}{2}\right)^{1\ \ \alpha}}{\left(\frac{R_c}{R_0}\right)^\alpha\left(\alpha - 1\right)} + \frac{3}{\left(\frac{\sqrt{3}R_c}{R_0}\right)^\alpha}\end{aligned}$$

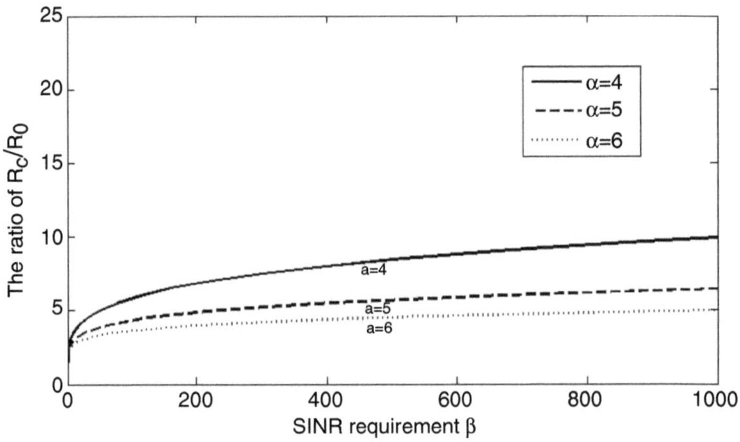

Fig. 7.2 Variation of the ratio $\frac{R_c}{R_0}$ with the SINR requirement β when the path loss exponent α is equal to 4, 5, 6, respectively

$$+\frac{3\left(\frac{5}{4}\right)^{1-\alpha}(3\alpha-1)}{(\alpha-1)(\alpha-2)\left(\sqrt{3}\frac{R_c}{R_0}\right)^{\alpha}} \tag{7.2.11}$$

Figure 7.2 shows the ratio $\frac{R_c}{R_0}$ as a function of the SINR requirement β. Different curves represent different choices of the path loss exponent α. For instance, when $\beta = 20$ and $\alpha = 4$, we have $\frac{R_c}{R_0} = 4$.

7.2.2 A Sufficient Condition for Connectivity

Based on the transmission range R_0 derived in Corollary 125, we obtain another main result of this section:

Theorem 126 *Consider a CSMA network with a total of n nodes identically and independently distributed on a square* $\left[-\frac{\sqrt{n}}{2}, \frac{\sqrt{n}}{2}\right]^2$ *following a uniform distribution. A pair of nodes are directly connected if and only if both (7.1.1) and (7.1.2) ($\gamma = 1$ and $N_0 = 0$ in (7.1.1) and (7.1.2)) are satisfied. Furthermore, the attenuation function* $l : \Re^+ \to \Re^+$ *in (7.1.1) and (7.1.2) is a power-law function given in (7.1.3) and the carrier-sensing condition is given in (7.1.4). As* $n \to \infty$*, the above network is* asymptotically almost surely *connected if the transmission power*

$$P = P_{th} b_1^{\alpha} (\log n + c(n))^{\frac{\alpha}{2}} \tag{7.2.12}$$

where $b_1 = b'/\sqrt{\pi}$, $c(n) \to \infty$ *as* $n \to \infty$ *and* $\infty > b' > 1$ *is the solution to the following equation:*

$$\frac{1}{\beta} = \frac{4\left(\frac{5\sqrt{3}}{4}x - 1\right)^{1-\alpha}\left(\frac{\sqrt{3}}{4}(3\alpha - 1)x - 1\right)}{x^2(\alpha - 1)(\alpha - 2)}$$
$$+ \frac{3}{(x-1)^{\alpha}} + \frac{3}{\left(\sqrt{3}x - 1\right)^{\alpha}} + \frac{3\left(\frac{3}{2}x - 1\right)^{1-\alpha}}{(\alpha - 1)x}$$
$$+ \frac{3}{x^{\alpha}} + \frac{3(\frac{3}{2})^{1-\alpha}}{x^{\alpha}(\alpha - 1)} + \frac{3}{\left(\sqrt{3}x\right)^{\alpha}}$$
$$+ \frac{3\left(\frac{5}{4}\right)^{1-\alpha}(3\alpha - 1)}{(\alpha - 1)(\alpha - 2)\left(\sqrt{3}x\right)^{\alpha}} \tag{7.2.13}$$

Proof We first show that there is a unique solution to (7.2.13) for any value of $\beta>0$. Let $f(x)$ be a function of x and $f(x)$ is equal to the right-hand side of (7.2.13). It can be shown that $f(x)$ is a monotonically decreasing function of x for $x > 1$. Then, noting that $f(x) \to \infty$ as $x \to 1$ and $f(x) \to 0$ as $x \to \infty$, it follows that for any value of $\beta > 0$, there exists a unique $\infty > b' > 0$ such that $f(b') = \frac{1}{\beta}$.

Using the results in Chap. 3 with proper scaling and coupling, it can be shown that for a network with a total of n nodes identically and independently distributed on a $\sqrt{n}\times\sqrt{n}$ square following a uniform distribution and a pair of nodes are directly connected if and only if their Euclidean distance is smaller than or equal to a given threshold $r(n)$ (i.e., the unit disk connection model), the network is *asymptotically almost surely* connected as $n \to \infty$ if and only if $r(n) = \sqrt{\frac{\log n + c(n)}{\pi}}$ where $c(n) \to \infty$ as $n \to \infty$. Using the above result, (7.2.11) (letting $b' = \frac{R_c}{R_0}$), Corollary 125 and Theorem 122, the result in the theorem follows. □

The implication of Theorem 126 is that in CSMA networks, even when the impact of mutual interference is considered, there exists a scheduling algorithm that allows as many as possible concurrent transmissions, and meanwhile, allows any pair of nodes in the network to be connected, in the sense that they can exchange packets, under an arbitrarily high SINR requirement. This result is in contrast to the ALOHA networks considered in [57, 58] in which percolation only occurs for a sufficiently small γ. Furthermore, it is evident from (7.2.12) that the transmission power only needs to be increased by a constant factor to combat interference and maintain connectivity compared with that in the unit disk connection model, in which no interference is considered.

7.3 Necessary Condition for Almost Surely Connected CSMA Networks

Section 7.2 derived a sufficient condition for a connected CSMA network as $n \to \infty$ in the presence of interference. A logical question after the results in Sect. 7.2 is: what is the necessary condition for a connected CSMA network as $n \to \infty$ in the presence of interference.

In a CSMA network, any set of nodes can transmit simultaneously as long as the carrier-sensing constraints are satisfied. Furthermore, in a large network, scheduling is often performed in a distributed manner. In the absence of accurate global knowledge on which particular set of nodes are simultaneously transmitting at a particular time instant, it is natural that a node sets its transmission power to be above *the minimum transmission power* required for a CSMA network to be connected under *any* scheduling algorithm. Denote that minimum power by P'_Ω where Ω represents the set of all scheduling algorithms satisfying the carrier-sensing constraints. In this section, we investigate P'_Ω, i.e., a necessary condition required for the same CSMA network considered in Sect. 7.2 to be connected as $n \to \infty$. This is done by analyzing the transmission power required for the above CSMA network to have no isolated node, where having no isolated node is a necessary condition for a connected network.

With a bit abuse of terminology, we call a set of nodes that can simultaneously transmit while satisfying the carrier-sensing constraints as an *independent set*. Obviously the independent set depends on the transmission power of nodes. As the transmission power decreases, other things being equal, R_c will decrease and the number of nodes that can simultaneously transmit will increase.

Denote by ϕ' a set of nodes that are scheduled to transmit simultaneously in the CSMA network. It follows that ϕ' must be an independent set. *Given* ϕ', a node $v \in \phi'$ is *isolated* if there is no node in the network that can successfully receive from it when the set of nodes in ϕ' are simultaneously transmitting. Furthermore, as explained in the last paragraph, the independent set depends on the transmission power. Let ϕ' be an independent set when the transmission power is P_1. When the transmission power is reduced from P_1 to P_2, where $P_2 \leq P_1$, if ϕ' is an independent set at power level P_1, it will also be an independent set at power level P_2. Based on the above observation and using (7.1.1) and (7.1.2), which suggests that when the set of active transmitters remain the same, a reduction in the transmission power will cause a decrease in the SINR, it readily follows that if a node $v \in \phi'$ is isolated at power level P_1 when the set of active transmitters is ϕ', it will also be isolated at power level P_2 when the set of active transmitters is ϕ'. Therefore, the probability that an arbitrary node is isolated is a nonincreasing function of the transmission power. It then follows that the probability that a network has no isolated node is a nonincreasing function of the transmission power.

Denote by P_Ω (respectively, P_ω) the minimum transmission power required for the network to have no isolated node under any scheduling algorithm (respectively, under a particular scheduling algorithm ω). Based on the above discussions, we have

$$P'_\Omega \geq P_\Omega = \max_{\omega \in \Omega} P_\omega \text{ and } P_\Omega \geq P_{\omega,\, \omega \in \Omega} \tag{7.3.1}$$

Now the task becomes constructing a particular algorithm which gives as large P_ω as possible, i.e., a tight lower bound on P'_Ω. In the following, we construct such scheduling algorithm ω heuristically.

7.3.1 *Construction of Scheduling Algorithm for CSMA Networks*

It is obvious that such a scheduling algorithm ω first needs to satisfy the constraint on the minimum separation distance between simultaneously active transmitters imposed by the carrier-sensing requirement. Meanwhile, the algorithm ω needs to schedule as many concurrent transmissions as possible to maximize interference, hence P_ω.

We first start with a technical lemma that is required for the construction of the scheduling algorithm ω.

Lemma 127 *Partition the square* $\left[-\frac{\sqrt{n}}{2}, \frac{\sqrt{n}}{2}\right]^2$ *into nonoverlapping hexagons of equal side length* s_n *such that the origin* o *coincides with the center of a hexagon and two diagonal vertices of this hexagon, whose Euclidean distance is* $2s_n$, *are located on* y *axis, as shown in Fig. 7.3. Furthermore, we call a hexagon an* interior hexagon *if it is entirely contained in the square* $\left[-\frac{\sqrt{n}}{2}, \frac{\sqrt{n}}{2}\right]^2$. *When*

$$s_n = \sqrt{\frac{2\log n}{5}} \tag{7.3.2}$$

asymptotically almost surely each interior hexagon is occupied by at least one node as $n \to \infty$.

Proof Because nodes are identically and independently distributed following a uniform distribution, for an arbitrary interior hexagon, the probability that it is empty is given by $\left(1 - \frac{3\sqrt{3}s_n^2}{2n}\right)^n$. Let ξ_i be the event that an interior hexagon i is empty, where $i \in \Xi$ and Ξ denotes the set of indices of all interior hexagons. Since each interior hexagon occupies an area of $\frac{3\sqrt{3}s_n^2}{2}$, there are at most $\frac{2n}{3\sqrt{3}s_n^2}$ interior hexagons. Therefore, $|\Xi| \leq \frac{2n}{3\sqrt{3}s_n^2}$.

Denote by A_n the event that there is at least one empty interior hexagon in $\left[-\frac{\sqrt{n}}{2}, \frac{\sqrt{n}}{2}\right]^2$. It follows that

$$\Pr(A_n) = \Pr\left(\bigcup_{i \in \Xi} \xi_i\right)$$

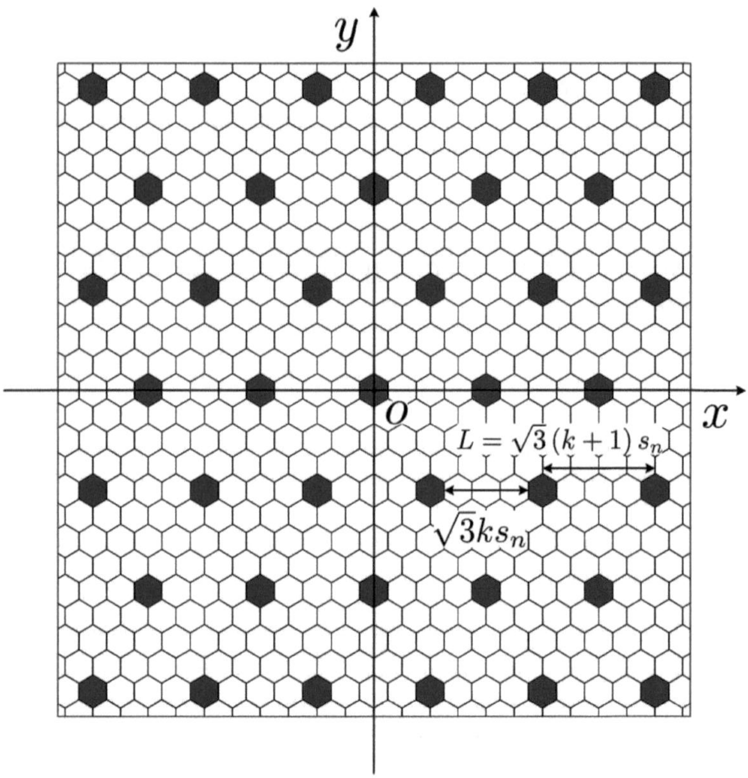

Fig. 7.3 An illustration of the hexagonal partition of the network area. The *shaded hexagons* represent simultaneously active hexagons, where $k = 3$

Using union bound, we have

$$\Pr\left(\bigcup_{i\in\Xi}\xi_i\right) \le \sum_{i\in\Xi}\Pr(\xi_i) \le \frac{2n\left(1-\frac{3\sqrt{3}s_n^2}{2n}\right)^n}{3\sqrt{3}s_n^2}$$

Using the fact that $1 - x \le \exp(-x)$ and $s_n = \sqrt{\frac{\log n}{\sqrt{3}}}$, we have

$$\begin{aligned}\lim_{n\to\infty}\Pr(A_n) &\le \lim_{n\to\infty}\frac{2ne^{-\frac{3\sqrt{3}s_n^2}{2}}}{3\sqrt{3}s_n^2} \\ &= \lim_{n\to\infty}\frac{5n}{3\sqrt{3}n^{\frac{3\sqrt{3}}{5}}\log n} \\ &= 0\end{aligned}$$

Therefore, it can be concluded that *asymptotically almost surely* every interior hexagon is occupied by at least one node as $n \to \infty$. □

Hereinafter, we declare a hexagon to be *active* if there is a node transmitting in it. In the following, we consider a scheduling algorithm ω that uses the hexagons as the basic unit for scheduling and allows at most one node in a hexagon to be active at any time instant. Due to the minimum separation distance constraint between simultaneously active transmitters in CSMA networks, any two simultaneously active hexagons should be separated by a minimum Euclidean distance (depending on the carrier-sensing range given in (7.1.4)). Let k be an integer, which represents the minimum number of inactive hexagons between two closest simultaneously active hexagons (see Fig. 7.3). Any two nodes inside the two active hexagons are separated by a Euclidean distance of at least $\sqrt{3}ks_n$. With a bit twist of terminology, we further define a *maximal independent set* for scheduling to be the set of hexagons that (a) includes as many hexagons as possible; and (b) closest hexagons in the set are separated by exactly k adjacent hexagons. Figure 7.3 illustrates such a maximal independent set with $k = 3$. Note that according to the earlier definition of maximal independent set, the number of maximal independent sets is finite and depends on k only.

The *scheduling algorithm* ω is defined such that only hexagons belonging to the same maximal independent set can be active at the same time. No nodes in the same hexagon can be scheduled to transmit simultaneously. Note that if a hexagon intersecting the border of $\left[-\frac{\sqrt{n}}{2}, \frac{\sqrt{n}}{2}\right]^2$ has node(s) in it, it is also included into the maximal independent set and its node(s) are treated in the same way as other nodes in interior hexagons. As a consequence of the CSMA constraint and the definition of k:

$$\sqrt{3}ks_n \geq R_c \geq \sqrt{3}\,(k-1)\,s_n \tag{7.3.3}$$

7.3.2 *Probability of Having No Isolated Node*

In this subsection, we derive a lower bound on P_ω for the scheduling algorithm ω defined in the previous subsection. This is done by analyzing the event that the network has no isolated node under the scheduling algorithm ω.

We first give a lemma, which will be used in the later analysis.

Lemma 128 *Consider a triangular lattice with unit side length and having a vertex located at the origin o. Define the* 1st *tier of points to be the six points placed at the vertices of the triangular lattice at a distance of 1 to the origin o. Let the* m^{th} *tier of points be the* $6m$ *points placed at the vertices of the triangular lattice located at distances within* $(m-1, m]$ *from the origin o, as shown in Fig. 7.4. The total number of points from the* 1st *tier to the* m^{th} *tier then is equal to* $j = 3m\,(1+m)$. *Let* $\boldsymbol{v}_1, \boldsymbol{v}_2, \ldots \boldsymbol{v}_j$ *be the location vectors of these* j *points and the points are ordered*

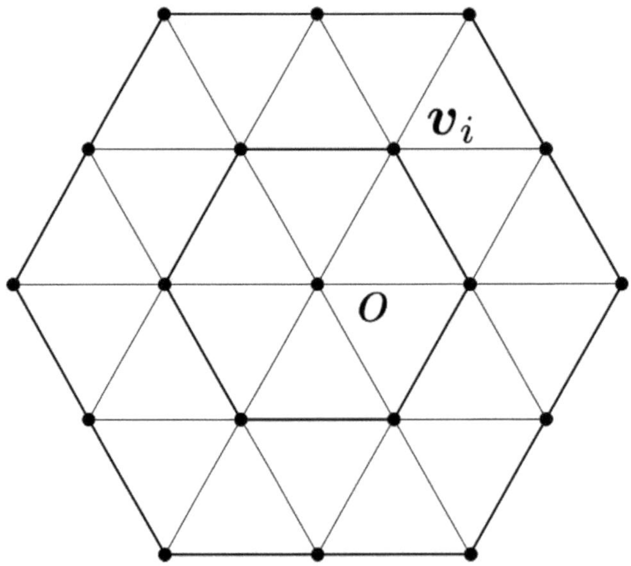

Fig. 7.4 An illustration of a triangular lattice

according to their distances to the origin o in a nondecreasing order. For an arbitrary point z located inside the hexagon formed by the 1st *tier of six points, the following holds:* $\sum_{i=1}^{j} \|\boldsymbol{v}_i - z\|^{-\alpha}$ *is minimized when z is located at the origin o, or equivalently when*

$$z = \frac{1}{j}\sum_{i=1}^{j} \boldsymbol{v}_i$$

where α *is the path loss exponent.*

The proof of Lemma 128 is based on the following theorem.

Theorem 129 ([6, Theorem 1]) *Let* $\boldsymbol{v}_1, \boldsymbol{v}_2, \ldots, \boldsymbol{v}_j$ *be j arbitrary points in* $\Re^2$. *Let* $w_1, w_2, \ldots w_j$ *be j positive numbers regarded as weights attached to these points, and define a position vector* **c** *by*

$$\sum_{i=1}^{j} w_i \boldsymbol{v}_i = W\boldsymbol{c} \ \text{ where } W = \sum_{i=1}^{j} w_i$$

Then, for an arbitrary point z, the following holds:

$$\sum_{i=1}^{j} w_i \|\boldsymbol{v}_i - z\|^2 = \sum_{i=1}^{j} w_i \|\boldsymbol{v}_i - \boldsymbol{c}\|^2 + W\|z - \boldsymbol{c}\|^2$$

Now we use Theorem 129 to prove Lemma 128.

Proof Letting all attached weights w_i equal to 1 and using Theorem 129, for an arbitrary point z located inside the hexagon formed by the 1st tier of six points, it can be shown that

$$\sum_{i=1}^{6} \|\boldsymbol{v}_i - \boldsymbol{z}\|^2 = \sum_{i=1}^{6} \|\boldsymbol{v}_i - \boldsymbol{c}\|^2 + 6\|\boldsymbol{z} - \boldsymbol{c}\|^2 \tag{7.3.4}$$

where $\boldsymbol{c}$ is given by

$$\sum_{i=1}^{6} \boldsymbol{v}_i = 6\boldsymbol{c} \tag{7.3.5}$$

It is clear that $\boldsymbol{c}$ is the centroid of the six points. Since the hexagon has a unit side length, $\|\boldsymbol{v}_i - \boldsymbol{c}\|$ is equal to 1. Let $x_i = \|\boldsymbol{v}_i - \boldsymbol{z}\|$ and $y = \|\boldsymbol{z} - \boldsymbol{c}\|$. The problem in Lemma 128 can then be converted to the following constrained minimization problem:

$$\begin{aligned} &\text{minimize} \;\; f(x_1, \ldots, x_6) = \sum_{i=1}^{6} x_i^{-\alpha} \\ &\text{subject to} \;\; h(x_1, \ldots, x_6) = \sum_{i=1}^{6} x_i^2 - 6 - 6y^2 = 0 \end{aligned}$$

where the constraint is due to (7.3.4). Using the method of Lagrange multipliers, we first construct the Lagrangian in the following:

$$F(x_1, \ldots x_6, \Lambda) = f(x_1, \ldots, x_6) + \Lambda h(x_1, \ldots, x_6)$$

where the parameter Λ is known as the Lagrange multiplier. Then, find the gradient and set it to zero:

$$\begin{aligned} &\nabla F(x_1, \ldots x_6, \Lambda) \\ &= \begin{pmatrix} -\alpha x_1^{-\alpha-1} + 2\Lambda x_1 \\ \vdots \\ -\alpha x_6^{-\alpha-1} + 2\Lambda x_6 \\ h(x_1, x_2, \ldots, x_6) \end{pmatrix}^T = \boldsymbol{0} \end{aligned}$$

Solving the above equation, it is obtained that

$$\Lambda = \frac{\alpha}{2}\left(1 + y^2\right)^{-\frac{\alpha+2}{2}}$$

$$\text{and } x_1 = x_2 \ldots = x_6 = \left(\frac{2\Lambda}{\alpha}\right)^{\frac{-1}{\alpha+2}} = \left(1 + y^2\right)^{\frac{1}{2}}$$

Since $x_i = \|\boldsymbol{v}_i - \boldsymbol{z}\|$ denotes the Euclidean distance from $\boldsymbol{v}_i$ to $\boldsymbol{z}$, only when $\boldsymbol{z} = \boldsymbol{c}$, we can have $x_1 = x_2 = \ldots = x_6 = 1$. It follows that the minimum of $f(x_1, x_2, \ldots, x_6)$ is obtained only when $\boldsymbol{z}$ is located at the origin o. Furthermore, for the $6m$ points of the mth tier, using the same method, it can be shown that $\sum_{i=1}^{6m} \|\boldsymbol{v}_i - \boldsymbol{z}\|^{-\alpha}$ is minimized only when $\boldsymbol{z}$ is located at the origin o. Hence, the result follows. □

On the basis of the above lemma, the following theorem, which summarizes a major outcome of the chapter, can be obtained:

Theorem 130 *Under the same setting in Theorem 126 and assuming the scheduling algorithm ω, a necessary condition on P_ω for the CSMA network to have no isolated node asymptotically almost surely as $n \to \infty$ is*

$$P_\omega \geq P_{th} b_2^\alpha (\log n)^{\frac{\alpha}{2}} \tag{7.3.6}$$

where $b_2 = \sqrt{\frac{6}{5}}(b-1)$ and b is the smallest integer satisfying the inequality:

$$\frac{2\left(\sqrt{3}(b+1)+1\right)^{1-\alpha}\left(\sqrt{3}(\alpha-1)(b+1)+1\right)}{(b+1)^2(\alpha-1)(\alpha-2)} \leq \frac{1}{\beta}\left(\frac{2\pi}{5}\right)^{\frac{\alpha}{2}} \tag{7.3.7}$$

Proof The main strategy used in proving Theorem 130 is to couple the network under the SINR connection model with the associated network under the unit disk connection model. On that basis, an upper bound on the probability of having no isolated node in the network under the SINR connection model can be found by using existing results in the literature obtained for the network under the unit disk connection model.

Denote the Euclidean distance between the centers of two closest hexagons in a maximal independent set by

$$L = \sqrt{3}(k+1)s_n \tag{7.3.8}$$

See Fig. 7.3 for an illustration. We divide the hexagons belonging to the same maximal independent set as a hexagon h_i into tiers of increasing Euclidean distance from the center of h_i using a similar strategy as that in Lemma 128. The Euclidean distances between the center of h_i and the centers of its 1st tier are $L = \sqrt{3}(k+1)s_n$. The centers of the hexagons in the mth tier of h_i are located within $((m-1)L, mL]$ from the center of h_i. With this definition, the mth tier of h_i has at most $6m$ hexagons. Furthermore, we declare that the mth tier of h_i is *complete in a given area* if all the $6m$ hexagons are entirely enclosed in this given area. Denote by C_A a square $\left[-\frac{\sqrt{cn}}{2}, \frac{\sqrt{cn}}{2}\right]^2$ ($0 < c < 1$ and the exact value of c will be decided later in this paragraph). The hexagon containing the origin o has a number of $t = \left\lfloor \frac{\frac{c\sqrt{n}}{2} - \frac{\sqrt{3}s_n}{2}}{L} \right\rfloor$ complete tiers in C_A. As the value of c increases, t increases as well. For the hexagons located inside C_A but near the border of C_A, the number of complete tiers in the square $\left[-\frac{\sqrt{n}}{2}, \frac{\sqrt{n}}{2}\right]^2$ decreases with an increase in c however. In the following analysis we choose the value of c such that each hexagon inside C_A has at least t complete tiers in the square $\left[-\frac{\sqrt{n}}{2}, \frac{\sqrt{n}}{2}\right]^2$, and the value of t is maximized. Let C'_A be the union of hexagons entirely contained in C_A. With a little bit abuse of terminology, we use C_A (respectively C'_A) to denote both the area itself and the size of the area. It can be readily shown that

$$\lim_{n\to\infty} \frac{C'_A}{C_A} = 1 \tag{7.3.9}$$

Consider an arbitrarily node i transmitting inside a hexagon h_i in C'_A. If there is no node that can receive from node i, then node i is isolated. Let I_{min} be the *minimum* interference that could possibly be experienced by a potential receiver of node i under the scheduling algorithm ω. Note that the Euclidean distance between the transmitter inside a hexagon in the mth tier of h_i and the center of hexagon h_i is less than $mL + s_n$ (see Fig. 7.3). Using (7.1.3) and Lemma 128, it can be shown that

$$\begin{aligned}
I_{min} & \\
&\geq \sum_{m=1}^{t} 6m\,(mL + s_n)^{-\alpha} P \\
&= 6P \sum_{m=1}^{t} m \left(\sqrt{3}m\,(k+1)\,s_n + s_n\right)^{-\alpha} \\
&= 6P s_n^{-\alpha} \sum_{m=1}^{t} m \left(\sqrt{3}m\,(k+1) + 1\right)^{-\alpha} \\
&= 6P s_n^{-\alpha} \int_1^t \lfloor x \rfloor \left(\sqrt{3}\lfloor x \rfloor (k+1) + 1\right)^{-\alpha} dx && (7.3.10) \\
&\geq 6P s_n^{-\alpha} \int_1^t x \left(\sqrt{3}x\,(k+1) + 1\right)^{-\alpha} dx && (7.3.11)
\end{aligned}$$

where $\lfloor x \rfloor$ denotes the largest integer smaller than or equal to x. Equation (7.3.11) is obtained due to the fact that the function $x\left(\sqrt{3}x\,(k+1)+1\right)^{-\alpha}$ is a decreasing function when $x > \frac{1}{\sqrt{3}(k+1)(\alpha-1)}$ and $\sqrt{3}\,(k+1)\,(\alpha-1) > 1$ for $\alpha > 2$ and $k \geq 1$. Therefore, $x\left(\sqrt{3}x\,(k+1)+1\right)^{-\alpha}$ is a decreasing function when $x > 1$.

Furthermore, noting that $\lim_{n\to\infty} t = \lim_{n\to\infty} \left\lfloor \frac{\frac{\sqrt{cn}}{2} - \frac{\sqrt{3}s_n}{2}}{L} \right\rfloor = \infty$, it follows that

$$\begin{aligned}
&\lim_{n\to\infty} 6 \int_1^t x\left(\sqrt{3}x\,(k+1)+1\right)^{-\alpha} dx \\
&= \frac{2\left(\sqrt{3}\,(k+1)+1\right)^{1-\alpha}\left(\sqrt{3}\,(\alpha-1)\,(k+1)+1\right)}{(k+1)^2\,(\alpha-1)\,(\alpha-2)} \\
&\triangleq f\,(k)
\end{aligned}$$

The above equation implies that for an arbitrarily small positive constant ε, there exists a positive integer n_ε such that when $n \geq n_\varepsilon$

$$\begin{aligned}
&6Ps_n^{-\alpha} \int_1^t \lfloor x \rfloor \left(\sqrt{3}\,\lfloor x \rfloor\,(k+1)+1\right)^{-\alpha} dx \\
&\geq Ps_n^{-\alpha}\,(f\,(k)-\varepsilon) \triangleq J_n
\end{aligned} \tag{7.3.12}$$

Let d be the Euclidean distance between node i and its receiver. Using (7.1.1), (7.1.2) and (7.1.3), it follows that *only* when $\frac{Pd^{-\alpha}}{J_n} \geq \beta$, the transmission from node i to its receiver could *possibly* be successful. In other words, if there is no node within a Euclidean distance of $\left(\frac{1}{P}\beta J_n\right)^{-\frac{1}{\alpha}}$ to node i, then node i is isolated. Denote this distance by

$$R = \left(\frac{1}{P}\beta J_n\right)^{-\frac{1}{\alpha}} \tag{7.3.13}$$

Denote by M and M^{SINR} the (random) number of isolated nodes in the CSMA network in the square $\left[-\frac{\sqrt{n}}{2}, \frac{\sqrt{n}}{2}\right]^2$ and in $C'_A \subset \left[-\frac{\sqrt{n}}{2}, \frac{\sqrt{n}}{2}\right]^2$, respectively. Denote by M^{UDM} the (random) number of isolated nodes in an area $C'_A \subset \left[-\frac{\sqrt{n}}{2}, \frac{\sqrt{n}}{2}\right]^2$ in a network with a total of n nodes identically, independently, and uniformly distributed on the square $\left[-\frac{\sqrt{n}}{2}, \frac{\sqrt{n}}{2}\right]^2$ and a pair of nodes are directly connected if and only if their Euclidean distance is smaller than or equal to R. Based on the discussion in the last paragraph and using the coupling technique, it can be shown that

$$\Pr\,(M \geq 1) \geq \Pr\left(M^{\text{SINR}} \geq 1\right) \geq \Pr\left(M^{\text{UDM}} \geq 1\right)$$

Consequently,

$$\Pr\,(M = 0) \leq \Pr\left(M^{\text{UDM}} = 0\right) \tag{7.3.14}$$

It remains to find the value of $\Pr\left(M^{\mathrm{UDM}} = 0\right)$.

We first consider a network with a total of n nodes distributed on a square $\left[-\frac{\sqrt{n}}{2}, \frac{\sqrt{n}}{2}\right]^2$ and a pair of nodes are directly connected if and only if their Euclidean distance is smaller than or equal to $r(n)$. It is well known that when $r(n) = \sqrt{\frac{\log n+\gamma(n)}{\pi}}$ and $\lim_{n\to\infty}\gamma(n) = \gamma$ where γ is a constant ($\gamma = \infty$ is allowed), the probability that there is no isolated node in the above network asymptotically converges to $e^{-e^{-\gamma}}$ as $n \to \infty$ (See Theorem 70). Furthermore, it was shown in Chap. 2 that boundary effect has an asymptotically vanishingly small impact on the number of isolated nodes. Let Z be a random integer representing the number of nodes located inside $C_A \subset \left[-\frac{\sqrt{n}}{2}, \frac{\sqrt{n}}{2}\right]^2$. Let $M^{r(n)}$ be the number of isolated nodes within an area C_A in the above network with a transmission range $r(n)$. Based on the above results, it follows that conditioned on that there are cn number of nodes in the subarea $C_A \subset \left[-\frac{\sqrt{n}}{2}, \frac{\sqrt{n}}{2}\right]^2$, i.e., when $Z = cn$ (Here we have omitted some trivial discussions involving the situation that cn is not an integer.),

$$\lim_{n\to\infty}\Pr\left(M^{r(n)} = 0 \middle| Z = cn\right) = e^{-ce^{-\gamma}} \tag{7.3.15}$$

It can also be shown that $E(Z) = cn$ and $\mathrm{Var}(z) = cn(1-c)$. Using Chebyshev's inequality, for $0 < \delta < \frac{1}{2}$, we obtain that

$$\begin{aligned}
&\lim_{n\to\infty}\Pr\left(|Z-cn| \geq (cn)^{\frac{1}{2}+\delta}\right)\\
&\leq \lim_{n\to\infty}\frac{\mathrm{Var}(Z)}{\left((cn)^{\frac{1}{2}+\delta}\right)^2}\\
&= 0
\end{aligned} \tag{7.3.16}$$

Let $f(n) = (cn)^{\frac{1}{2}+\delta}$. Using the following two equations:

$$\log(n+f(n)) + \gamma(n) = \log n + \log\left(1+\frac{f(n)}{n}\right) + \gamma(n)$$

and

$$\lim_{n\to\infty}\log\left(1+\frac{f(n)}{n}\right) + \gamma(n) = \lim_{n\to\infty}\gamma(n) = \gamma$$

and (7.3.15), it can be shown that

$$\lim_{n\to\infty}\Pr\left(M^{r(n)} = 0 \middle| Z = cn + f(n)\right) = e^{-ce^{-\gamma}}$$

Following the above procedure, it can be established that for any integer m satisfying $-f(n) \le m \le f(n)$,

$$\lim_{n\to\infty} \Pr\left(M^{r(n)} = 0 \middle| Z = cn + m\right) = e^{-ce^{-\gamma}}$$

The above equation, together with (7.3.16), allows us to conclude that when $r(n) = \sqrt{\frac{\log n + \gamma(n)}{\pi}}$

$$\lim_{n\to\infty} \Pr\left(M^{r(n)} = 0\right) = e^{-ce^{-\gamma}} \tag{7.3.17}$$

We are now ready to discuss $\Pr\left(M^{\mathrm{UDM}} = 0\right)$ and $\Pr(M = 0)$. As a result of (7.3.14), a necessary condition for $\lim_{n\to\infty} \Pr(M = 0) = 1$ is that $\lim_{n\to\infty} \Pr\left(M^{\mathrm{UDM}} = 0\right) = 1$. Using (7.3.9) and (7.3.17), it follows that a necessary condition for the network under the SINR connection model to have no isolated node asymptotically almost surely is that $R \ge \sqrt{\frac{\log n + \gamma(n)}{\pi}}$ and $\gamma(n) \to \infty$ as $n \to \infty$. As a consequence of (7.3.13), the above requirement on R means that

$$\left(\frac{1}{P}\beta J_n\right)^{-\frac{1}{\alpha}} \ge \sqrt{\frac{\log n + \gamma(n)}{\pi}}$$

as $n \to \infty$. Substituting the value of J_n in (7.3.12) and the value of s_n in (7.3.2) into the above equation, it can be further obtained that

$$f(k) \le \frac{1}{\beta}\left(\frac{2\pi}{5}\frac{\log n}{\log n + \delta}\right)^{\frac{\alpha}{2}} + \varepsilon$$

Letting $n \to \infty$ and then $\varepsilon \to 0$ in the above inequality, it can be obtained that

$$f(k) \le \frac{1}{\beta}\left(\frac{2\pi}{5}\right)^{\frac{\alpha}{2}}$$

Based on the above equation, together with (7.1.4) and (7.3.3), Theorem 130 results. □

The following corollary can be obtained as a ready consequence of Theorem 130 and (7.3.1).

Corollary 131 *A necessary condition required for CSMA networks to be asymptotically almost surely connected as $n \to \infty$ under* any *scheduling algorithm, i.e., a lower bound on P'_Ω, is given by*

$$P'_\Omega \ge P_{\mathrm{th}} b_2^{\alpha} (\log n)^{\frac{\alpha}{2}} \tag{7.3.18}$$

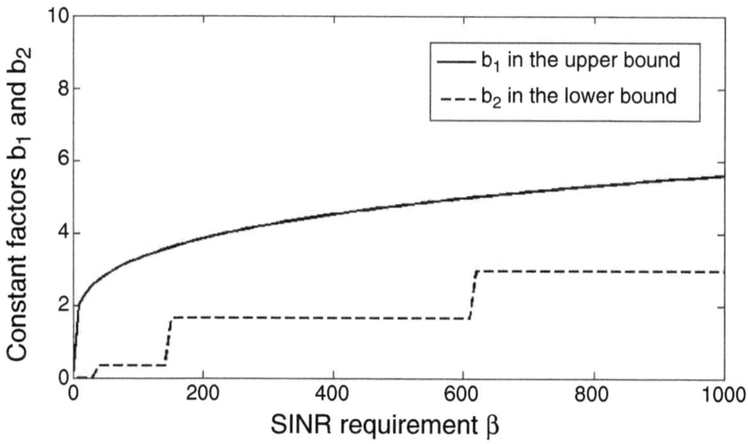

Fig. 7.5 A plot of the two constant factors b_1 and b_2 in the upper bound (7.2.12) and in the lower bound (7.3.18) when $\alpha = 4$

Comparing the lower bound on P'_Ω in (7.3.18) with the upper bound in (7.2.12) (i.e., a sufficient condition), it can be shown that, given an arbitrary SINR requirement β, the lower bound and the upper bound differ by a constant factor only as $n \to \infty$. Figure 7.5 shows a plot of the two constant factors, viz., b_1 and b_2, in the upper bound and in the lower bound, respectively, as a function of β when the path loss exponent $\alpha = 4$. The curve representing b_2 is a step function due to the granularity caused by the integer k in the scheduling algorithm ω.

7.4 Notes and Further Readings

In this chapter, we studied connectivity of wireless CSMA networks considering the impact of interference. We showed that, different from an ALOHA network, the aggregate interference experienced by any receiver in CSMA networks is upper bounded even when the coefficient γ in (7.1.1) and (7.1.2) is equal to 1.

An upper bound and a lower bound were obtained on the critical transmission power required for having an asymptotically almost surely connected CSMA network. These two bounds are tight and differ from each other by a constant factor only for an arbitrary SINR requirement. The results suggested that any pair of nodes can be connected (via multi-hop paths) under an arbitrarily high SINR requirement so long as the carrier-sensing capability is available. Compared with that considering a unit disk connection model without interference, the transmission power only needs to be increased by a constant factor to combat interference and maintain connectivity. This is a much more optimistic result compared with previous results considering connectivity of ALOHA networks under the SINR connection model.

As shown in Fig. 7.5, the upper and lower bounds on the critical transmission power for having an asymptotically connected CSMA network are close. The gap between the two bounds can be further narrowed by considering more intricate geometric shapes than hexagons. However, such improvement is possibly of minor importance. The implication of the results in this chapter is that there exists a spatial and temporal scheduling algorithm in a large CSMA network that allows as many as possible concurrent transmissions, and meanwhile, allows any pair of nodes in the network to be connected, in the sense that they can exchange packets, under an arbitrarily high SINR requirement.

Despite the significant impact of mutual interference caused by multiple concurrent transmissions on network connectivity, limited work has been done on analyzing network connectivity under the SINR connection model. In [10, 124], the authors studied connectivity from the perspective of channel assignment. Specifically, channels/time slots are assigned to each link in order for all active links to be simultaneously transmitting while satisfying the SINR requirement. The two papers [57, 58] discussed in Sect. 7.1 studied the impact of interference on connectivity of ALOHA networks from the percolation perspective. Some other work exists on modeling the point process formed by the location of simultaneously active transmitters in CSMA networks. Nguyen et al. [157] and Busson et al. [34] proposed using the Matern point process and Simple Sequential Inhibition point process, respectively, to model the set of active transmitters in CSMA networks.

Part II
Connectivity of Highly Dynamic Networks

Chapter 8
Connectivity of Dynamic Networks

Abstract In this chapter, we study connectivity of networks with dynamically changing network topology. In wireless networks, a change in topology can be caused by node mobility, by node or link failure, or by nodes being switched on or off for energy saving purposes. Dynamic networks offer many interesting and unique challenges than its static counterparts. Particularly, a dynamic network may be disconnected at any time instant but a message transmitted by any node can still reach any other node over time. That is, a dynamic network may become *connected over time*. In this chapter, we introduce those properties of dynamic networks that are different from statistic networks from the connectivity perspective. We extend and develop the concepts of route matrix, connectivity matrix, and probabilistic connectivity matrix as convenient tools to characterize and investigate properties of dynamic networks. The properties of these matrices are established and their relevance to properties of dynamic networks are introduced.

In the previous chapters, we have studied connectivity of large static wireless networks where nodes in the network are not moving or the network topology is time-invariant. In the next three chapters, we shall investigate connectivity of dynamic networks with dynamically changing network topology. In wireless networks, a change in topology can be caused by node mobility, by node or link failure, or by nodes being switched on or off for energy saving purposes.

8.1 Challenges and Opportunities in Dynamic Networks

From graph theoretic point of view, there are two major differences between connectivity of a static network and connectivity of a dynamic network:

- In dynamic networks, it is possible that two nodes may never be part of the same connected component but they are still able to communicate with each other within a finite time interval. This is illustrated in Fig. 8.1. In the figure, the network of six mobile nodes is disconnected at any time instant but there is a path from any node to any other node during the four time slots in the sense that a message transmitted by any node can reach any other node within the four

G. Mao, *Connectivity of Communication Networks*,
DOI 10.1007/978-3-319-52989-9_8

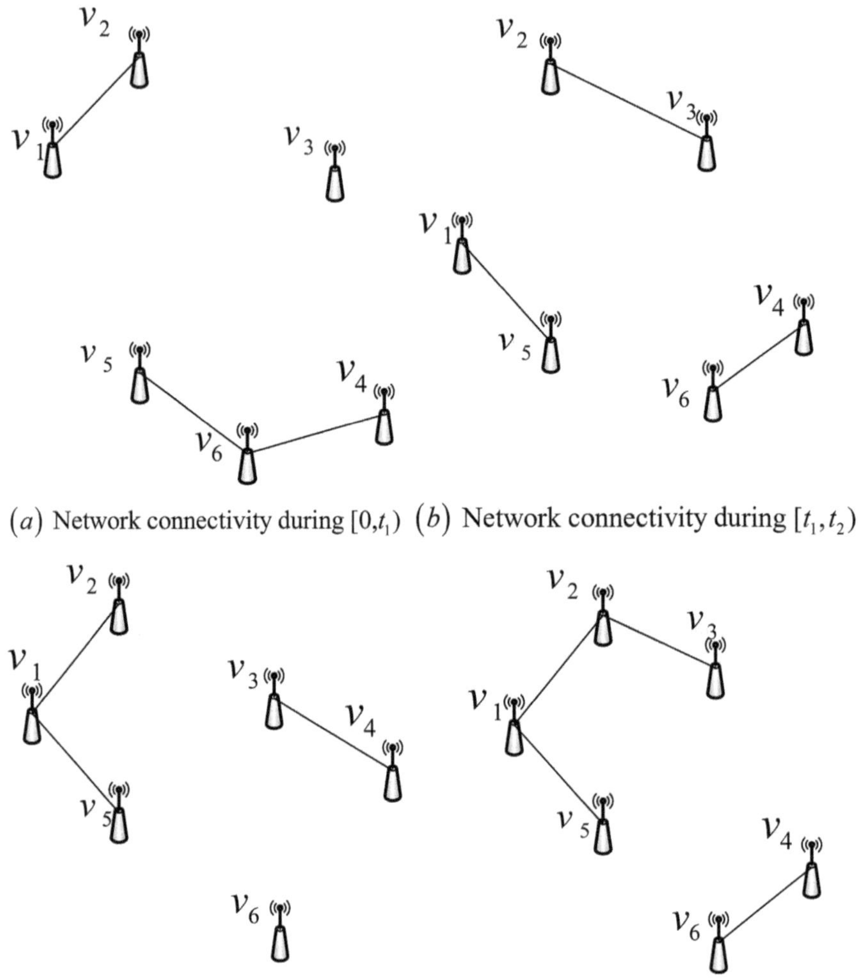

(a) Network connectivity during $[0, t_1)$ (b) Network connectivity during $[t_1, t_2)$

(c) Network connectivity during $[t_2, t_3)$ (d) Network connectivity during $[t_3, t_4)$

Fig. 8.1 An illustration of connectivity in a mobile network with six nodes. A *solid line* represents a direct connection between two nodes. The network is disconnected at any time instant but there is a path from any node to any other node in the network over the four time slots. For example, nodes v_1 and v_6 are never part of the same connected component but a message from v_1 can still reach v_6 through the following path: $[0, t_1) : v_1 \rightarrow v_2$, $[t_1, t_2) : v_2 \rightarrow v_3$, $[t_2, t_3) : v_3 \rightarrow v_4$, $[t_3, t_4) : v_4 \rightarrow v_6$

time slots. For example, a message transmitted by node v_1 can reach node v_6 through the following path: in the first time slot, the message may be transmitted from v_1 to v_2; in the second time slot, the message may be transmitted from v_2 to v_3; in the third time slot, the message may be transmitted from v_3 to v_4; while in the fourth time slot, the message reaches v_6 from v_4. Conversely, a message transmitted by node v_6 can reach node v_1 through the following path: in the first time slot, the message may be transmitted from v_6 to v_5 and in the second time slot, the message reaches v_1 through v_5.

- In dynamic networks, while any wireless link may be (or may be assumed to be) undirectional, the path connecting any two nodes must be regarded as directional, i.e., the fact that there is a path from node v_i to node v_j within a designated time period does not necessarily mean there is a path from v_j to v_i within the same period. For example, in Fig. 8.1 a message from v_6 can reach v_1 at t_2 but a message from v_1 can only reach v_6 at t_4. Therefore, there is a path from v_6 to v_1 during time interval $[0, t_2)$ does not imply that a path from v_1 to v_6 also exists within the same time interval $[0, t_2)$. This has been caused by the fact that in dynamic networks, the time sequence following which links appear affects the existence of a path from a node to another node.

Consequently, many established concepts in static networks must be revisited for dynamic networks. For example, a static network is said to be *connected* if there is a path between any pair of nodes in the network. A static network is *k-connected* if there are k mutually independent paths, i.e., paths not having any nodes in common except the starting and ending nodes, between any pair of nodes in the network. However a more meaningful definition of connectivity in dynamic networks is to say that a dynamic network is *connected in time period* $[0, T]$ if there is a path *from* any node *to* any other node within $[0, T]$. A more rigorous definition of connectivity in dynamic networks is deferred to later in the chapter. The above definition implies that the tradeoff between connectivity, mobility (or active period in networks with stationary nodes but nodes may be switched on or off) and delay must be properly considered in the analysis of dynamic networks. Particularly, the example shown in Fig. 8.1 reveals that node mobility in dynamic networks can be explored to improve connectivity.

In addition to connectivity, mobility has also been found to improve network capacity. It was shown in [87] that in a mobile network *with no delay requirement*, mobility of nodes can be explored to significantly increase network capacity [87, 91]. This can be understood intuitively as follows. In static networks, if the wireless channel is being occupied, a node has to wait whereas in mobile networks the node can carry the message over a physical distance, hopefully in the direction of the destination, until a better transmission opportunity arises. In other words, information propagates in static networks via a store-and-forward process whereas information propagates in mobile networks via a store-*carry*-forward process. It is this possibility that the information may be *carried*, instead of just being transmitted wirelessly, over a physical distance towards its destination in mobile networks that makes it possible for mobility to be explored to increase network capacity.

In this chapter, we use and extend a recently proposed graph theoretic model, i.e., *evolving graphs*, to capture the characteristics, in particular connectivity, of dynamic wireless networks. We extend and develop a set of tools, i.e., route matrix, connectivity matrix and probabilistic connectivity matrix, to investigate properties of dynamic wireless networks. The properties of these matrices related to dynamic wireless networks will be established. Note that in addition to the algebraic tools introduced in this chapter, branching process has also been widely used to model information propagation in dynamic networks. Please refer to Sect. 1.3.2 for a brief introduction to the branching process.

8.2 Connectivity Matrix and Probabilistic Connectivity Matrix for Dynamic Networks

A major challenge in the analysis of dynamic networks is choosing a proper graphic model that can both capture the dynamics of the network and be tractable for analytical studies. We use and extend a class of graphs, *evolving graphs* [24, 66, 192], as a formal abstraction of dynamic networks for our analysis. There has been research in the literature on the evaluation of computational complexity of algorithms in evolving graphs [24, 192] and on simulation based studies on the performance of routing algorithms in dynamic networks using evolving graphs [151].

An *evolving graph* $\mathcal{G} = (G, S_G)$ consists of a given graph $G(V, E)$, along with an ordered sequence of its subgraphs, $S_G = G_1(V_1, E_1), G_2(V_2, E_2), \ldots, G_\tau(V_\tau, E_\tau)$, $\tau \in \mathbb{N}$. Let the vertex set and the edge set of $\mathcal{G}$ be $V_\mathcal{G} = \cup V_i$ and $E_\mathcal{G} = \cup E_i$, respectively. If we assign a physical meaning to subgraphs such that the subgraph $G_i(V_i, E_i)$ at a given index i is the *underlying graph* of the network during the time interval $[t_{i-1}, t_i)$ and $t_0 < t_1 < \cdots < t_\tau$, the time domain is then incorporated into the model. Let P_i be a path in G_i. Let $F(P_i)$ and $L(P_i)$ be the first vertex and the last vertex of P_i, respectively, and $|P_i|$ be its length. We define a *journey* in $\mathcal{G} = (G, S_\mathcal{G})$ from vertex u to vertex v of $V_\mathcal{G}$ as a sequence $J(u, v) = P_{\lambda 1}, P_{\lambda 2}, \ldots, P_{\lambda l}$, with $1 \leq \lambda 1 < \lambda 2 < \cdots < \lambda l \leq \tau$, such that (a) $P_{\lambda i}$ is a path in $G_{\lambda i}$ with $F(P_{\lambda 1}) = u$, $L(P_{\lambda l}) = v$; (b) for all $1 \leq i \leq l$, it holds that $F(P_{\lambda i}) \neq L(P_{\lambda i})$ and (c) for all $i < l$, it holds that $L(P_{\lambda i}) = F(P_{\lambda i+1})$. The length of the journey is $\sum_{i=1}^{l} |P_{\lambda i}|$. The restriction $F(P_{\lambda i}) \neq L(P_{\lambda i})$ has limited the journey to consist of non-cycle paths only. Nevertheless a vertex may still appear more than once inside a journey, viz., a vertex might appear both in a path $P_{\lambda i} \in J(u, v)$ and in a path $P_{\lambda j} \in J(u, v)$, where $\lambda i \neq \lambda j$. The starting and the ending vertices of a journey may also be the same. We define an *evolving path* in $\mathcal{G} = (G, S_\mathcal{G})$ from vertex u to vertex v of $V_\mathcal{G}$ as a journey $J(u, v)$ in which all vertices are distinct except that the ending vertex of the journey may be the same as the starting vertex. The term *evolving path* is intended to differentiate a path in an evolving graph from a path in one of the subgraphs $G_{\lambda i}$. A vertex v is said to be k-hops away from another vertex u if the length of the shortest journey (or equivalently the shortest evolving path) from vertex u to vertex v is k. The evolving graph $\mathcal{G} = (G, S_\mathcal{G})$ is said to be a *connected* graph if there exists a journey (or equivalently an evolving path) from any vertex to any other vertex of $V_\mathcal{G}$.

To facilitate later discussions, we call a dynamic network *deterministic* if all entries of its adjacency matrix, as defined in Sect. 1.3.3, during each time interval are either 1 or 0, representing the fact that there are or there are no direct connections between the corresponding pairs of vertices, respectively. We call a dynamic network *probabilistic* if during some time intervals only statistical information on the probabilities that there are direct connections between certain pairs of nodes is known.

8.2.1 Connectivity Matrix of Deterministic Dynamic Networks

Let $0, \tau_1, \tau_2, \ldots, \tau_k, \ldots, 0 < \tau_1 < \tau_2 < \cdots < \tau_k \cdots$ be the embedding points of the network such that network topology, defined by a collection of all direct connections between nodes in the network, only changes at these discrete time instants. We define the *route matrix* H_k of a deterministic dynamic network of n nodes during time interval $[\tau_{k-1}, \tau_k)$ whose underlying graph is G_k as an $n \times n$ square matrix such that its (i,j)-th entry is the number of distinct paths in G_k from vertex i to another *distinct* vertex j. If there is no path from i to j, $(H_k)_{ij} = 0$ where $(H_k)_{ij}$ is the (i,j)-th entry of H_k. Furthermore, the diagonal entries of H_k *is 1 by definition.* As will be shown later, this definition of the diagonal entries properly reflects the fact that in dynamic networks routes and connections are constantly changing, when a node either becomes isolated or cannot find a (better) route to the destination, it has to store the message until a (better) route is available. For convenience, we assume that $V_1 = V_2 = \cdots = V_k = \cdots$ such that the route matrices during different time intervals have the same dimension. If a node disappears and then reappears in the network, the situation can be readily accommodated by considering that node as an isolated node during some time intervals.

Based on the above definition of the route matrix, we can establish some properties of the route matrix and their relevance to connectivity of a dynamic network.

Lemma 132 *There are a total of H^k_{ij} journeys from node i to node j in the evolving graph $\mathcal{G} = (G, S_G)$, $S_G = G_1(V_1, E_1), G_2(V_2, E_2), \ldots, G_k(V_k, E_k)$, $k \in \mathbb{N}$ within $[0, \tau_k)$, where H^k_{ij} is the (i,j)-th entry of $H_1 H_2 \cdots H_k - I$.*

Proof We prove Lemma 132 by recursion.

When $k = 1$, the lemma is obviously correct. Note that the diagonal entries of $H_1 - I$ is 0, i.e., there is no journey from a node to itself within $[0, \tau_1)$.

When $k = 2$,

$$H^2_{ii} = h^1_{ii} h^2_{ii} + \sum_{l \neq i} h^1_{il} h^2_{li} - 1 = \sum_{l \neq i} h^1_{il} h^2_{li}$$

is the total number of journeys from node i to itself within $[0, \tau_1) \bigcup [\tau_1, \tau_2)$, where $h_{ij}^p, p = 1, 2$ is the (i, j)-th entry of H_p. Each journey is made of two non-cycle paths, one from i to $l \neq i$ in G_1 and the other from l to i in G_2. For distinct i and j,

$$H_{ij}^2 = \sum_l h_{il}^1 h_{lj}^2 = h_{ij}^1 + h_{ij}^2 + \sum_{l \neq i,j} h_{il}^1 h_{lj}^2$$

The first term in the above equation is the number of journeys from node i to node j made of paths in G_1 only; the second term is the number of journeys from i to j made of paths in G_2 only; and the third term is the number of journeys from node i to node j, where each journey is made of a non-cycle path in G_1 and a non-cycle path in G_2. Therefore, the lemma is true for $k = 2$.

Assuming that the lemma is correct for $k = T$, let us consider $k = T + 1$ and $i \neq j$:

$$H_{ij}^{T+1} = \sum_l H_{il}^T h_{lj}^{T+1} = H_{ij}^T + H_{ii}^T h_{ij}^{T+1} + \sum_{l \neq i,j} H_{il}^T h_{lj}^{T+1}$$

In the above equation, the first term is the number of journeys from i to j within $[0, \tau_T)$. The sum of the second and the third terms is the number of newly added journeys from i to j where each journey consists of a non-cycle path in G_{T+1}. The second term is the sum of $\left(H_{ii}^T - 1\right) h_{ij}^{T+1}$, representing the number of journeys that (a) start at i before τ_T ; (b) end at i at τ_T; and (c) reach j at τ_{T+1}, and h_{ij}^{T+1}, representing the number of journeys from i to j made of non-cycle paths in G_{T+1} only. The third term is the number of journeys that end at an intermediate node at τ_T and reach j at τ_{T+1}. The sum of the three terms gives the total number of journeys from node i to node j within $[0, \tau_{T+1})$. It is trivial to show that $H_{ii}^{T+1} - 1$ is the total number of journeys that start and end at i within $[0, \tau_{T+1})$. □

Remark 133 Denote by H_k' the route matrix of G_k except that the diagonal entries of H_k' are zeros, i.e., $H_k' = H_k - I$. It can be shown that

$$H_1 H_2 \cdots H_k - I = \sum_i H_i' + \sum_{i<j} H_i' H_j' + \cdots + H_1' H_2' \cdots H_k'.$$

The r-th sum on the right-hand side of the above equation represents the number of journeys with exactly r non-cycle paths in r distinct subgraphs of $\mathcal{G}$, respectively. Therefore, the earlier definition of the route matrix H_k and Lemma 132 allow an easy computation of the total number of journeys ever occurred in the evolving graph $\mathcal{G}$.

Lemma 134 *A dynamic network is connected in* $[0, \tau_k)$ *if and only if* $H_1 H_2 \cdots H_k \geq J$ *element wise, where* J *is an* $n \times n$ *matrix with all entries equal to* 1.

The proof of Lemma 134 is straightforward and hence omitted.

Lemma 135 *H_{ij}^k is a nondecreasing function of k.*

This lemma is an easy consequence of Lemma 132 and the fact that H^k is the sum of I and a nonnegative matrix. An implication of Lemma 135 is that in dynamic networks the number of journeys from any node to any other node will only increase over time.

We define the *connectivity matrix* Q_k of a deterministic dynamic network of n nodes whose underlying graph is G_k during time interval $[\tau_{k-1}, \tau_k)$ as an $n \times n$ square matrix such that its (i,j)-th entry is 1 if there exists a path in G_k from vertex i to another distinct vertex j. If there is no path from i to j, $(Q_k)_{ij} = 0$. We also set $(Q_k)_{ii} \triangleq 1$. Alternatively, Q_k can be considered as the sign matrix of H_k. Using the properties of nonnegative matrices [177] and the above lemmas, we can establish the following properties of the connectivity matrix.

Lemma 136 *A dynamic network is connected in $[0, \tau_k)$ if and only if $Q_1 Q_2 \cdots Q_k \geq J$ element wise. There is a journey from vertex i to j if and only if $(Q_1 Q_2 \cdots Q_k)_{ij} \geq 1$.*

It can also be established that $(Q_1 Q_2 \cdots Q_k)_{ij}$ is a nondecreasing function of k. This, together with Lemma 136, suggests for a mobile network in which mobile nodes travel according to a random mobility model within a bounded area, e.g., random walk model [90], the network will become connected eventually. The interesting question is thus the rate at which the network becomes connected or equivalently the rate at which all entries of $Q_1 Q_2 \cdots Q_k$ (or equivalently $H_1 H_2 \cdots H_k$) become greater than or equal to 1.

The connectivity matrix and the route matrix defined earlier, although are convenient tools to investigate connectivity, do not capture the fact that in a communications network, a message can only travel a limited distance within a limited amount of time. In view of this, we extend the earlier definitions to account for message propagation delays. We call a path P_i in G_i *feasible* if a message of a unit length transmitted by the starting point of P_i at τ_{i-1} can reach the end point of P_i before τ_i in ideal conditions, i.e., no traffic congestion. We call a journey *feasible* if it is made entirely of feasible paths. We can also define the concept of *feasible connectivity matrix* and *feasible route matrix* analogously. Obviously the above lemmas on the properties of connectivity matrix and route matrix still hold if the relevant terms are replaced by their "*feasible*" counterparts. By a direct application of Lemma 132, we obtain the following Lemma that relates the feasible connectivity matrix to end-to-end throughput (or latency).

Lemma 137 *Assuming that node i starts to transmit a message of unit length to another distinct node j at time 0, there are M_{ij}^k feasible journeys from i to j that take less than or equal to τ_k time for the unit message to reach j, where M_{ij}^k is the (i,j)-th entry of $M_1 M_2 \cdots M_k$. M_k is the feasible route matrix of graph G_k during $[\tau_{k-1}, \tau_k)$.*

Lemma 137 can be used to derive an upper bound on the end-to-end throughput in a dynamic network. Furthermore, let τ be the time required to transmit a message of unit length from a node to its immediate neighbors. We choose the embedding points of the network $0, \tau_1, \tau_2, \ldots, \tau_k, \ldots$ to be $\tau_k = k\tau$ and assume that a node can only start transmitting a message at τ_k, which accords with a synchronized network. If a direct connection between two nodes is broken during the interval $[(i-1)\tau, i\tau)$, it is considered that there is no direct connection between the two nodes during $[(i-1)\tau, i\tau)$. Under the above conditions, a *feasible connectivity matrix* becomes the sum of the corresponding adjacency matrix and I. We can obtain the following lemma:

Lemma 138 *Assuming that node i starts to transmit a unit message to another distinct node j at time* 0*, the least amount of time for the message to arrive at node j is* $k\tau$ *if and only if* $M_{ij}^{k-1} = 0$ *and* $M_{ij}^{k} \geq 1$*. The number of messages that can be transmitted from i to j within* $[0, k\tau)$ *is less than or equal to* M_{ij}^{k}*.*

Lemmas 137 and 138 give an easy way to study the *foremost* journeys [24, 192] in an evolving graph. A slight modification of the two lemmas can be used to investigate the *shortest* journeys and the *fastest* journeys [24, 192] in an evolving graph algebraically.

8.2.2 Probabilistic Connectivity Matrix of Probabilistic Dynamic Networks

Let $0, \tau_1, \tau_2, \ldots, \tau_k, \ldots$, $0 < \tau_1 < \tau_2 < \cdots < \tau_k \cdots$ be the embedding points of the network such that network topology only changes at these discrete time points. We define the *probabilistic connectivity matrix* Q_k of a probabilistic dynamic network of n nodes during time interval $[\tau_{k-1}, \tau_k)$, whose underlying graph is G_k, as an $n \times n$ square matrix such that its (i,j)-th entry is the probability that there is a path in G_k from vertex i to another distinct vertex j *and* in addition for all i, $(Q_k)_{ii} \triangleq 1$. We further assume that

(1) the events whose probabilities are the entries in the same probabilistic connectivity matrix are independent;
(2) the events whose probabilities are the entries in probabilistic connectivity matrices during different time intervals are independent.

An example in which these assumptions hold is wireless sensor networks with uncoordinated power saving mechanisms [59]. We can establish the following properties of probabilistic connectivity matrix.

Lemma 139 *The probability that there exists at least one journey from node i to another distinct node j in the evolving graph* $\mathcal{G} = (G, S_G)$, $S_G = G_1(V_1, E_1), G_2(V_2, E_2), \ldots, G_k(V_k, E_k)$, $k \in \mathbb{N}$ *is* Q_{ij}^k, *where* Q_{ij}^k *is the* (i,j)*-th entry of* $Q_1 \otimes Q_2 \cdots \otimes Q_k$*. Here,* $C = A \otimes B$ *means* $C_{ij} = 1 - \prod_l \left(1 - A_{il}B_{lj}\right)$, *and* A_{ij}, B_{ij} *and* C_{ij} *are the* (i,j)*-th entries of* $n \times n$ *square matrices A, B, and C respectively.*

Proof We prove this lemma by recursion and omit the proof that the defined product operation is associative.

The lemma is obviously correct for $k = 1$ and the diagonal elements $Q^1_{ii} = 1$, $1 \leq i \leq n$.

Assuming the lemma is correct for $k = T$, let us consider $k = T + 1$. First, it can be shown that the diagonal elements of $Q_1 \otimes Q_2 \cdots \otimes Q_{T+1}$ are always 1. Let us now consider Q^{T+1}_{ij} where $i \neq j$,

$$Q^{T+1}_{ij} = 1 - \left(1 - q^{T+1}_{ij}\right)\left(1 - Q^T_{ij}\right)\prod_{l \neq i,j}\left(1 - Q^T_{il}q^{T+1}_{lj}\right)$$

In the above equation, q^{T+1}_{ij} is the (i,j)-th entry of Q_{T+1} and represents the probability of having a journey consisting of a path from i to j in G_{T+1} only. Q^T_{ij} is the probability of having a journey from i to j in the evolving graph $\mathcal{G} = (G, S_G)$, $S_G = G_1, G_2, \ldots, G_T$. The existence of such a journey necessarily means the existence of a journey in $\mathcal{G} = (G, S_G)$, $S_G = G_1, G_2, \ldots, G_{T+1}$. The term $Q^T_{il}q^{T+1}_{lj}$, $l \neq i, j$ is the probability of having a journey in $\mathcal{G} = (G, S_G)$, $S_G = G_1, G_2, \ldots, G_{T+1}$ that ends at an intermediate node l at τ_T and reaches j at τ_{T+1}. Using the assumption on the independence of events represented by entries in the probabilistic connectivity matrix, we can obtain that the above three events are independent. Therefore, $\prod_l \left(1 - Q^T_{il}q^{T+1}_{lj}\right)$ is the probability that none of the above three events occurs. Finally, $1 - \prod_l \left(1 - Q^T_{il}q^{T+1}_{lj}\right)$ gives the probability that at least one of the above three events occurs, which is the probability of having a journey from i to j in the evolving graph $\mathcal{G} = (G, S_G)$, $S_G = G_1, G_2, \ldots, G_{T+1}$. □

Lemma 140 *Q^k_{ij} is a nondecreasing function of k.*

This lemma can be proved using the definition of Q^k_{ij}.

Suppose that the probabilistic connectivity matrix is the same on every interval, so that the network simply resets certain connections, with the same probabilities, at each clock time. Then, this lemma can be used to show that in the limit as k tends to infinity, either Q^k_{ij} tends to 1, or it is zero for all k, i.e., there is never a connection from node i to node j.

Lemma 141 *A network of n nodes is connected with probability* 1 *in* $[0, \tau_k)$ *if and only if* $Q^k_{ij} = 1$ *for all pairs of i and j.*

This lemma is an easy consequence of Lemma 139.

Similarly by defining the concept of *feasible probabilistic connectivity matrix*, we can use the probabilistic connectivity matrix to investigate end-to-end delay. We define the *feasible probabilistic connectivity matrix* Q_k of a probabilistic dynamic network of n nodes during time interval $[\tau_{k-1}, \tau_k)$ as an $n \times n$ square matrix such that its (i,j)-th entry is the probability that there is a *feasible* path in G_k from vertex i to another distinct vertex j and $(Q_k)_{ij} \triangleq 1$. The above lemmas on the properties of probabilistic connectivity matrix still hold if the relevant terms are replaced by their "*feasible*" counterparts.

Furthermore, let τ be the time required to transmit a message of unit length from a node to its immediate neighbors. We choose the embedding points of the network $0, \tau_1, \tau_2, \ldots, \tau_k, \ldots$ to be $\tau_k = k\tau$ and assume a node can only start transmitting a message at τ_k. If a direct connection between two nodes is broken during the interval $[(i-1)\tau, i\tau)$, it is considered that there is no connection between the two nodes during $[(i-1)\tau, i\tau)$. Under the above conditions, a *feasible probabilistic connectivity matrix* becomes the sum of the corresponding probabilistic adjacency matrix, as defined in [32], and I. We can establish the following lemma:

Lemma 142 *Assuming that node i starts to transmit a unit message to a distinct node j at time* 0*, the probability that the message can reach j in less than* τ_k *time is* Q^k_{ij}*, where* Q^k_{ij} *is the* (i,j)*-th entry of* $Q_1 \otimes Q_2 \cdots \otimes Q_k$ *and the mean end-to-end delay from i to j is given by:*

$$\sum_{k=1}^{\infty} \left(1 - Q^{k-1}_{ij}\right) Q^k_{ij} k\tau.$$

This lemma can be obtained from Lemma 139. It can be used to study the mean end-to-end delay and the delay distribution.

8.3 Notes and Further Readings

In this chapter, we used and extended a recently proposed graph theoretic model, i.e., evolving graphs, as a formal abstraction of dynamic wireless networks. We extended and developed a set of tools, i.e., route matrix, connectivity matrix, and probabilistic connectivity matrix, to investigate properties of dynamic wireless networks. The properties of these matrices were investigated in relation to properties of the corresponding dynamic networks. We expect them to provide a convenient tool to formally investigate properties of dynamic networks.

The most related work to ours includes the use of *adjacency matrix* to study the connectivity of networks by Cvetkovic et al. [48] and Brooks et al. [32]. In [48], the *adjacency matrix* of a static network of n nodes is defined to be an $n \times n$ square matrix M whose (i,j)-th entry $m_{ij} = 1$ if there is a direct connection between distinct nodes i and j, otherwise $m_{ij} = 0$. Furthermore m_{ii} is defined to be 0, i.e., no self-loop is allowed. It is shown that the number of *walks* of length z between nodes i and j (with $j = i$ permitted) is m^z_{ij}, where m^z_{ij} is the (i,j)-th entry of M^z. A walk of length z between nodes i and j is a sequence of $z+1$ nodes $i, v_1, v_2, \ldots, v_{z-1}, j$ which starts at i and ends at j and there is a direct connection between adjacent nodes. A *path* of length z between nodes i and j is a walk of length z between nodes i and j in which the vertices $i, v_1, v_2, \ldots, v_{z-1}$ are distinct, save that $i = j$ is permitted. If $i = j$, the path is called a *cycle*. In [32] Brooks et al. considered a probabilistic version of the adjacency matrix for wireless networks and defined a *probabilistic adjacency matrix* as an $n \times n$ square matrix M whose (i,j)-th entry m_{ij} represents the

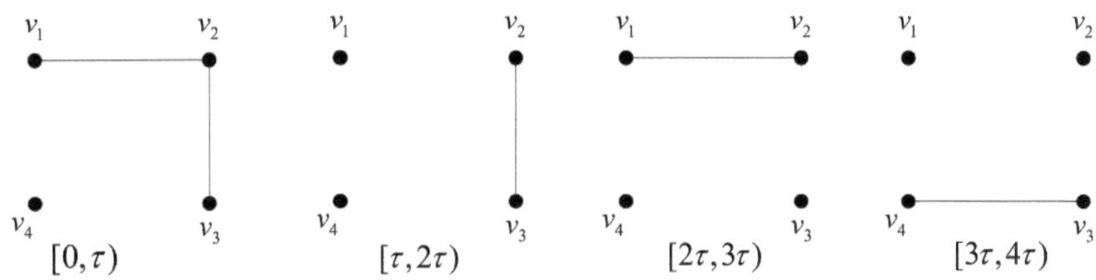

Fig. 8.2 An illustration of the connectivity of a dynamic network in different time intervals. Each subfigure represents the connectivity of a four-node network in a corresponding time interval. A *solid line* represents a direct connection between two nodes

probability of having a direct connection between distinct nodes i and j, and $m_{ii} = 0$. They established that the probability that there exists at least one walk of length z between nodes i and j is m_{ij}^z, where m_{ij}^z is the (i,j)-th entry of $M \otimes M \otimes \cdots \otimes M$ (z times). $C \triangleq A \otimes B$ if and only if $C_{ij} = 1 - \prod_{l \neq i,j} (1 - A_{il}B_{lj})$ where A_{ij}, B_{ij}, and C_{ij} are the (i,j)-th entries of the $n \times n$ square matrix A, B, and C, respectively. We remark that the product defined this way has the associativity property, allowing the computation of a product of three or more matrices with an arbitrary sequence of single product operations.

The above adjacency matrix and probabilistic adjacency matrix, however, is not suitable for investigating dynamic networks. This is illustrated through the example dynamic network of four nodes shown in Fig. 8.2. Denote the adjacency matrix of the network during $[(i-1)\tau, i\tau)$ by M_i. If one were to follow the above definition of adjacency matrix, it would be concluded that there is no connection from v_1 to v_4 in the network within $[0, 4\tau)$ because the $(1,4)$-th entries in M_1, M_1M_2, $M_1M_2M_3$ and $M_1M_2M_3M_4$ are all zero. This is obviously incorrect because a message from v_1 can reach v_4 using the following journey: $[0,\tau) : v_1 \to v_2$, $[\tau, 2\tau) : v_2 \to v_3$, $[3\tau, 4\tau) : v_3 \to v_4$. If the 1s in M_i, $1 \le i \le 4$ are replaced by values between 0 and 1 representing the probabilities that there are direct connections between the corresponding pairs of vertices, M_i, $1 \le i \le 4$ becomes the probabilistic adjacency matrix of the network during $[(i-1)\tau, i\tau)$ as defined in [32]. It can be obtained that the probability that there exists a connection from v_1 to v_4 in the network within $[0, 4\tau)$ is zero because the $(1,4)$-th entries in M_1, $M_1 \otimes M_2$, $M_1 \otimes M_2 \otimes M_3$ and $M_1 \otimes M_2 \otimes M_3 \otimes M_4$ are all zero. Apparently, this is not true. Therefore, neither the adjacency matrix defined in [48] and the probabilistic adjacency matrix defined in [32] is suitable for investigating the connectivity of dynamic networks. The problem with the definition of Brooks et al. [32] is that it fails to capture the notion that in a wireless network, it is legitimate for a message to remain at a node over one or more time intervals.

Chapter 9
Information Propagation in One-Dimensional Dynamic Networks

Abstract In this chapter, we study the information propagation process in one-dimensional dynamic networks with vehicular ad hoc networks being the motivating example. Particularly, we consider a vehicular ad hoc network formed by vehicles Poissonly distributed on a highway. Corresponding to different lanes of the highway and different types of vehicles, we consider that vehicles in the network can be categorized into a number of traffic streams, where vehicles in the same traffic stream have the same speed distribution while the speed distributions of vehicles in different traffic streams are different. We analyze the information propagation process of the above vehicular network and characterize the information propagation speed. The results established in this chapter allow one to study the impact of parameters such as radio range, vehicle traffic density, vehicle speed distribution, and the temporal variation of vehicle speed on the information propagation speed.

In this chapter, we study information propagation in one-dimensional dynamic networks with vehicular ad hoc networks (VANETs) being a typical example. As mentioned in the last chapter, a mobile network in which mobile nodes move randomly and independently will become connected eventually, in the sense that information from a node can reach another node eventually. The interesting question is thus the rate at which the network becomes connected. This motivates us to study the information propagation process in dynamic networks.

A VANET is a mobile multi-hop network formed by vehicles traveling on the road. As a new way of communication, VANETs have attracted significant interest not only in academia but also in industry. IEEE has taken up working on new standards for VANETs, such as the IEEE 1609 Family of Standards for Wireless Access in Vehicular Environments (WAVE). Furthermore, there are many projects on VANETs such as InternetITS in Japan, Network on Wheels in Germany, PReVENT project in Europe, etc. [198]. In this chapter, we study the expected propagation speed for a piece of information to be broadcast along the road in a VANET, referred to as the *information propagation speed (IPS)*. Due to the mobility of vehicles, the topology of a VANET is changing over time. Furthermore, traffic density on a road can vary significantly depending on time-of-day or day-of-week. Therefore, the information propagation process in a VANET can be quite different from that in a static network.

G. Mao, *Connectivity of Communication Networks*,
DOI 10.1007/978-3-319-52989-9_9

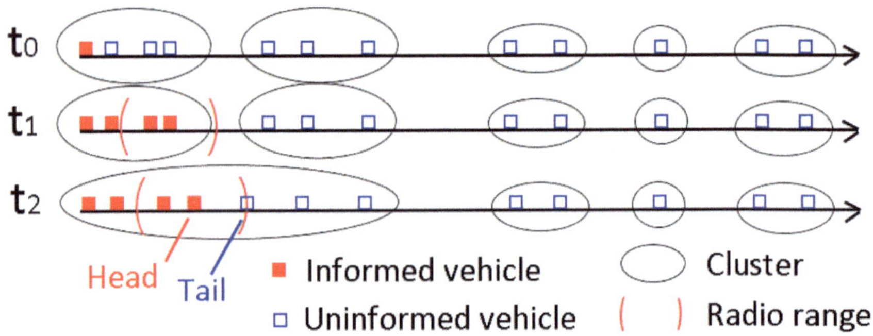

Fig. 9.1 An illustration of the topology of a VANET at different time instants. In the figure, the positive direction of the axis is the direction of information propagation

A VANET is often naturally partitioned into a number of clusters [190], where a *cluster* is a maximal set of vehicles in which every pair of vehicles is connected by at least one multi-hop path. Due to the mobility of vehicles, the clusters are splitting and merging over time. Therefore, information propagation in a VANET is typically based on a store-carry-forward scheme alike that in a delay-tolerant network. Considering the example illustrated in Fig. 9.1, a piece of information starts to propagate from the origin toward the positive direction of the axis at time t_0. The vehicles that have received this piece of information are referred to as the *informed vehicles*, where other vehicles are *uninformed*. As indicated by the left most ellipse, the first informed vehicle is inside a cluster of four vehicles at time t_0. At time t_1, the message is forwarded, in a multi-hop manner, to the foremost vehicle in its cluster. The propagation of the message within a cluster, which begins at t_0 and ends at t_1, is called a *forwarding process*. In a forwarding process the information propagation speed is determined by the per-hop delay and the length of the cluster. The *per-hop delay* β is the time required for a vehicle to receive and process a message before it is available for further retransmission [191]. The value of β depends on the practical implementation, and a common assumption for the value of β reflecting typical technology is 4ms [191]. We shall show that the per-hop delay has a significant impact on the IPS, especially when the vehicle density is high.

Define the *head* at time t to be the informed vehicle with the largest coordinate at time t. Define the *tail* at time t to be the uninformed vehicle with the smallest coordinate at time t. Two vehicles can directly communicate with each other if and only if their Euclidean distance is smaller than or equal to the transmission range r_0. Although this *unit disk connection model* is a simplified model, it can be indicative for real world scenarios. A realistic radio model usually takes into account statistical variations of the received signal power around its mean value. It is shown in [100] that these variations actually can increase the connectivity of a network. Therefore, the analysis under the unit disk connection model provides a conservative estimate on the performance of a VANET. As shown in Fig. 9.1, at time t_1 the tail is outside the transmission range of the head. Then, a *catch-up process* begins, during which the informed vehicles hold the information until the head catches up the tail (at time t_2). We study both the forwarding process and the catch-up process to characterize the information propagation process in VANET. We start by considering single traffic stream, i.e., vehicles traveling in the same direction and with the same speed distribution, and then move to examine the situation with multiple traffic streams.

9.1 Information Propagation Process in VANETs with Single Traffic Stream

In this section, we study information propagation process in one-dimensional VANETs with one traffic stream only. We start by describing the network model and mobility model for the VANET.

A synchronized random walk mobility model is used for modeling movement of vehicles. Specifically, time is divided into time slots with equal length τ. Each vehicle randomly chooses its new speed at the beginning of each time slot, independent of other vehicles and its own speed in other time slots, according to a certain distribution with a mean value $E[v]$. It is shown later that the constant speed model forms a special case in the above mobility model when $\tau \to \infty$. Due to the limited speed acceleration, the vehicle speed in the real world does not change as rapidly as in the above mobility model. Therefore, as will be shown later in Sect. 9.1.7.3, the results based on the above model provide a conservative estimate on the IPS. Furthermore, we also discuss the impact of a non-synchronized mobility model at the beginning of Sect. 9.1.4.

In the above model, the speed of a vehicle can be considered as having a constant component $E[v]$ and a variable component with a zero mean. Accordingly, the vehicular network can be decomposed into two components: a network in which all vehicles travel at a constant speed and a network in which vehicles travel at speeds following the same prescribed distribution $f_v(v)$ with a zero mean. Our analysis focuses on the IPS in the second network component, where $f_v(v)$ is the probability density function of the speed distribution. The first network component is considered separately and is combined into the result at the end of the analysis. Furthermore, a positive (negative) value of the speed means that the vehicle is traveling in the same (opposite) direction as the direction of information propagation. Therefore, when $E[v]$ is a positive (negative) value, our analysis provides the results for a VANET in which a message propagates in the same (opposite) direction as the direction of the vehicle traffic flow. In this section, the analysis is first performed for a general speed distribution. Then, detailed analytical results are given for the Gaussian speed distribution with standard deviation σ, which is commonly used for VANETs on a highway [126, 198].

The aforementioned speed-change time interval τ depends on practical conditions, e.g., a sports car may change its speed more frequently than a heavy truck.

Reasonable values for the time interval can be from 1s to 25s [18]. The vehicle mobility parameters, i.e., $E[v]$, σ and τ, are taken from practical measurements. Typical values for $E[v]$ and σ are given in [176], where the usual record time intervals for a vehicle speed monitor are $\tau = 1s, 5s$. We conduct our analysis in the discrete time domain ($t = i\tau$). Extension to the continuous time domain is straightforward following the procedure outlined later in the chapter.

We adopt a commonly used traffic model in vehicular traffic theory [126] in which vehicles travel independently in the same direction on a one-dimensional infinite line and follow the Poisson arrival model with a rate λ veh/s. The Poisson arrival model is a commonly used traffic model in vehicular traffic theory based on real world measurements. Furthermore, the Poisson arrival model and the Poisson distribution of the vehicles are also commonly used traffic models in the studies of vehicular ad hoc networks [68, 198]. The following lemma relates the spatial distribution of the vehicles on the road to the Poisson arrival model of the vehicles. The result on the spatial distribution of vehicles is used in the rest of this section.

Lemma 143 *If the traffic in a VANET follows the Poisson arrival model with rate λ and the speed of each vehicle changes at the beginning of each time slot, independent of other vehicles, according to $f_v(v)$, then at any time instant, the spatial distribution of the vehicles on the road follows a homogeneous Poisson point process with density $\rho = \lambda \int_{-\infty}^{\infty} \frac{f_v(v)}{v} dv$.*

Proof It has been shown in [198] that if the vehicle speeds do not change over time, then at any time instant the distances between adjacent vehicles (inter-vehicle distance l) are independent and follow an exponential distribution with rate parameter $\rho = \lambda \int_{-\infty}^{\infty} \frac{f_v(v)}{v} dv$. It follows that the spatial distribution of the vehicles follows a homogeneous Poisson point process with density ρ.

Next we apply the mathematical induction to study the spatial distribution of vehicles under our mobility model, in which vehicles are allowed to change their speeds from one time slot to another. In the first time slot $[0,\tau)$, it is straightforward to show that the spatial distribution of the vehicles follows a homogeneous Poisson point process with density ρ, because the speed does not change during a time slot.

Assume that in the ith time slot $[(i-1)\tau, i\tau)$, the spatial distribution of the vehicles still follows a homogeneous Poisson point process with density ρ. When the next time slot begins, each vehicle chooses its new speed, independent of other vehicles, according to $f_v(v)$. Then, according to the random splitting property of a Poisson point process ([155], see also Sect. 1.2), the spatial distribution of the vehicles traveling at the speed $v \in [v_m, v_m + dv_m)$ during the $i + 1^{th}$ time slot $[i\tau,(i + 1)\tau)$ follows a homogeneous Poisson sub-process with density $\rho f_v(v_m)dv_m$. These vehicles have the same speed so that at any time instant during the $i + 1^{th}$ time slot, the spatial distribution of these vehicles does not change. Then, according to the random coupling property of a Poisson point process [155], the spatial distribution of all vehicles in the $i + 1^{th}$ time slot follows a homogeneous Poisson point process, which is the sum of all the sub-processes, with the density $\int_{-\infty}^{\infty} \rho f_v(v_m)dv_m = \rho$. □

Note that previous research [198] only considered that the vehicle speeds do not change over time. In Lemma 143, we consider the time variation of vehicle speeds in the analysis of vehicle distribution on the road.

9.1.1 Catch-Up Process for a General Speed Distribution

In this subsection, we study the catch-up process in a VANET where the vehicle speeds follow a general probability density function $f_v(v)$. Without loss of generality, it is assumed that the catch-up process starts at time 0. The *displacement* x of a vehicle at time t is defined as the difference between the position of the vehicle at time 0 and its position at time t.

9.1.2 Modeling the Movement of Single Vehicle

Denote by $p(x, t)$ the probability that the displacement of a vehicle is x at time t. Because the speed does not change during a time slot, $p(x, \tau)$ can be readily obtained from $f_v(v)$.

Due to the independence of the vehicle speeds in different time slots, hence the displacements, we have for $t = i\tau$:

$$p(x, t) = p(x, i\tau) = \overbrace{(p * p * \ldots * p)}^{\text{i-fold convolution}}(x, \tau) \tag{9.1.1}$$

The calculation of the above i-fold convolution can be simplified by using the Fourier and inverse Fourier transform.

9.1.3 Modeling the Movement of the Head and Tail

Denote by H_m (P_m) the mth vehicle to the left of the head H_0 (to the right of the tail P_0) at time 0, as shown in Fig. 9.2. Define w_m to be the Euclidean distance between H_m and H_0 at time 0.

Let us first consider the movement of the head. Define $x_m(t)$ to be the displacement of H_m at time t. Define $y(t)$ to be the displacement of the head at time t. Note that the head vehicle at time 0 is not necessarily the head vehicle at time t because the original head may be overtaken by another informed vehicle during time $(0, t]$. It follows that:

$$y(t) = \max\{x_0(t), x_1(t) - w_1, x_2(t) - w_2, \ldots, x_n(t) - w_n\} \tag{9.1.2}$$

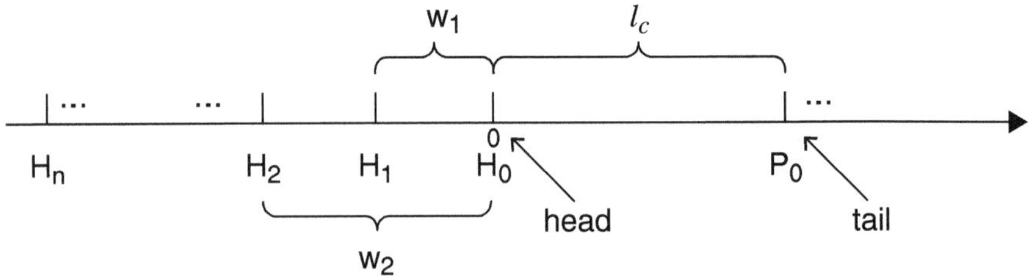

Fig. 9.2 An illustration of a VANET at the beginning of a catch-up process with gap l_c, where l_c is the Euclidean distance between the head and the tail at time 0. Hereinafter, a catch-up process where the initial distance between the head and tail is l_c at time 0 is referred to as a *catch-up process with gap* l_c

where n is the number of vehicles, to the left of the head, that have the potential to overtake the head vehicle at time 0.

Because the movement of a vehicle is independent of other vehicles, $x_m(t)$ and $x_j(t)$ are independent for any $m \neq j$. Therefore, the cumulative distribution function of the displacement of the head at time t is:

$$\begin{aligned} \Pr(y(t) \leq y) &= \prod_{m=0}^{n} \Pr(x_m(t) - w_m \leq y) \\ &= \prod_{m=0}^{n} \left(\int_0^{\infty} \int_{-\infty}^{y+w_m} p(x,t) f_{w_m}(w_m) dx dw_m \right) \end{aligned} \tag{9.1.3}$$

where $p(x,t)$ is given by (9.1.1), w_m is the distance between H_m and H_0 at time 0, $w_0 = 0$ and $f_{w_m}(w_m)$ is the probability density function of w_m. As an easy consequence of the Poisson distribution of the vehicles, proved in Lemma 143, the inter-vehicle distance follows an exponential distribution. Therefore, we have:

$$f_{w_m}(w_m) = \frac{\rho e^{-\rho w_m} (\rho w_m)^{m-1}}{(m-1)!} \text{ for } m \geq 1 \tag{9.1.4}$$

Define $p_h(y,t)$ to be the probability that the displacement of the head is y at time t. Then:

$$p_h(y,t) = \frac{\partial \Pr(y(t) \leq y)}{\partial y} \tag{9.1.5}$$

The calculation is however tedious for a general speed distribution. Therefore, only the methodology for the analysis of the catch-up process is shown in this subsection. A detailed analytical result for the catch-up process under the Gaussian speed distribution is shown in Sect. 9.1.4.

Denote by $p_g(\tilde{y}, t)$ the probability that the displacement of the tail is $\tilde{y}$ at time t. The analysis for the movement of the tail is similar to the one for the head and therefore is omitted.

9.1.3.1 Catch-Up Delay

As illustrated in Fig. 9.2, consider a catch-up process where the Euclidean distance between the head and the tail is l_c at the beginning of the catch-up process (designated as time 0), which is referred to as *a catch-up process with gap* l_c. Define the *catch-up delay* t_c to be the time taken from time 0 till the time when the head and tail move into the transmission range of each other for the first time, i.e., $t_2 - t_1$ in Fig. 9.1. We do not consider the rare event that the distance between the head and tail becomes larger than r_0 again during $(t_c, t_c+\beta)$, which may cause the transmission of a packet being interrupted, because the per-hop delay β (e.g., 4ms [191]) is usually much smaller than the time interval for a vehicle to change speed (typically longer than a second [18]). It is worth noting that there is a first passage phenomenon in the catch-up process, i.e., the catch-up process finishes as soon as the head and the tail move into the transmission range of each other. Therefore, the catch-up delay is t_c *if and only if* the distance between the head and tail reduces from l_c at time 0 to r_0 *for the first time* at time t_c. This first passage phenomenon is important for the analysis of the catch-up process.

Denote by $p_H(z, t)$ the probability that the reduction of the Euclidean distance between head and tail is z at time t, with regard to their original distance at time 0. As illustrated in Fig. 9.3, it can be shown that:

$$p_H(z, t) = \int_{-\infty}^{\infty} p_h(y, t) p_g(y - z, t) dy \tag{9.1.6}$$

Note that the above equation can be converted into convolution if $p_g(\tilde{y}, t) = p_g(-\tilde{y}, t)$, which is the case to be introduced in the next subsection when the vehicle speed follows a Gaussian distribution.

Denote by $h(z, i)$ the probability that the reduction of the distance between head and tail reaches z in the ith time slot $[(i-1)\tau, i\tau)$. Therefore,

$$h(z, i) = \int_{(i-1)\tau}^{i\tau} p_H(z, t) dt \tag{9.1.7}$$

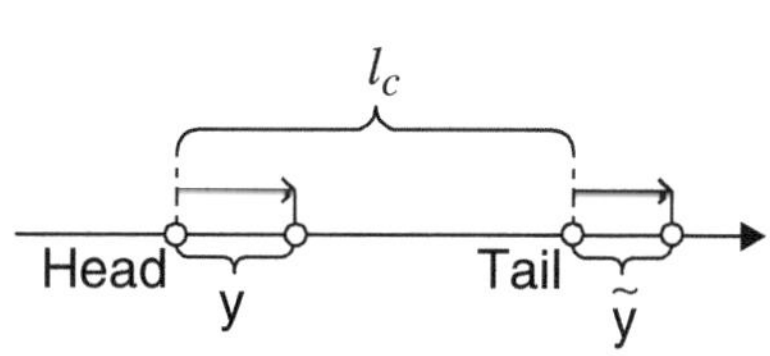

Fig. 9.3 An illustration of the displacements of the head and tail at time t in a catch-up process with gap l_c. The reduction of distance is $z = y - \tilde{y}$

In order to obtain a closed-form result, the probability of the reduction of the distance between head and tail being z at time $t \in [(i-1)\tau, i\tau)$ is considered to be approximately equal to the probability of the reduction of the distance between head and tail being z at time $t = i\tau$. This approximation provides a fairly accurate result when τ is small (e.g., τ =1s, 5s) as shown in Sect. 9.1.7.1 and Sect. 9.1.7.3. Therefore, from (9.1.7),

$$h(z,i) = \int_{(i-1)\tau}^{i\tau} p_H(z,t)dt \approx \tau p_H(z,i\tau) \tag{9.1.8}$$

Define $\xi(z,i)$ to be the first passage probability [174] of $h(z,i)$, viz., the probability that the reduction of the distance between head and tail reaches z in the ith time slot $[(i-1)\tau, i\tau)$ *for the first time* since time 0. The relationship between $\xi(z,i)$ and $h(z,i)$ can be studied as the first passage time in a stochastic process [152]. The *first passage time* of a diffusing particle or a random walker is the time at which the particle or the random walker *first* reaches a specified site.

A standard procedure is applied to determine the first passage probability $\xi(z,i)$ [152, 174]. As illustrated in Fig. 9.4, consider a class of random walks starting at time 0 and walking from point 0 to z' must proceed by going through a point z. The transition from 0 to z' can be decomposed into two independent stages: in the first stage an agent walks from 0 to z for the first time in the ith time slot; in the second stage the agent walks from z to z' in the $i' - i^{th}$ time slot, not necessarily for the first time. Then, we can obtain the following equation [152, 174]:

$$h(z',i') = \sum_{i=1}^{i'} \xi(z,i)h(z'-z,i'-i) \tag{9.1.9}$$

The convolution can be simplified by using the Z-transform with regard to i, which is denoted by the operator $\mathcal{Z}$. According to the convolution theorem, (9.1.9) becomes

$$(\mathcal{Z}h)(z',s) = (\mathcal{Z}\xi)(z,s)\ (\mathcal{Z}h)(z'-z,s) \tag{9.1.10}$$

$$\text{Thus: } (\mathcal{Z}\xi)(z,s) = \frac{(\mathcal{Z}h)(z',s)}{(\mathcal{Z}h)(z'-z,s)} \tag{9.1.11}$$

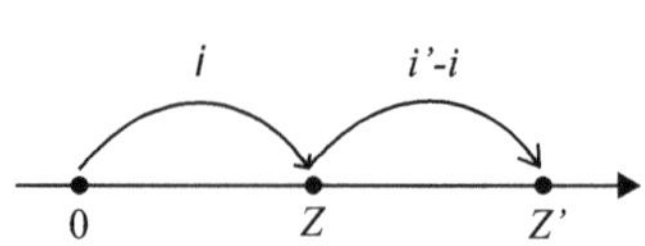

Fig. 9.4 An illustration of a class of random walks taking i' time slots to walk from 0 to z' through an intermediate point z

Then, by inverse Z-transform we can obtain $\xi(z,i)$. Denote by $F_\xi(z,i)$ the cumulative distribution function of $\xi(z,i)$ with regard to i, i.e., the probability that the reduction of the distance between head and tail has reached z during time $(0, i\tau]$. It follows that $F_\xi(l_c - r_0, i)$ is the probability that the head and tail have moved into the transmission range of each other during time $(0, i\tau]$. Therefore, the expected catch-up delay (t_c) for a catch-up process with gap l_c is

$$E[t_c|l_c] = \tau \sum_{i=1}^{\infty} (1 - F_\xi(l_c - r_0, i)) \tag{9.1.12}$$

9.1.3.2 Distribution of the Gaps l_c

Denote by $f_l(l)$ the probability density function of the Euclidean distance between any two adjacent vehicles. Due to the Poisson distribution of vehicles, it is evident that $f_l(l) = \rho e^{-\rho l}$. Denote by $f_{l_c}(l_c)$ the probability density function of the Euclidean distance between any two adjacent but disconnected vehicles. It is straightforward that for $l_c > r_0$:

$$f_{l_c}(l_c) = \frac{f_l(l_c)}{1 - \int_0^{r_0} f_l(l)dl} = \rho e^{-\rho(l_c - r_0)} \tag{9.1.13}$$

9.1.4 Catch-Up Process for a Gaussian Speed Distribution

In this subsection, we provide detailed analytical results on the catch-up process assuming a Gaussian speed distribution, which is a commonly used assumption for VANETs on a highway [126, 176, 198]. The procedure of the analysis is the same as the one introduced in the previous subsection, except for some adjustments in order to obtain a simpler result.

For a zero mean Gaussian speed distribution with standard deviation σ, the probability density function of the vehicle speed is:

$$f_v(v) = \frac{1}{\sigma\sqrt{2\pi}} \exp\left(\frac{-v^2}{2\sigma^2}\right) \tag{9.1.14}$$

At the end of the first time slot, i.e., $t = \tau$, it is straightforward to show that:

$$p(x,\tau) = \frac{1}{\sigma\tau\sqrt{2\pi}} \exp\left(\frac{-x^2}{2(\sigma\tau)^2}\right) \tag{9.1.15}$$

Furthermore, because the convolution of two Gaussian functions yields another Gaussian function [85], using (9.1.1) we can obtain:

$$p(y, i\tau) = \frac{1}{\sigma_i\sqrt{2\pi}} \exp\left(\frac{-y^2}{2\sigma_i^2}\right) \tag{9.1.16}$$

where $\sigma_i^2 = i(\sigma\tau)^2$.

Next, we consider the situation in which vehicles are allowed to change speed at different time instants. Without loss of generality, consider that a vehicle changes its speed at time τ_0 for the first time since time 0, where τ_0 is uniformly distributed in $(0, \tau]$. For $t = i\tau$, (9.1.1) becomes:

$$\begin{aligned} p(x,t) &= p(x, i\tau) \\ &= p(x, \tau_0) * \overbrace{(p * p * \ldots * p)(x, \tau)}^{\text{(i-1)-fold convolution}} * p(x, \tau - \tau_0) \\ &= \frac{1}{\sigma_i\sqrt{2\pi}} \exp\left(\frac{-x^2}{2\sigma_i^2}\right) \end{aligned} \tag{9.1.17}$$

where $\sigma_i^2 = (i-1)\sigma^2\tau^2 + \sigma^2\tau_0^2 + \sigma^2(\tau - \tau_0)^2 = i\sigma^2\tau^2 + 2\sigma^2\tau_0^2 - 2\sigma^2\tau\tau_0$.

Compared with the result obtained using a synchronized mobility model (i.e., (9.1.16)), the additional terms are $2\sigma^2\tau_0^2 - 2\sigma^2\tau\tau_0$. To simplify the analysis, we ignore the additional terms and consider the synchronized mobility model only. The error caused by ignoring these additional terms and assuming a synchronized mobility model is given by $\int_0^\tau (2\sigma^2\tau_0^2 - 2\sigma^2\tau\tau_0)\frac{1}{\tau}d\tau_0 = -\frac{\sigma^2\tau^2}{3}$. It is obvious that the error is small compared with the dominant term $i\sigma^2\tau^2$, particularly when i is large. Furthermore, the accuracy of this approximation is verified in Sect. 9.1.7.3.

9.1.4.1 Catch-Up Delay in a Basic Catch-Up Process

In this subsection, we temporarily ignore the possibility of overtaking, i.e., we consider *a basic catch-up process* involving only the vehicle, which is the head at time 0, catching up with the vehicle, which is the tail at time 0. We have the following lemma for the basic catch-up process.

Lemma 144 *In a basic catch-up process where vehicle speed follows a zero mean Gaussian distribution with standard deviation σ, the probability that the reduction of distance between the head and tail is z for the first time during the ith time slot is:*

$$\xi(z, i) = \frac{z}{2i\tau\sigma\sqrt{\pi i}} \exp\left(-\frac{z^2}{4i\tau^2\sigma^2}\right) \tag{9.1.18}$$

Proof Because the Gaussian speed distribution is symmetric with respect to the mean, for the displacement of the tail we have: $p_g(\tilde{y}, t) = p(\tilde{y}, t) = p(-\tilde{y}, t)$. Using (9.1.6), it can be shown that:

$$\begin{aligned} p_H(z, i\tau) &= \int_{-\infty}^{\infty} p(y, i\tau) p_g(z - y, i\tau) dy \\ &= (p * p)(z, i\tau) \\ &= \frac{1}{\tilde{\sigma}_i \sqrt{2\pi}} \exp\left(\frac{-z^2}{2\tilde{\sigma}_i^2}\right) \end{aligned} \tag{9.1.19}$$

where $\tilde{\sigma}_i^2 = 2i(\sigma\tau)^2$. Therefore:

$$h(z, i) = \tau p_H(z, i\tau) = \frac{\tau}{\tilde{\sigma}_i \sqrt{2\pi}} \exp(\frac{-z^2}{2\tilde{\sigma}_i^2}) \tag{9.1.20}$$

Inspired by [152, Eq. 6.4], $h(z, i)$ in (9.1.20) can be rewritten in the following form in order to calculate the Z-transform:

$$h(z, i) = \frac{\tau}{2\pi} \int_{-\infty}^{\infty} \exp(-jz\alpha - \frac{\tilde{\sigma}_i^2}{2}\alpha^2) d\alpha \tag{9.1.21}$$

where j denotes $\sqrt{-1}$.

Then, perform the Z-transform on (9.1.21) with regard to i:

$$\begin{aligned} (\mathcal{Z}h)(z, s) &= \sum_{i=1}^{\infty} e^{-si} h(z, i) \\ &= \frac{\tau}{2\pi} \int_{-\infty}^{\infty} \exp(-jz\alpha) \sum_{i=1}^{\infty} \exp(-si) \exp(-\frac{\tilde{\sigma}_i^2}{2}\alpha^2) d\alpha \end{aligned}$$

With $\tilde{\sigma}_i^2 = 2i(\sigma\tau)^2$, there holds:

$$\begin{aligned} (\mathcal{Z}h)(z, s) &= \frac{\tau}{2\pi} \int_{-\infty}^{\infty} \exp(-jz\alpha) \sum_{i=1}^{\infty} \exp(-si) \exp(-i(\sigma\tau)^2\alpha^2) d\alpha \\ &= \frac{\tau}{2\pi} \int_{-\infty}^{\infty} \exp(-jz\alpha)(s + \sigma^2\tau^2\alpha^2)^{-1} d\alpha \\ &= \frac{\tau}{2} \frac{\exp(-z\sqrt{s/(\sigma^2\tau^2)})}{\sqrt{s\sigma^2\tau^2}} \end{aligned} \tag{9.1.22}$$

Then, according to (9.1.11) we have:

$$\begin{aligned}(\mathcal{Z}\xi)(z,s) &= \frac{(\mathcal{Z}h)(z',s)}{(\mathcal{Z}h)(z'-z,s)} \\ &= \frac{\exp(-z'\sqrt{s/(\sigma^2\tau^2)})}{\exp(-(z'-z)\sqrt{s/(\sigma^2\tau^2)})} \\ &= \exp(-z\sqrt{s/(\sigma^2\tau^2)}) \end{aligned} \tag{9.1.23}$$

Finally, using the inverse Z-transform, it can be obtained that:

$$\xi(z,i) = \frac{z}{2i\tau\sigma\sqrt{\pi i}}\exp(-\frac{z^2}{4i\tau^2\sigma^2}) \tag{9.1.24}$$

□

We say that one vehicle *catches up* another vehicle if and only if the Euclidean distance between them reduces to the transmission range r_0. Then, using Lemma 144, one can readily obtain the following result.

Theorem 145 *Consider two vehicles separated by a Euclidean distance l_c at time 0. The probability that one vehicle catches up the other for the first time in the ith time slot is $\frac{l_c-r_0}{2i\tau\sigma\sqrt{\pi i}}\exp(-\frac{(l_c-r_0)^2}{4i\tau^2\sigma^2})$.*

Proof It is straightforward that the reduction of distance is $z = l_c - r_0$. Then, using Lemma 144, the theorem is readily proved. □

9.1.4.2 Catch-Up Delay with Overtaking Permitted

In the previous subsection, the possibility of overtaking is not included in the calculation of the first passage probability, in order to obtain a closed-form result in (9.1.24). In this subsection, the possibility of overtaking is considered to provide a more accurate result on the catch-up delay. Recall that H_m denotes the mth vehicle to the left of the head H_0 at time 0 and $P_{m'}$ denotes the m'^{th} vehicle to the right of the tail P_0 at time 0. Note that all the vehicles H_m ($P_{m'}$) for $m, m' \in [1,\infty)$ can possibly overtake the head H_0 (the tail P_0).

Lemma 146 *Denote by $q_{mm'}(i|l_c)$ the probability that H_m catches up $P_{m'}$ ($m, m' \in [0,\infty)$) for the first time in the ith time slot, in a catch-up process with gap l_c. Then, using Lemma 144, we have*

$$q_{mm'}(i|l_c) \approx \xi\left(l_c - r_0 + m/\rho + m'/\rho, i\right) \tag{9.1.25}$$

Proof Recall that w_m is the distance between H_m and H_0 at time 0, which follows an exponential distribution. It follows that the expected distance between H_m and

H_0 at time 0 is $\int_0^\infty w_m f_{w_m}(w_m)dw_m = m/\rho$, where $f_{w_m}(w_m)$ is given by (9.1.4). Similarly, the expected distance between $P_{m'}$ and P_0 at time 0 is m'/ρ. Therefore, in a catch-up process with gap l_c, the expected distance between H_m and $P_{m'}$ at time 0 is $l_c + m/\rho + m'/\rho$. In order for H_m to catch up with $P_{m'}$, the reduction of distance should be $z = l_c - r_0 + m/\rho + m'/\rho$. □

Remark 147 In Lemma 146, only the mean value of the distance between vehicles is required. This provides us with the flexibility to extend the analysis from the Poisson distribution model to another vehicle distribution model, i.e., we only need to replace m/ρ in (9.1.25) by the corresponding average inter-vehicle distance if a different vehicle distribution model is used. The rest of the analysis on the catch-up process does not depend on the particular vehicle distribution model being used. However, the accuracy of using mean value approximation for another vehicle distribution needs to be validated.

Denote by $H(i|l_c)$ the probability that none of the H_m-$P_{m'}$ pairs $(m, m' \in [0, \infty))$ catches up in the ith time slot, in a catch-up process with gap l_c. Due to the independence of the movements of vehicles, we have:

$$H(i|l_c) = \prod_{m,m' \in [0,\infty)} (1 - q_{mm'}(i|l_c)) \tag{9.1.26}$$

where $q_{mm'}(i|l_c)$ is given by Lemma 146. During numerical evaluation, finite values of m, m' can provide fairly accurate results, which is discussed later in Sect. 9.1.7.1.

Denote by $h(i|l_c)$ the probability that at least one pair of H_m-$P_{m'}$ catches up in the ith time slot *and* none of them has caught up before the ith time slot, in a catch-up process with gap l_c. It is straightforward that:

$$h(i_c|l_c) = (1 - H(i_c|l_c)) \prod_{i=1}^{i_c-1} H(i|l_c) \tag{9.1.27}$$

Finally, the expected delay for a catch-up process with gap l_c is:

$$E[t_c|l_c] = \sum_{i=1}^{\infty} i\tau h(i|l_c) \tag{9.1.28}$$

9.1.5 Analysis of the Forwarding Process

In a forwarding process, the packet is forwarded in a multi-hop manner between vehicles inside a cluster. We start with an analysis on the distribution of the length of the cluster.

9.1.5.1 Cluster Length

Define the cluster length x_0 to be the diameter of a cluster, which is the Euclidean distance between the vehicles at the two ends of a cluster. Define $f_{x_0}(x_0)$ to be the probability density function of the cluster length, which has often been studied as the probability density function of the busy period in queuing theory by drawing an analogy between the inter-vehicle distance and the inter-arrival time in a queue. In previous research only numerical solutions [198] or approximate results [190] were provided for the cluster length distribution. In this subsection, we provide a closed-form formula for the probability density function of the cluster length using a different method inspired by the study on the connectivity of random interval graph [79] and theory of coverage processes [96].

Theorem 148 *In a VANET where the spatial distribution of vehicles follows a homogeneous Poisson point process with density ρ, the probability density function of the cluster length is:*

$$f_{x_0}(x_0) = \frac{\rho}{(e^{\rho r_0} - 1)} \sum_{m=0}^{\lfloor x_0/r_0 \rfloor} \frac{(-\rho(x_0 - m r_0))^{m-1}}{-m!}(\rho(x_0 - m r_0) + m)e^{-\rho m r_0}$$

where m is an integer and $\lfloor . \rfloor$ is the floor function.

Proof Place the origin of the axis at the position of the leftmost vehicle of a cluster. Let N be the random integer representing the number of vehicles in the cluster. The cluster ends in $[x_0, x_0 + dx_0)$ *and* there are n vehicles in this cluster if and only if:

- $\mathcal{E}_1$: There is a vehicle in $[x_0, x_0 + dx_0)$; and
- $\mathcal{E}_2$: There is no vehicle in $[x_0 + dx_0, x_0 + r_0)$; and
- $\mathcal{E}_3$: There are $n - 2$ vehicles in $(0, x_0)$; and
- $\mathcal{E}_4$: The inter-vehicle distance between any two adjacent vehicles for those n vehicles in $[0, x_0 + dx_0)$ is smaller than or equal to r_0.

Denote by $\Pr(\mathcal{E}_m)$ the probability of event $\mathcal{E}_m, m = 1, 2, 3, 4$ in the above list. Due to the Poisson distribution of vehicles, it is straightforward to show that:

$$\Pr(\mathcal{E}_1) = \rho dx_0 \tag{9.1.29}$$

$$\Pr(\mathcal{E}_2) = e^{-\rho r_0} \tag{9.1.30}$$

$$\Pr(\mathcal{E}_3) = \frac{(\rho x_0)^{n-2} e^{-\rho x_0}}{(n-2)!} \tag{9.1.31}$$

Furthermore, $\Pr(\mathcal{E}_4)$ can be studied using [79, Lemma 1], which provides results on the connectivity of random interval graph. In [79], vertices are randomly and uniformly distributed on a unit interval. Due to the Poisson distribution of the vehicles in our case, conditioned on that there are n vehicles in a cluster with length x_0, these n vehicles follow a uniform distribution. Therefore, by scaling the cluster length x_0 to 1 and consequently the transmission range to $\frac{r_0}{x_0}$, we have the following equation obtained from [79, Lemma 1]:

$$\Pr(\mathcal{E}_4) = \sum_{m=0}^{\min\{n-1,\lfloor x_0/r_0 \rfloor\}} \binom{n-1}{m} (-1)^m (1 - m\frac{r_0}{x_0})^{n-2} \tag{9.1.32}$$

where m is an integer. For the convenience of the following calculation, let $\binom{n-1}{m} = 0$ for $m > n-1$. Thus the above summation is from $m = 0$ to $\lfloor x_0/r_0 \rfloor$.

Events $\mathcal{E}_1$, $\mathcal{E}_2$, and $\mathcal{E}_3$ are independent of each other. Event $\mathcal{E}_4$, which is conditioned on event $\mathcal{E}_3$, is independent of events $\mathcal{E}_1$ and $\mathcal{E}_2$. Define $f(x_0, N = n)$ to be the joint probability that the cluster ends in $[x_0, x_0 + dx_0)$ *and* there are n vehicles in this cluster. It is evident that $f(x_0, N = n) = \Pr(\mathcal{E}_1) \Pr(\mathcal{E}_2) \Pr(\mathcal{E}_3) \Pr(\mathcal{E}_4)$. Next we derive $f_{x_0}(x_0)$ using $f(x_0, N = n)$.

If the cluster only consists of one vehicle, then the cluster length is 0 and the probability of this event is $e^{-\rho r_0}$. If the cluster consists of more than one vehicles, then the probability density function of the cluster length x_0 is:

$$\begin{aligned}
f_{x_0}(x_0) &= \frac{\sum_{n=2}^{\infty} f(x_0, N = n)}{\Pr(N \geq 2)} \\
&= \sum_{n=2}^{\infty} \left\{ \frac{\rho e^{-\rho r_0} (\rho x_0)^{n-2} e^{-\rho x_0}}{(n-2)!(1 - e^{-\rho r_0})} \right. \\
&\quad \left. \left[\sum_{m=0}^{\lfloor x_0/r_0 \rfloor} \binom{n-1}{m} (-1)^m (1 - m\frac{r_0}{x_0})^{n-2} \right] \right\} \\
&= \frac{\rho e^{-\rho r_0} e^{-\rho x_0}}{(1 - e^{-\rho r_0})} \sum_{n=2}^{\infty} \left\{ \frac{(\rho x_0)^{n-2}}{(n-2)!} \right. \\
&\quad \left. \left[\sum_{m=0}^{\lfloor x_0/r_0 \rfloor} \binom{n-1}{m} (-1)^m (1 - m\frac{r_0}{x_0})^{n-2} \right] \right\}
\end{aligned} \tag{9.1.33}$$

Using $\binom{n}{m} = \frac{n!}{m!(n-m)!}$, the summation terms in (9.1.33) can be simplified as:

$$\begin{aligned}
&\sum_{n=2}^{\infty} \frac{(\rho x_0)^{n-2}}{(n-2)!} \sum_{m=0}^{\lfloor x_0/r_0 \rfloor} \binom{n-1}{m} (-1)^m (1 - m\frac{r_0}{x_0})^{n-2} \\
&= \sum_{m=0}^{\lfloor x_0/r_0 \rfloor} \frac{(-1)^m}{m!} \sum_{n=2}^{\infty} \frac{(n-1)!}{(n-2)!} \frac{(\rho x_0(1 - m\frac{r_0}{x_0}))^{n-2}}{(n-m-1)!}
\end{aligned}$$

$$= \sum_{m=0}^{\lfloor x_0/r_0 \rfloor} \frac{(-1)^m}{m!} \sum_{n=2}^{\infty} \frac{(n-1)(\rho(x_0 - mr_0))^{n-2}}{(n-m-1)!}$$

$$= \sum_{m=0}^{\lfloor x_0/r_0 \rfloor} \left[\frac{(-1)^m}{m!} (\rho(x_0 - mr_0))^{m-1} \right.$$
$$\left. \times \sum_{n=2}^{\infty} \frac{(n-1)(\rho(x_0 - mr_0))^{n-m-1}}{(n-m-1)!} \right]$$

$$= \sum_{m=0}^{\lfloor x_0/r_0 \rfloor} \left[\frac{(-1)^m}{m!} (\rho(x_0 - mr_0))^{m-1} \right.$$
$$\left. \times \sum_{n=m+1}^{\infty} \frac{(n-1)(\rho(x_0 - mr_0))^{n-m-1}}{(n-m-1)!} \right]$$

$$= \sum_{m=0}^{\lfloor x_0/r_0 \rfloor} \left[\frac{(-1)^m}{m!} (\rho(x_0 - mr_0))^{m-1} \right.$$
$$\left. \times \sum_{\alpha=0}^{\infty} \frac{(\alpha + m)(\rho(x_0 - mr_0))^{\alpha}}{\alpha!} \right] \quad (9.1.34)$$

Using $xe^x = \sum_{\alpha=0}^{\infty} \frac{\alpha x^\alpha}{\alpha!}$, (9.1.34) can be written as:

$$\sum_{m=0}^{\lfloor x_0/r_0 \rfloor} \frac{(-1)^m}{m!} \left[\rho(x_0 - mr_0)\right]^{m-1} (\rho(x_0 - mr_0) + m)\, e^{\rho(x_0 - mr_0)}$$

Substituting the above equation into (9.1.33), one can obtain:

$$f_{x_0}(x_0) = \frac{\rho e^{-\rho r_0} e^{-\rho x_0}}{(1 - e^{-\rho r_0})} \sum_{m=0}^{\lfloor x_0/r_0 \rfloor} \frac{(-1)^m}{m!} (\rho(x_0 - mr_0))^{m-1}$$
$$(\rho(x_0 - mr_0) + m)\, e^{\rho(x_0 - mr_0)}$$
$$= \frac{\rho e^{-\rho r_0}}{(1 - e^{-\rho r_0})} \sum_{m=0}^{\lfloor x_0/r_0 \rfloor} \frac{(-1)^m}{m!} (\rho(x_0 - mr_0))^{m-1}$$
$$(\rho(x_0 - mr_0) + m)\, e^{-\rho m r_0}$$
$$= \frac{\rho}{(e^{\rho r_0} - 1)} \sum_{m=0}^{\lfloor x_0/r_0 \rfloor} \frac{(-1)^m}{m!} (\rho(x_0 - mr_0))^{m-1}$$

$$(\rho(x_0 - mr_0) + m)\, e^{-\rho m r_0}$$

$$= \frac{\rho}{(e^{\rho r_0} - 1)} \sum_{m=0}^{\lfloor x_0/r_0 \rfloor} \frac{(-\rho(x_0 - mr_0))^{m-1}}{-m!}$$

$$(\rho(x_0 - mr_0) + m)\, e^{-\rho m r_0} \tag{9.1.35}$$

□

Theorem 148 gives a closed-form formula for the probability density function of the cluster length, which is important for analytical study on the performance of VANETs.

9.1.5.2 Hop Count Statistics in a Cluster

In order to study the information propagation speed in the forwarding process, the number of hops between the leftmost vehicle and the rightmost vehicle in the same cluster needs to be calculated. Two vehicles are said to be k hops apart if the shortest path between them, measured by the number of hops, is k. Define $\phi_k(x_0)$ to be the probability that two vehicles separated by Euclidean distance x_0 are k hops apart. It is assumed that the positions of the vehicles do not change during the forwarding process since the forwarding delay is relatively small, which is also confirmed in Fig. 9.7. Therefore, the probability $\phi_k(x_0)$ can be calculated by the result introduced in [63] for a static one-dimensional multi-hop network.

Define $\Pr_s(x_0)$ to be the probability of successful transmissions between any pair of vehicles separated by a Euclidean distance x_0. An end-to-end packet transmission is successful if a packet can reach the destination, by any number of hops. Therefore,

$$\Pr_s(x_0) = \sum_{k=1}^{\infty} \phi_k(x_0)$$

Define $\phi_{ks}(x_0)$ to be the conditional probability that a packet reaches its destination at the k_s^{th} hop, conditioned on the transmission being successful *and* the Euclidean distance between the source and the destination being x_0. It is trivial to see that $\phi_k(x_0) = \phi_{ks}(x_0) \Pr_s(x_0)$. Therefore, the expected number of hops between two vehicles separated by distance x_0, given that they are connected, is:

$$E_s[k_s|x_0] = \frac{\sum_{k=1}^{\infty} k\phi_k(x_0)}{\Pr_s(x_0)} \tag{9.1.36}$$

Define the *forwarding delay* to be the time required for a packet to be forwarded from the leftmost vehicle to the rightmost vehicle in the cluster, which is $t_1 - t_0$ in the example illustrated in Fig. 9.1. Then, the expected forwarding delay in a cluster with length x_0 is $E[t_f|x_0] = \beta E_s[k_s|x_0]$.

Remark 149 The hop count statistics for a one-dimensional network with inhomogeneous Poisson distribution of nodes is studied in [201], which provides us with the required methodology for extending our analysis on the forwarding process from the Poisson distribution model to another vehicle distribution model.

9.1.6 Information Propagation Speed

The entire information propagation process can be considered as a renewal reward process [175, Chapter 7.4] where each cycle consists of a catch-up process followed by a forwarding process and the reward is the information propagation distance during each cycle. As mentioned in Sect. 9.1, $E[v]$ is the constant component of the vehicle speed. It can be shown that [190]:

$$E[v_{ip}] \approx \frac{\text{expected length of one cycle}}{\text{expected time duration of one cycle}} + E[v]$$
$$= \frac{\int_{r_0}^{\infty} l_c f_{l_c}(l_c) dl_c + \int_0^{\infty} x_0 f_{x_0}(x_0) dx_0}{\int_{r_0}^{\infty} E[t_c|l_c] f_{l_c}(l_c) dl_c + \beta + \frac{1}{1-p_c} \int_0^{\infty} E[t_f|x_0] f_{x_0}(x_0) dx_0} + E[v]$$

where $p_c = \frac{2W_{min}N_b}{(W_{min}+1)^2+2W_{min}N_b}$ is the probability of collision given in [127], W_{min} is the minimum contention window size and $N_b = \rho 2\pi r_0$ is the average node degree. Packet collision can be shown to have negative impact on the forwarding process, i.e., reducing the IPS, when the vehicle density is high. To illustrate this effect, we conduct simulations using the parameters shown in [36], i.e., $W_{min} = 32$.

9.1.7 Simulation Results

A number of approximations were used in the earlier analysis. In this subsection, we report on simulations to validate the accuracy of the analytical results. Each point shown in the figures is the average value from 2000 simulations. The confidence interval is too small to be distinguishable and hence is ignored in the following plots. The transmission range is $r_0 = 250m$ [190]. The mobility parameters are $E[v] = 25m/s, \sigma = 7.5m/s$ [176]. In order to distinguish the impact on the IPS of packet collision and other parameters, we let $p_c = 0$ except in Fig. 9.11.

9.1.7.1 Catch-Up Process

As mentioned earlier, we use $\tau = 1s, 5s$. Only the results for $\tau = 5s$ are shown in this subsection since the results for $\tau = 1s$ have a similar (and slightly better)

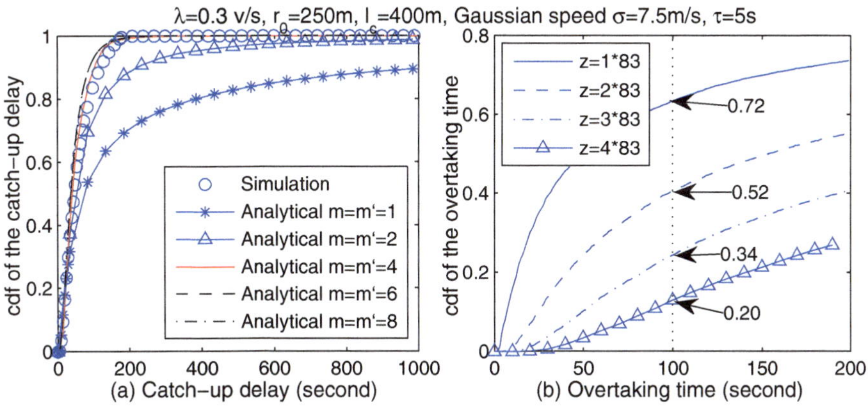

Fig. 9.5 (**a**) Cumulative distribution function of the catch-up delay for a catch-up process with gap $l_c = 400$. (**b**) Simulation results on the probability that a randomly chosen vehicle overtakes another vehicle within time t, where their initial Euclidean distance is $z = m \times 83$ at time 0

accuracy. The traffic density is $\lambda = 0.3$ veh/s. It follows that the spatial distribution of the vehicles follows a homogeneous Poisson point process with density $\rho = \int_0^\infty \frac{f_v(v)}{v} dv = 0.012$ veh/m, which is a low traffic density resulting in a large number of catch-up processes. The results for other densities are quite similar hence not shown in this subsection.

Figure 9.5a shows the probability that a catch-up process with gap $l_c = 400$ finishes within time t. It can be seen that when $m = m' = 4$, the analytical result gives a good approximation. Moreover, considering more vehicles in the overtake process, e.g., $m = 6$ or $m = 8$, has marginal impact on the results. This is because as the distance between vehicles increases, the probability of overtaking decreases rapidly, which can be seen in Fig. 9.5b. Figure 9.5b shows the simulation result of the probability that a randomly chosen vehicle overtakes another vehicle within time $t = i\tau$, where their distance is z at time 0. Because the average inter-vehicle distance is $1/\rho = 1/0.012 \approx 83$ m, the curve of $z = m \times 83$ in Fig. 9.5b approximately illustrates the probability that vehicle H_m overtakes H_0 before time t. As can be seen in the figure, the probability that H_4 overtakes H_0 within 100 s while none of H_1, H_2, H_3 overtakes H_0 within 100 s is approximately given by $0.2(1-0.72)(1-0.52)(1-0.34) = 0.01774$. It can be understood that the probability that H_0 is overtaken by another vehicle, e.g., H_5, H_6 ..., is very small. Therefore, considering $m = m' = 4$ can provide a good approximation.

Figure 9.6a shows the expected catch-up delay for a catch-up process with gap l_c. It can be seen that the analytical result, which considers $m = m' = 4$, provides a good approximation. The discrepancy between the simulation result and the analytical result is caused by the approximations used during the analysis. Specifically, the first passage analysis is only applied to the analysis of the catch-up process between a pair of vehicles which are the head and the tail at the start of the catch-up process. However, the first passage analysis does not consider the

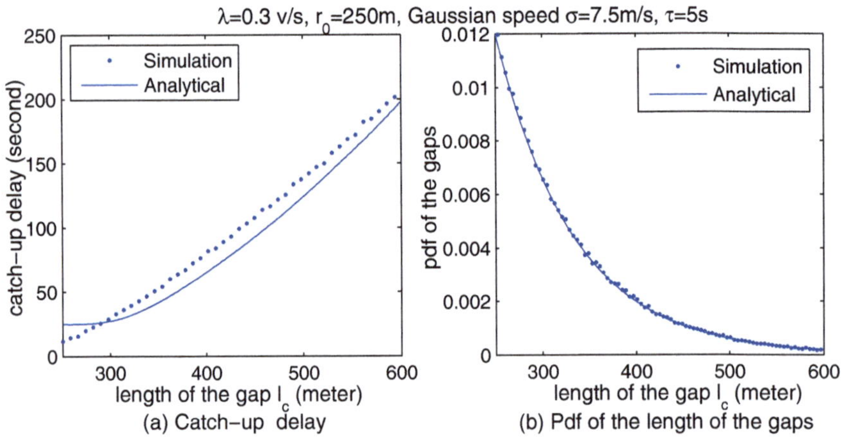

Fig. 9.6 (**a**) Expected catch-up delay, (**b**) probability density function of the length of the gap (l_c)

possibility that the head (the tail) may be overtaken by other vehicles during the catch-up process. Furthermore, Fig. 9.6b also verifies that the inter-vehicle distance, under our network model and the Gaussian speed distribution, still follows an exponential distribution with $\rho = 0.012$. This property is also expected to hold in some other speed distributions, although it needs to be analyzed and verified separately.

9.1.7.2 Forwarding Process

In addition to the simulation settings introduced earlier, the per-hop delay is set to be $\beta = 4ms$ [190]. Figure 9.7a shows the expected forwarding delay in a cluster with a given length. Figure 9.7b shows the probability density function of the cluster length. It can be seen that analytical results match simulation results very well. The results for other values of the parameters have a similar accuracy and so are omitted. Furthermore it is interesting to note that in Fig. 9.7b, the probability density function of the cluster length is a constant for $x_0 \in [0, 250]$. This is because within the transmission range ($r_0 = 250$) of the first vehicle, $\Pr(\mathcal{E}_4) = 1$ according to (9.1.32) and the cluster length is x_0 if and only if there is a vehicle in $[x_0, x_0 + dx_0)$ *and* there is no vehicle in $[x_0 + dx, x_0 + r_0)$. It follows that the probability density function of the cluster length is a constant for $x_0 \in [0, r_0]$.

9.1.7.3 Information Propagation Speed

In addition to the simulation settings introduced earlier, the Poisson arrival rate λ is varied from 0 to 1.5 veh/s. With $E[v] = 25$, the spatial distribution of the vehicles follows a homogeneous Poisson point process with density ρ ranging from 0 to

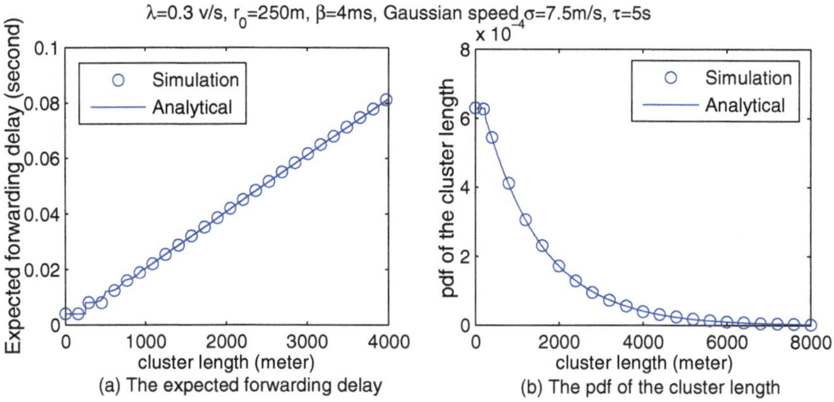

Fig. 9.7 The expected forwarding delay and the probability density function of the cluster length

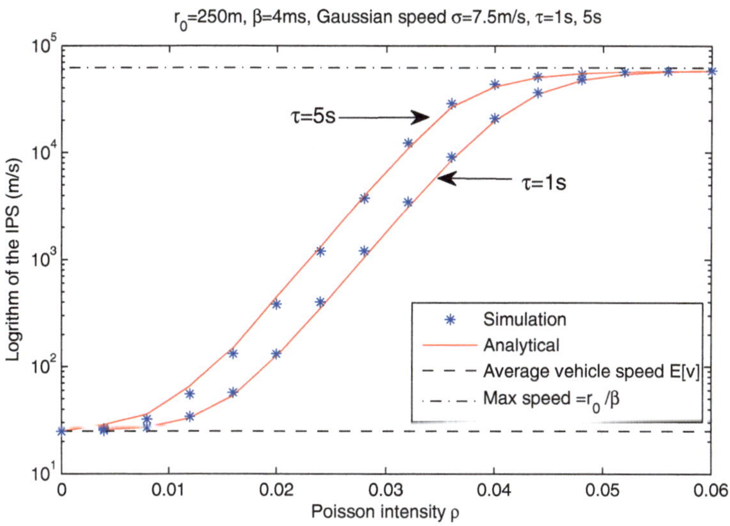

Fig. 9.8 The expected information propagation speed for $\tau = 1s, 5s$

0.06. For completeness of the plot, $\rho = 0$ is included which means there is only one vehicle on the road. Therefore, the average number of neighbors (average node degree) varies from 0 to 30, which represents a large range of traffic densities.

Figure 9.8 shows the expected IPS for $\tau = 1s, 5s$. It can be seen that when the vehicle density is low, the IPS is determined by vehicle speeds because there is little packet forwarding in the network. When the vehicle density increases, small clusters are formed and the IPS is determined by the catch-up delay, which is further determined by the mobility of vehicles. It can be seen that the more frequently the speed changes, the slower the information propagates. This is mainly because changing speed has the potential to interrupt the catch-up process, i.e., during a catch-up process, the tail may speed up and the head may slow down. An intuitive explanation can be provided by considering an extreme case. In the extreme case that the speed-change time interval tends to 0, it can be shown using the central limit theorem that the average vehicle speed in any specified time interval converges to the mean speed $E[v]$. Hence the network topology becomes static and the expected IPS is equal to the mean speed $E[v]$ because there is no catch-up process to bridge the gaps. Finally, as the vehicle density further increases, clusters become larger and the forwarding process starts to dominate. Therefore, the IPS increases until it reaches the maximum value, which is determined by the per-hop delay in the forwarding process. The maximum IPS is obviously equal to r_0/β, where β is the per-hop delay. See also the analysis in [3], which formally characterizes this phase transition phenomenon in the IPS.

Figure 9.9 shows the expected IPS under the constant speed model, i.e., the vehicle speed does not change over time. The result *Analytical-wu* is calculated based on [190] for comparison. The constant speed model is a special case of the mobility model considered in this chapter, i.e., when $\tau \to \infty$. In Fig. 9.9, we choose a fairly large value of τ to obtain analytical result under our mobility model and use the result as an approximation of the result under the constant speed model. The result *Analytical-wu*, which does not consider first passage phenomenon, provides a good match with the simulations. This is because the first passage phenomenon does not have a significant impact when the vehicle speed does not change over time. Finally, it can be seen that the constant speed model used in previous research, e.g.,

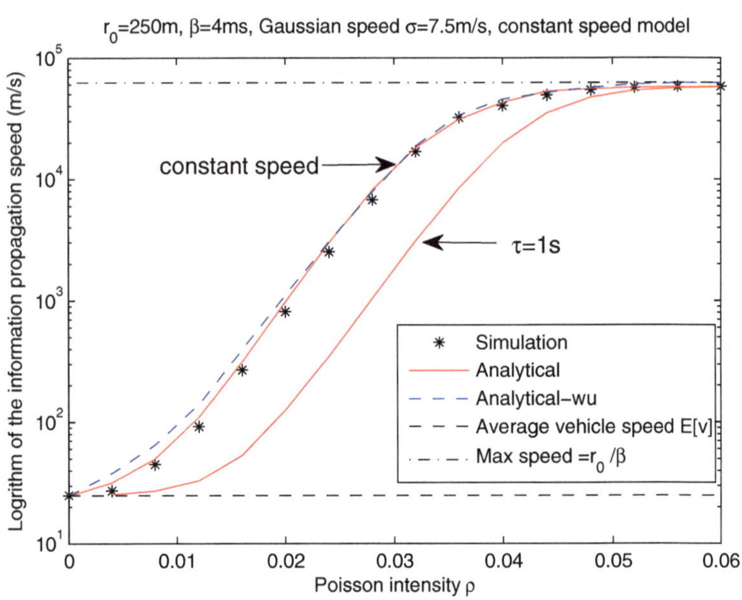

Fig. 9.9 The expected information propagation speed under the constant speed model, together with the curve for $\tau = 1$ for comparison. The result *Analytical-wu* is calculated based on [190]

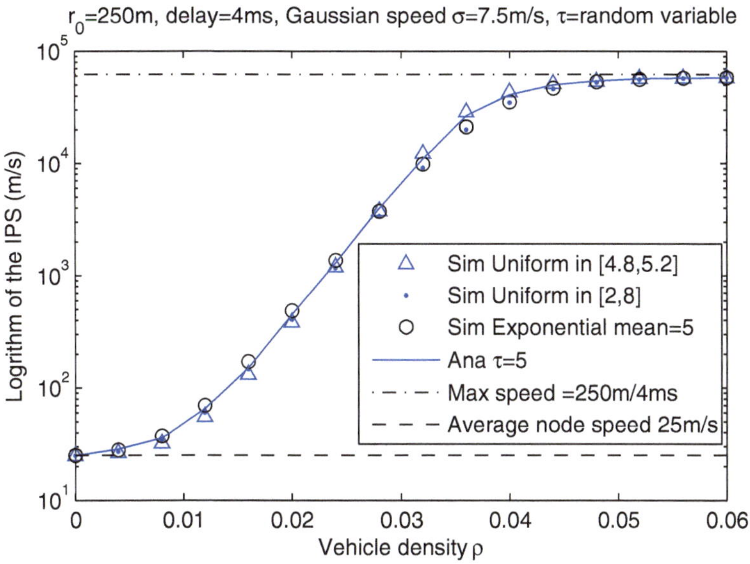

Fig. 9.10 The expected IPS under non-synchronized mobility models

[190], causes serious over-estimation of the IPS by almost an order of magnitude. Therefore, time variation of vehicle speed is an important factor affecting the IPS.

Figure 9.10 shows the expected IPS under non-synchronized mobility models. In the analysis we choose the synchronized random walk mobility model in order to study the impact of the time variation of vehicle speed on the IPS. In order to verify the general applicability of the analytical study assuming the simplified mobility model, more simulations are conducted. As shown in Fig. 9.10, three different mobility models are evaluated, i.e., the speed-change time interval τ of each vehicle is uniformly selected from $[4.8, 5.2]$ (i.e., around 5) or from $[2, 8]$ (i.e., within a larger range around 5); and for τ following an exponential distribution with mean 5. Under all three mobility models, vehicles change speed at different time instants (non-synchronized) and the average speed-change time interval is 5 s. It can be seen that the IPSs under non-synchronized mobility models are very close (almost indistinguishable) to each other and our analysis assuming the simplified (synchronized) mobility model provides a good estimation on the IPS.

We consider the unit disk connection model in the analysis. The unit disk communication model is constructed based on the path loss attenuation model, which is suitable to model the radio environment in free space without clutters [172]. Therefore, the unit disk connection model is suitable for the VANET on the freeway. In order to study the impact of clutters such as road-side buildings, simulation results of the expected IPS under the log-normal shadowing model [172] are shown in Fig. 9.11. The results are compared under the condition that the average node degrees (i.e., the average number of neighbors per node) under the

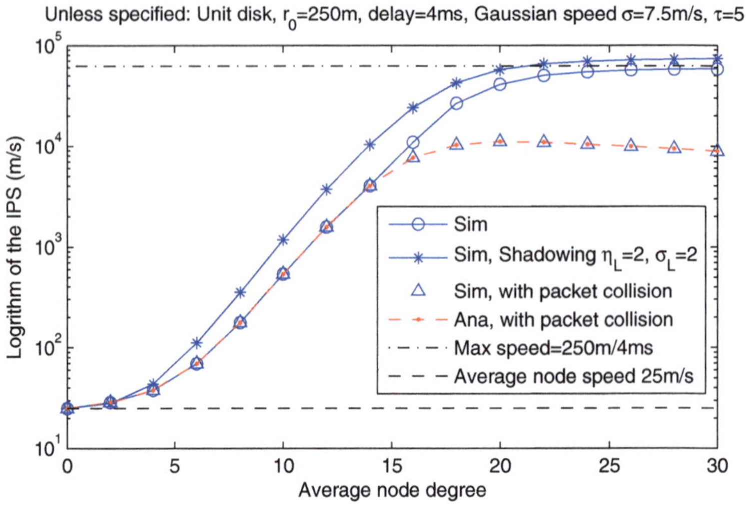

Fig. 9.11 The expected IPS in a VANET subject to log-normal shadowing and packet collision, where η_L is the path loss exponent and σ_L is the standard deviation of the log-normal shadowing. Furthermore, when a head cannot transmit a packet to any uninformed vehicle to the right of itself, the head keep trying to re-transmit the packet after every 0.9s time delay. The value is chosen according to the real world measurement that the channel coherence time of a VANET on the freeway is around $0.3s - 1.5s$

log-normal model and under the unit disk model, respectively, are the same. In the log-normal shadowing model, the received signal strength (RSS) attenuation (in dB) follows a normal distribution with a mean value equal to the RSS under the path loss attenuation model. This random variation on the RSS attenuation provides a higher chance for a node to find a next-hop neighbor. Hence even with the same average node degree, the IPS under the log-normal shadowing model is faster than that under the unit disk model. A similar observation is obtained in the study of network connectivity in [100] and [71, Theorem 2.5.2]. Therefore, the IPS under the unit disk model can be considered as a lower bound on the IPS of a VANET in the real world.

In addition, the third and forth curves in Fig. 9.11 show the IPS subject to packet collision with collision probability p_c given in Sect. 9.1.6. It can be seen that the packet collision has a significant impact on the IPS, especially when the vehicle density is high.

9.2 Information Propagation Process in VANETs with Multiple Traffic Streams

On the basis of previous section which studied the information propagation process in a VANET with single traffic stream, in this section, we continue to investigate the information propagation process in a VANET with multiple traffic streams as illustrated in Fig. 9.12.

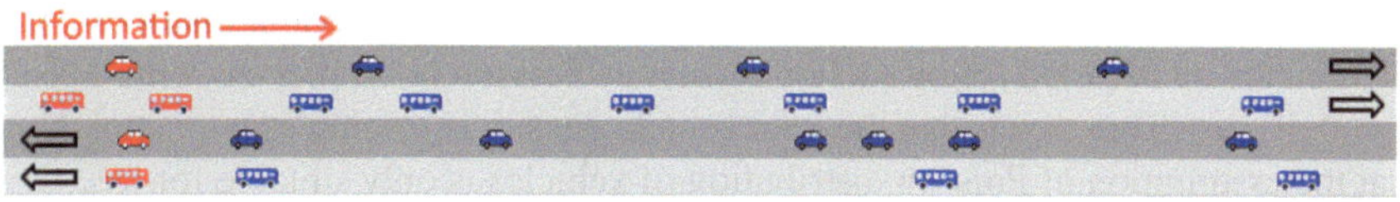

Fig. 9.12 An illustration of a VANET with four traffic streams. Information is propagating in the positive direction of the axis

Recent research has shown that downstream traffic, i.e., a set of vehicles traveling in the opposite direction of information propagation, can be explored to improve the IPS [1, 2, 11]. Furthermore, real world measurements show that vehicles traveling in different lanes (e.g., bus lane or heavy truck lane) often have different speed distributions. In view of these observations, this section further investigates information propagation in VANETs with multiple traffic streams, where a *traffic stream* is a set of vehicles following the same speed distribution. Traffic streams can be used to represent vehicles traveling in different lanes, or different types of vehicles (e.g., sports cars or heavy trucks).

Consider a VANET with a total of N traffic streams and a synchronized random walk mobility model. Each vehicle in the nth, $n \in \{1, \ldots, N\}$, traffic stream chooses its speed randomly at the beginning of each time slot, independent of the speeds of other vehicles and its own speeds in other time slots, according to a probability density function $f_n(v)$. A non-synchronized mobility model, i.e., each vehicle changes its speed at different time instants, is evaluated by simulations later in Sect. 9.2.6 and shown to have little impact on the IPS. We consider the Gaussian speed distribution, i.e., $f_n(v) \sim \mathcal{N}(\mu_n, \sigma_n^2)$, where μ_n and σ_n^2 are the mean speed and variance of the nth traffic stream, respectively. Though the mobility model allows arbitrary values of μ_n, in real world [176], typical values of μ_n can be $\pm 25m/s$ or $\pm 10m/s$ corresponding to the speeds of vehicles traveling in fast lane or slow lane, respectively. Note that a positive (negative) value of the vehicle speed means that the vehicle is moving in the positive (negative) direction of the axis. When a vehicle changes its speed from a positive value to a negative one, it means that the vehicle changes its moving direction. This does not occur very often when μ_n is much greater than σ_n. Furthermore, another interesting scenario is a traffic stream with $\mu_n = \sigma_n = 0$, corresponding to the vehicles stopped at roadside or some roadside units without (expensive) wired connection; these roadside units are shown later to have great potential of improving the IPS.

Same as in the previous section, we adopt the unit disk connection model with a transmission range r_0 and consider a per-hop delay β where the value of β is chosen to be 4ms.

At the beginning of the information propagation process, vehicles in the nth traffic stream follow a homogeneous Poisson spatial distribution with density ρ_n. Then, these vehicles start to move according to the random walk mobility model. As an easy consequence of Lemma 143, the spatial distribution of vehicles in the nth traffic stream remains a homogeneous Poisson distribution with density ρ_n at any time instant. Then, according to the superposition property of Poisson point processes ([155], see also Sect. 1.2), at any time instant the spatial distribution of all vehicles on the road follows a homogeneous Poisson point process with density $\rho = \sum_{n=1}^{N} \rho_n$. This means that the total vehicle density is ρ vehicles per meter. Note that the assumption of Poisson distribution of vehicles is only suitable for free-flow traffic [198], i.e., the movements of vehicles are independent. Therefore, the model is applicable to VANET scenarios where the vehicle density is low or moderate.

In the following three subsections, we study the information propagation process by separately analyzing its two major constituent sub-processes: the forwarding process and the catch-up process. On the basis of these studied, simplified results are provided in Sect. 9.2.4 for both the forwarding process and the catch-up process.

9.2.1 Forwarding Process

Define the *forwarding delay* as the time required for a packet to be forwarded from the leftmost vehicle in a cluster to the rightmost vehicle in the cluster. Note that we consider that the radio propagation speed is sufficiently large in a forwarding process, because the information propagation delay is usually small. For example, a typical value of per-hop-delay in vehicle-to-vehicle communication is $\beta = 4ms$ [191], which means that the radio propagation speed can be as large as $250/0.004 = 62500m/s$, which is much faster than the moving speed of vehicles (typical speed is 25m/s [176]). Therefore, it is assumed that a cluster does not become disconnected during the forwarding process. Then, the expected forwarding delay in a cluster with length x is $E[t_f|x] = \beta E[k|x]$, where $E[k|x]$ (given in [63], see also Sect. 11.1) is the expected number of hops between two vehicles separated by distance x and t_f is the forwarding delay.

Denote by $X(\rho)$ the length of an arbitrary cluster in a VANET with density ρ, i.e., it is the Euclidean distance between the leftmost vehicle and the rightmost vehicle in a cluster. Using Theorem 148, the probability density function of the cluster length is:

$$f_{X(\rho)}(x) = \frac{\rho}{(e^{\rho r_0} - 1)} \sum_{m=0}^{\lfloor x/r_0 \rfloor} \left[\frac{(-\rho(x - mr_0))^{m-1}}{-m!} (\rho(x - mr_0) + m)e^{-\rho m r_0} \right]$$

It follows that the cumulative distribution function of the cluster length is:

$$\Pr(X(\rho) \le x) = \int_0^x f_{X(\rho)}(y)\, \mathrm{d}y \tag{9.2.1}$$

9.2.2 *Catch-Up Process*

We study the catch-up process in this subsection. Without loss of generality, it is assumed that the catch-up process starts at time 0. The *displacement* γ of a vehicle at time t is defined to be the distance between the vehicle position at time t and its position at time 0. Note that $\gamma \in (-\infty, \infty)$, where a positive (negative) value of γ means that the position of the vehicle at time t is on the right (left) of its position at time 0.

This subsection is organized as follows: in Lemma 150, we study the movement of single vehicle. Then, the result on the distance between two vehicles at a given time is summarized in Lemma 151 and Lemma 152, followed by the study on the so-called pseudo catch-up process in Lemma 153. After considering the overtaking of vehicles in Sect. 9.2.2.4, we obtain the delay of a catch-up process in Sect. 9.2.2.5.

9.2.2.1 Modeling the Movement of Single Vehicle

Consider single vehicle first. Suppose that a vehicle is located at the origin ($\gamma = 0$) at time 0. Denote by $p_n(\gamma, t)$ the probability density function that the displacement of a vehicle in the nth traffic stream is γ at time t. We have the following result:

Lemma 150 *Suppose that a vehicle in the nth traffic stream follows a Gaussian speed distribution with mean μ_n and variance σ_n^2. Under the synchronized random walk mobility model, there holds*

$$p_n(\gamma, t) = \frac{1}{\tilde{\sigma}_i \sqrt{2\pi}} \exp\left(\frac{-(\gamma - \mu_n i\tau - \mu_n(t - i\tau))^2}{2\tilde{\sigma}_i^2}\right) \tag{9.2.2}$$

where $\tilde{\sigma}_i^2 = i(\sigma_n \tau)^2 + \sigma_n^2(t - i\tau)^2$ and $i = \lfloor \frac{t}{\iota} \rfloor$.

Proof According to the Gaussian speed distribution model, the probability density function of the speed of a vehicle in the nth traffic stream is

$$f_n(v) = \frac{1}{\sigma_n \sqrt{2\pi}} \exp\left(\frac{-(v - \mu_n)^2}{2\sigma_n^2}\right) \tag{9.2.3}$$

Because the speed does not change during a time slot, it is straightforward that $p_n(\gamma, \tau)$ is also a Gaussian function:

$$p_n(\gamma, \tau) = \frac{1}{\sigma_n \tau \sqrt{2\pi}} \exp\left(\frac{-(\gamma - \mu_n \tau)^2}{2(\sigma_n \tau)^2}\right) \tag{9.2.4}$$

Recall that we consider that the speed of a vehicle at a time slot is independent of that in another time slot. Hence the displacements of a vehicle are independent across time slots. Then, there holds:

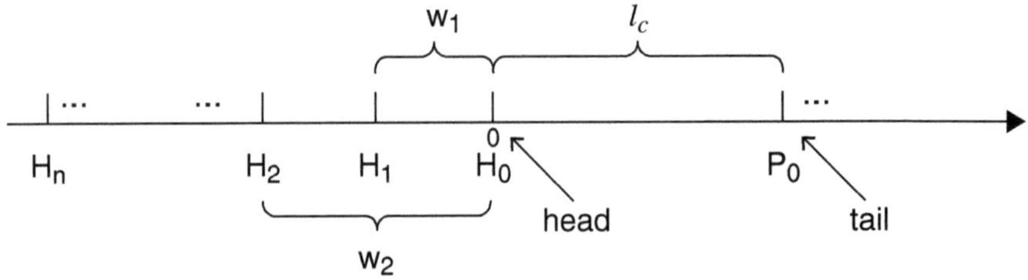

Fig. 9.13 An illustration of a VANET at the beginning of a catch-up process

$$p_n(\gamma, t) = \overbrace{(p_n * p_n * \ldots * p_n)(\gamma, \tau)}^{\text{i-fold convolution}} * p_n(\gamma, t - i\tau) \tag{9.2.5}$$

where $i = \lfloor \frac{t}{\tau} \rfloor$.

Using the property that the convolution of two Gaussian functions is a Gaussian function [85], the Gaussian function after i-fold convolution has mean $\mu_n i\tau$ and variance $i(\sigma_n \tau)^2$. Then, the lemma is readily proved. □

9.2.2.2 The Distance Between a Pair of Vehicles

As shown in Fig. 9.13, denote by H_ζ (P_ζ) the ζth vehicle to the left of the head H_0 (to the right of the tail P_0) at the beginning of a catch-up process (viz., at time 0). If H_ζ happens to be in the nth traffic stream, then an additional label n (e.g., H_ζ^n) is used to indicate that the vehicle is in the nth traffic stream. In this subsection, we study the Euclidean distance between H_ζ^n and P_η^m, where $n, m \in \{1, \ldots, N\}$, n may or may not be equal to m, and parameters ζ and η are nonnegative integers. Note that we need to consider arbitrary nonnegative integer values of ζ, η rather than $\zeta = \eta = 0$ due to the fact that at time $t > 0$ the head vehicle H_0 (the tail vehicle P_0) can be possibly overtaken by another informed vehicle (uninformed vehicle). Consequently, the information may be forwarded from H_ζ to P_ζ for any nonnegative integer values of ζ and η.

The information can be forwarded from H_ζ^n to P_η^m as soon as their distance reduces to the transmission range r_0. Next, we characterize the reduction in the distance between H_ζ^n and P_η^m compared with their initial distance at the beginning of the catch-up process.

Lemma 151 *Denote by $g_{\zeta\eta}^{nm}(z, t)$ the probability density function of the reduction in the distance between H_ζ^n and P_η^m being z at time t, with regard to their distance at time 0 (viz., the start-time of the catch-up process). Then, there holds*

$$g_{\zeta\eta}^{nm}(z, t) = \frac{1}{\sqrt{2\pi\sigma_t^2}} \exp\left(\frac{-(z - \mu_t)^2}{2\sigma_t^2}\right) \tag{9.2.6}$$

where $\mu_t = \mu_n i\tau + \mu_n(t - i\tau) - \mu_m i\tau - \mu_m(t - i\tau)$, $\sigma_t^2 = i\sigma_n^2\tau^2 + \sigma_n^2(t - i\tau)^2 + i\sigma_m^2\tau^2 + \sigma_m^2(t - i\tau)^2$ *and* $i = \lfloor \frac{t}{\tau} \rfloor$.

The proof is straightforward using the same technique as that used in the proof of Lemma 150, hence omitted here.

Note that vehicle H_ζ^n *catches up* another vehicle P_η^m when their Euclidean distance reduces to the transmission range r_0 *for the first time*. However, during a sufficiently long time interval, the reduction in distance between H_ζ^n and P_η^m can reach a given value z at several occasions with nonzero probabilities. We are only interested in the event that the reduction in the distance between H_ζ^n and P_η^m reaches z *for the first time*, i.e., the first passage phenomenon. A thorough understanding of the first passage phenomenon is therefore essential for the analysis of the catch-up process.

Lemma 152 *Denote by* $G_{\zeta\eta}^{nm}(z, i)$ *the probability density function of the event that the reduction in the distance between* H_ζ^n *and* P_η^m*, with regard to their distance at time 0, reaches* z $(z > 0)$ for the first time *in the* i^{th} $(i \geq 1)$ *time slot. Then:*

(a) When $\mu_n = \mu_m$*:*

$$G_{\zeta\eta}^{nm}(z, i) = \frac{z}{i\sqrt{2\pi i\tau^2(\sigma_n^2 + \sigma_m^2)}} \exp\left(\frac{-z^2}{2i\tau^2(\sigma_n^2 + \sigma_m^2)}\right) \quad (9.2.7)$$

(b) When $\mu_n \neq \mu_m$ *and* $i = 1$,

$$G_{\zeta\eta}^{nm}(z, i) \leq \frac{1}{2}\left(1 - erf(\frac{z - \hat{\mu}}{\sqrt{2\sigma_i^2}})\right) \quad (9.2.8)$$

(c) When $\mu_n \neq \mu_m$ *and* $i \geq 2$,

$$G_{\zeta\eta}^{nm}(z, i) \leq \frac{1}{4}(1 + erf(\frac{z - \mu_{i-1}}{\sqrt{2\sigma_{i-1}^2}}))(1 - erf(\frac{z - \hat{\mu}}{\sqrt{2\sigma_i^2}})) \quad (9.2.9)$$

where $\hat{\mu} = \mu_n i\tau - \mu_m i\tau$, $\sigma_i^2 = (\sigma_n^2 + \sigma_m^2)\tau^2 i$ *and* $erf(.)$ *is the error function.*

Proof We first consider the case that two traffic streams have the same mean speed, i.e., $\mu_n = \mu_m$. In this case, we have $\mu_t = 0$ in (9.2.6).

It is straightforward that in the ith time slot, the probability that the reduction in the distance between H_ζ^n and P_η^m reaches z is $\int_{(i-1)\tau}^{i\tau} g_{\zeta\eta}^{nm}(z, t)dt$. According to the mean value theorem, there exists $\hat{t} \in ((i-1)\tau, i\tau]$ such that $\int_{(i-1)\tau}^{i\tau} g_{\zeta\eta}^{nm}(z, t)dt = \tau g_{\zeta\eta}^{nm}(z, \hat{t})$. To facilitate the first passage analysis, we approximate $\hat{t}$ by $i\tau$ and apply a standard procedure, as shown in [174], to determine the first passage probability $G_{\zeta\eta}^{nm}(z, i)$. Then, there holds

$$\tau g_{\zeta\eta}^{nm}(z', i'\tau) = \tau \sum_{i=1}^{i'} G_{\zeta\eta}^{nm}(z, i) g_{\zeta\eta}^{nm}\left(z' - z, (i' - i)\tau\right) \tag{9.2.10}$$

The convolution can be simplified by performing the Z-transform on $g_{\zeta\eta}^{nm}(z, i\tau)$ with regard to i:

$$(\mathcal{Z}g)(z, s) = \sum_{i=1}^{\infty} e^{-si} g_{\zeta\eta}^{nm}(z, i\tau) \tag{9.2.11}$$

$$= \frac{1}{2} \frac{\exp(-z\sqrt{2s/((\sigma_n^2 + \sigma_m^2)\tau^2)})}{\sqrt{s(\sigma_n^2 + \sigma_m^2)\tau^2/2}} \tag{9.2.12}$$

The derivation from (9.2.11) to (9.2.12) is similar to the proof of Lemma 144, hence omitted here. Then, according to (9.2.10), there holds

$$(\mathcal{Z}G)(z, s) = \frac{(\mathcal{Z}g)(z', s)}{(\mathcal{Z}g)(z' - z, s)}$$

$$= \exp\left(-z\sqrt{2s/((\sigma_n^2 + \sigma_m^2)\tau^2)}\right)$$

Finally, using the inverse Z-transform operation, it is straightforward to obtain that for $\mu_n = \mu_m$,

$$G_{\zeta\eta}^{nm}(z, i) = \frac{z}{i\sqrt{2\pi i\tau^2(\sigma_n^2 + \sigma_m^2)}} \exp\left(\frac{-z^2}{2i\tau^2(\sigma_n^2 + \sigma_m^2)}\right) \tag{9.2.13}$$

When $\mu_n \neq \mu_m$, we use a method different from the complicated first passage analysis to calculate $G_{\zeta\eta}^{nm}(z, i)$ from $g_{\zeta\eta}^{nm}(z, t)$. Define $\mathcal{A}$ to be the event that the reduction in distance between H_ζ^n and P_η^m is smaller than z at time $(i - 1)\tau$, with regard to time 0. Let $\mathcal{B}$ be the event that the reduction in distance between H_ζ^n and P_η^m is larger than z at time $i\tau$, with regard to time 0. Then, for $\mu_n \neq \mu_m$ and $i \geq 2$:

$$G_{\zeta\eta}^{nm}(z, i)$$

$$\leq \Pr(\mathcal{A}) \Pr(\mathcal{B}|\mathcal{A})$$

$$\leq \Pr(\mathcal{A}) \Pr(\mathcal{B}) \tag{9.2.14}$$

$$= \int_0^z g_{\zeta\eta}^{nm}(z_0, (i-1)\tau) \mathrm{d}z_0 \int_z^{\infty} g_{\zeta\eta}^{nm}(z_0, i\tau) \mathrm{d}z_0 \tag{9.2.15}$$

$$= \frac{1}{2}\left(1 + erf(\frac{z - \mu_{i-1}}{\sqrt{2\sigma_{i-1}^2}})\right) \frac{1}{2}\left(1 - erf(\frac{z - \hat{\mu}}{\sqrt{2\sigma_i^2}})\right) \tag{9.2.16}$$

where $\hat{\mu} = \mu_n i\tau - \mu_m i\tau$, $\sigma_i^2 = (\sigma_n^2 + \sigma_m^2)\tau^2 i$. The first inequality is due to the fact that we do not consider the first passage phenomenon. Hence $\Pr(\mathcal{A})\Pr(\mathcal{B}|\mathcal{A})$ is larger than or equal to the probability that the reduction in distance between H_ζ^n and P_η^m reaches z *for the first time* in the ith time slot. When $\mu_n \neq \mu_m$, $g_{\zeta\eta}^{nm}(z,t)$ becomes a Gaussian function with a nonzero mean according to Lemma 151. It follows that the mean value of the distance between H_ζ^n and P_η^m is either a decreasing (when $\mu_n < \mu_m$) or an increasing (when $\mu_n > \mu_m$) function of t. Therefore, this bound is tight when the difference between μ_n and μ_m is large, which is usually satisfied in situations when different traffic streams are formed by vehicles traveling in fast lane, slow lane or even vehicles stopped at roadside. Secondly, the inequality $\Pr(\mathcal{B}|\mathcal{A}) \leq \Pr(\mathcal{B})$ in (9.2.14) is due to the observation that $\mathcal{A}$ and $\mathcal{B}$ are negatively correlated. Events $\mathcal{A}$ and $\mathcal{B}$ are negatively correlated because conditioned on the occurrence of event $\mathcal{A}$, that the reduction in distance is less than z at time $(i-1)\tau$, the occurrence of event $\mathcal{B}$, that the reduction in distance is larger than z at time $i\tau$, is less likely due to the bounded value of vehicle speed.

Furthermore, when $\mu_n \neq \mu_m$ and $i = 1$, it is straightforward that:

$$G_{\zeta\eta}^{nm}(z,i) \leq \Pr(\mathcal{B}) = \frac{1}{2}\left(1 - erf\left(\frac{z-\hat{\mu}}{\sqrt{2\sigma_i^2}}\right)\right) \tag{9.2.17}$$

□

The results on $G_{\zeta\eta}^{nm}(z,i)$ are important steps in the analysis of the catch-up delay and will be used in the following subsections.

9.2.2.3 Pseudo Catch-Up Events

Another major challenge in the analysis of the catch-up process in a VANET with multiple traffic streams is the study of the *pseudo catch-up event*, denoted by Ψ. The pseudo catch-up event occurs when a packet is forwarded from a head (say H_0) to a tail (P_0) that is in another traffic stream different from the head, but later H_0 overtakes P_0 (and all other informed vehicles) and becomes the head again owing to the mobility of vehicles. Such event often occurs when the head and the tail are in two traffic streams with very different mean vehicle speeds, e.g., two traffic streams in opposite directions.

An example of the pseudo catch-up event is illustrated in Fig. 9.14. Suppose that the distance between the head (H_0) and the tail (P_0^+) is l_c at time 0, as shown in Fig. 9.14a. After some time, the information is forwarded from H_0^+ to an uninformed vehicle P_0 as shown in Fig. 9.14b, i.e., a catch-up event occurs. However, due to the higher speed of H_0^+ compared with P_0, later H_0^+ may overtake P_0 and all vehicles in the cluster containing P_0 and become the head again. In this case, the previous catch-up event becomes a pseudo catch-up event in the sense that the previous catch-up does not advance the propagation of the information. The pseudo

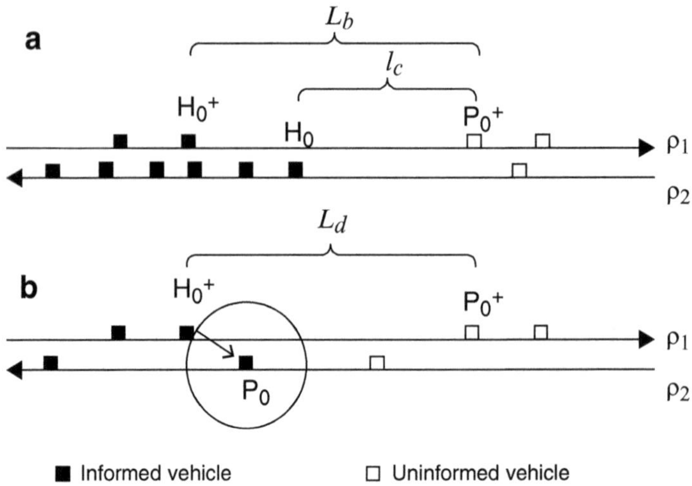

Fig. 9.14 An illustration of a pseudo catch-up event in a VANET with two traffic streams traveling in opposite directions. (**a**) Suppose that at time 0, the distance between the head (H_0) and the tail (P_0^+) is l_c and the distance between H_0^+ and P_0^+ is L_b. (**b**) Suppose that at time $t > 0$, the distance between H_0^+ and P_0^+ becomes L_d. Furthermore, suppose that the information is forwarded from H_0^+ to P_0 at time t. However, due to the higher speed of H_0^+ compared with P_0, H_0^+ may overtake P_0 later and become the head again, resulting in a pseudo catch-up event

catch-up event was firstly analyzed in [11], and has been ignored in other previous work. Indeed, as shown later, under certain conditions, the pseudo catch-up event can have significant impact on the information propagation process. This subsection quantitatively characterizes the impact of the pseudo catch-up events on the IPS and provides a more accurate estimate on the IPS, as shown in Sect. 9.2.6.

Next, we present an analytical result on the probability that a pseudo catch-up event occurs. For simplicity, hereafter a catch-up process, where the initial distance between the head H_0 and tail P_0 is l_c at time 0, is referred to as a *catch-up process with gap* l_c. Denote by $f_c(l_c)$ the probability density function of l_c. Because of the Poisson distribution of vehicles, it is straightforward that for $l_c > r_0$,

$$f_c(l_c) = \frac{\rho \exp(-\rho l_c)}{\int_{r_0}^{\infty} \rho \exp(-\rho l) \mathrm{d}l} = \rho \exp(-\rho(l_c - r_0)) \tag{9.2.18}$$

Lemma 153 *Denote by* $\Pr(\Psi)$ *the probability that a pseudo catch-up event occurs. There holds*

$$\Pr(\Psi) \le \int_{r_0}^{\infty} f_b(l_b) \Pr\left(X(\rho - \rho_{n_{\max}}) \le l_b - 2r_0\right) \mathrm{d}l_b \tag{9.2.19}$$

where $n_{\max} \triangleq \arg\max_n\{\mu_n\}$, f_b *is given by (9.2.22) and* $\Pr(X(\rho) \le x)$ *is given by (9.2.1).*

Proof To characterize pseudo catch-up events, we first consider the distance between the head and the tail. Consider a catch-up process with gap l_c, i.e., the Euclidean distance between the head and the tail is l_c at the beginning of the catch-up process. Denote by H_0^+ (P_0^+) the rightmost informed (leftmost uninformed) vehicle in the $n_{\max}^{th}$ traffic stream at the beginning of a catch-up process, where the $n_{\max}^{th}$ traffic stream is the traffic stream with the largest mean speed, i.e., $n_{\max} = \arg\max_n\{\mu_n\}$. As illustrated in Fig. 9.14, denote by L_b (L_d) the Euclidean distance between H_0^+ and P_0^+ at the beginning (at time $t > 0$) of a catch-up process. Suppose that at time $t > 0$, a message is forwarded from H_0^+ to P_0. If the length of the cluster following P_0 is shorter than $L_d - 2r_0$, then the information cannot be forwarded to P_0^+ via multi-hop forwarding. Furthermore, as mentioned earlier, H_0^+ and P_0 are in different traffic streams and have a nonzero mean relative speed, hence H_0^+ may overtake P_0 and all vehicles in the cluster containing P_0 and become the head again. Therefore, a sufficient condition, which gives an upper bound on $\Pr(\Psi)$, for a catch-up process to be successful is that the cluster containing P_0 is long enough to bridge the gap L_d, such that the catch-up process moves the message from H_0^+ to P_0^+.

Furthermore, there holds $L_b \geq_{st} L_d$ in stochastic ordering, due to the fact that the distance between H_0^+ and P_0^+ tends to reduce over time, because vehicles with a fast (slow) speed may overtake other vehicles and become the new head H_0^+ (new tail P_0^+) after a certain time period. Using stochastic ordering, we say $L_b \geq_{st} L_d$ if and only if $\Pr(L_b > c) \geq \Pr(L_b > c)$ for any number $c \in (-\infty, \infty)$. In view of the analytical tractability of L_b, the following analysis considers L_b instead of L_d to obtain an upper bound on $\Pr(\Psi)$.

Define $f_l(l_b|l_c)$ as the probability density function of L_b given a catch-up process with gap l_c. Because the distribution of vehicles in each traffic stream follows a Poisson point process independent of other traffic streams, there holds

$$f_l(l_b|l_c) = \rho_{n_{\max}} \exp\left(-\rho_{n_{\max}}(l_b - l_c)\right) \tag{9.2.20}$$

Then, using the total probability theorem, there holds

$$\begin{aligned} f_l(l_b) &= \int_{r_0}^{l_b} f_l(l_b|l_c)\frac{\rho\exp(-\rho l_c)}{\int_{r_0}^{l_b}\rho\exp(-\rho l)\mathrm{d}l}\mathrm{d}l_c \\ &= \int_{r_0}^{l_b} \rho_{n_{\max}}\exp(-\rho_{n_{\max}}(l_b - l_c))\frac{\rho\exp(-\rho l_c)}{\int_{r_0}^{l_b}\rho\exp(-\rho l)\mathrm{d}l}\mathrm{d}l_c \\ &= \frac{\rho_{n_{\max}}\rho\exp(-\rho_{n_{\max}}l_b))}{\exp(-\rho r_0) - \exp(-\rho l_b)}\int_{r_0}^{l_b}\exp(\rho_{n_{\max}}l_c - \rho l_c)\mathrm{d}l_c \\ &= \frac{\rho\rho_{n_{\max}}}{\rho - \rho_{n_{\max}}} \times \frac{\exp(-\rho_{n_{\max}}l_b + (\rho_{n_{\max}} - \rho)r_0) - \exp(-\rho l_b)}{\exp(-\rho r_0) - \exp(-\rho l_b)} \end{aligned} \tag{9.2.21}$$

Define $f_b(l_b)$ as the probability density function of l_b. Then, through normalization, there holds

$$f_b(l_b) = \frac{f_l(l_b)}{\int_{r_0}^{\infty} f_l(l_b)\mathrm{d}l_b} \tag{9.2.22}$$

Given the gap l_b, we next study the pseudo catch-up event. As discussed at the beginning of this lemma, to obtain an upper bound on $\Pr(\Psi)$, we consider that the pseudo catch-up event occurs if the cluster containing P_0 is shorter than $l_b - 2r_0$, which happens with probability $\Pr(X(\rho - \rho_{n_{\max}}) \leq l_b - 2r_0)$, where $\Pr(X(\rho) \leq x)$ is the cumulative distribution function of the cluster length given in (9.2.1).

It follows that the probability that a pseudo catch-up event occurs satisfies

$$\Pr(\Psi) \leq \int_{r_0}^{\infty} f_b(l_b) \Pr\left(X(\rho - \rho_{n_{\max}}) \leq l_b - 2r_0\right) \mathrm{d}l_b$$

□

Figure 9.15 shows the probability of pseudo catch-up events, viz., $\Pr(\Psi)$. It can be seen that the probability of pseudo catch-up events is affected by the vehicle densities in both traffic streams. When the density of traffic stream $n_{\max}$ is small (say 0.001 veh/m), a large percentage of catch-up processes are pseudo catch-up events. On the other hand, when the density of the traffic stream $n_{\max}$ is large (say 0.01 veh/m), another traffic stream with a small density (say 0.001 veh/m) can significantly boost the catch-up process, in the sense of bridging the gap in the $n_{\max}^{th}$ traffic stream. Note that the analysis provides a fairly tight bound on $\Pr(\Psi)$.

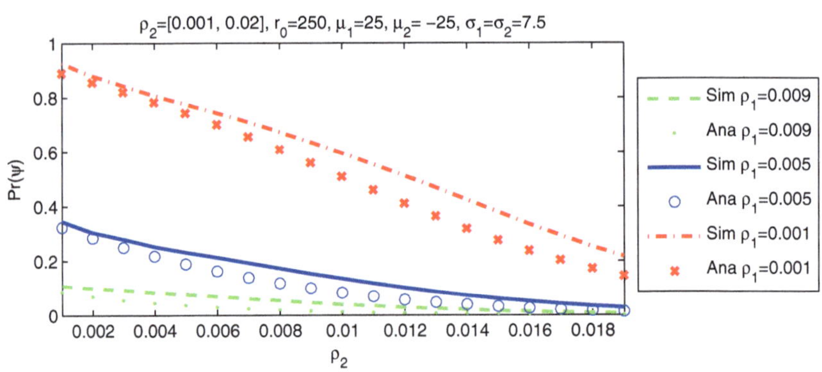

Fig. 9.15 Simulation (Sim) and analytical (Ana) results on the probability of pseudo catch-up events, viz., $\Pr(\Psi)$, in a VANET with two traffic streams. Detailed simulation parameters are introduced in Sect. 9.2.6

9.2.2.4 The Catch-Up Process Between H_ζ and P_η

Recall that H_ζ (P_ζ) is the ζth vehicle to the left of the head H_0 (to the right of the tail P_0) at the beginning of a catch-up process. viz., at time 0. It is worth noting that the head vehicle at time $t > 0$ is not necessarily the head vehicle at time 0 because the previous head vehicle may be overtaken by another informed vehicle, either in the same traffic stream or in a different stream, during time $(0, t]$. In view of this, we study the catch-up process between vehicles H_ζ and P_η in this subsection.

Lemma 154 *Denote by $q_{\zeta\eta}(i|l_c)$ the probability that H_ζ catches up P_η for the first time in the ith time slot, in a catch-up process with gap l_c, where ζ and η are nonnegative integers. Then, there holds*

$$
\begin{aligned}
q_{\zeta\eta}(i|l_c) \approx & \sum_{n=m\in\{1,\ldots,N\}} \mathcal{G}_{\zeta\eta}^{nm}(z_{nm}, i)\frac{\rho_n\rho_m}{\rho^2} \\
& + \sum_{n,m\in\{1,\ldots,N\}\ for\ n\neq m} \mathcal{G}_{\zeta\eta}^{nm}(z_{nm}, i)\frac{\rho_n\rho_m}{\rho^2} \\
& \int_{l_c}^{\infty} (1 - \Pr(X(\rho - \rho_{n_{\max}}) \leq l_b - 2r_0)) f_l(l_b|l_c, \rho_{n_{\max}})\mathrm{d}l_b
\end{aligned}
\tag{9.2.23}
$$

where $n_{\max} \triangleq \arg\max_n\{\mu_n\}$, $z_{nm} = l_c - r_0 + \zeta/\rho_n + \eta/\rho_m$, $f_b(l_b|l_c, \rho_{n_{\max}}) = \rho_{n_{\max}} \exp(-\rho_{n_{\max}}(l_b - l_c))$, $\Pr(X(\rho) \leq x)$ is given by (9.2.1) and $\mathcal{G}_{\zeta\eta}^{nm}$ is given by (9.2.24).

Proof Denote by ϕ_{nm} the probability that a randomly chosen pair of vehicles belong to the nth and the mth traffic stream, respectively. It is straightforward that $\phi_{nm} = \frac{\rho_n\rho_m}{\rho^2}$ owing to the Poisson distribution of vehicles. Furthermore, due to the effect of pseudo catch-up processes studied in the previous subsection, the probability that the head and the tail vehicles of a catch-up process belong to the nth and the mth traffic stream, respectively, can be different from $\phi_{nm} = \frac{\rho_n\rho_m}{\rho^2}$. This is one of the major challenges in the study of the catch-up process in a VANET with multiple traffic streams. In the following analysis, we take this effect into account.

Let $z_{nm} = l_c - r_0 + w_\zeta^n + w_\eta^m$, where w_ζ^n (w_η^m) is the distance between H_ζ^n and H_0 (P_η^m and P_0) at time 0. Due to the Poisson distribution of vehicles, the inter-vehicle distance in the nth traffic stream follows an exponential distribution with mean $1/\rho_n$. Denote by $f(w_\zeta^n)$ the probability density function of w_ζ^n, then $f(w_\zeta^n) = \text{Erlang}(\zeta, \rho_n)$ is an Erlang function because the sum of ζ independent exponentially distributed random variables follows an Erlang distribution. Denote by $\mathcal{G}_{\zeta\eta}^{nm}(i|l_c)$ the probability that H_ζ^n catches up P_η^m for the first time in the ith time slot in a catch-up process with gap l_c. Then, it is straightforward that

$$
\mathcal{G}_{\zeta\eta}^{nm}(i|l_c) = \int_0^\infty \int_0^\infty G_{\zeta\eta}^{nm}(z_{nm}, i) f(w_\zeta^n) f(w_\eta^m)\mathrm{d}w_\zeta^n \mathrm{d}w_\eta^m \tag{9.2.24}
$$

where $G_{\xi\eta}^{nm}(z_{nm}, i)$ is given by Lemma 152, which is the probability that the distance between H_{ξ}^{n} and P_{η}^{m} reduces from z_{nm} to r_0 for the first time in the ith time slot. Furthermore, denote by $\psi(l_c)$ the probability that a catch-up event between H_{ξ} and P_{η} is *not* a pseudo catch-up event in a catch-up process with gap l_c. Then, according to the total probability theorem, it is evident that

$$q_{\xi\eta}(i|l_c) = \sum_{n,m\in\{1,\dots,N\}} \mathcal{G}_{\xi\eta}^{nm}(i|l_c)\psi(l_c)\phi_{nm} \tag{9.2.25}$$

Note that $\psi(l_c)$ can be seen as a weighting factor for ϕ_{nm} that addresses the bias of ϕ_{nm} introduced at the beginning of this proof. Next, we calculate $\psi(l_c)$. To take the effect of pseudo catch-up events into account without making the calculation too complicated, we approximately consider that a catch-up event between two vehicles is a pseudo catch-up event if these vehicles are in different traffic streams *and* the cluster containing the tail is shorter than $L_b - 2r_0$, as shown in Fig. 9.14. The accuracy of this approximation has been discussed in the previous subsection and verified by simulations. Then, (9.2.25) becomes

$$\begin{aligned} q_{\xi\eta}(i|l_c) \approx & \sum_{n=m\in\{1,\dots,N\}} \mathcal{G}_{\xi\eta}^{nm}(i|l_c)\phi_{nm} \\ & + \sum_{n,m\in\{1,\dots,N\} \text{ for } n\neq m} \mathcal{G}_{\xi\eta}^{nm}(i|l_c)\phi_{nm} \\ & \int_{l_c}^{\infty} (1 - \Pr(X(\rho - \rho_{n_{\max}}) \le l_b - 2r_0)) f_b(l_b|l_c, \rho_{n_{\max}}) \mathrm{d}l_b \end{aligned} \tag{9.2.26}$$

where $f_b(l_b|l_c, \rho_{n_{\max}})$ is the probability density function of l_b given that the distance between the head and the tail at time 0 is l_c. Because the vehicles in each traffic stream follow a homogeneous Poisson distribution independent of other traffic streams, there holds

$$f_b(l_b|l_c, \rho_{n_{\max}}) = \rho_{n_{\max}} \exp(-\rho_{n_{\max}}(l_b - l_c)) \tag{9.2.27}$$

□

9.2.2.5 Delay of a Catch-Up Process

Define the *catch-up delay* t_c to be the time interval from the beginning of a catch-up process till the time when the head and the tail move into the transmission range of each other for the first time, i.e., $t_2 - t_1$ in Fig. 9.1. This subsection studies the catch-up delay. Note that we do not consider the rare event that the distance between the head and tail becomes larger than r_0 again during the transmission between head and tail (from time t_c to $t_c + \beta$) due to changes of vehicle speeds, which may cause the transmission to be interrupted, because the per-hop delay β (e.g., 4ms) is usually much smaller than the time interval for a vehicle to change speed (typically longer than a second [18]).

Denote by $H(i|l_c)$ the probability that none of the H_ζ-P_η vehicle pairs for $\zeta, \eta \in \{0, 1, 2, \ldots\}$ catches up in the ith time slot, in a catch-up process with gap l_c. We next derive $H(i|l_c)$. Denote by $\Xi_{\zeta\eta}$ the event that a pair of vehicles H_ζ-P_η does *not* catch-up in the ith time slot. It is worth noting that the events $\Xi_{\zeta\eta}$, for $\zeta, \eta \in \{0, 1, 2, \ldots\}$, are *not* independent of each other. Specifically, the event $\Xi_{0\eta}$ is positively correlated [103, Section 1.2] with events $\Xi_{\zeta\eta}$, for any given $\eta \in \{0, 1, 2, \ldots\}$ and all $\zeta \in \{1, 2, \ldots\}$, because the occurrence of event $\Xi_{0\eta}$ means that the occurrences of events $\Xi_{\zeta\eta}$ are more likely. In other words, if the rightmost informed vehicle, viz., the head H_0 as shown in Fig. 9.13, does not catch-up to an uninformed vehicle, other things being equal, then other informed vehicles are less likely to catch up the uninformed vehicle. Similarly, the event $\Xi_{\zeta 0}$ is positively correlated with events $\Xi_{\zeta\eta}$, for any given $\zeta \in \{0, 1, 2, \ldots\}$ and all $\eta \in \{1, 2, \ldots\}$, because if the leftmost uninformed vehicle, viz., P_0, was not caught up by an informed vehicle, then other uninformed vehicles are less likely to be caught up by the informed vehicle. Then, according to the FKG inequality [103, Section 1.2], there holds:

$$H(i|l_c) = \Pr\Big(\bigcap_{\zeta,\eta\in\{0,1,2,\ldots\}} \Xi_{\zeta\eta}\Big) \geq \prod_{\zeta,\eta\in\{0,1,2,\ldots\}} (1 - q_{\zeta\eta}(i|l_c)) \tag{9.2.28}$$

where $q_{\zeta\eta}(i|l_c)$ is given by Lemma 154. Furthermore, the above inequality is tight in VANETs with a low density, where the distance between vehicle pair H_ζ-P_η increases rapidly as either ζ or η increases, so that the effect of the aforementioned correlation issue reduces rapidly as either ζ or η increases. This is exactly the situation where this analysis on the catch-up process focuses on.

Denote by $h(i|l_c)$ the probability that one vehicle pair (H_ζ and P_η) catches up in the ith time slot *and* no vehicle pair has caught up before the ith time slot, in a catch-up process with gap l_c. Assume that the catch-up event in the ith time slot is independent of those in the jth time slots for $0 < j < i$, which is an accurate approximation when the duration of a time slot is large, e.g., $\tau = 1$s or 5s as shown in Sect. 9.2.6. Then:

$$h(i_c|l_c) = (1 - H(i_c|l_c)) \prod_{i=1}^{i_c-1} H(i|l_c) \tag{9.2.29}$$

Finally, the expected delay of a catch-up process with gap l_c is $E[t_c|l_c] = \sum_{i=1}^{\infty} i\tau h(i|l_c)$.

9.2.3 Information Propagation Speed

Similarly as that in the previous section, the information propagation process is modeled by a renewal reward process. Each cycle in the renewal reward process starts at the time when a catch-up process starts and consequently each cycle consists of a catch-up process followed by a forwarding process. The reward of each cycle is the distance traveled by a packet in the catch-up process and forwarding process during the cycle. Then, the expected IPS ($E[v_{ip}]$) is:

$$\begin{aligned} E[v_{ip}] &= \frac{\text{expected length of one cycle}}{\text{expected time duration of one cycle}} \\ &\approx \frac{E[l_c] + E[X(\rho)]}{E[t_c] + \beta + E[t_f]} + \mu_{n_{\max}} \\ &= \frac{\int_{r_0}^{\infty} l_c f_c(l_c) \mathrm{d}l_c + \int_0^{\infty} x \Pr(X(\rho) = x) \mathrm{d}x}{\int_{r_0}^{\infty} E[t_c | l_c] f_c(l_c) \mathrm{d}l_c + \beta + \int_0^{\infty} E[t_f | x] \Pr(X(\rho) = x) \mathrm{d}x} \\ &\quad + \mu_{n_{\max}} \end{aligned} \tag{9.2.30}$$

where the term $E[l_c] + E[X(\rho)]$ is the expected distance traveled by a packet during a catch-up process and a forwarding process, without considering the distance traveled when the packet is carried by the head vehicle during the catch-up process. The contribution of the movement of the head vehicle to the information propagation speed is considered separately by the term $\mu_{n_{\max}}$. Furthermore, the term $E[t_c]+\beta+E[t_f]$ is the expected time duration of a catch-up process and a forwarding process.

9.2.4 Simplified Results Charactering the Information Propagation Process

In the previous three subsections, we studied in detail the information propagation process of a VANET. To provide a more insightful understanding of the information propagation process, this subsection presents some simplified results to reveal more clearly the interactions among the main parameters that determine the performance of the information propagation process. In this subsection, we first provide simplified results on the number of hops traveled by a packet in a cluster and the length of a cluster, followed by the results on the catch-up delay and the IPS.

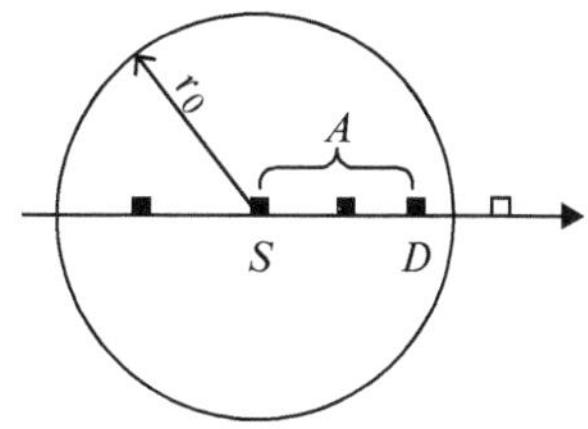

Fig. 9.16 An illustration of the per-hop progress, marked by A, which is the distance between a node (say S) and the rightmost node in the transmission range r_0 of S, which is node D

9.2.4.1 Simplified Hop Count

Denote by A (where $0 < A \leq r_0$) the *per-hop progress*, which is the Euclidean distance between a randomly chosen vehicle and its rightmost directly connected neighbor at a randomly chosen time instant, as illustrated in Fig. 9.16. Denote by Ψ_1 the event that there is no node in a given road segment with length $r_0 - a$, where $0 < a \leq r_0$. Because vehicles follow a Poisson distribution with density ρ at any time instant, it is evident that $\Pr(\Psi_1) = \exp(-\rho(r_0 - a))$. Denote by Ψ_2 the event that there is at least one node in a given road segment with length a. It is straightforward that $\Pr(\Psi_2) = 1 - \exp(-\rho a)$. Denote by C_r the event that a progress is made at a given node (say S), i.e., there is at least one node to the right of node S within its transmission range r_0, as illustrated in Fig. 9.16. Furthermore, denote by $\Pr(A \leq a, C_r)$ the probability that a progress is made at a given node *and* the progress is not larger than a. It is evident that $\Pr(A \leq a, C_r) = \Pr(\Psi_1)\Pr(\Psi_2)$. Finally, denote by $\Pr(A \leq a|C_r)$ the cumulative distribution function of per-hop progress A, conditioned on the event that a progress is made. It is evident that $\Pr(A \leq a|C_r) = \frac{\Pr(A \leq a, C_r)}{\Pr(C_r)}$, where $\Pr(C_r) = 1 - \exp(-\rho r_0)$. Then, the expected per-hop progress is:

$$\begin{aligned}\bar{a} \triangleq E[A|C_r] &= \int_0^{r_0} (1 - \Pr(A \leq a|C_r))\mathrm{d}a \\ &= \int_0^{r_0} \left(1 - \frac{(1 - \exp(-\rho a))\exp(-\rho(r_0 - a))}{1 - \exp(-\rho r_0)}\right) \mathrm{d}a \\ &= r_0 - \frac{1}{\rho} + \frac{r_0 \exp(-\rho r_0)}{1 - \exp(-\rho r_0)} \end{aligned} \tag{9.2.31}$$

Furthermore, the expected number of hops from the left-most vehicle to the right-most vehicle in a cluster with length x can be approximated by:

$$E[k|x] \approx \frac{x}{\bar{a}} \tag{9.2.32}$$

Figure 9.17a shows the expected number of hops from the left-most vehicle to the right-most vehicle in clusters with different lengths. The exact analytical result (denoted by *Ana*) is given by [63] and the approximate analytical result (denoted by *Ana-simple*) is given by (9.2.32). It is interesting to observe that (9.2.32), which

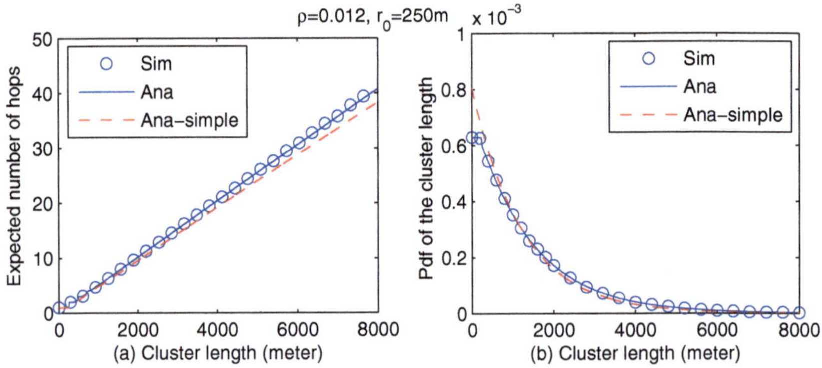

Fig. 9.17 Simulation (Sim) and analytical (Ana) results for (**a**) the expected number of hops from the leftmost vehicle to the rightmost vehicle in a cluster; (**b**) probability density function of the cluster length

is much simpler than the recursive formula presented in [63], provides a fairly accurate estimate on the hop count in VANETs. The accuracy of the approximation reduces as the cluster length increases (say longer than 6000m), however, as shown in Fig. 9.17b and studied in the next subsection, the probability of forming a long cluster decreases exponentially.

9.2.4.2 Simplified Cluster Length

In the previous subsection, we have shown that the expected per-hop progress $\bar{a}$ is a performance-determining factor which can be used to characterize the forwarding process. Next, we consider that a multi-hop forwarding process can continue as long as there is at least one vehicle within distance $\bar{a}$ to the right of an arbitrary vehicle, which happens with probability $p_s = 1 - \exp(-\rho\bar{a})$, because of the Poisson distribution of vehicles. Then, the number of successful multi-hop forwarding progresses, denoted by κ, follows a geometric distribution with the following probability density function

$$f_\kappa(\kappa) = (1 - p_s)p_s^{\kappa-1} \tag{9.2.33}$$

The expected time delay of a forwarding process is

$$E[t_f] \approx \beta E[\kappa] = \frac{\beta}{1 - p_s} = \frac{\beta}{\exp(-\rho\bar{a})} \tag{9.2.34}$$

Furthermore, the expected cluster length in a VANET with vehicle density ρ is

$$E[X(\rho)] \approx \bar{a}E[\kappa] = \frac{\bar{a}}{1 - p_s} = \frac{\bar{a}}{\exp(-\rho\bar{a})} \tag{9.2.35}$$

Next we present an approximation of the distribution of the cluster length. Note that κ is a discrete random variable, whereas the cluster length x is a continuous random variable. To facilitate the analysis, let κ_c be the continuous analogue of κ, where κ_c follows an exponential distribution with parameter p_c. Then, the probability of having at least κ consecutive successful progresses can be expressed as $p_s^{\kappa} = \exp(-p_c\kappa)$, from which we can obtain $p_c = -\log(p_s)$. Therefore, the probability density function of the cluster length can be approximated by

$$
\begin{aligned}
f_{X(\rho)}(x) &\approx \frac{p_c \exp(-p_c \frac{x}{\bar{a}})}{\int_0^{\infty} p_c \exp(-p_c \frac{x_0}{\bar{a}}) \mathrm{d}x_0} \\
&= \frac{p_c \exp(-p_c \frac{x}{\bar{a}})}{\bar{a}} \\
&= -\frac{\log(1-\exp(-\rho\bar{a}))}{\bar{a}} \exp(\frac{x \log(1-\exp(-\rho\bar{a}))}{\bar{a}})
\end{aligned} \tag{9.2.36}
$$

Furthermore, its cumulative distribution function is

$$
\begin{aligned}
\Pr(X(\rho) \le x) &\approx \frac{1-\exp(-\frac{p_c x}{\bar{a}})}{\bar{a}} \\
&= \frac{1-\exp(\frac{\log(1-\exp(-\rho\bar{a}))}{\bar{a}} x)}{\bar{a}}
\end{aligned} \tag{9.2.37}
$$

Figure 9.17b shows the probability density function of the cluster length. The exact analytical result is given by (9.2.1) and the approximate analytical result is given by (9.2.36). It is interesting to note that the accuracy of the approximation is good, while at the same time the analytical formula of the approximate result is much simpler than that for the exact result.

9.2.4.3 Simplified Catch-Up Delay

Recall that the $n_{\max}^{th}$ traffic stream is the traffic stream with the largest mean speed, i.e., $n_{\max} = \arg\max_n\{\mu_n\}$. Furthermore, let the $n_{\min}^{th}$ traffic stream be the traffic stream with the smallest mean speed, i.e., $n_{\min} = \arg\min_n\{\mu_n\}$. Define t_ζ to be the expected time interval from the beginning of an arbitrary catch-up process to the time when a vehicle in traffic stream $n \in \{1, \ldots, N\} \setminus n_{\max}$ comes into the transmission range of a vehicle in traffic stream $n_{\max}$. Then, it can be shown that $t_m \approx \frac{D_e(\rho - \rho_{n_{\max}})}{\mu_{n_{\max}} - \mu_{n_{\min}}}$, where $D_e(\rho)$ is the expected distance between two adjacent but disconnected vehicles. Because vehicles follow a homogeneous Poisson distribution with density ρ, there holds

$$
D_e(\rho) = \frac{\int_{r_0}^{\infty} l_c \rho \exp(-\rho l_c) \mathrm{d}l_c}{\int_{r_0}^{\infty} \rho \exp(-\rho l_c) \mathrm{d}l_c}
$$

$$= \frac{\exp(-\rho r_0)(\rho r_0 + 1)}{\exp(-\rho r_0)\rho}$$

$$= \frac{(\rho r_0 + 1)}{\rho} \tag{9.2.38}$$

Next, we take the pseudo catch-up events into consideration. Let Υ be the event that a vehicle in traffic stream $n \in \{1, \ldots, N\} \setminus n_{\max}$ comes into the transmission range of a vehicle in traffic stream $n_{\max}$ in an arbitrary catch-up process. Recall that Υ is a pseudo catch-up event if the length of the cluster containing the tail vehicle is shorter than $L_b - 2r_0$, where L_b is the Euclidean distance between H_0^+ and P_0^+. Denote by p_p the probability that Υ is a pseudo catch-up event. Then

$$\begin{aligned} p_p &= E[\Pr(X(\rho - \rho_{n_{\max}}) \le L_b - 2r_0)] \\ &\approx \Pr(X(\rho - \rho_{n_{\max}}) \le E[L_b - 2r_0]) \end{aligned}$$

where

$$\begin{aligned} E[L_b - 2r_0] &= \int_{2r_0}^{\infty} (l_b - 2r_0)\rho_{n_{\max}} \exp(-\rho_{n_{\max}} l_b) \mathrm{d}l_b \\ &= \frac{\exp(-\rho_{n_{\max}} 2r_0)}{\rho_{n_{\max}}} \end{aligned} \tag{9.2.39}$$

Recognizing that an approximation is involved, assume that the catch-up events are independent of each other. Then, the number of pseudo catch-up events before a successful one follows a geometric distribution with mean $\frac{1}{1-p_p}$. Therefore, the expected catch-up delay is:

$$E[t_c] \approx \frac{t_m}{1-p_p} \approx \frac{(\rho-\rho_{n\max})r_0+1}{(\mu_{n\max}-\mu_{n\min})(\rho-\rho_{n\max})(1-p_p)} \tag{9.2.40}$$

9.2.5 *Simplified IPS*

According to (9.2.30) and the results obtained in the previous subsections, the expected IPS satisfies

$$\begin{aligned} E[v_{ip}] &= \frac{\text{expected length of one cycle}}{\text{expected time duration of one cycle}} \\ &\approx \frac{r_0 + E[X(\rho)]}{E[t_c] + E[t_f]} + \mu_{n_{\max}} \end{aligned}$$

$$\approx \frac{r_0 + \frac{\bar{a}}{\exp(-\rho\bar{a})}}{E[t_c] + \frac{\beta}{\exp(-\rho\bar{a})}} + \mu_{n_{\max}}$$

$$= \frac{r_0 + \frac{\rho r_0 - 1}{\rho \exp(1-\rho r_0)}}{E[t_c] + \frac{\beta}{\exp(1-\rho r_0)}} + \mu_{n_{\max}} \qquad (9.2.41)$$

where $E[t_c]$ is given by (9.2.40).

Note that the result is much simpler than that given by (9.2.30) and in Sect. 9.1, where the accuracy and applicability of the simplified result are discussed in the next subsection. Furthermore, the impact of main parameters on the performance of the information propagation process is clearly shown. For example, a larger difference between mean vehicle speeds ($\mu_{n_{\max}} - \mu_{n_{\min}}$ in (9.2.40)) leads to a faster IPS. More observations dealing with the impact of the fundamental parameters are discussed in the next subsection.

9.2.6 Simulation Results

In this subsection, we report on simulations to validate the accuracy of the analytical results. Each point shown in the figures is the average value from 1000 simulations. The confidence interval is too small to be distinguishable and hence is ignored in the following plots. Similar as in Sect. 9.1.7, the transmission range is $r_0 = 250m$. The typical values of the mean and standard deviation of vehicle speed are 25m/s and 7.5m/s. The vehicle mobility parameters, i.e., μ_n, σ_n and τ, are taken from real measurements, where the usual record time intervals for a vehicle speed monitor are $\tau = 1s, 5s$. The traffic density is varied so that the density ρ of the homogeneous Poisson point process governing the spatial distribution of the vehicles varies from 0 to 0.06 veh/m. For completeness of the plot, $\rho = 0$ is included which means that there is only one vehicle in each traffic stream. Consequently, the average number of neighbors varies from 0 to 30, which represents a large range of traffic densities.

Firstly, Fig. 9.18 validates our mobility model. The analytical results are given by (9.2.30), which is derived under the synchronized random walk model, whereas the simulation results show the expected IPS under non-synchronized mobility models. Three different mobility models are evaluated: the speed-change time interval τ of each vehicle is uniformly selected from $[4.8, 5.2]$ (i.e., around 5s) or from $[2, 8]$ (i.e., within a larger range around 5s); and for τ following an exponential distribution with mean 5s. Under all three mobility models, vehicles change speed at different time instants (non-synchronized) and the average speed-change time interval is 5s. It can be seen that the IPSs under non-synchronized mobility models are very close (almost indistinguishable) to each other and our analysis using the simplified (synchronized) mobility model provides a good estimation on the IPS.

Moreover, it can be observed in Fig. 9.18 that when the vehicle density is low, the IPS is close to the vehicle moving speed because there is little packet forwarding in

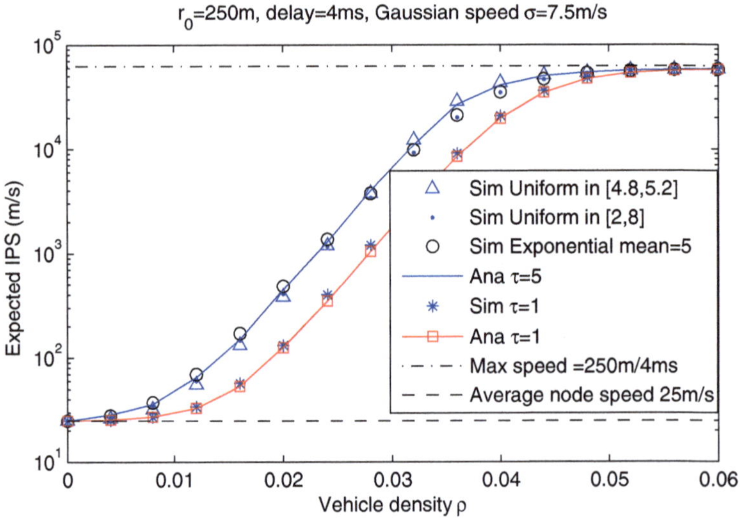

Fig. 9.18 Simulation (Sim) and analytical (Ana) results of the expected information propagation speed in a VANET with one traffic stream ($N = 1$)

the network. Furthermore, when the vehicle density increases, small clusters are formed and the IPS increases. As the vehicle density further increases, clusters become larger and the forwarding process starts to dominate. Therefore, the IPS increases until it reaches the maximum value, which is determined by the per-hop delay in the forwarding process. The maximum IPS is obviously equal to r_0/β, where β is the per-hop delay. When the vehicle density is moderate, the IPS is determined by the catch-up delay, which is further determined by the mobility of the vehicles. It can be seen from Fig. 9.18 that the more frequently the speed changes, the slower the information propagates. This is mainly because changing speed has the potential to interrupt the catch-up process, i.e., during a catch-up process, the tail may speed up and the head may slow down.

Moreover, vehicle speed distribution also has a significant impact on the catch-up process, hence the IPS. Figure 9.19 shows the IPS in a VANET with two traffic streams with equal vehicle density but different vehicle speed distributions. Specifically, the mean speed of one traffic stream is fixed at 25 and the mean speed of another traffic stream is varied. It can be seen that our analytical results have a better match with the simulation results than that in [11]. An interesting observation is that the IPS increases when the average speed of the vehicles in one of the traffic streams is reduced from 25m/s to 0m/s. Furthermore, an even faster IPS is observed when the average speed of the vehicles in one of the traffic streams is −25m/s, i.e., two traffic streams traveling in opposite directions. The reason behind this interesting observation is that a larger relative speed between vehicles results in faster catch-up processes, hence a faster IPS. This can also be seen from the analytical results of $G_{\xi\eta}^{nm}(z, i)$, e.g., applying the fact that the error function is an increasing function

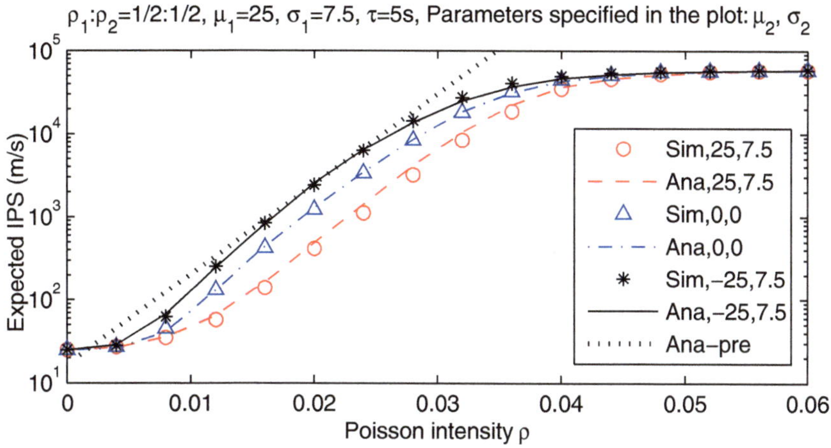

Fig. 9.19 The expected IPS in a VANET with different vehicle speeds in two traffic streams ($N = 2$). The plot *Ana-pre* shows the IPS derived in [11] for a VANET with vehicle speed 25 and −25 in two streams respectively

to (9.2.8). This observation tells us that making use of the vehicles traveling in the opposite direction of information propagation can increase the IPS. Moreover, some (stationary) roadside units without (expensive) wired connections can also significantly increase the IPS in a VANET. The optimum placement of the roadside units is not the focus of this chapter and will be studied in Chap. 11.

Figure 9.20 shows the expected IPS in a VANET with different vehicle densities in two traffic streams. Specifically, the total traffic density ρ is fixed while the fractions of traffic density in two traffic streams are varied to examine the impact of traffic densities in different traffic streams on the IPS. Firstly, it can be seen that an uneven distribution of the vehicle densities between traffic streams (e.g., $\rho_1 = \rho/10, \rho_2 = 9\rho/10$) results in a slower IPS compared to the IPS in a VANET with evenly distributed traffic densities (e.g., $\rho_1 = \rho_2 = \rho/2$). This is because an uneven distribution of the vehicle densities between traffic streams results in a smaller number of catch-ups between vehicles in different traffic streams (as manifested by the factor $\frac{\rho_n \rho_m}{\rho^2}$ in (9.2.23)). Hence there is less improvement on the IPS provided by the large relative vehicle speed, compared with a VANET with evenly distributed vehicle densities between traffic streams. This situation can be observed in the real world on freeways connecting the business district and residential district. Densities of the traffic streams in opposite directions vary a great deal depending on time-of-day when people go to work or come back home.

On the other hand, it can be seen from Fig. 9.20 that a small amount (e.g., $\rho/10$) of vehicle traffic in an opposite direction can still significantly increase the IPS. Therefore, it is meaningful to explore the traffic streams with different vehicle moving speeds. This phenomenon can also be observed in Fig. 9.21, owing to the fact that large relative speed between vehicles reduces the catch-up delay, as

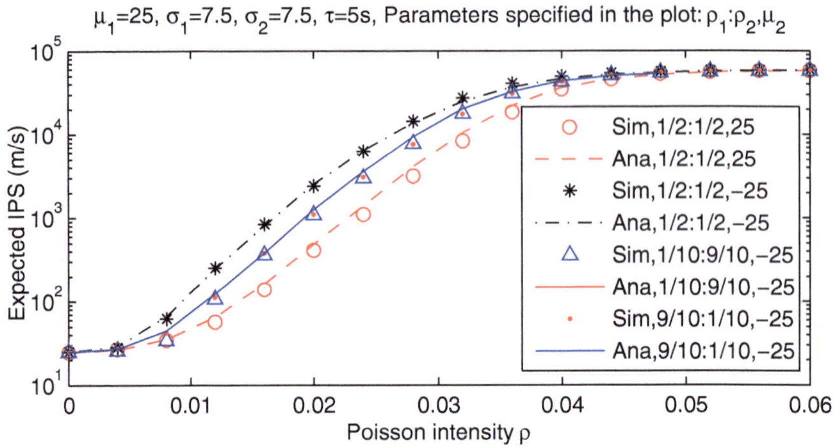

Fig. 9.20 The expected IPS in a VANET with different vehicle densities in two traffic streams. The first four curves are kept from Fig. 9.19 for comparison

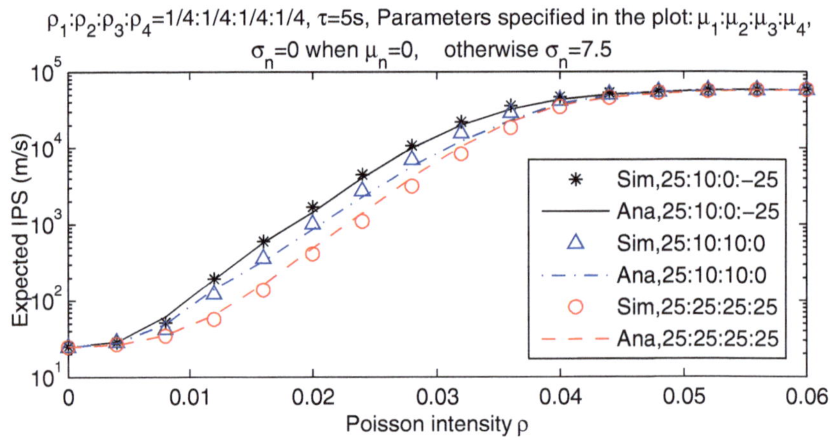

Fig. 9.21 The expected IPS in a VANET with four traffic streams

discussed earlier. It is worth noting that the IPS can still be large in a VANET where most, or even all, vehicles are in the same traffic stream, which is different from previous studies, e.g., [1, 11], considering that all vehicles traveling at the same and time-invariant speed where the catch-up events can only occur between vehicles in different traffic streams.

Figure 9.22 compares the results on the IPS given in Sect. 9.2.1 with previous results of related studies. In Fig. 9.22a, the transmission range is 250m and the vehicle density is varied from 0 to 0.015, whereas in Fig. 9.22b the vehicle density is 0.036 and the transmission range is varied from 0 to 90m. Consider a common case that the mean distance between two consecutive vehicles on the same lane is

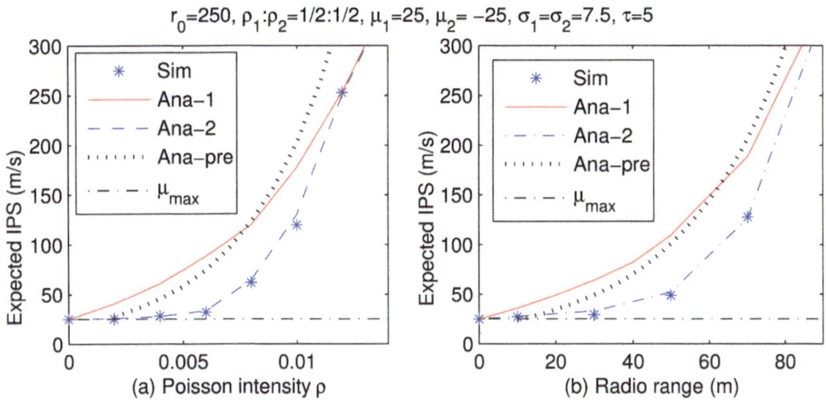

Fig. 9.22 The expected information propagation speed in a VANET with two traffic streams, where *Ana-1* is the previous result given in Sect. 9.1, *Ana-2* is the result given in this chapter, and *Ana-pre* is the result given by [11]

around 110 m [65], leading to a mean vehicle density of around 9 veh/km/lane. Consider two lanes in each direction, leading to an overall mean vehicle density of around 36 veh/km or 0.036 veh/m [65]. It can be seen that the results presented in this subsection, which consider the effects of pseudo catch-up events in detail, have better accuracy compared with previous results such as [11] and our own result in Sect. 9.1. Moreover, it can also be seen that the impact of pseudo catch-up events can be significant.

Figure 9.23 shows the simplified results of the IPS, derived in Sect. 9.2.4. Without detailed analysis of the catch-up process as shown in Sect. 9.2.2, the simplified results are only valid for VANETs with more than one traffic streams, i.e., vehicles have different mean speeds. Nevertheless, it can be seen that the simplified results, though do not match exactly with simulation results, capture the impact of fundamental parameters, e.g., vehicle density and speed, on the IPS.

9.3 Notes and Further Readings

In this chapter, we studied the information propagation process in one-dimensional dynamic networks with vehicular networks being the motivating example. Different from static networks in which information propagates via a store-and-forward process, in mobile networks, information propagates via a store-*carry*-forward process. Accordingly, the information propagation process in mobile networks can be divided into two sub-processes: the forwarding process and the catch-up process. We presented analysis for both sub-processes and the results were combined using the renewal theory to form a complete picture on the information propagation

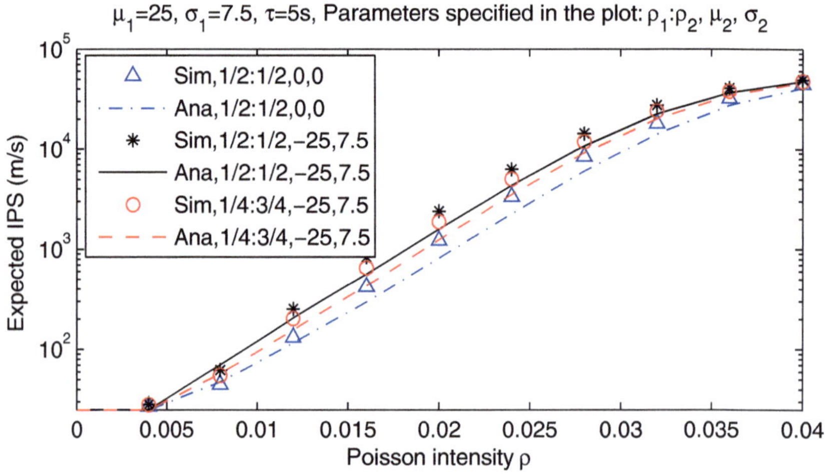

Fig. 9.23 Simplified results of the expected IPS, where *Ana* is given by (9.2.41)

process. Both vehicular networks with single traffic stream and vehicular networks with multiple traffic streams were investigated.

In recent years, VANET has attracted significant interest. In [68], Fracchia *et al.* introduced the design of a warning delivery service in VANETs. They studied the propagation of a warning message in a one-dimensional VANET where vehicles move in the opposite direction of the propagation direction of the warning message. Their analysis is based on an overly simplified assumption that the vehicular network topology does not change over time during the information propagation process. In [191], Wu *et al.* studied the IPS through simulations. They used a commercial microscopic traffic simulator, CORSIM, to simulate the traffic on a highway. Then, the topology data from CORSIM were imported into a wireless communication simulator to study the properties of the information propagation process. They showed that the IPS varies significantly with vehicle densities.

There are analytical studies on VANETs based on the assumption that vehicle speed does not change over time, referred to as the *constant speed model*. In [198], Yousefi *et al.* provided analytical results on the distribution of the inter-vehicle distance in a one-dimensional VANET under the constant speed model and the Poisson arrival model: in the *Poisson arrival model*, the number of vehicles passing an observation point on the road during any time interval follows a homogeneous Poisson point process with density λ. By applying the results used in studying the busy period in queueing theory, they further analyzed the connectivity distance, a measure similar to the cluster length in this chapter. However, they did not provide a closed-form formula for the distribution of the cluster length. In [2], Agarwal *et al.* studied the IPS in a one-dimensional VANET where vehicles are Poissonly distributed and move at the same speed but either in the positive or negative direction of the axis. They derived upper and lower bounds for the IPS, which provided a hint on the impact of vehicle density on the IPS. However the bounds are not tight and many factors, e.g., temporal variation of speeds and propagation delay, were ignored in their analysis. In [190], Wu *et al.* considered a one-dimensional VANET where vehicles are Poissonly distributed and vehicle speeds are uniformly distributed in a designated range. They provided a numerical method to compute the IPS under two special network models, i.e., when the vehicle density is either very low or very high. The aforementioned studies [2, 190, 198] were all based on the constant speed model.

In [3], Ashish, Starobinski, and Little analyzed the phase transition phenomenon in the IPS. They found that below a certain critical threshold in vehicle density, the average information propagation speed is the same as the average vehicle speed; on the other hand, above another higher critical threshold, the average information propagation speed quickly increases as a function of vehicle density and approaches the speed of wireless communications. In this chapter, we also found and discussed this phase transition phenomenon.

Chapter 10
Information Propagation in Two-Dimensional Dynamic Networks

Abstract In this chapter, we investigate the information propagation process in two-dimensional dynamic networks with mobile ad hoc networks being the motivating example. Particularly, we consider a two-dimensional mobile ad hoc network where nodes are initially randomly distributed and then move following a random direction mobility model. Adopting a popularly used Susceptible-Infectious-Recovered epidemic broadcast scheme, we analyze performance measures such as the fraction of nodes that can receive the information and the delay of the information dissemination process. The accuracy of analytical results is verified using simulations driven by both the random direction mobility model and a real world trace.

In this chapter, we continue to investigate the information propagation process in two-dimensional dynamic networks with mobile ad hoc networks being the motivating example.

A mobile ad hoc network (MANET) is a self-organizing network comprising mobile devices like smart phones, tablet PCs, or connected and automated vehicles (CAVs). In a MANET, information dissemination often relies on local ad hoc connections that emerge opportunistically as mobile devices move and meet each other. The main challenge of data communication in a MANET is the time-varying nature of ad hoc connections, which is attributable to two major factors: dynamic network topology and channel randomness.

The dynamic topology of a MANET is caused by the mobility of wireless communication devices. Specifically, as mobile users or CAVs move over time, the distances between mobile devices are changing constantly, resulting in time-varying wireless connections. This makes many commonly used routing protocols, e.g., the well-known Ad hoc On Demand Distance Vector (AODV) [166] or a basic flooding broadcast algorithm [189], less effective for MANETs, because they can only disseminate information to the node(s) that is connected to the source by at least one (multi-hop) path at the time instant when the source node transmits. Furthermore, differently from the store-forward pattern that AODV relies on to disseminate information, information dissemination in MANETs has to propagate by way of store-*carry*-forward that allows a node to carry the information over a physical distance and forward to other nodes over time. It has been widely

G. Mao, *Connectivity of Communication Networks*,
DOI 10.1007/978-3-319-52989-9_10

recognized that the dynamic topology of a MANET often resembles the topology of a human network [183, 202], in the sense that the movement of mobile devices in a MANET is not only similar to, but often governed by, the movement of their human owners. In view of this, epidemic schemes [183] have been proposed as a fast and efficient approach to broadcast information in MANETs. On the other hand, unlike the spreading of epidemic disease in human networks, information dissemination schemes in MANETs can often be carefully designed to meet certain goals in addition to information delivery, such as achieving energy efficiency or fulfilling certain delay constraint, as shown in this chapter.

In addition to the fast-changing network topology, channel randomness also has a significant impact on the performance of information broadcast in MANETs. It has been shown that channel shadowing has negative impacts on information dissemination in MANETs employing traditional routing algorithms like AODV [169]. Moreover, the wireless connection between two devices can also be affected by the availability of spectrum resource. Due to scarcity of the radio frequency spectrum, a frequency band is usually shared by more than one devices. Consequently, wireless communication between two mobile devices is affected by temporal availability of spectrum band in the vicinity of the devices. Due to these factors, i.e., the uncertainty in the availability of spectrum band as well as shadowing and fading effects, wireless connections between nodes are random and time-varying, which often need to be considered in the design of information dissemination schemes for MANETs. Therefore, it is challenging to establish a distributed information dissemination scheme for MANETs that is adaptive to both dynamic topology and channel randomness, while meeting the performance objectives, i.e., delay constraint and energy efficiency.

This chapter presents the analysis and design of a broadcast scheme for information dissemination in two-dimensional MANETs based on an in-depth analysis of the advantages and inadequacies of the widely used epidemic broadcast schemes.

10.1 Information Dissemination Scheme and Network Model

This section introduces the information dissemination model, i.e., the Susceptible-Infectious-Recovered (SIR) model, and the network model used in the chapter.

10.1.1 Network Model

Suppose that at some initial time instant ($t = 0$), a set of N nodes are independently, randomly and uniformly placed on a torus $(0, L]^2$. It follows that the node density is $\lambda = N/L^2$. Then, these nodes start to move according to a random direction model (RDM). Specifically, at time $t = 0$, each node chooses its direction independently and uniformly in $[0, 2\pi)$, and then moves in this direction thereafter at a constant speed V. It has been established in [154] that under the aforementioned model, at any time instant $t \geq 0$, the spatial distribution of nodes is stationary and still follows the uniform distribution. Note that the uniform spatial distribution and random direction mobility are both simplified but widely used models in the field [35, 71, 154]. Furthermore, we also evaluate the applicability of our analysis in networks whose node distribution and mobility deviate from the above assumption using a real world mobility trace, which is described in detail in Sect. 10.3.1.

A commonly used radio propagation model in this field is the *unit disk connection model*, under which two nodes are directly connected if and only if the Euclidean distance between them is not larger than the transmission range r_0. More specifically, under the unit disk connection model, the received signal strength (RSS) at a receiver separated by distance x from the transmitter is $P_u(x) = Cp_t x^{-\eta}$, where C is a constant, p_t is the transmission power common to all nodes, and η is the path loss exponent [172]. A transmission is successful if and only if the RSS exceeds a given threshold $p_{\min}$. Therefore, the required transmission power p_t allowing a transmission range r_0 is $p_t = \frac{p_{\min}}{C} r_0^{\eta}$. Refer to Sect. 1.1 for more detailed discussions.

The unit disk connection model is simple but unrealistic. In reality, the RSS may have significant variations around the mean value. This is typically taken into account in the log-normal shadowing model. Under the log-normal shadowing model, the RSS attenuation (in dB) follows a Gaussian distribution, i.e., there is $10\log_{10}(P_l(x)/Cp_t x^{-\eta}) \sim Z$, where $P_l(x)$ is the RSS under the log-normal shadowing model and Z is the *shadowing factor*, which is a zero-mean Gaussian distributed random variable with standard deviation σ. When $\sigma = 0$, the model reduces to the unit disk connection model. Denote by $q(z)$ the probability density function of the shadowing factor Z; then:

$$q(z) = \frac{1}{\sigma\sqrt{2\pi}} \exp\left(-\frac{z^2}{2\sigma^2}\right) \tag{10.1.1}$$

Furthermore, we consider that the shadowing factors Z between all pairs of transmitter and receiver are independent and all links are symmetric.

In addition to the above large-scale fading, we consider a general model of small-scale fading, i.e., the *Nakagami-m* fading model [180]. Assuming the Nakagami-m fading, the RSS per symbol, $\hat{\omega}$, is distributed according to a Gamma distribution with the following probability density function [180]:

$$\hat{\xi}(\hat{\omega}) = \frac{m^m \hat{\omega}^{m-1}}{(E[\hat{\omega}])^m \Gamma(m)} \exp\left(-\frac{m\hat{\omega}}{E[\hat{\omega}]}\right), \quad \hat{\omega} \geq 0 \tag{10.1.2}$$

where $\Gamma(.)$ is the standard Gamma function and $E[\hat{\omega}] = P_l(x)$ is the mean RSS (over time), which is determined by the path loss and shadowing effects. By choosing different values for the parameter m, the Nakagami-m fading model easily includes several widely used fading distributions as its special cases [180].

To incorporate both shadowing and fading effects, we adopt a wireless connection model, named the *general connection model*, which is built on the basis of the random connection model introduced in Chap. 1. Specifically, let $P_g(p_t x^{-\eta}, Z, \Omega)$: $\Re^+ \times \Re \times \Re \to [0,1]$ be the RSS at a receiver separated by distance x from the transmitter, where Z and Ω (called *channel factors*) are random variables representing the random variation of the RSS caused by shadowing effect and small-scale fading effect, respectively. Note that the analysis in this chapter allows a general form of the RSS function $P_g(p_t x^{-\eta}, Z, \Omega)$, and the analytical results under the log-normal shadowing model with Nakagami fading (called the *Log-normal-Nakagami model*) are provided in Sect. 10.2 as a typical example. Lastly, to be practically meaningful in modeling the RSS attenuation, it is assumed that the RSS $P_g(p_t x^{-\eta}, Z, \Omega)$ is a nondecreasing function of $p_t x^{-\eta}$, Z and Ω, respectively.

10.1.2 Information Dissemination Scheme

Suppose that a piece of information is broadcast from an arbitrary node. Once a node receives the information for the first time, it becomes *infectious*. The infectious node holds the information for a fixed amount of time τ_s, called the *sleep time interval*, followed by a random amount of time τ_r, to be specified in the next paragraph, then re-transmits the information once to all nodes directly connected to the infectious node. Such a *sleep-active cycle* repeats for a fixed number of times, denoted by a positive integer β, after which the node *recovers*. A *recovered node* stops transmitting the information and will ignore all future transmissions of the same information. The information dissemination process naturally stops, i.e., reaches the *steady state*, when there is no infectious node in the network. The nodes that have received the information are referred to as the *informed nodes*. It is obvious that the fraction of informed nodes is a key performance metric of information dissemination.

Note that the time interval between two consecutive transmissions is determined by two additive components: a pre-designated sleep time interval τ_s and a random time interval τ_r. The pre-designated waiting period τ_s is chosen to allow sufficient time, e.g., $V\tau_s \geq 2r_0$, for a node to move away from the location of its previous broadcast, thereby reducing redundant transmissions to the same nodes. The random time interval τ_r introduces randomness in the transmitting time instants, which can reduce collisions and contention between nodes caused by simultaneous transmissions. Furthermore, τ_r also reflects the channel access time in some practical scenarios. For example, using carrier sense multiple access (CSMA), if a node finds the channel busy at the end of the pre-designated sleep time interval τ_s, then, the node needs to wait a random time interval, viz., random back-off time, for the next transmission opportunity. We include τ_r in our broadcast scheme to introduce flexibility in determining the transmitting time of each infectious node, so that a node can transmit at its convenience, e.g., when the node using CSMA senses the channel idle, in a decentralized manner while the performance of broadcast in the whole network, e.g., measured by the fraction of informed nodes, is still guaranteed. These features are valuable for a MANET subject to dynamic topological changes and channel randomness.

Note that when τ_r takes a constant value 0, $\tau_s \rightarrow 0$ and $\beta\tau_s$ is a positive value, the above broadcast scheme becomes a traditional SIR epidemic scheme [183, 202] with an active period $\beta\tau_s$ seconds. Furthermore, when $\beta\tau_s \rightarrow \infty$, the broadcast scheme becomes a Susceptible-Infectious (SI) epidemic scheme. The advantage of the considered broadcast scheme compared with the traditional SIR scheme is discussed in detail later in Sect. 10.2.5.

The network described above is hereafter denoted by $\mathcal{G}(\eta, \sigma, \lambda, V, r_0, \beta, \tau_s, \tau_r)$. Furthermore, we assume a sufficiently large network, i.e., $L \gg \beta(\tau_s + \max\{\tau_r\})V$, so that a node will not be wrapped, through its motion in the torus, back to the point where it became infected before it recovers.

10.2 Analytical Characterization of the Information Propagation Process

In this section, we analytically study the information propagation process in a MANET adopting the information dissemination scheme introduced in the previous section.

10.2.1 The Probability of Direct Connection

In order to incorporate channel randomness into the analysis, we need the following lemma.

Lemma 155 *Consider a set of transmitter-receiver pairs where the wireless channel between each pair has shadowing factor z and fading factor ω. Then, there exists a constant value $r_N(z, \omega)$, called the* transmission range given the channel factors z and ω, *such that an arbitrary transmitter-receiver pair in the aforementioned set of transmitter-receiver pairs is directly connected if and only if the Euclidean distance between the transmitter and receiver is less than or equal to $r_N(z, \omega)$. Furthermore, $r_N(z, \omega)$ is the solution of r to the equation $P_g(p_t r^{-\eta}, z, \omega) = p_{\min}$.*

Given the values of the channel factor z and ω, the randomness in the RSS disappears. Noting that $P_g(p_t x^{-\eta}, z, \omega)$ is a nonincreasing function of the distance x between transmitter and receiver, the conclusion readily follows. Hence the proof of the lemma is omitted.

Next, we consider a typical wireless channel model, i.e., the Log-normal-Nakagami model, and obtain the associated value of the transmission range.

Lemma 156 *Suppose that the wireless channel between a transmitter and a receiver is subject to the Log-normal-Nakagami model with channel factors $Z = z$ and $\Omega = \omega$ respectively, then, the* transmission range given the channel factors z *and* ω *is:*

$$r_N(z, \omega) \triangleq r_0 \omega^{1/\eta} \exp\left(\frac{z \log 10}{10\eta}\right) \tag{10.2.1}$$

where r_0 is the transmission range under the unit disk connection model, the probability density function of Z is given by (10.1.1) and Ω follows a Gamma distribution with mean 1.

Proof Denote by x the Euclidean distance between a transmitter and a receiver. Subject to Nakagami-m fading, the RSS varies around its mean value over time according to a Gamma distribution with mean $P_l(x)$, where $P_l(x) = Cp_t x^{-\eta} 10^{Z/10}$ is the RSS under the log-normal shadowing model.

To facilitate the analysis, we introduce a random variable Ω which follows a Gamma distribution with mean 1. Therefore, the probability density function of Ω is

$$\zeta(\omega) = \frac{m^m \omega^{m-1}}{\Gamma(m)} \exp(-m\omega), \quad \omega \geq 0 \tag{10.2.2}$$

Furthermore, it can be shown that for any constant $C_1 \in \Re$, the random variable $C_1\Omega$ follows a Gamma distribution with mean C_1. Then, under the Log-normal-Nakagami model, the RSS at a receiver at distance x from the transmitter is $P_g(p_t x^{-\eta}, Z, \Omega) = P_l(x)\Omega = Cp_t x^{-\eta} 10^{Z/10}\Omega$.

Recall that two nodes are directly connected if and only if the RSS exceeds a given threshold $p_{\min}$. Without shadowing and fading effects, i.e., considering path loss only, there holds $p_{\min} = Cp_t r_0^{-\eta}$ (see Sect. 10.1). With shadowing and fading, there holds:

$$\begin{aligned}
&\Pr\left(P_g(p_t x^{-\eta}, Z, \Omega) \geq p_{\min}\right) \\
&= \Pr\left(Cp_t x^{-\eta} 10^{Z/10}\Omega \geq Cp_t r_0^{-\eta}\right) \\
&= \Pr\left(x \leq r_0 \Omega^{\frac{1}{\eta}} \exp(\frac{Z \log 10}{10\eta})\right)
\end{aligned}$$

Conditioned on the channel factors $Z = z$ and $\Omega = \omega$ between two nodes, the two nodes are directly connected if and only if their distance is not larger than $r_0 \omega^{1/\eta} \exp(\frac{z \log 10}{10\eta})$. □

10.2.2 The Effective Node Degree

In this subsection, we analyze the effective node degree, a key parameter characterizing the information propagation process.

Definition 157 The *effective node degree* R_0 of an infectious node is the expected number of nodes that are directly connected to the infectious node in at least one of the β transmissions.

Note that R_0 is the same for all nodes due to the stationarity and homogeneity of node distribution on the torus. To compute the effective node degree, we further need to calculate the clustering factor as defined in the following.

Definition 158 The *clustering factor* $\phi(\tau_s)$ is the expected number of nodes that are directly connected to an infectious node in *both* of two consecutive transmissions when the sleep time interval is τ_s.

Using Lemma 156, we have the following results.

Lemma 159 *In network* $\mathcal{G}(\eta, \sigma, \lambda, V, r_0, \beta, \tau_s, \tau_r)$*, the clustering factor satisfies*

$$\phi(\tau_s) = \int_0^{\infty}\int_0^{\infty}\int_{-\infty}^{\infty}\int_{-\infty}^{\infty}\int_0^{\infty}\int_0^{\pi} A_p(\theta, \tau_s, r_N(z_1, \omega_1), r_N(z_2, \omega_2)) \times \frac{\lambda}{\pi} p_\tau(\tau_r) q(z_1) q(z_2) \zeta(\omega_1) \zeta(\omega_2) d\theta d\tau_r dz_1 dz_2 d\omega_1 d\omega_2 \qquad (10.2.3)$$

where $A_p(\theta, \tau_s, r_1, r_2)$ *is given by (10.2.4).*

Proof Denote by Θ the angle measured counterclockwise from the moving direction of an infectious node to the moving direction of an arbitrary node. Recall that the direction of a node is randomly and uniformly chosen in $[0, 2\pi)$, independent of the directions of other nodes. Consequently, it can be shown that the angle Θ is also uniformly distributed in $[0, 2\pi)$.

Suppose that an infectious node transmits once at point S_1, then it moves by distance $(\tau_s + \tau_r)V$ to point S_2 and transmits again, as shown in Fig. 10.1.

Next, we focus on a subset of nodes that fulfill the following three conditions: 1) they move in direction $\Theta \in (\theta, \theta + d\theta)$; *and* 2) their RSS from the infectious node has shadowing factors $Z_1 \in (z_1, z_1 + dz_1)$ and $\Omega_1 \in (\omega_1, \omega_1 + d\omega_1)$ when the infectious node transmits at S_1; *and* 3) their RSS from the infectious node has shadowing factors $Z_2 \in (z_2, z_2 + dz_2)$ and $\Omega_2 \in (\omega_2, \omega_2 + d\omega_2)$ when the infectious node transmits at S_2. Due to the independence of shadowing factors and the mobility of nodes, these nodes are uniformly distributed with density $\frac{\lambda}{2\pi} q(z_1) q(z_2) \zeta(\omega_1) \zeta(\omega_2) d\theta dz_1 dz_2 d\omega_1 d\omega_2$. Among this subset of nodes, the nodes that are connected to the infectious node in the first transmission are in a disk centered at point S_1 with radius $r_N(z_1, \omega_1)$, denoted by $C(S_1, r_N(z_1, \omega_1))$. Furthermore, when the infectious node transmits at S_2, these nodes move by distance $(\tau_s + \tau_r)V$ from being contained in $C(S_1, r_N(z_1, \omega_1))$ to being contained

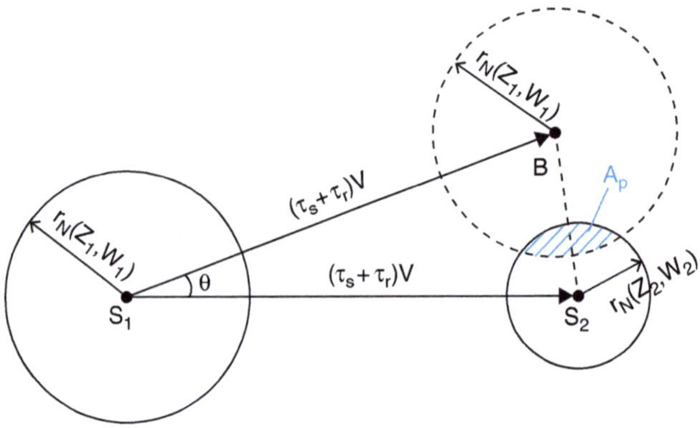

Fig. 10.1 An illustration of the nodes (in the shaded area A_p) that are directly connected to an infectious node in both of the two consecutive transmissions (occurring at S_1 and S_2). Symbols are defined in Lemma 159

in a new disk $C(B, r_N(z_1, \omega_1))$ as shown in Fig. 10.1. Then, the nodes that are connected to the infectious node in both transmissions are in the intersectional area $C(S_2, r_N(z_2, \omega_2)) \cap C(B, r_N(z_1, \omega_1))$. Denote by $A_p(\theta, \tau_s, r_N(z_1, \omega_1), r_N(z_2, \omega_2))$ the size of the intersectional area $C(S_2, r_N(z_2, \omega_2)) \cap C(B, r_N(z_1, \omega_1))$. It can be readily calculated using the following formula:

$$A_p(\theta, \tau_s, r_1, r_2) = \begin{cases} \min(\pi r_1^2, \pi r_2^2) & \text{for } \psi(\theta, \tau_s) \le |r_1 - r_2| \\ r_1^2 \arccos(\frac{\psi^2(\theta,\tau_s)+r_1^2-r_2^2}{2r_1\psi(\theta,\tau_s)}) + r_2^2 \arccos(\frac{\psi^2(\theta,\tau_s)+r_2^2-r_1^2}{2r_2\psi(\theta,\tau_s)}) & \\ -\frac{1}{2}\sqrt{[(r_1+r_2)^2 - \psi^2(\theta,\tau_s)][\psi^2(\theta,\tau_s) - (r_1-r_2)^2]} & \text{for } |r_1 - r_2| < \psi(\theta, \tau_s) \\ & \quad < r_1 + r_2 \\ 0 & \text{otherwise} \end{cases} \tag{10.2.4}$$

where $\psi(\theta, \tau_s) = 2(\tau_s + \tau_r)V \sin\frac{\theta}{2}$ is the length of BS_2 (see Fig. 10.1).

Next we consider all subsets of nodes. Note that only the cases for $\theta \in [0, \pi)$ need to be calculated due to symmetry. Then, it can be shown that the clustering factor satisfies $\phi(\tau_s) = \lambda E[A_p(\theta, \tau_s, r_N(z_1, \omega_1), r_N(z_2, \omega_2))]$. □

Remark 160 The results of Lemma 159 can be extended to different channel models by replacing the probability density functions of the channel factors z_1, z_2, ω_1, and ω_2.

Finally, we have the following theorem for the value of R_0.

Theorem 161 *In network $\mathcal{G}(\eta, \sigma, \lambda, V, r_0, \beta, \tau_s, \tau_r)$, under the Log-normal-Nakagami model, the effective node degree satisfies*

$$R_0 \leq \beta\lambda\pi r_0^2 \exp\left(\frac{(\sigma \log 10)^2}{50\eta^2}\right) \frac{m^{\frac{-2}{\eta}}\Gamma(m+\frac{2}{\eta})}{\Gamma(m)} - (\beta - 1)\phi(\tau_s) \qquad (10.2.5)$$

where $\Gamma(.)$ *is the standard Gamma function and* $\phi(.)$ *is given by Lemma 159.*

Proof We first consider two consecutive transmissions, as illustrated in Fig. 10.1. Along the same lines as Lemma 159, consider that the shadowing factors of the ith transmission, $i = \{1, 2, \ldots, \beta\}$, is $Z_i \in [z_i, z_i + dz_i]$ and $\Omega_i \in [\omega_i, \omega_i + d\omega_i]$. Define $\phi_{1,2}(\tau_s)$ to be the clustering factor of the two consecutive transmissions. It is straightforward that $\phi_{1,2}(\tau_s)$ is given by the (10.2.3) without integrating over z_1, z_2, ω_1, and ω_2.

Note that the size of the area covered by the transmission range in the ith transmission is $\pi(r_N(z_i, \omega_i))^2$. Furthermore, the size of the area wherein the nodes receive both the 1st and the 2nd transmissions is $\phi_{1,2}(\tau_s)/\lambda$. Denote by A_R the total (union) size of the area wherein the nodes receive either the 1st or the 2nd transmissions. Some straightforward geometric analysis leads to the conclusion that $A_R = \pi(r_N(z_1, \omega_1))^2 + \pi(r_N(z_2, \omega_2))^2 - \phi_{1,2}(\tau_s)/\lambda$.

Then, the expected number of nodes that are directly connected to an infectious node in either of the two transmissions can be calculated using (10.1.1) and Lemma 156:

$$\int_0^\infty \int_0^\infty \int_{-\infty}^\infty \int_{-\infty}^\infty \Big(\lambda\pi(r_N(z_1, \omega_1))^2 + \lambda\pi(r_N(z_2, \omega_2))^2$$
$$- \phi_{1,2}(\tau_s)\Big) q(z_1) q(z_2) \zeta(\omega_1) \zeta(\omega_2) dz_1 dz_2 d\omega_1 d\omega_2$$
$$= 2 \int_0^\infty \int_{-\infty}^\infty \lambda\pi \left(r_0 \omega^{1/\eta} \exp(\frac{z \log 10}{10\eta})\right)^2$$
$$\frac{1}{\sigma\sqrt{2\pi}} \exp\left(-\frac{z^2}{2\sigma^2}\right) dz \zeta(\omega) d\omega - \phi(\tau_s)$$
$$= 2 \int_0^\infty \lambda\pi r_0^2 \omega^{2/\eta} \exp\left(\frac{(\sigma \log 10)^2}{50\eta^2}\right) \zeta(\omega) d\omega - \phi(\tau_s)$$
$$= 2\lambda\pi r_0^2 \exp\left(\frac{(\sigma \log 10)^2}{50\eta^2}\right) \frac{m^{\frac{-2}{\eta}}\Gamma(m+\frac{2}{\eta})}{\Gamma(m)} - \phi(\tau_s) \qquad (10.2.6)$$

Using the inclusion-exclusion principle, (10.2.5) can be readily obtained. More specifically, if the number of transmissions is $\beta = 1$, then R_0 is equal to the first term in (10.2.5). If the number of transmissions is $\beta = 2$, then R_0 is equal to the two terms in (10.2.5). Furthermore, when $\beta \geq 3$, the analysis involves the intersectional area of more than two circles, as shown in Fig. 10.2. To avoid complicated formulas, we only calculate the intersectional area of two circles (viz., the term $\phi(\tau_s)$) and provide an upper bound on R_0 as shown in (10.2.5). □

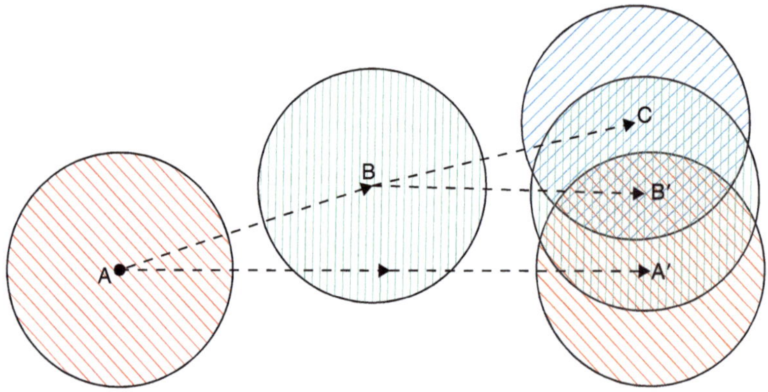

Fig. 10.2 An illustration of the area covered by the transmission range of an infectious node during three transmissions at points A, B, and C sequentially. Consider a set of nodes moving in the direction AA'. When the infectious node transmits the third time (at point C), the nodes that have received the first and second transmissions are in the disks centered at points A' and B' respectively

10.2.3 Percolation Probability

We first study the fraction of informed nodes from a percolation perspective *asymptotically*, viz., we increase the network area towards infinity, i.e., letting $L \to \infty$, while keeping other parameters, i.e., $\lambda, V, r_0, \beta, \tau_s$ and τ_r, unchanged.

Consider a realization of a random network on a torus $(0, L]^2$. Denote by a constant $N(L)$ the total number of nodes in this network. Further denote by a random integer $N_R(L)$ the number of informed nodes of a packet in this network, i.e., the nodes that have received the packet at the end of the information dissemination process. Then, the fraction of informed nodes of the packet in this network can be calculated by $N_R(L)/N(L)$. Note that the fraction of informed nodes is calculated for each information dissemination process in each realization of a random network.

Recall that in an infinite network, the network is said to *percolate* if there exists a component of infinite size in the network. Percolation probability is the probability that the network percolates. As widely used in the study of percolation in random networks, we consider asymptotic networks in this chapter. More specifically, we consider asymptotic networks where $L \to \infty$ while keeping the node density unchanged. It follows that the total number of nodes in each realization of the network becomes $N(L) \to \infty$.

Further note that, a series a_n depending on n is said to be *nonvanishingly small* if there exists a sufficiently small positive constant ε and a positive integer n_0 such that for all $n > n_0$, $a_n > \varepsilon$. Here, we are interested in the probability that the random number $N_R(L)/N(L)$ is nonvanishingly small as $N(L)$ approaches infinity, i.e., the probability that the fraction of informed nodes is nonvanishingly small. With a bit twist of its standard definition, we call this probability the *percolation probability* in this chapter.

Definition 162 The *percolation probability* p_c of a MANET is the probability that a piece of information broadcast from an arbitrary node can be received by a *nonvanishingly small* fraction of nodes asymptotically.

The main result of this subsection is as follows:

Theorem 163 *In network $\mathcal{G}(\eta, \sigma, \lambda, V, r_0, \beta, \tau_s, \tau_r)$ with effective node degree R_0. The percolation probability p_c satisfies $p_c \leq 1 + \frac{1}{R_0} W(-R_0 e^{-R_0})$, where $W(.)$ is the Lambert W Function.*

Proof We first model the information dissemination process using a Galton-Watson branching process ([110], see also Sect. 1.3.2). The root, viz., the 0th generation, is the source node. The expected number of children per node is given by R_0. Denote by $\chi(k)$ the number of individuals in the kth generation of the branching process.

Define $q \triangleq \Pr(\lim_{k\to\infty} \chi(k) \to 0)$ to be the *extinction probability* of the branching process, viz., the probability that the number of individuals in the kth generation diminishes to zero as $k \to \infty$. It has been shown in [110, Theorem 6.5.1] and Sect. 1.3.2 that the extinction probability q is the smallest nonzero solution of $q = \exp((1-q)R_0)$ as the number of nodes $N \to \infty$. Solving the equation, it can be obtained that $q = \frac{W(-R_0 e^{-R_0})}{-R_0}$, where $W(.)$ is the Lambert W Function [47].

Next we establish the connection between the branching process and the information dissemination process in a MANET. Denote by $\chi_N(k)$ the number of nodes in the kth generation of the information dissemination process, where the kth generation of the information dissemination process consists of the nodes that receive the information *for the first time* from a node belonging to the $(k-1)^{th}$ generation, and the 0th generation is the source node. Similarly as above, define $q_N \triangleq \Pr(\lim_{k\to\infty} \chi_N(k) \to 0)$ to be the extinction probability of the information dissemination process. Because some of the nodes connected to an infectious node may have already received the information from other infectious nodes, the number of children per node of the information dissemination process is stochastically less than that in the branching process introduced earlier. Therefore, there holds $\chi_N(k) \leq_{st} \chi(k)$. Recall that in stochastic ordering, we say $\chi_N(k) \leq_{st} \chi(k)$ if and only if $\Pr(\chi_N(k) > c) \leq \Pr(\chi(k) > c)$ for any constant c. Then, there holds $q_N \geq q = \frac{W(-R_0 e^{-R_0})}{-R_0}$.

Denote by p'_c the probability that the information dissemination process does not become extinct. It is clear that $p'_c = 1 - q_N$. Furthermore, it is obvious that the information dissemination process does not become extinct is a necessary but not a sufficient condition for having an nonvanishingly small fraction of informed nodes. Therefore, there holds $p_c \leq p'_c$. □

10.2.4 Expected Fraction of Informed Nodes

Define z_0 as the expected fraction of informed nodes in the steady state of a percolated network. Then, we report the following two results:

Theorem 164 *Consider a network ($L \to \infty$), whose effective node degree is R_0. The expected fraction of informed nodes in the steady state of a percolated network satisfies $z_0 \leq 1 + \frac{1}{R_0} W(-R_0 e^{-R_0})$, where $W(.)$ is the Lambert W Function.*

Theorem 164 can be readily proved using a set of ordinary differential equations and a mean field limit theorem that is commonly used in the analysis of epidemic broadcast schemes, see [106] and the reference herein. The proof is therefore omitted here and the accuracy of the result is verified by simulation in Sect. 10.3.

Theorems 163 and 164 show that the effective node degree determines the performance of broadcast scheme. We shall present further analysis quantifying the value of the effective node degree in the later subsections.

10.2.5 Energy and Bandwidth Efficiency

In a communication network, the energy and bandwidth consumed in disseminating information can be important considerations in the design of the information dissemination scheme. We next introduce the energy and bandwidth consumption metrics. Specifically, assume that the time spent on transmitting a packet of unit size over single hop is a constant T_t. Therefore, the energy consumed in transmitting a packet is $T_t p_t$, denoted via a constant $E_1 = T_t p_t$. Similarly, denote by constant E_2 the energy consumed when receiving a packet at a node.

Denote by random variable $\mathcal{D}_i$ the node degree of a randomly chosen node at a random time instant. Then, the sum of the energy consumption for a randomly chosen node broadcasting a packet and the energy consumed by all its neighbors in receiving the packet is evidently $E_1 + \mathcal{D}_i E_2$. It follows that the expected energy consumption for a randomly chosen node broadcasting a packet and all its neighbors receiving the packet is $E_1 + E[\mathcal{D}_i]E_2$, where $E[\mathcal{D}_i]$ is the average node degree.

Therefore, we can combine the energy consumption at single node, including energy consumed by other nodes in receiving its packet, by a constant $E_c = E_1 + E[\mathcal{D}_i]E_2$. As manifested in the above equation, the overall energy consumption is directly related to the number of transmissions and is equal to the number of transmissions times E_c.

Similarly, if each transmission occupies, e.g., using CSMA, the frequency band in an area whose expected size is A_c, then the expected size of the area where the frequency band is occupied by an infectious node during its β transmissions is βA_c. Therefore, the consumption of bandwidth is also an increasing function of the number of transmissions. To save energy and bandwidth, we need to reduce the number of transmissions β.

On the other hand, to meet pre-designated broadcast performance objectives, measured by the percolation probability and expected fraction of informed nodes, a certain number of transmissions are required. Given the dependence of the performance objectives on the effective node degree, which is determined by the number of transmissions β, and the reliance of the energy and bandwidth consumption on the number of transmissions β, we propose using the following ratio to measure the *energy and bandwidth efficiency* of the proposed broadcast scheme:

$$Y \triangleq \frac{R_0}{\beta} \tag{10.2.7}$$

which is the average effective node degree achieved per transmission.

To improve energy and bandwidth efficiency, it is obvious that the clustering factor, which characterizes the amount of overlap between two transmissions, needs to be reduced. The following lemma reports a useful property of the clustering factor $\phi(\tau_s)$.

Lemma 165 *The clustering factor $\phi(\tau_s)$ is a monotone nonincreasing function of τ_s, when all other parameters, i.e., $\eta, \sigma, \lambda, V, r_0, \beta$, and τ_r, are fixed.*

Proof In (10.2.3), the only term determined by τ_s is the size of the intersectional area $\int_0^{\pi} A_p(\theta, \tau_s, r_N(z_1, \omega_1), r_N(z_2, \omega_2))d\theta$. It can be shown using (10.2.4) that for any given value of θ, $A_p(\theta, \tau_s, r_N(z_1, \omega_1), r_N(z_2, \omega_2))$ is a nonincreasing function of τ_s. Then, the conclusion follows that the clustering factor $\phi(\tau_s)$ is a monotone nonincreasing function of τ_s. □

Introducing the sleep time interval τ_s, the proposed broadcast scheme separates consecutive transmissions in both space and time, allowing each transmission to be received by more new nodes that have not received the information during the previous transmission(s). This is more energy and bandwidth efficient than the traditional forwarding schemes, e.g., SI or SIR scheme, where an infectious node transmits the information to every susceptible node coming into the transmission range. To quantitatively compare the energy and bandwidth efficiencies between a traditional SIR scheme and our scheme, we consider a SIR scheme with a perfect neighbor discovery mechanism where an infectious node only wakes up when there is a new node coming into the transmission range. Then, the energy and bandwidth efficiency of the SIR scheme can be calculated in the same way as (10.2.7):

$$\hat{Y} \triangleq \frac{R_0}{R_0 - \lambda\pi r_0^2 + 1} \tag{10.2.8}$$

where $\lambda\pi r_0^2$ is the expected number of nodes receiving the first transmission and $R_0 - \lambda\pi r_0^2 + 1$ is the expected number of transmissions required by an infectious node using an ideal SIR scheme to transmit a piece of information to R_0 nodes.

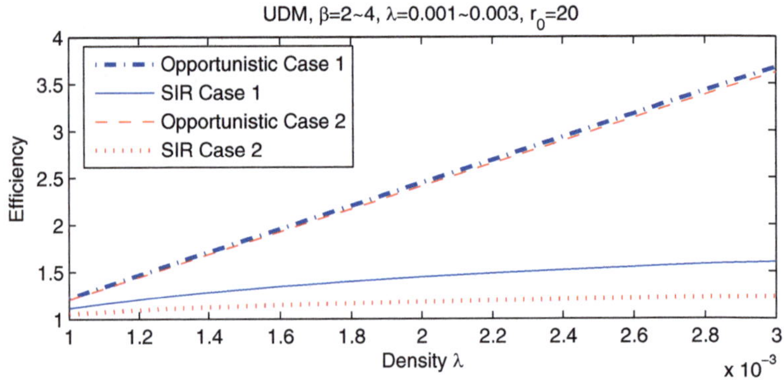

Fig. 10.3 The energy and bandwidth efficiency metrics under the proposed scheme (given by (10.2.7)) and the traditional SIR scheme (given by (10.2.8)). Note that R_0 in (10.2.8) takes the same value as R_0 in (10.2.7). In Case 1 and Case 2, the proposed scheme transmits $\beta = 2$ and $\beta = 4$ times, respectively. Recall that the efficiency metric is a measure of the average effective node degree achieved per transmission

As can be seen in Fig. 10.3, the proposed scheme has a higher energy and bandwidth efficiency than the SIR scheme, especially when the density is large. This is because when the density is large, the traditional method needs to transmit frequently to every new neighbor whereas the proposed scheme can wait until the infectious node moves to a new region and it will then transmit to a set of new nodes at the same time.

10.2.6 Information Dissemination Delay

Suppose that a piece of information is broadcast from an arbitrary node at time $t = 0$ using the proposed broadcast scheme. Let $T(z)$ be the expected time when the fraction of informed nodes reaches z, for $0 < z < 1$.

Theorem 166 *In network $\mathcal{G}(\eta, \sigma, \lambda, V, r_0, \beta, \tau_s, \tau_r)$, whose effective node degree is R_0, as $L \to \infty$ and $N = \lambda L^2 \to \infty$, there holds*

$$T(z) \geq \tau_s \left\lfloor 1 + \frac{\log(Nz)}{\log\left(1 + \frac{R_0}{\beta}\right)} \right\rfloor \tag{10.2.9}$$

Proof Recall that each infectious node has β transmissions separated by a random time interval $\tau_s + \tau_r$. In this proof, we obtain a lower bound on the delay $T(z)$ by constructing a new network, denoted by $\mathcal{G}_L$, where the time interval between any two consecutive transmissions of each node is a constant τ_s and the infectious nodes never recover. Then, in the new network $\mathcal{G}_L$, all infectious nodes transmit simultaneously at time $\tau_s, 2\tau_s, 3\tau_s, \ldots$. It can be established that the delay of information dissemination in the new network is a lower bound on the delay in the original network because 1) each node retransmits the information sooner because sleep time interval is τ_s in the new network compared with $\tau_s + \tau_r$ in the original network, and 2) the number of transmissions of each node is greater than or equal to that in the original network, and 3) the number of infectious nodes created by each infectious node is greater than or equal to that in the original network (to be described in the next paragraph).

Let a_k be the total number of infectious nodes at time $k\tau_s$ in the new network $\mathcal{G}_L$, for $k \geq 0$. Initially, there is $a_0 = 1$. Denote by $Q_i \in \{0, 1, 2, \ldots\}$ the number of nodes directly connected to node i at a randomly chosen time instant. Due to the uniform distribution of nodes and the fact that nodes move independently of one another, it can be shown that q_i has an identical distribution across all nodes. Furthermore, $E[Q_i] = R_0$. Note that some of the nodes directly connected to an infectious node, say node i, may have already received the information. Hence the number of new infectious nodes created by infectious node i at each transmission is not larger than Q_i. To obtain a lower bound on the delay, we consider that the number of new infectious nodes created by an infectious node i at a transmission is equal to Q_i.

It can be readily shown that the total number of infectious nodes at time $k\tau_s$ satisfies the relationship $a_k = a_{k-1} + \sum_{i=1}^{a_{k-1}} Q_i$. However, an explicit and non-recursive form for a_k is difficult to find. To simplify the analysis, an asymptotic model is used. Specifically, as we let $L \to \infty$ while keeping the density $\frac{N}{L^2}$ unchanged, the expected number of children $E[Q_i]$ does not vary as $N \to \infty$. Furthermore, because a_k only depends on a_{k-1} and the distribution of Q_i, according to the mean field limit [106], as $N \to \infty$, the number of nodes in the kth generation converges almost surely to the deterministic form $a_k = a_{k-1} + a_{k-1}E[Q_i] = a_{k-1}\left(1 + \frac{R_0}{\beta}\right)$.

Further noting that $a_0 = 1$, it follows that for $N \to \infty$, $a_k = \left(1 + \frac{R_0}{\beta}\right)^k$ and

$$
\begin{aligned}
\lim_{n\to\infty} T(z) &\geq E\left[\tau_s \arg\max_k \left(a_k \leq Nz\right)\right] \\
&= \tau_s E\left[\arg\max_k \left((1 + \frac{R_0}{\beta})^k \leq Nz\right)\right] \\
&= \tau_s\left[1 + \frac{\log(Nz)}{\log\left(1 + \frac{R_0}{\beta}\right)}\right]
\end{aligned}
\tag{10.2.10}
$$

□

10.2.7 Optimization of the Inter-Transmission Time Interval

In the previous subsections, we postulated that consecutive transmissions of an infectious node are separated by a random time interval $\tau_s + \tau_r \triangleq \tau$. The randomly distributed time interval τ reflects the uncertainty in the channel access time in practical scenarios. This subsection disregards the technology limitation, i.e., we assume that channel access time can be pre-determined and there is no contention or collision between concurrent transmissions of different nodes, and investigates the optimal probability distribution of τ that maximizes the effective node degree when all other parameters, i.e., $\eta, \sigma, \lambda, V, r_0$ and β, are fixed.

Note that this subsection conducts optimization under the unit disk connection model for $\beta = 2$ only, where the R_0 is equal to the expression given in (10.2.5). The same method can be applied to other scenarios which, however, involve nontrivial analysis of the intersectional area of multiple disks. The performance of the information dissemination process using the optimal broadcast scheme is shown and discussed later in Sect. 10.3.

Theorem 167 *Denote by $p_\tau(\tau)$ the probability density function of the random time interval τ. In networks assuming the unit disk connection model, consider a set of proposed broadcast schemes with $\beta = 2$ but using different sleep strategies, i.e., different values of τ_s and different distributions for τ_r. Among all sleep strategies with the same mean sleep time interval $E[\tau]$, the optimal one that maximizes the effective node degree R_0 is a strategy using a constant sleep time interval with length $E[\tau]$.*

Proof Note that under the unit disk connection model $r_1 = r_2 = r_0$. Taking the second derivative of (10.2.4) with regard to τ, there results

$$\frac{\partial^2}{\partial\tau^2}A_p(\theta,\tau,r_0,r_0)$$
$$= \begin{cases} 0, & \text{for } \psi(\theta,\tau)=0 \\ \frac{\partial^2}{\partial\tau^2}2r_0^2\arccos(\frac{\psi(\theta,\tau)}{2r_0}) - \frac{\partial^2\frac{1}{2}\sqrt{[(2r_0)^2-\psi^2(\theta,\tau)][\psi^2(\theta,\tau)]}}{\partial\tau^2}, & \text{for } 0<\psi(\theta,\tau)<2r_0 \\ 0, & \text{otherwise} \end{cases} \tag{10.2.11}$$

First consider the second derivative of (10.2.4) for $0 < a\tau < 2r_0$. Let $a = 2V\sin\frac{\theta}{2}$, then $\psi(\theta,\tau) = a\tau$. There holds

$$\begin{aligned} \mathcal{A}_1 &\triangleq \frac{\partial^2}{\partial\tau^2}2r_0^2\arccos\left(\frac{\psi(\theta,\tau)}{2r_0}\right) \\ &= \frac{\partial^2}{\partial\tau^2}2r_0^2\arccos\left(\frac{a\tau}{2r_0}\right) \\ &= -\frac{2a^3\tau r_0^2}{\sqrt{4r_0^2 - a^2\tau^2}(4r_0^2 - a^2\tau^2)} \end{aligned} \tag{10.2.12}$$

Furthermore,

$$\mathcal{A}_2 \triangleq \frac{\partial^2}{\partial \tau^2}\frac{1}{2}\sqrt{[(2r_0)^2 - \psi^2(\theta,\tau)][\psi^2(\theta,\tau)]}$$
$$= \frac{a^5\tau^3 - 6a^3r_0^2\tau}{(4r_0^2 - a^2\tau^2)\sqrt{4r_0^2 - a^2\tau^2}} \quad (10.2.13)$$

Then, the second derivative of (10.2.4) for $0 < a\tau < 2r_0$ is given by

$$\mathcal{A}_1 - \mathcal{A}_2 = \frac{-2a^3\tau r_0^2 - a^5\tau^3 + 6a^3r_0^2\tau}{(4r_0^2 - a^2\tau^2)\sqrt{4r_0^2 - a^2\tau^2}}$$
$$= a^3\tau\frac{4r_0^2 - a^2\tau^2}{(4r_0^2 - a^2\tau^2)\sqrt{4r_0^2 - a^2\tau^2}} > 0 \quad (10.2.14)$$

It is evident that $A_p(\theta,\tau,r_0,r_0)$ is a convex function of τ. Then, according to the Jensen's inequality, there holds $E[A_p(\theta,\tau,r_0,r_0)] \geq A_p(\theta,E[\tau],r_0,r_0)$.

Using Lemma 159, the clustering factor is

$$\phi(\tau) = \int_0^\infty\int_0^\pi A_p(\theta,\tau,r_0,r_0)\frac{\lambda}{\pi}p_\tau(\tau)d\theta d\tau_r$$
$$\geq \int_0^\pi \frac{\lambda}{\pi}A_p(\theta,E[\tau],r_0,r_0)d\theta \quad (10.2.15)$$

Finally, according to Theorem 161, under the unit disk connection model when $\beta = 2$, $R_0 = 2\lambda\pi r_0^2 - \phi(\tau)$.

This means that among all distributions of the sleep time interval τ with mean $E[\tau]$, the case where consecutive transmissions are separated by a constant $E[\tau]$ minimizes the clustering factor $\phi(\tau)$, consequently maximizes the effective node degree. □

Note that only Theorem 167 has the limitation of $\beta = 2$, while other results presented in the chapter can be applied to any positive integer value of β.

Remark 168 Note that the result of Theorem 167 only applies to the situation where nodes move at the same constant speed. In the situation where nodes' speeds are randomly distributed, the optimal sleep interval depends on the speeds of the active nodes as well as the speeds of other nodes, or more precisely, depends on the relative speeds between the active nodes and other nodes. Specifically, when an active node moves slower than V and other nodes move at speed V, the active node needs to sleep for a longer time interval in order to move away from its previous transmission location. On the other hand, if the active node moves slower than V but other nodes move faster than V, then the active node may not necessarily need a longer sleep time interval because other nodes can move away from the active node. Therefore, in the situation where node speed is randomly distributed, there does not exist a simple optimal sleep time interval that maximizes the effective node degree of every node. An algorithm needs to be designed to adjust the sleep time interval of a node dynamically according to the relative speeds of the active node and all its neighbors.

Remark 169 The randomly distributed time interval reflects the uncertainty in the channel access time. Theorem 167 disregards this limitation, i.e., assuming that channel access time can be pre-determined and there is no contention or collision between concurrent transmissions, and investigates the optimal distribution that maximizes the effective node degree.

10.3 Simulation Studies of the Information Propagation Process

In this section we use simulations to further investigate the information propagation process and establish the accuracy of the analysis presented in the previous sections.

Initially, $N = 1280$ nodes are randomly, independently, and uniformly deployed on a torus $(0, 800m]^2$, i.e., density is $\lambda = 0.002$ nodes/m^2. After deployment of the nodes, they start to move according to the mobility model introduced in Sect. 10.1. The speed V is set to 10 m/s. The sleep time interval is a constant, i.e., $\tau_r = 0$, except for Fig. 10.7.

Figure 10.4 shows the percolation probability and the expected fraction of informed nodes. It can be seen that both metrics improve as either τ_s or σ increases, owing to the reduction of the clustering factor as shown in Sect. 10.2.5. Our analytical bounds are close to the simulation results as shown in Fig. 10.4a, while the discrepancy in Fig. 10.4b is caused by the well-mixing assumption used in deriving the fraction of informed nodes, which however requires sophisticated analysis to adjust. More specifically, in the analysis it was assumed that informed and unformed nodes are very well mixed but in reality the informed nodes may be clustered together.

Note that the percolation probability is an asymptotic figure of merit, whereas any simulation can only be conducted on a finite network. The simulation in this section is conducted in a finite but large network with 1280 nodes. The simulation result for the percolation probability shows the probability of having at least $x = 10\%$ informed nodes in the steady state. Note that we have tried some different values for x ranging from 5% to 20% and the plots are similar. It can be seen that the percolation model, though constructed for an infinite network, is of predictive value for networks likely to be encountered in real world.

Figure 10.5 illustrates the impact of Rayleigh fading on the percolation probability and the expected fraction of informed nodes. It can be seen that Rayleigh fading has a negative impact on the information dissemination process, which can also be seen from the analysis in Lemma 165.

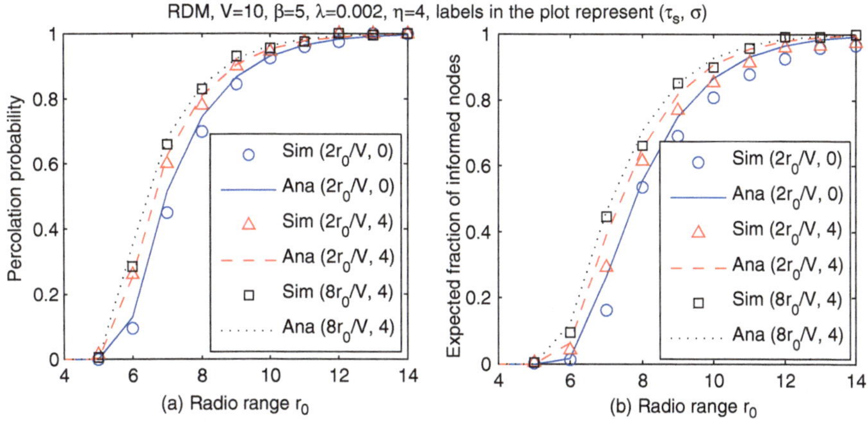

Fig. 10.4 Analytical (Ana) and simulation (Sim) results of the percolation probability and the expected fraction of informed nodes, using different network parameters. Analytical results are the upper bounds obtained by combining Theorems 163, 164 and 161. Simulation result for the percolation probability shows the probability of having at least 10% informed nodes in the steady state

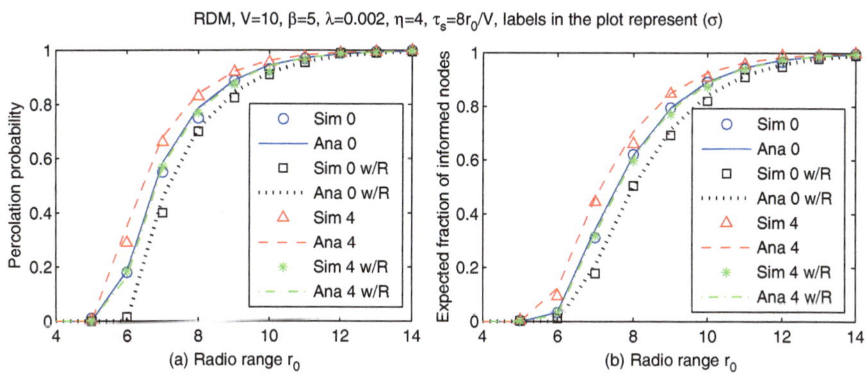

Fig. 10.5 The percolation probability and the expected fraction of informed nodes, using different network parameters, where "w/R" represents "with Rayleigh fading." Note that the curves of "Ana 0" and "Ana 4 w/R" almost overlap hence hard to distinguish

Figure 10.6 shows the lower bound of the delay for a piece of information to be received by 50% of nodes. It can be seen that the length of the sleep time interval has a significant impact on the delay and our analytical results provide valid lower bounds on the delay. It is clear that a longer sleep time interval can cause a longer delay. A network designer needs to consider the trade-off between delay and resource consumption.

Furthermore, Figs. 10.4, 10.5, and 10.6 suggest that shadowing effects benefit the information dissemination process in terms of the percolation probability, the expected fraction of informed nodes and delay, because an increase in σ leads to

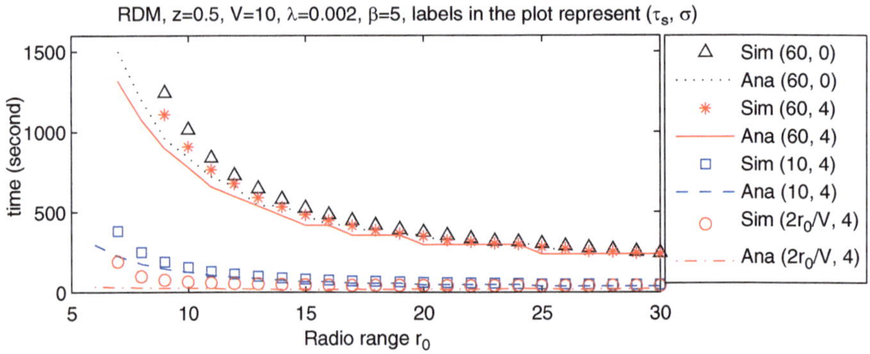

Fig. 10.6 The lower bound on the delay for a piece of information to be received by 50% of nodes using different network parameters. For comparison, the value of τ_s is kept constant (i.e., 10 or 60 s) while the value of r_0 being increased in the first six curves

an increase in the effective node degree R_0 as shown in Theorem 161. This is in sharp contrast with previous conclusions, e.g., [169], because traditional routing algorithms like AODV need to establish a route before sending data, whilst the route is unstable due to dynamic topology and channel randomness.

The proposed scheme and the analysis presented can also be useful for unicast. For example, consider that a source node sends a unicast packet to a randomly chosen destination node using the proposed broadcast scheme. It is interesting to note that the probability that the packet reaches the randomly chosen destination at time $T(z)$ is z as shown in Theorem 166. Moreover, the probability that the packet reaches the randomly chosen destination at the steady state is given by the expected fraction of informed node z_0 as shown in Theorem 164.

Figure 10.7a verifies the results of Theorem 167 that a constant time interval between consecutive transmissions leads to the largest effective node degree, which further leads to the largest percolation probability as shown in Fig. 10.7b. Note that in a practical scenario where concurrent transmissions of different nodes can cause wireless channel contention and transmission failure, further analysis is required to characterize the effective node degree and the percolation probability.

Now we present simulation results evaluating the energy and bandwidth efficiency of the proposed scheme. The energy and bandwidth consumed in the information dissemination is an increasing function of the number of transmissions of the packets containing the information. According to a study by Feeney *et al.* [64], using 11Mbps wireless transmission, the energy consumed by a node to broadcast a packet of 1K bytes is 482μJ, and the energy consumed by a node to receive a broadcast packet of size 1K bytes is 76μJ. Therefore, the overall energy consumption, including the transmitting and receiving energy, is $E_c = 482 + E[\mathcal{D}_i]76$, where $E[\mathcal{D}_i]$ is the average node degree, which is equal to the right-hand side of (10.2.5) (i.e., it is no longer an upper bound) by letting $\beta = 1$.

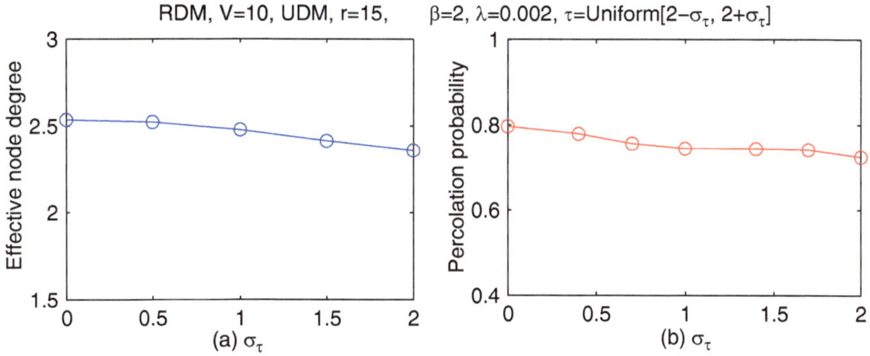

Fig. 10.7 Simulation results of (**a**) the effective node degree and (**b**) the percolation probability in networks using the proposed broadcast scheme where the time interval between consecutive transmissions of an infectious node is uniformly distributed in $[2 - \sigma_\tau, 2 + \sigma_\tau]$ and σ_τ is varied from 0 to 2

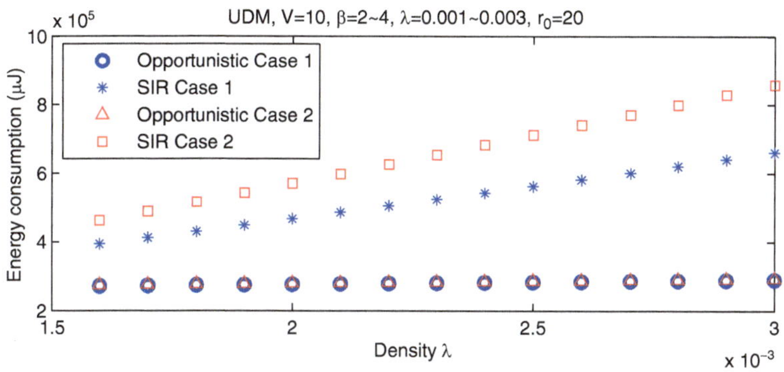

Fig. 10.8 Simulation results of the overall energy consumption using the proposed scheme and the traditional SIR scheme for a packet to be received by 99% of nodes in a network. In Case 1 and Case 2, $\beta = 2$ and 4, respectively

Figure 10.8 shows simulation results of the overall energy consumption for the dissemination of a piece of information using the proposed scheme and the traditional SIR scheme. Firstly, it can be seen that the proposed scheme has a significantly smaller energy consumption compared with the traditional SIR scheme. Furthermore, the increase in node density has much less impact on the overall energy consumption than the SIR scheme. This is because a larger node density causes more overlap between consecutive transmissions, which leads to a larger clustering factor and consequently more energy consumption for the traditional SIR scheme. On the other hand, the enforced gaps between transmissions under the proposed scheme reduce the aforementioned overlap. Hence the proposed scheme is more energy efficient.

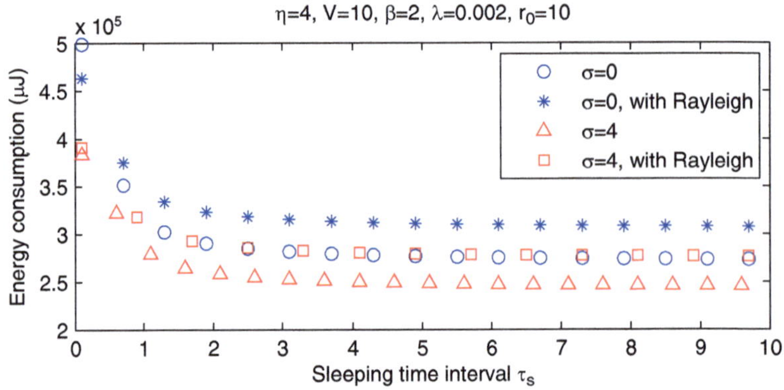

Fig. 10.9 Simulation results of the overall energy consumption using the proposed scheme and the traditional SIR scheme for a packet to be received by 99% of nodes is a network. The impact of the sleep time interval τ_s on the overall energy consumption is evaluated by letting the random component τ_r take a constant value 0

Figure 10.9 evaluates the impact of the sleep time interval τ_s on the energy consumption. Note that other things being equal, the energy and bandwidth consumption is a linearly increasing function of the number of transmissions. It can be seen that increasing the sleep time interval reduces the energy consumption. This is because a larger gap between transmissions leads to less overlap and the same reason applies to the network under the unit disk connection model (labeled $\sigma = 0$), the network under the log-normal shadowing model (labeled $\sigma = 4$) and the networks subject to Rayleigh fading (labeled *with Rayleigh*).

Figure 10.10 compares the proposed scheme with two classic schemes in the literature: the SI scheme and the SIR scheme. A large number of studies in this area considered the SI scheme, where every node carries the received information and forwards it to all nodes coming into the transmission range. In particular, Thedinger *et al.* proposed an SI-based broadcast scheme in [182], where each node broadcasts the received information to its directly connected neighbors repeatedly and consecutive re-broadcasts are separated by a fixed time interval. Using an SI-based scheme, every node needs to re-broadcast the received packet. For a fair comparison, we stop the re-broadcast when the packet is received by 99% of nodes in the network. It can be seen in Fig. 10.10 that the SIR based model requires a significantly smaller overall energy consumption compared with the SI based model. This is because the recovery mechanism in the SIR model limits the number of transmissions of each node to save energy. Moreover, as described earlier, the scheme presented in this chapter further reduces the energy consumption by reducing the overlap between consecutive transmissions and thereby reducing the number of transmissions to nodes already having a copy of the packet. Therefore, the proposed scheme requires less number of transmissions for a packet to be received by the same percentage of nodes in a network.

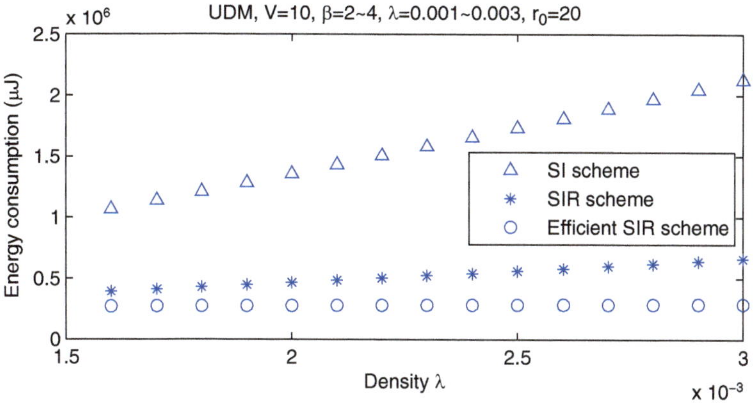

Fig. 10.10 Simulation results of the overall energy consumption using the proposed scheme and the other classic schemes for a packet to be received by 99% of nodes in a network

10.3.1 Real Mobility Trace Simulation

In this subsection, we consider a MANET driven by a real world trace of cabs in the San Francisco Bay Area [168]. The cab trace contains GPS coordinates of 536 cabs over approximately 30 days. Linear interpolation was used to increase the trace granularity. We focus on the downtown area (3.5km×3.5km) as indicated by the square in Fig. 10.11. A piece of information is broadcast from an arbitrary cab in the downtown area using the scheme proposed in Sect. 10.1. When information dissemination stops, we count the fraction of cabs in the downtown area that have received the information.

It is obvious that the cabs do not follow the mobility model used in the analysis. Actually there has been no analytical model that can fully characterize the movement of real world devices. The analysis based on analytical models under some ideal assumptions is used to predict the real world situations. We use the cab trace to evaluate the analytical results in a large network when the node mobility and distribution deviate from assumptions described in Sect. 10.1, and to identify the limitations of the analysis and present possible usage in estimating the performance of information dissemination in real world dynamic networks. It is interesting to see in the following figures that the impact of varying each network parameters on the network performance predicted by the analysis highly coincides with the simulations results.

Figure 10.12a and b shows the percolation probability and the expected fraction of informed nodes in the steady state. It can be seen that though the analytical result and simulation result exhibit the same trend, the fraction of informed nodes in a MANET driven by the actual cab trace is lower than that predicted by analysis. The discrepancy is mainly caused by the inhomogeneity of the node distribution.

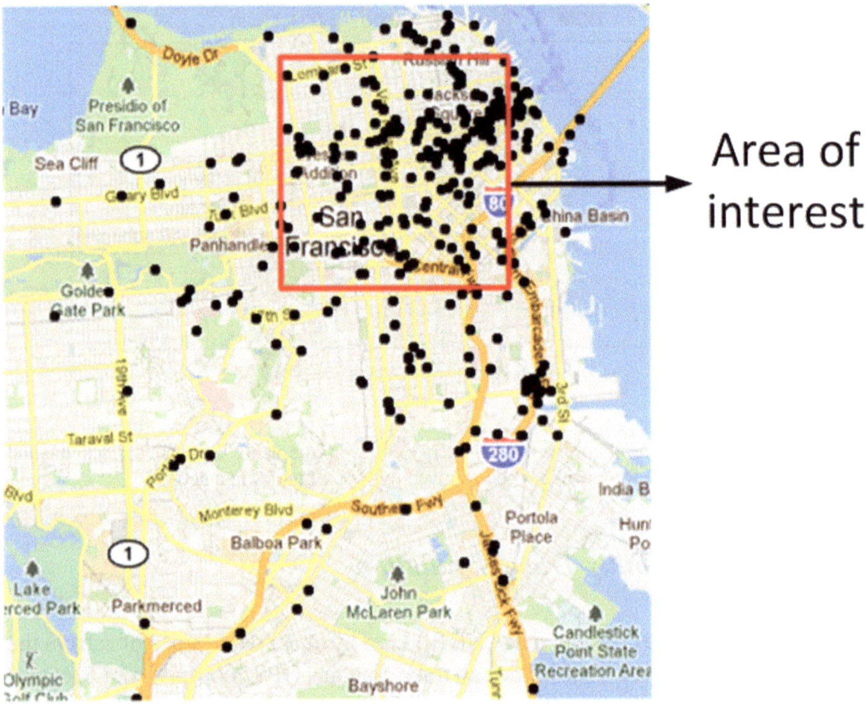

Fig. 10.11 A snapshot of the San Francisco Bay Area. The *dots* represent the positions of the cabs at a particular time instant. The *square* represents the downtown area we are interested in

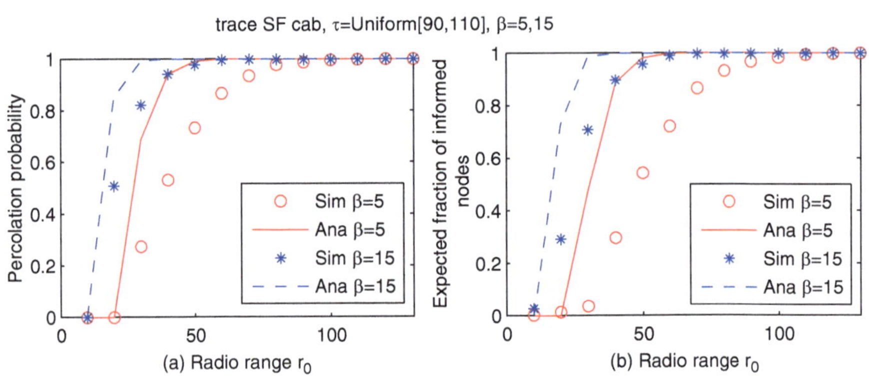

Fig. 10.12 The percolation probability and the expected fraction of informed nodes for the MANET driven by cab trace

Specifically, as can be seen in Fig. 10.11, the density of cabs in suburbs is lower than that in the downtown area. Therefore, when an infectious node moves to the suburbs, it has little chance to inform other nodes.

10.4 Notes and Further Readings

This chapter presented an information dissemination scheme for dynamic networks subject to channel randomness and with consideration of the energy and bandwidth consumed in the information propagation process. The information propagation process in dynamic networks adopting the information dissemination scheme was studied. The relevant parameters characterizing the information propagation process, i.e., the percolation probability, the expected fraction of informed nodes and delay, were analytically investigated. The accuracy of the analytical results was validated using simulations.

A number of existing studies on information dissemination in wireless networks, e.g., [115], focused on connected networks, where a network is said to be *connected* if and only if there is at least one multi-hop path connecting any pair of nodes all the time. As pointed out in Chap. 8, in mobile networks, it is often unnecessary or impractical to require a network to be *always* connected, due to fast-changing network topology or channel randomness.

Epidemic broadcast schemes are popularly used for information dissemination in MANETs. In [40], Chen *et al.* studied the information dissemination process using a Susceptible-Infectious (SI) epidemic scheme, where every node carries the received information and forwards it to all nodes coming into the transmission range. The SI epidemic scheme is a reliable but costly scheme due to a lack of a proper mechanism to stop the transmission.

In [43], Clementi *et al.* studied the speed of information propagation in a mobile network where nodes move independently at random over a square area. They obtained an upper bound on the flooding time, which is the maximum time required for all nodes of the network to be informed. In [109], Jacquet *et al.* studied the information propagation speed in mobile networks where nodes are uniformly distributed in a bounded area. The nodes were assumed to move following identically and independently distributed random trajectories. An upper bound on the information propagation speed, viz., ratio of the distance traveled by information over a given amount of time, was obtained.

There are other broadcast schemes for MANETs beside epidemic schemes. In [73], Friedman *et al.* reviewed some gossip-based algorithms that can be suitable candidates for information dissemination in MANETs. They pointed out that the design of energy and bandwidth efficient information dissemination schemes for MANETs is a challenging and open problem. A more recent work [188] of Gao *et al.* proposed a novel data forwarding strategy which explores the transient social contact patterns in social networks. They identified that low node density and a lack of global information are two major challenges in effective data forwarding in delay tolerant networks.

Part III
Connectivity of Small to Medium Sized Networks

Chapter 11
Connectivity of One-Dimensional Small to Medium Sized Networks

Abstract In this chapter, we study the connectivity of one-dimensional small to medium sized networks with and without infrastructure support. Different from the connectivity of large static networks and dynamic networks where asymptotic analysis is a major tool used in the analysis, for small to medium sized networks, the number of nodes in the network is not necessarily large enough to warrant the use of asymptotic analysis. Therefore, for small to medium sized networks, our focus is on characterizing three related connectivity measures, collectively called the probabilities of k-hop connection or hop count statistics, viz., 1) the probability that a randomly selected node is k-hops apart from another randomly selected node, i.e., the length of the shortest path from the first node to the second node measured by the number of hops is k; 2) the probability that a node $\boldsymbol{x}$ apart from another node is connected to that node in exactly k hops; and 3) the spatial distribution of the nodes k-hops apart from another designated node. In this chapter, we investigate the aforementioned three connectivity measures for one-dimensional small to medium sized networks with and without infrastructure support.

In the previous chapters, we have studied the connectivity of large static networks and dynamic networks. Asymptotic analysis is a major tool used in the study of large networks. In the next several chapters, we continue to study the connectivity of small to medium sized networks whose number of nodes is not necessarily large enough for asymptotic results to apply.

In a very general setting, we consider a wireless network with nodes identically and independently distributed following a homogeneous Poisson distribution with known density ρ in a given d-dimensional area ($d = 1, 2$), denoted by S. A node at $\boldsymbol{x}_2 \in S$ is directly connected to a node at $\boldsymbol{x}_1 \in S$ with probability $g(\boldsymbol{x}_2 - \boldsymbol{x}_1)$, i.e., following the random connection model. There are three related probabilities characterizing the connectivity properties of such a network: 1) $\Pr(k)$ the probability that a randomly selected node is k-hops apart from another randomly selected node, i.e., the length of the shortest path from the first node to the second node measured by the number of hops is k; 2) $\Pr(k|\boldsymbol{x})$ the probability that a node $\boldsymbol{x}$ apart from another node is connected to that node in exactly k hops; and 3) $\Pr(\boldsymbol{x}|k)$ the spatial distribution of the nodes k-hops apart from another designated node. These three probabilities are related through Bayes' formula. Given the probability density $p(\boldsymbol{x})$

G. Mao, *Connectivity of Communication Networks*,
DOI 10.1007/978-3-319-52989-9_11

for the displacement $\boldsymbol{x}$ between two randomly chosen node [144] and if one is computable, the other two will be computable using similar techniques. Therefore, we call these three probabilities collectively *the probabilities of k-hop connection* or *hop count statistics*.

As special cases of the above general setting, we study, respectively, the cases when $d = 1$, i.e., one-dimensional networks, or when $d = 2$, i.e., two-dimensional networks; when the connection model is a unit disk connection model or when the connection model becomes a log-normal connection model. When the probabilistic measures $\Pr(k|\boldsymbol{x})$ and $\Pr(\boldsymbol{x}|k)$ depend only on the Euclidean distance x between nodes, we can write $\Pr(k|x)$ for $\Pr(k|\boldsymbol{x})$ and $\Pr(x|k)$ for $\Pr(\boldsymbol{x}|k)$, instead of their displacement. Obviously, $\sum_{k=1}^{\infty} \Pr(k)$ gives the probability that two randomly selected nodes are connected (via a multi-hop path). Assuming the unit disk connection model, $\sum_{k=1}^{\infty} \Pr(k|x)$ gives the probability that a one-dimensional network with two nodes placed at both ends of a line segment of length x and other nodes in between is connected. We shall also show that the three measures and their variants can be used to derive other connectivity related properties of the network.

Solutions to the probability of k-hop connection problems can be used in a number of areas in wireless networks. The probability $\Pr(k)$ is useful in estimating the overall energy consumption, lifetime, and capacity of a wireless sensor network [204, 205] because these measures are tightly related to the number of hops required to transmit a packet from its source to its destination. The probability $\Pr(k|\boldsymbol{x})$ can be used in the analysis of end-to-end delay, energy consumption, and reliability of packet transmission [54, 203] and the probability in vehicular networks that a vehicle can access the base station within a designated number of hops [196]. The probability $\Pr(\boldsymbol{x}|k)$ is useful in estimating the distance between two nodes from their neighborhood information and obtaining variance of such estimate, which can then be used in localization [62].

A focus of our study on the connectivity of small to medium networks will be on the characterization of the three related probabilistic measures.

11.1 Probabilities of k-Hop Connection in One-Dimensional Ad Hoc Networks

In this section, we consider the special case of a one-dimensional network with nodes distributed on a semi-infinite line extending towards $+x$ axis following a homogeneous Poisson distribution with density ρ. A pair of nodes are directly connected following the unit disk connection model with a transmission range r_0. Without loss of generality, place the origin on the leftmost node and designate the node as the source node. We study the conditional probability $\Pr(k|x)$, i.e., the probability that a node located at x is k-hop apart from the source node.

Let ξ_k, $k = 1, \ldots$, be the furthest Euclidean distance between the source node and its k-hop neighbors, if exist. We first study the distribution of ξ_k, $k = 1, \ldots$ and then use the distribution of ξ_k, $k = 1, \ldots$ to derive $\Pr(k|x)$.

We derive the distribution of ξ_k, $k = 1, \ldots$, recursively and start with ξ_1. Due to the Poisson distribution of nodes, it can be readily shown that for $x \leq r_0$

$$\Pr(\xi_1 \leq x) = \left(1 - e^{-\rho x}\right) e^{-\rho(r_0 - x)} \tag{11.1.1}$$

where the term $1 - e^{-\rho x}$ is the probability that there is at least one node in $[0, x]$ and the term $e^{-\rho(r_0 - x)}$ is the probability that there is no node in $(x, r_0]$. It follows that the probability density function of ξ_1, denoted by $f_{\xi_1}(x)$, is given by

$$f_{\xi_1}(x) = \frac{d \Pr(\xi_1 \leq x)}{dx} = \rho e^{-\rho(r_0 - x)} \tag{11.1.2}$$

Due to the unit disk connection model being considered, it follows that $f_{\xi_1}(x) = 0$ for $x > r_0$ and $\Pr(\xi_1 \leq r_0) = 1 - e^{-\rho r_0}$ is the probability that there exists a 1-hop neighbor of the source.

Now we move on to consider the distribution of ξ_2. Obviously ξ_2 has to satisfy the condition that $r_0 < \xi_2 \leq \xi_1 + r_0$, it follows that conditioned on ξ_1, the cumulative distribution function of ξ_2 is given by

$$\Pr(\xi_2 \leq x | \xi_1) = \begin{cases} 0 & \text{for } x \leq r_0 \\ \left(1 - e^{-\rho(x - r_0)}\right) e^{-\rho(\xi_1 + r_0 - x)} & \text{for } r_0 < x \leq \xi_1 + r_0 \\ 1 - e^{-\rho \xi_1} & \text{for } x > \xi_1 + r_0 \end{cases} \tag{11.1.3}$$

The probability density function of ξ_2, conditioned on ξ_1, is given by

$$f_{\xi_2}(x | \xi_1) = \begin{cases} 0 & \text{for } x \leq r_0 \\ \rho e^{-\rho(\xi_1 + r_0 - x)} & \text{for } r_0 < x \leq \xi_1 + r_0 \\ 0 & \text{for } x > \xi_1 + r_0 \end{cases} \tag{11.1.4}$$

Supposing that the (conditional) distributions of $\xi_1, \xi_2, \ldots, \xi_k$ are known, we now consider the conditional distribution of ξ_{k+1}. First note that under the unit disk connection model, the distribution of ξ_{k+1} only depends on the distributions of ξ_k and ξ_{k-1}, i.e., $f_{\xi_{k+1}}(x | \xi_k, \ldots, \xi_1) = f_{\xi_{k+1}}(x | \xi_k, \xi_{k-1})$ because a node is a $k+1$-hop neighbor of the source if and only if its location x satisfies $\xi_k + r_0 < x \leq \xi_{k-1} + r_0$. This is also illustrated in Fig. 11.1.

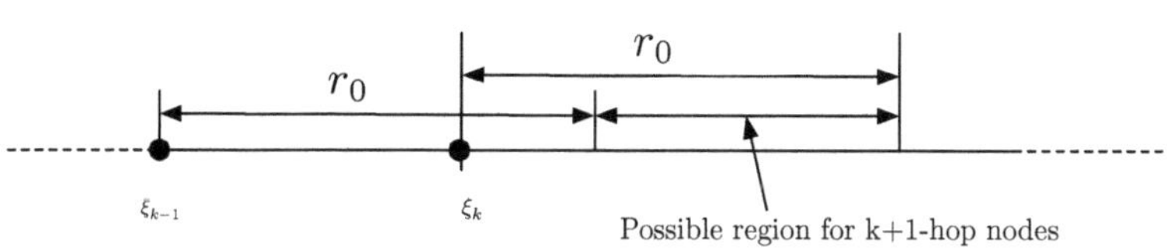

Fig. 11.1 An illustration of possible region for $k + 1$-hop neighbors

Similarly as before, it can be shown that

$$\begin{aligned}
&\Pr\left(\xi_{k+1} \le x \middle| \xi_k, \ldots, \xi_1\right) \\
&\quad = \Pr\left(\xi_{k+1} \le x \middle| \xi_k, \xi_{k-1}\right) \\
&\quad = \begin{cases} 0 & \text{for } x \le \xi_{k-1} + r_0 \\ \left(1 - e^{-\rho(x-\xi_{k-1}-r_0)}\right) e^{-\rho(\xi_k + r_0 - x)} & \text{for } \xi_{k-1} + r_0 < x \le \xi_k + r_0 \\ 1 - e^{\rho(\xi_k - \xi_{k-1})} & \text{for } x > \xi_k + r_0 \end{cases}
\end{aligned} \tag{11.1.5}$$

and the probability density function of ξ_{k+1}, conditioned on $\xi_k, \ldots, \xi_1$, is given by

$$f_{\xi_{k+1}}\left(x \middle| \xi_k, \ldots, \xi_1\right) = \begin{cases} 0 & \text{for } x \le \xi_{k-1} + r_0 \\ \rho e^{-\rho(\xi_k + r_0 - x)} & \text{for } \xi_{k-1} + r_0 < x \le \xi_k + r_0 \\ 0 & \text{for } x > \xi_k + r_0 \end{cases} \tag{11.1.6}$$

Remark 170 It is evident from (11.1.5) that, under the unit disk connection model, the distribution of the furtherest distance of $k + 1$-hop nodes depends only on the distributions of the furtherest distances of the previous two hop nodes. The same conclusion may not be true for other connections models, e.g., the log-normal connection model and the random connection model.

It follows that the joint distribution of $\xi_{k+1}, \ldots, \xi_1$ can be obtained as

$$\begin{aligned}
f\left(\xi_{k+1}, \ldots, \xi_1\right) &= f\left(\xi_{k+1} \middle| \xi_k, \ldots, \xi_1\right) f\left(\xi_k \middle| \xi_{k-1}, \ldots, \xi_1\right) \cdots f\left(\xi_1\right) \\
&= f\left(\xi_{k+1} \middle| \xi_k, \xi_{k-1}\right) f\left(\xi_k \middle| \xi_{k-1}, \xi_{k-2}\right) \cdots f\left(\xi_1\right)
\end{aligned} \tag{11.1.7}$$

and the probability density function (and the cumulative distribution function) of ξ_k, $k = 1, \ldots$, can be obtained from (11.1.7) using the total probability theorem.

Given the distribution of ξ_k, $k = 1, \ldots$, $\Pr(k|x)$ can be readily obtained:

$$\begin{aligned}
\Pr(k|x) &= \Pr\left(\xi_{k-1} < x \le \xi_{k-1} + r_0\right) \\
&= \Pr\left(x - r_0 \le \xi_{k-1} < x\right)
\end{aligned} \tag{11.1.8}$$

11.2 Connectivity of One-Dimensional Infrastructure Based Networks

In this section, we continue to investigate the connectivity of one-dimensional wireless networks with infrastructure support. Examples of such networks can be found in vehicular networks and wireless sensor networks, see [44, 76] for more examples. In vehicular networks, roadside infrastructure plays an important role in the reliable and timely distribution of critical information to vehicles on the road [14]. In wireless sensor networks, data sinks gather useful information collected by the sensors via multi-hop paths. Then, these data sinks may either store the data for later retrieval or aggregate and transfer the data immediately via a backbone network to the remote base station or Internet. An example is the sensor network deployed on Great Duck Island for habitat monitoring [136]. From the above examples, it can be summarized that an infrastructure-based wireless network has the following characteristics: (a) the communication between "ordinary" nodes (vehicles / sensors) and "powerful" nodes (roadside infrastructure / data sinks) is important for the core functions of the networks to be carried out properly; (b) the powerful nodes are inter-connected, either by wired or by wireless links and their location is usually deterministic; (c) the location of ordinary nodes is often random. In this section, we investigate the connectivity of this type of one-dimensional infrastructure based networks, which is also referred to as hybrid networks in the literature.

To characterize connectivity of one-dimensional infrastructure based networks, we propose a new concept of connectivity, which we term *type-II connectivity*, to distinguish it from connectivity of homogeneous networks studied earlier. We say that a network is *type-II connected* if every ordinary node in the network is connected, directly or via multi-hop paths, to at least one of a small subset of powerful nodes. Type-II connectivity problem is a broad topic. In addition to connectivity probability, we also analyze the average number of clusters. The average number of clusters can be considered as an alternative measure of network connectivity, which measures how fragmented a network is if it is not connected. It tells us how many additional powerful nodes are required for all ordinary nodes in a network to be connected to at least one powerful node with high probability. Assuming these additional powerful nodes are mobile, then there exist a number of ways [16] that they can connect the ordinary nodes together to achieve certain purposes such as maximizing the communication reliability between nodes [113] or balancing the traffic load among the nodes [128]. Such problems are important in network topology control and routing. Based on the above connectivity and clustering results, we obtain the optimum powerful node distribution that minimizes the average number of clusters and maximizes the asymptotic connectivity probability of the network.

More specifically, the network model studied in this section is defined as follows.

Definition 171 Denote by $G(\lambda, n_p; L; r_o, r_p)$ a one-dimensional wireless network with two types of nodes: ordinary nodes and powerful nodes. Ordinary nodes are identically, independently, and Poissonly distributed with a known density λ in the interval $[0, L]$. There are $n_p \geq 2$ powerful nodes in the network, where two of them are placed at both ends of the interval and the rest are arbitrarily distributed in the interior of the same interval. A direct connection between two ordinary nodes (respectively, between an ordinary node and a powerful node) exists if their Euclidean distance is smaller than or equal to r_o (respectively, r_p). Furthermore, all powerful nodes are assumed to be inter-connected to each other.

An example of our model is illustrated in Fig. 11.2.

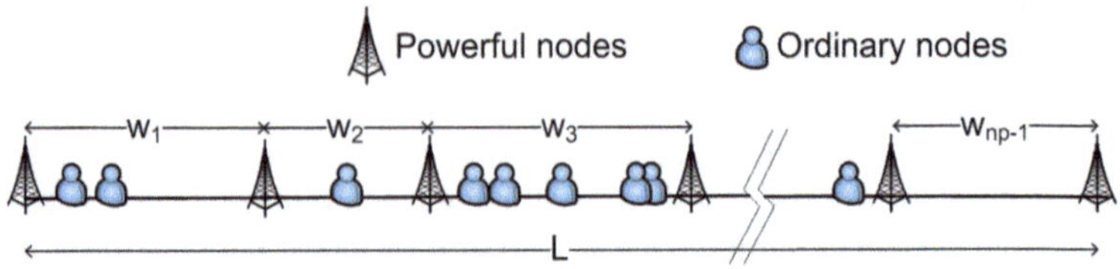

Fig. 11.2 An example of a wireless network with a mixture of ordinary nodes and powerful nodes. The distances between adjacent powerful nodes are denoted by w_i, for $1 \leq i \leq n_p - 1$

In the model, the powerful nodes divide the interval $[0, L]$ into $n_p - 1$ sub-intervals and each sub-interval i has length w_i for $1 \leq i \leq n_p - 1$.

In general, we assume that $r_p \geq r_o$. This assumption is justified because it is often the case that a powerful node can not only transmit at a larger transmission power than an ordinary node,but it may also be equipped with more sophisticated antennas, which make it more sensitive to the transmitted signal from an ordinary node [172]. As will be shown later, in this section we mostly focus on the network with $L = 1$, i.e., on the unit interval. Using the scaling technique, the results for the network on the unit interval can be easily applied to a network on the interval $[0, L]$ where $L \neq 1$.

11.2.1 Characterization of Type-II Connectivity Probability

In this subsection we investigate the type-II connectivity probability of a network $G(\lambda, n_p; 1; r_o, r_p)$, i.e., on the unit interval. Under the unit disk connection model, the connectivity probability can be derived by first examining each sub-interval bounded by two consecutive powerful nodes.

Let $\mathcal{A}_i(w_i)$ be the event that sub-interval i with length w_i is type-II connected under the assumption that $r_p \geq r_o$. It is trivial to show that $\Pr(\mathcal{A}_i(w_i))$, the probability that $\mathcal{A}_i(w_i)$ occurs, is 1 when $w_i \leq 2r_p$. For $w_i > 2r_p$, a realization of a type-II connected sub-interval i for $r_p \geq r_o$ *remains* type-II connected if and only if after removing an interval of length $r_p - r_o$ (and the ordinary nodes within that interval) from the left end and right end of sub-interval i respectively, the resulting sub-interval with $r_p = r_o$ is still type-II connected. Hence for $w_i > 2r_p$,

$$\Pr(\mathcal{A}_i(w_i)) = \Pr\left(\mathcal{A}_i^{eq}(w_i - 2(r_p - r_o))\right) \tag{11.2.1}$$

where $\mathcal{A}_i^{eq}(x)$ is the event that sub-interval i with length x is type-II connected under the situation that $r_p = r_o$. In the next subsection we provide the derivation of $\Pr\left(\mathcal{A}_i^{eq}(x)\right)$.

A network is type-II connected if and only if each sub-interval is type-II connected. Under the unit disk connection model, the event that one sub-interval is type-II connected is independent of the event that another sub-interval is type-II connected. Hence, the probability that a network with each sub-interval i having length w_i is type-II connected (say, event $\mathcal{B}(w_1, \cdots, w_{n_p-1})$) is

$$\Pr\left(\mathcal{B}(w_1, \cdots, w_{n_p-1})\right) = \prod_{i=1}^{n_p-1} \Pr\left(\mathcal{A}_i(w_i)\right) \tag{11.2.2}$$

Based on (11.2.2), we can obtain the following theorem.

Theorem 172 *Denote by $\mathcal{B}$ the event that a random instance of $G(\lambda, n_p; 1; r_o, r_p)$ is type-II connected. Then, the probability that event $\mathcal{B}$ occurs is*

$$\Pr(\mathcal{B}) = \int_{\mathbb{D}} \left(\prod_{i=1}^{n_p-1} \Pr\left(\mathcal{A}_i(w_i)\right) \right) f(\mathbf{w}) d\mathbf{w} \tag{11.2.3}$$

where $\mathbb{D} = \{(w_1, \cdots, w_{n_p-1}) : \sum_{i=1}^{n_p-1} w_i = 1\}$; $f(\mathbf{w}) = f(w_1, \cdots, w_{n_p-1})$ is the joint probability density function of the distances between adjacent powerful nodes; $\Pr(\mathcal{A}_i(w_i))$ *is given in (11.2.1) and* $\Pr\left(\mathcal{A}_i^{eq}(w_i)\right)$ *is given in (11.2.10).*

Using (11.2.3), we can calculate the type-II connectivity probability of a network with any distribution of powerful nodes as long as $f(\mathbf{w})$ of that distribution is known. For example, if the powerful nodes are uniformly distributed, then $f(\mathbf{w}) = (n_p - 2)!$ [53]. If the powerful nodes are placed in an equidistant fashion, then (11.2.3) simplifies into

$$\Pr(\mathcal{B}) = \left(\Pr\left(\mathcal{A}_i(w)\right)\right)^{n_p-1} \tag{11.2.4}$$

with $w = \frac{1}{n_p-1}$. In the following subsections, we derive $\Pr\left(\mathcal{A}_i^{eq}(x)\right)$ and its asymptotic approximation where the superscript eq is used to highlight that the power nodes are placed in an equidistant fashion.

11.2.1.1 Exact Probability that a Sub-Interval Is Type-II Connected for $r_p = r_o$

As mentioned earlier, the result for $r_p = r_o$ can be used to obtain the result for the general case where $r_p \geq r_o$. Let $\mathcal{A}_i^{eq}(m_i, w_i)$ be the event that sub-interval i with length $0 < w_i \leq 1$ is type-II connected given that there are m_i ordinary nodes in the sub-interval. Denote the common transmission range by r, i.e., $r = r_p = r_o$. The derivation of $\Pr\left(\mathcal{A}_i^{eq}(m_i, w_i)\right)$ relies on the following lemma from [79].

Lemma 173 *Let $[x, x + y]$ be a sub-interval of length y within $[0, 1]$. Assume two of k given vertices have been placed at the borders of this sub-interval. Define two vertices to be neighbors if and only if they are at distance r or less apart, let $\mathcal{Z}_{k,y,r}$ be* *the event that $k-2$ vertices, corresponding to the remaining vertices and uniformly placed in $[0,1]$, are inside $[x, x+y]$ and "join" the borders, that is, the k vertices form a connected subgraph of length y; and let $P(k,y,r) = \Pr(\mathcal{Z}_{k,y,r})$. Then, for $k \geq 2$,*

$$P(k,y,r) = \sum_{j=0}^{\min(k-1,\lfloor y/r \rfloor)} \binom{k-1}{j} (-1)^j (y-jr)^{k-2} \tag{11.2.5}$$

A sub-interval is type-II connected if all ordinary nodes within the sub-interval are connected to at least one of the two powerful nodes located at both ends of the sub-interval. Hence, event $\mathcal{A}_i^{eq}(m_i, w_i)$ occurs with probability

$$\Pr\left(\mathcal{A}_i^{eq}(m_i, w_i)\right) = P(m_i+2, 1, \hat{r}) + m_i(m_i+1) \int_0^{1-\hat{r}} P(m_i+1, \hat{x}, \hat{r}) \mathrm{d}\hat{x} \tag{11.2.6}$$

where $\hat{r} = \frac{r}{w_i}$ is the normalized transmission range and $\hat{x} = \frac{x}{w_i}$ is the normalized distance of x. The two terms on the right-hand side of (11.2.6) represent the two possible cases of the event, as illustrated in Fig. 11.3. Figure 11.3a corresponds to the first term in (11.2.6), and Fig. 11.3b corresponds to the second term.

Figure 11.3a shows a possible case where all m_i ordinary nodes within sub-interval i are connected to both powerful nodes. That is, none of the $m_i + 1$ spacings between the adjacent ordinary nodes and the two powerful nodes is larger than r. In this case, all ordinary nodes and the two powerful nodes in sub-interval i form a connected "subgraph" of length w_i. From Lemma 173, the probability of this case is $P(m_i+2, 1, \hat{r})$, where $m_i + 2$ is the sum of the number of ordinary nodes and the two powerful nodes.

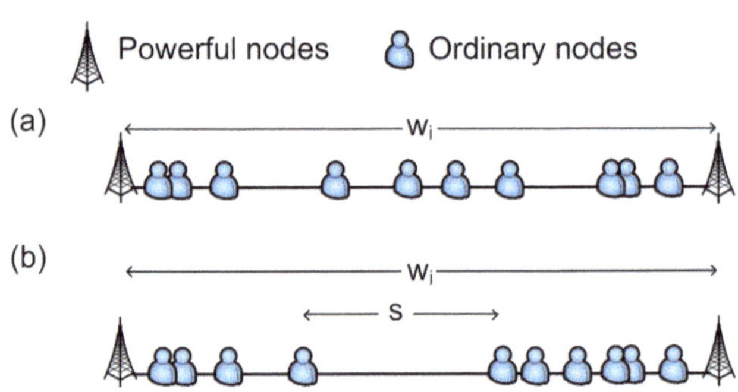

Fig. 11.3 An illustration that sub-interval i is type-II connected. Subgraph (**a**) shows that all ordinary nodes are connected to both powerful nodes. All spacings between any two adjacent nodes are not greater than the transmission range r. Subgraph (**b**) shows an example where there is a big spacing with length s in the sub-interval, and $s > r$. All ordinary nodes located to the left (right) of the big spacing are connected to the left (right) powerful node

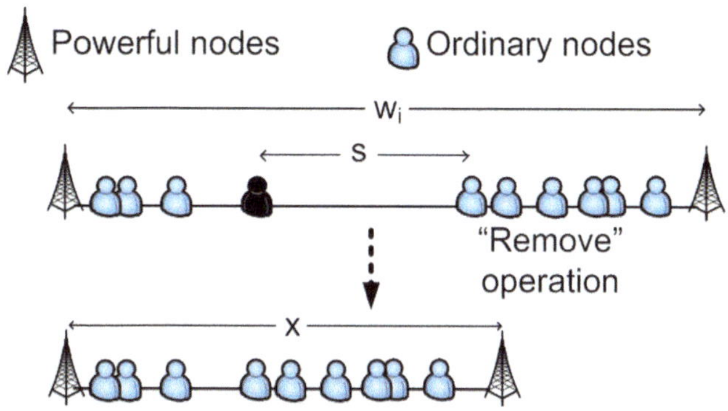

Fig. 11.4 An illustration of the scenario that if the big spacing with length s and the ordinary node attached to the left end of the big spacing are removed (the "remove" operation) from the sub-interval i (except for the case when the big spacing is the left most spacing, and then the ordinary node attached to the right of the big spacing is removed), then the $m_i - 1$ remaining ordinary nodes and the two powerful nodes form a connected "subgraph" of length $x = w_i - s$

Figure 11.3b shows the other possible case where the m_i ordinary nodes inside sub-interval i are connected to either one of the two powerful nodes but not both. Then, among the $m_i + 1$ spacings between adjacent nodes, there is exactly one spacing with length $s > r$. Suppose temporarily that the big spacing of length s and the ordinary node attached to the left end of the big spacing are removed from sub-interval i, as illustrated in Fig. 11.4. Then, the $m_i - 1$ remaining ordinary nodes and the two powerful nodes form a connected "subgraph" of length $x = w_i - s$.

A special case occurs when the big spacing is the leftmost spacing in the sub-interval. If this is the case, then we remove the ordinary node attached to the right end of the big spacing instead. The probability that the $m_i - 1$ remaining ordinary nodes and the two powerful nodes, with the sub-interval having the new length of x, form a connected interval is given by $P(m_i + 1, \hat{x}, \hat{r})$ where $\hat{x} = \frac{x}{w_i}$. Following the convention in [79] that nodes are treated as distinguishable, the event that a particular node i attached to the left end of the big spacing is removed together with the big spacing and the remaining nodes form a connected interval, and the event that a particular node j attached to the left end of the big spacing is removed together with the big spacing and the remaining nodes form a connected interval are treated as different events. Therefore, any of the m_i ordinary nodes can be attached to the left end of the big spacing (or attached to the right end for the special case), and the big spacing can be any of the $m_i + 1$ spacings in sub-interval i. As a result, the probability that events like Fig. 11.3b occur is then $m_i(m_i + 1)P(m_i + 1, \hat{x}, \hat{r})$, for $\hat{x}$ ranging from zero to $1 - \hat{r}$. Therefore, we obtain the second term in (11.2.6).

After applying (11.2.5) into the second term of (11.2.6), we can get rid of the integral in the second term by moving the inner sum outside the integral. With further changes to the range of summation and integral, we obtain

$$m_i(m_i+1)\int_0^{1-\hat{r}} P(m_i+1,\hat{x},\hat{r})\,\mathrm{d}\hat{x}$$

$$=(m_i+1)\sum_{j=0}^{\min(m_i,\lfloor 1/\hat{r}\rfloor-1)}\binom{m_i}{j}(-1)^j\left(1-(j+1)\hat{r}\right)^{m_i} \tag{11.2.7}$$

Using (11.2.7), and replacing the first term in (11.2.6) by (11.2.5), we can simplify (11.2.6):

$$\Pr\left(\mathcal{A}_i^{eq}(m_i,w_i)\right)$$

$$=\sum_{j=0}^{\min(m_i+1,\lfloor w_i/r\rfloor)}(1-j)\binom{m_i+1}{j}(-1)^j(1-j\frac{r}{w_i})^{m_i} \tag{11.2.8}$$

Since all sub-intervals bounded by powerful nodes are nonoverlapping segments with length w_i and ordinary nodes are Poissonly distributed, m_i is a Poisson random variable with mean $w_i\lambda$, and m_i and m_j are mutually independent for $i \neq j$. Let $\mathcal{A}_i^{eq}(w_i)$ be the event that sub-interval i with length $0 < w_i \leq 1$ is type-II connected. Then,

$$\Pr\left(\mathcal{A}_i^{eq}(w_i)\right)=\sum_{m_i=0}^{\infty}\Pr\left(\mathcal{A}_i^{eq}(m_i,w_i)\right)\frac{(w_i\lambda)^{m_i}}{m_i!}\exp(-w_i\lambda) \tag{11.2.9}$$

$$=\sum_{j=0}^{\lfloor w_i/r\rfloor}(1-j)(-1)^j\frac{1}{j!}(j+w_i\lambda-jr\lambda)$$

$$\times\,(w_i\lambda-jr\lambda)^{j-1}\exp(-jr\lambda) \tag{11.2.10}$$

where from (11.2.9) we first exchange the order of the inner sum and the outer sum after we substitute $\Pr\left(\mathcal{A}_i^{eq}(m_i,w_i)\right)$ by (11.2.8). Then, we substitute $\binom{m_i+1}{j}=\frac{1}{j!}\frac{\mathrm{d}^j}{\mathrm{d}t^j}t^{m_i+1}\Big|_{t=1}$, move the derivative outside the inner sum, and with some arithmetic steps we obtain (11.2.10). Note that (11.2.10) is still valid even when $w_i \leq 2r$. That is, $\Pr\left(\mathcal{A}_i^{eq}(w_i)\right)=1$ when $w_i \leq 2r$ as expected.

11.2.1.2 Asymptotic Probability That a Sub-Interval Is Type-II Connected for $r_p = r_o$

Equation (11.2.10) is in a very complicated form which may prevent us from obtaining in-depth understanding on the relations among parameters that determine $\Pr\left(\mathcal{A}_i^{eq}(w_i)\right)$. In the following we derive a simplified asymptotic approximation for $\Pr\left(\mathcal{A}_i^{eq}(w_i)\right)$.

Let $\hat{w} = w_i/r$ be the normalized length of sub-interval i by r, where $r = r_p = r_o$ as usual. Let $\mu = 2r\lambda$ be the average node degree ignoring the boundary effect. Let $\phi(\hat{w}) = \Pr\left(\mathcal{A}_i^{eq}(\hat{w}r)\right)$. Then, (11.2.10) implies the following difference-differential equation.

$$\frac{\mathrm{d}^2}{\mathrm{d}\hat{w}^2}\phi(\hat{w}) + 2\beta\frac{\mathrm{d}}{\mathrm{d}\hat{w}}\phi(\hat{w}-1) + \beta^2\phi(\hat{w}-2) = 0 \quad (11.2.11)$$

where $\beta = \frac{\mu}{2}\exp(-\frac{\mu}{2})$. Using (11.2.11), we can obtain the Laplace transform of $\phi(\hat{w})$ as

$$\Phi(s) = \frac{1}{s+\beta\exp(-s)} + \frac{\beta\exp(-s)}{(s+\beta\exp(-s))^2} \quad (11.2.12)$$

As $s \to 0$, $\exp(-s) \approx 1 - s$. Substituting this approximation into (11.2.12), the inverse Laplace transform of the approximated equation is then

$$\phi(\hat{w}) \approx \left[\frac{1-2\beta}{(1-\beta)^2} + \frac{\beta\hat{w}}{(1-\beta)^3}\right]\exp(\frac{-\beta\hat{w}}{1-\beta}) \quad (11.2.13)$$

Figure 11.5 shows that (11.2.13) serves as a good approximation for the exact result in (11.2.10) provided $\mu \geq 6$, and virtually all values of $\hat{w} \geq 2$, not just large values of $\hat{w}$.

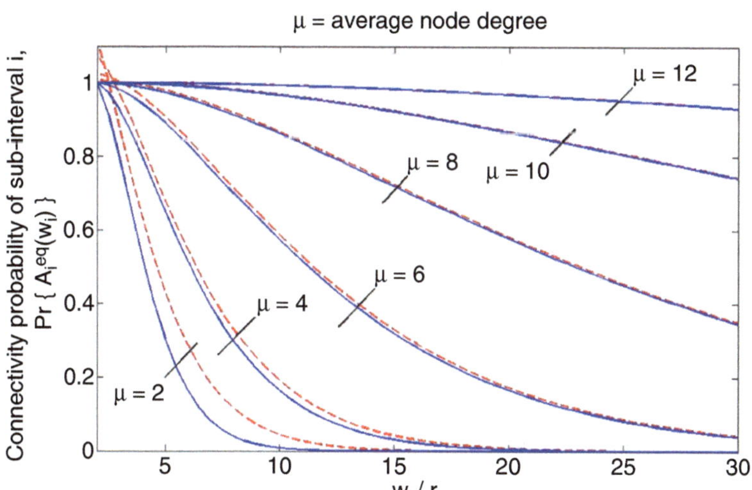

Fig. 11.5 The type-II connectivity probability of a sub-interval given different normalized length of the sub-interval w_i/r and the average node degree. The *solid lines* are the exact results, and the *dashed lines* are the asymptotic results

Solving (11.2.13) for $\hat{w}$ leads to

$$\hat{w} = \{-\mathrm{W}_{-1}[-(1-\beta)^2 \exp(-(1-2\beta))\phi(\hat{w})] - (1-2\beta)\} \frac{1-\beta}{\beta} \tag{11.2.14}$$

where $\mathrm{W}_{-1}[\cdot]$ is the real-valued, non-principal branch of the LambertW function [47]. Given the required connectivity probability $\phi(\hat{w})$, and the value of β, which is related to the ordinary node density λ and the transmission range r, we can use (11.2.14) to obtain $\hat{w}$, the maximum distance between two adjacent powerful nodes so that the designated connectivity probability requirement is fulfilled.

11.2.2 Average Number of Clusters

Besides connectivity probability, another measure of interest is the number of clusters in the network. It is an indicator of how fragmented a network is. In our network model, *all ordinary nodes that are connected to at least one of the powerful nodes belong to the same cluster*. Denote by "main cluster" the cluster formed by the ordinary nodes which are connected to at least one powerful node. Other ordinary nodes which are not connected to any powerful node, if exist, form one or more "secondary clusters." Therefore, there is always one main cluster and zero or more secondary clusters in a network. Note that a network is type-II connected if and only if there is no secondary cluster in the network. In this subsection we investigate the average number of clusters in a network $G(\lambda, n_p; 1; r_o, r_p)$ where the powerful nodes are placed in an equidistant fashion. The reason for focusing on equidistant powerful node distribution is that, as will be shown later, it gives the best performance in terms of minimizing the average number of clusters and maximizing the asymptotic type-II connectivity probability. Nevertheless, the analysis in the previous subsection has provided the conceptual basis of how the results can be generalized to having powerful nodes arbitrarily distributed.

Let $C_i(w)$ be the number of secondary clusters in sub-interval i with length w under the assumption that $r_p \geq r_o$. Note that $w = \frac{1}{n_p-1}$ for equidistant powerful node distribution. We only consider $w > 2r_p$ to avoid triviality. Also note that ordinary nodes in sub-interval i, which are at most $r_p - r_o$ Euclidean distance away from the powerful nodes, belong to the main cluster with probability 1 and any other ordinary nodes which are directly connected to these ordinary nodes are also directly connected to the powerful nodes. Consequently, the number of secondary clusters remains the same after we remove an interval of length $r_p - r_o$ (and the ordinary nodes within that interval) from the left end and right end of sub-interval i, respectively, and then assume that $r_p = r_o$. So, we have $C_i(w) = C_i^{eq}(w-2(r_p-r_o))$ where $C_i^{eq}(x)$ is the number of secondary clusters in sub-interval i with length x under the special assumption that $r_p = r_o$.

To obtain $C_i^{eq}(x)$, let $t_i(x)$ be the number of spacings with length greater than r in sub-interval i of length x where $r = r_p = r_o$ as usual. Then, $C_i^{eq}(x)$ and $t_i(x)$ have the following relationship:

$$C_i^{eq}(x) = \begin{cases} t_i(x) - 1 & \text{for } t_i(x) \geq 1 \\ 0 & \text{for } t_i(x) = 0 \end{cases} \tag{11.2.15}$$

Assume that there are m_i ordinary nodes in sub-interval i of length x and let $1_{i,j}(x)$ be an indicator function such that

$$1_{i,j}(x) = \begin{cases} 1 & \text{if the } j\text{-th spacing in sub-interval } i \text{ of length } x \text{ has length greater than } r \\ 0 & \text{otherwise} \end{cases}$$

where $1 \leq j \leq m_i + 1$. Then, the expected value of $t_i(x)$ given m_i ordinary nodes in sub-interval i with length x is

$$\begin{aligned} E[t_i(x)|\, m_i] &= E\left[\sum_{1\leq j\leq m_i+1} 1_{i,j}(x)|m_i\right] \\ &= (m_i + 1)E\left[1_{i,j}(x)|m_i\right] \quad \text{for any } j \qquad (11.2.16) \\ &= (m_i + 1)(1 - \frac{r}{x})^{m_i} \qquad (11.2.17) \end{aligned}$$

where $E\left[1_{i,j}(x)\middle|\, m_i\right]$ is equal to the probability that the j-th spacing in sub-interval i of length x has length greater than r. Since this probability is equal to the probability that the m_i ordinary nodes fall into a smaller interval of length $1 - \frac{r}{x}$ in sub-interval i, we obtain (11.2.17).

Since m_i is a Poissonly distributed random variable with mean $x\lambda$, it follows immediately that

$$\begin{aligned} E\left[t_i(x)\right] &= \sum_{m_i=0}^{\infty} E\left[t_i(x)|\, m_i\right] \frac{(x\lambda)^{m_i}}{m_i!} \exp(-x\lambda) \\ &= (x\lambda - r\lambda + 1)\exp(-r\lambda). \end{aligned}$$

From (11.2.15) we have

$$\begin{aligned} E\left[C_i^{eq}(x)\right] &= E[t_i(x)] - 1 + \Pr\left(t_i(x) = 0\right) \qquad (11.2.18) \\ &= \sum_{j=2}^{\lfloor x/r \rfloor} (-1)^j \frac{1}{j!}(j + x\lambda - jr\lambda)(x\lambda - jr\lambda)^{j-1}\exp(-jr\lambda) \qquad (11.2.19) \end{aligned}$$

where from (11.2.18) to (11.2.19) we apply

$$\Pr\left(t_i(x) = 0\right) = \sum_{j=0}^{\lfloor x/r \rfloor} (-1)^j \frac{1}{j!}(j + x\lambda - jr\lambda)(x\lambda - jr\lambda)^{j-1}\exp(-jr\lambda) \tag{11.2.20}$$

which is obtained from the first term in (11.2.6) and simplified using the same procedure as that resulting in (11.2.10). Finally, let $D(n_p)$ be the number of clusters in a network with n_p powerful nodes equally spaced and assume $r_p \geq r_o$. Then,

$$D(n_p) = \sum_{i=1}^{n_p-1} C_i(w) + 1 \tag{11.2.21}$$

$$= \sum_{i=1}^{n_p-1} C_i^{eq}(w - 2(r_p - r_o)) + 1 \tag{11.2.22}$$

where $w = \frac{1}{n_p-1}$. That is, we add up the number of secondary clusters in each sub-interval and one (and the only) main cluster in the whole network. Based on (11.2.22), we can obtain the following theorem.

Theorem 174 *For $G(\lambda, n_p; 1; r_o, r_p)$ with powerful nodes placed in an equidistant fashion, the expected number of clusters in the network is then*

$$E\left[D(n_p)\right] = (n_p - 1)E\left[C_i^{eq}(w - 2(r_p - r_o))\right] + 1 \tag{11.2.23}$$

where $E\left[C_i^{eq}(x)\right]$ is given in (11.2.19) and $w = \frac{1}{n_p-1}$.

11.2.3 The Optimal Distribution of Powerful Nodes

In this subsection, we shall show that the equidistant placement of powerful nodes will minimize the average number of clusters in a network and maximize the asymptotic type-II connectivity probability.

11.2.3.1 Minimizing the Average Number of Clusters

From (11.2.21), we have the average number of clusters in a network $G(\lambda, n_p; 1; r_o, r_p)$ given each sub-interval i has length w_i, is

$$E\left[D(w_1, w_2, \cdots, w_{n_p-1})\right] = \sum_{i=1}^{n_p-1} E\left[C_i(w_i)\right] + 1 \tag{11.2.24}$$

where $E[C_i(w_i)]$ is the average number of secondary clusters in sub-interval i. Finding the optimal powerful node placement to minimize the average number of clusters can be treated as a constrained optimization problem:

$$\begin{aligned} \text{minimize} \quad & E\left[D(w_1, w_2, \cdots, w_{n_p-1})\right] \\ \text{subject to} \quad & \sum_{i=1}^{n_p-1} w_i = 1 \end{aligned}$$

In the following we prove that $E[C_i(w_i)]$ is a convex function of w_i.

Recall that for $w_i > 2r_p$, we have $E[C_i(w_i)] = E[C_i^{eq}(w_i - 2(r_p - r_o))]$. From (11.2.18) we further have,

$$E[C_i^{eq}(x)] = (x\lambda - r_o\lambda + 1)e^{-r_o\lambda} - 1 + \Pr(t_i(x) = 0)$$

where $\Pr(t_i(x) = 0)$ is given by (11.2.20). With some arithmetic steps we can derive the second derivative of $E[C_i^{eq}(x)]$ and obtain

$$\begin{aligned}\frac{d^2}{dx^2}E[C_i^{eq}(x)] &= \frac{d^2}{dx^2}\Pr(t_i(x) = 0)\\ &= (\lambda e^{-r_o\lambda})^2 \Pr(t_i(x - 2r_o) = 0)\\ &\geq 0\end{aligned}$$

Hence, the second derivative of $E[C_i(w_i)]$ is also greater or equal to zero for $w_i > 2r_p$. It is trivial to show that the second derivative is zero for $w_i \leq 2r_p$ as $E[C_i(w_i)] = 0$ in that range.

Since $E[C_i(w_i)]$ is a convex function, using (11.2.24), $E[D(w_1, w_2, \cdots, w_{n_p-1})]$ is also a convex function. Hence the optimization problem is a convex optimization problem. It is then straightforward to prove, e.g., using the method of Lagrange multipliers, that the minimum of the average number of clusters is achieved when $w_1 = \cdots = w_{n_p-1} = \frac{1}{n_p-1}$ and by convexity it is a global minimum.

11.2.3.2 Maximizing the Asymptotic Type-II Connectivity Probability

Using (11.2.13) we approximate the type-II connectivity probability

$$\Pr\left(\mathcal{B}(w_1, \cdots, w_{n_p-1})\right) = \prod_{i=1}^{n_p-1} \Pr(\mathcal{A}_i(w_i)) \approx \prod_{i=1}^{n_p-1} \phi(x_i)$$

where $x_i = w_i - 2(r_p - r_o)$, and

$$\phi(x_i) = \begin{cases}\left[\frac{1-2\beta}{(1-\beta)^2} + \frac{\beta x_i/r_o}{(1-\beta)^3}\right]\exp(\frac{-\beta x_i/r_o}{1-\beta}) & \text{if } x_i > 2r_o\\ 1 & \text{otherwise}\end{cases}$$

Since both expressions $\frac{1-2\beta}{(1-\beta)^2} + \frac{\beta x_i/r_o}{(1-\beta)^3}$ and $\exp(\frac{-\beta x_i/r_o}{1-\beta})$ are log-concave on $x_i \geq 0$, and the product of log-concave functions is a log-concave function [30], we have $\phi(x_i)$ is log-concave on x_i and $\Pr\left(\mathcal{B}(w_1, \cdots, w_{n_p-1})\right)$ is also a log-concave function of the lengths of sub-intervals. Using this property, it can be readily shown that the maximum of the probability that the network $G(\lambda, n_p; 1; r_o, r_p)$ is type-II connected is also achieved when powerful nodes are distributed in an equidistant fashion.

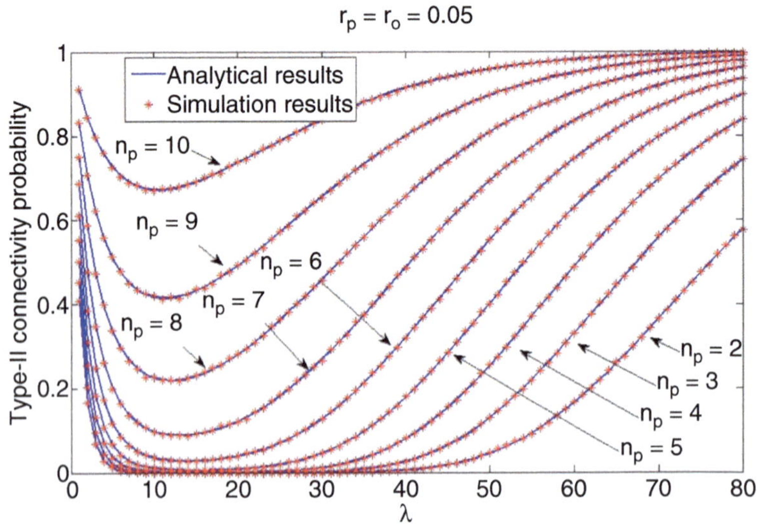

Fig. 11.6 The type-II connectivity probability given different values of λ, n_p, and $r_p = r_o = 0.05$ under the unit disk connection model. The *solid lines* are plotted using (11.2.4), verified by simulation results obtained from 40, 000 randomly generated network topologies

11.2.4 Impact of Different Parameters on Connectivity

In this subsection, we investigate the impact of different parameters on the performance of a network $G(\lambda, n_p; 1; r_o, r_p)$. Note that all figures are plotted under the condition that powerful nodes are placed in an equidistant fashion.

First, Fig. 11.6 shows the probability that a network is connected given different values of λ, n_p and $r_p = r_o = 0.05$ under the unit disk connection model. The accuracy of the analytical results is verified by simulation results obtained from 40, 000 randomly generated network topologies. As the numbers of instances of random networks used in the simulations are very large, the confidence interval is too small to be distinguishable and hence ignored in this plot and the latter plots. The number of powerful nodes has been varied from 2 to 10. With $r_p = 0.05$, the network will be fully covered by the powerful nodes for $n_p > 10$. It is shown that an increase in n_p significantly improves network connectivity probability. The impact of λ on connectivity is rather interesting. When λ is small, the network connectivity probability initially drops as λ increases. That is because when the number of ordinary nodes is small, the probability that an ordinary node is connected to a powerful node via a multi-hop path is small and can be almost neglected. Therefore, an ordinary node has to be close to a powerful node in order to be connected. Thus, when the number of ordinary nodes is small, an increase in the number of ordinary nodes causes a drop in the probability that *all* ordinary nodes are connected to at least one nearby powerful node. As the number of ordinary nodes further increases, the probability that an ordinary node far away from a powerful node can establish a multi-hop path to the powerful node increases, which consequently causes an increase in the probability of having a type-II connected network. Note that this phenomenon, i.e., the increase of node density will first reduce the connectivity probability and then improve it, can also be verified by examining the first derivative of $\Pr(\mathcal{B})$ from (11.2.4) with regard to λ. In addition, the value of node density which minimizes the connectivity probability can be obtained numerically using the classical Newton's method.

Note that the properties observed in Fig. 11.6, i.e., under the unit disk connection model, are also observed when the log-normal connection model is considered. In the log-normal connection model, two nodes separated by a Euclidean distance x are directly connected with probability

$$g(x) = Q\left(\frac{10\alpha}{\sigma}\log_{10}\frac{x}{r}\right) \tag{11.2.25}$$

where $Q(y) = \frac{1}{\sqrt{2\pi}}\int_y^{\infty}\exp(-\frac{z^2}{2})dz$ is the tail probability of the standard normal distribution, α is the path loss exponent, σ^2 is the shadowing variance, r is the transmission range ignoring shadowing effect. In order to further accurately model the direct connection between nodes, we consider channel correlation in our simulation. That is, we follow the approach in [78, 89, 194] and use an exponential model to model the fading correlations between wireless links. In the model, the received signals at two nearby nodes from the same transmitting node are correlated with correlation coefficient

$$\rho(x) = \exp\left(-\frac{x}{d_{corr}}\log_e 2\right) \tag{11.2.26}$$

where x is the Euclidean distance between two receiving nodes, d_{corr} is the decorrelation distance whose typical value is 20 m for the urban environment and 5 m for the indoor environment [194]. This is the well-known Gudmundson model. It works well for two-dimensional networks but it may not be able to accurately model some situations in one-dimensional networks. Particularly, consider a big obstacle located between a powerful node and two ordinary nodes in proximity where the wireless signals between the powerful node and the ordinary nodes cannot propagate through. One ordinary node cannot receive the signals from the powerful node implies that the other ordinary node also cannot receive the signals either. That is, two nodes "hiding" behind a big obstacle in a one-dimensional network are highly correlated compared to the two-dimensional case. Gudmundson model is less suitable in modeling such situation. Nevertheless, Gudmundson model is still used here due to its popularity in the literature.

Figure 11.7 is plotted using simulation results obtained from 10,000 randomly generated network topologies, and following the correlated log-normal model. To have a fairer comparison between different shadowing effect assumptions, we adjust

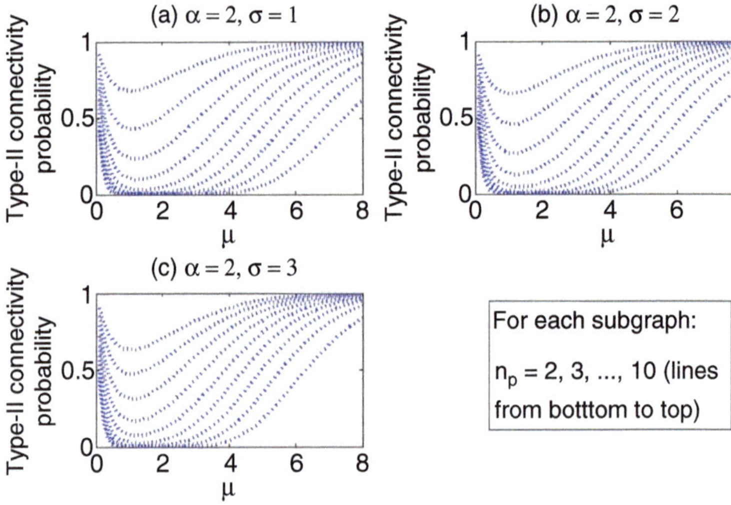

Fig. 11.7 The type-II connectivity probability given different values of the average node degree μ, n_p, $r_p = r_o = 0.05$, and considering the correlated log-normal model. The *dash lines* are plotted using simulation results obtained from 40000 randomly generated samples

the density λ of ordinary nodes in each simulation so that the average node degree μ of an arbitrary ordinary node is preserved under different path loss exponent α and shadowing variance σ^2 settings. Ignoring the boundary effect, we have $\mu = 2\lambda r_o \exp\left(\frac{1}{2}\left(\frac{\sigma}{10\alpha}\log_e 10\right)^2\right)$. The steps to derive the equation are omitted here. When $\sigma = 0$, the log-normal connection model reduces to the unit disk connection model and we have $\mu = 2\lambda r_o$. As a result, Fig. 11.6 can be directly compared with Fig. 11.7 as the former is also plotted with the average node degree ranging from 0 to 8. Figure 11.6 and 11.7 together show that the impact of the powerful nodes on the type-II connectivity probability under the correlated log-normal model has quantitatively little difference compared with the impact of the powerful nodes under the unit disk connection model. Furthermore, an increase in shadowing variance σ^2 will improve the connectivity probability even if the node density has been reduced to preserve the same average node degree. The better type-II connectivity probability observed under the log-normal model is consistent with the results in ad hoc networks without infrastructure support.

Next we investigate the impact of λ and n_p on the average number of clusters under the unit disk connection model. The analytical formula in (11.2.23) is verified by simulations obtained from 40,000 randomly generated network topologies.

Figure 11.8 shows virtually an exact match of (11.2.23) with the simulation results. In addition, the curves in the figure also agree with the curves in Fig. 11.6 and show that the connectivity probability reaches its minimum when the average number of clusters is maximized; conversely, the connectivity probability approaches one when the average number of clusters approaches one.

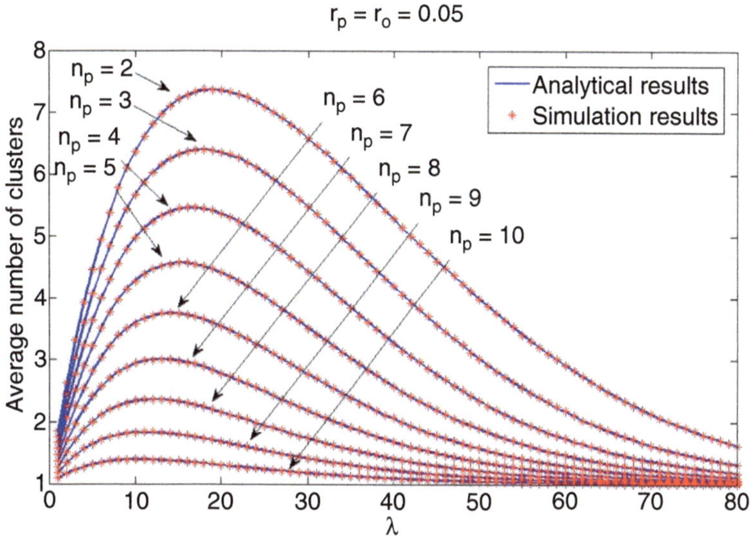

Fig. 11.8 The average number of clusters given different values of λ, n_p, and $r_p = r_o = 0.05$ under the unit disk model. The *solid lines* are plotted using (11.2.23), verified by simulation results obtained from 40, 000 randomly generated network topologies

11.3 Notes and Further Readings

In this chapter, we studied connectivity of one-dimensional small to medium sized networks with and without infrastructure support respectively. The results will be useful for many real world applications modelable by one-dimensional networks, e.g., a vehicular network built along a highway or a sensor network deployed along the border of a defined region for intrusion detection.

Connectivity of one-dimensional wireless ad hoc networks has been extensively studied [55, 67, 77, 147]. Among the studies, Miorandi and Altman [147] assumed that there is a pre-determined node located at the origin. They investigated the probability of other nodes, which are either arbitrarily or uniformly distributed along a semi-infinite line, being connected (either directly or via multi-hop paths) to the node at the origin. Both the unit disk connection model and a Boolean model with a random transmission range were considered. This scenario can be considered as a special case of type-II connectivity with only one powerful node placed at the origin.

Dousse *et al.* have conducted a study closely related to type-II connectivity in [61], considering one-dimensional networks under the unit disk connection model. The nodes are assumed to be Poissonly distributed on a line segment of length L with a known density. Two base stations are placed at both ends of the line segment. Based on the above model, they obtained analytically $p(x)$, the probability that a node at distance x from the left base station is connected to at least one base station.

Fig. 11.9 An illustration of the difficulty in analyzing connectivity of one-dimensional networks assuming a more general connection model than the unit disk connection model. Solid lines represent direct connections between nodes. A node may be connected to another node via a multi-hop path through a third node in the opposite direction of the other node

Based on $p(x)$, the authors concluded that the existence of base stations improves the probability that two arbitrary nodes are connected. The authors considered this line segment as a "reduced" version of a more general network with an infinite number of base stations placed every L units distance on an infinite line. Note that Dousse *et al.* analyzed the probability that *a node* at location x is connected to at least one base station, denoted by $p(x)$, whereas in this chapter we analyzed the probability that *all nodes* are connected to at least one powerful node (or base station). It is not trivial to derive the probability that a network is type-II connected using $p(x)$. The difficulty lies in the fact that the event that one node located at x is connected to a base station and the event that another node at y is connected to a base station are not independent, but correlated in a complicated way.

It is worth noting that despite numerous studies on network connectivity and advances in the field, analytical characterization of connectivity of one-dimensional networks assuming a more general connection model than the unit disk connection model (including the unit disk model where each node has a variable transmission range) remains an open problem. Figure 11.9 gives an illustration of the difficulty. Under a more general model, e.g., the log-normal connection model or the random connection model, a node may be connected to another node (via a multi-hop path) through a third node in the opposite direction of the other node. It becomes very difficult to enumerate and incorporate all such possibilities in the analysis of network connectivity.

Chapter 12
Connectivity of Two-Dimensional Small to Medium Sized Networks

Abstract In this chapter, we continue to investigate connectivity of two-dimensional small to medium sized networks by analyzing the three connectivity related measures: 1) the probability that a randomly selected node is k-hops apart from another randomly selected node, i.e., the length of the shortest path from the first node to the second node measured by the number of hops is k; 2) the probability that a node $\boldsymbol{x}$ apart from another node is connected to that node in exactly k hops; and 3) the spatial distribution of the nodes k-hops apart from another designated node, which are collectively called the probabilities of k-hop connection or hop count statistics. We start with networks assuming the unit disk connection model and then move to networks assuming the random connection model and the log-normal connection model.

In the previous chapter, we have studied connectivity of one-dimensional small to medium sized networks. In this chapter, we continue to investigate connectivity of two-dimensional small to medium sized networks by analyzing the three connectivity related measures: 1) $\Pr(k)$ the probability that a randomly selected node is k-hops apart from another randomly selected node, i.e., the length of the shortest path from the first node to the second node measured by the number of hops is k; 2) $\Pr(k|\boldsymbol{x})$ the probability that a node $\boldsymbol{x}$ apart from another node is connected to that node in exactly k hops; and 3) $\Pr(\boldsymbol{x}|k)$ the spatial distribution of the nodes k-hops apart from another designated node, which are collectively called *the probabilities of k-hop connection* or *hop count statistics*. We start with networks assuming the unit disk connection model and then move to networks assuming the random connection model and the log-normal connection model. Under the random connection model, a node located at $\boldsymbol{x}_2$ is directly connected to a node at $\boldsymbol{x}_1$ with probability $g(\boldsymbol{x}_2 - \boldsymbol{x}_1)$. Therefore, $\Pr(\boldsymbol{x}|k)$ and $\Pr(\boldsymbol{x}|k)$ are functions of the displacement $\boldsymbol{x}$ between nodes. When the connection model is a unit disk connection model or a log-normal connection model, the probabilistic measures $\Pr(k|\boldsymbol{x})$ and $\Pr(\boldsymbol{x}|k)$ depend on the Euclidean distance x between nodes only, which allows to write $\Pr(k|x)$ for $\Pr(k|\boldsymbol{x})$ and $\Pr(x|k)$ for $\Pr(\boldsymbol{x}|k)$, instead of their displacement.

G. Mao, *Connectivity of Communication Networks*,
DOI 10.1007/978-3-319-52989-9_12

12.1 Probabilities of k-Hop Connection Under the Unit Disk Connection Model

Consider a wireless network whose nodes are identically and independently distributed in a bounded region in $\Re^2$ according to a homogeneous Poisson point process with density ρ. Each node has an identical transmission range of r_0. Denote by $G = (V, E)$ the underlying graph of the network, where V is the vertex set and E is the edge set.

In this section we investigate the conditional probability $\Pr(k|x)$ that two randomly selected nodes separated by a known Euclidean distance x are k-hop neighbors for some positive integer k. Two nodes being k-hop neighbors means that the length of the shortest path between the two nodes, measured in the number of hops, is k. A recursive analytical equation embodying an approximation is given for computing this probability. As mentioned in Chap. 11, knowledge of $\Pr(k|x)$ can be used to compute two other related connectivity measures $\Pr(k)$ and $\Pr(x|k)$.

In the following analysis, we first derive the probability that any two nodes separated by a known distance x are two-hop neighbors. Then, we further develop a recursive analytical equation for the probability that any two nodes separated by a known distance x are k-hop neighbors for $k > 2$.

In what follows, the conditional probability $\Pr(k|x)$ is denoted by $\Phi_k(x)$ for convenience. Denote by $D(s, l)$ a disk of radius l centered at s.

Obviously, $\Phi_1(x) = 1$ when $x \leq r_0$ and $\Phi_1(x) = 0$ when $x > r_0$. For $k = 2$, it means that the source s and the destination d have no direct link between each other but can communicate through at least one intermediate node. Therefore, at least one node must lie in the intersectional area $D(s, r_0) \cap D(d, r_0)$, i.e., the shaded area **A** in Fig. 12.1.

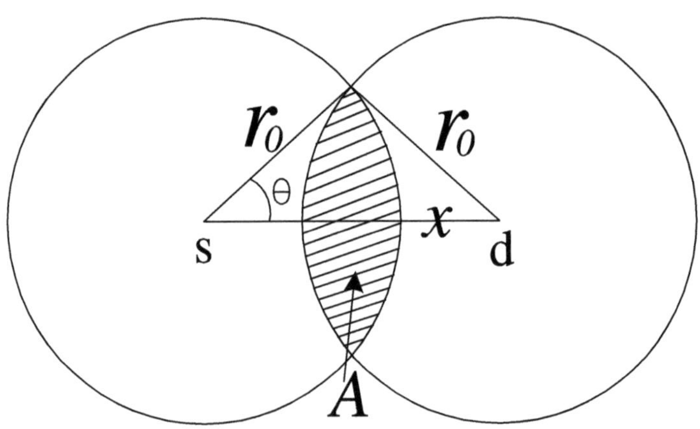

Fig. 12.1 An illustration of two-hop neighbors. Nodes s and d are two randomly selected nodes separated by a distance x ($r_0 < x \leq 2r_0$)

Ignoring the boundary effect, the probability $\Phi_2(x)$ can be found as the probability that there is at least one node located in **A**.

Since nodes are Poissonly distributed, the probability that there is no node located in **A** is $\exp(-\rho A)$, where A is the size of the area **A**, given by

$$A = 2r_0^2 \arcsin\left(\sqrt{1-\frac{x^2}{4r_0^2}}\right) - xr_0\sqrt{1-\frac{x^2}{4r_0^2}},\ r_0 < x \leq 2r_0 \tag{12.1.1}$$

Hence, the probability $\Phi_2(x)$ can be readily obtained:

$$\Phi_2(x) = 1 - \Pr(\{\texttt{no node in } \mathbf{A}\}) = 1 - e^{-\rho A} \tag{12.1.2}$$

When $x \leq r_0$, the two nodes can connect directly with each other, so $\Phi_2(x) = 0$; when $x > 2r_0$, we have $A = 0$, so that $\Phi_2(x) = 0$. Therefore,

$$\Phi_2(x) = \begin{cases} 1 - e^{-\rho A},\ r_0 < x \leq 2r_0 \\ 0, \quad \texttt{otherwise} \end{cases} \tag{12.1.3}$$

Now we shall further evaluate the probability $\Phi_k(x)$ for $k > 2$, continuing to ignore the boundary effect. Consider two random nodes s and d separated by distance x, as shown in Fig. 12.2. Node d is a k-hop neighbor of s if and only if node d is not an m-hop neighbor of s for any $m < k$ *and* there is at least one node within $D(d, r_0)$ which is a $k-1$ hop neighbor of s.

Let us first consider the probability that there is at least one node within $D(d, r_0)$ which is a $k-1$ hop neighbor of s. An approximation, termed the *independence assumption* has to be used in order to obtain an analytical solution, i.e., the event that an arbitrary node located within $D(d, r_0)$ is a $k-1$ hop neighbor of s is independent

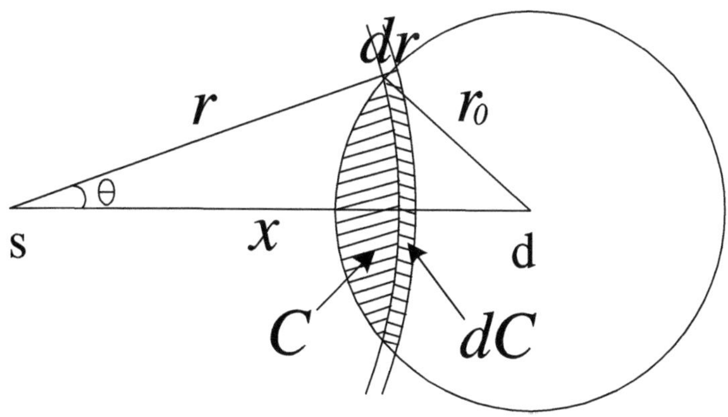

Fig. 12.2 An illustration of k-hop ($k > 1$) neighbors. Nodes s and d are two random nodes separated by a known distance x ($x > r_0$)

of the event that another arbitrary node located within $D(d, r_0)$ is a $k-1$ hop neighbor of s. The performance impact of the independence assumption will be dealt with separately in Sect. 12.3. Considering the area C in Fig. 12.2, i.e., $C = D(s, r) \cap D(d, r_0)$, a differential increment dr on r gives a differential area dC and the size of the differential area dC is $dC = 2r\theta dr$, where θ is given by

$$\theta = \arccos \frac{x^2 + r^2 - r_0^2}{2xr} \tag{12.1.4}$$

Since dr is a very small value, the probability that there exist more than one nodes within dC can be ignored and the probability that there exists a node in the differential area dC is given by $2\rho r\theta dr$. Given $\Phi_{k-1}(r)$, the probability that there is a node within dC which is also a $k-1$ hop neighbors of s is given by $2\Phi_{k-1}(r)\rho r\theta dr$.

Let $f(C)$ denote the probability that there is *no* node in C which is a $k-1$ hop neighbor of s. Then

$$f(C + dC) = f(C)(1 - 2\Phi_{k-1}(r)\rho r\theta dr) \tag{12.1.5}$$

Equation (12.1.5) readily leads to the conclusion that

$$df(C) = -2\Phi_{k-1}(r)\rho r\theta f(C)dr \tag{12.1.6}$$

Therefore, the probability that there is *no* node within $D(d, r_0)$ which is a $k-1$ hop neighbor of s at distance x is given by

$$g(x) = \exp\left(\int_{x-r_0}^{x+r_0} -2\Phi_{k-1}(r)\rho r\theta dr\right) \tag{12.1.7}$$

The probability that a node d at a distance x to s is not an m-hop neighbor of s for any $m < k$ is given by:

$$1 - \sum_{i=1}^{k-1} \Phi_i(x) \tag{12.1.8}$$

Therefore, the probability that node d is a k-hop neighbor of s is given recursively as:

$$\Phi_k(x) = \left(1 - \sum_{i=1}^{k-1} \Phi_i(x)\right)\left[1 - \exp\left(-\int_{x-r_0}^{x+r_0} 2\Phi_{k-1}(r)\rho r\theta dr\right)\right] \tag{12.1.9}$$

For $k = 2$, when $x \le r_0$ or $x > 2r_0$, it can be readily shown that $\Phi_2(x) = 0$; when $r_0 < x \le 2r_0$, it can be shown that

$$-\int_{x-r_0}^{x+r_0} 2\Phi_1(r)\rho r\theta dr - \rho\int_{x-r_0}^{r_0} 2r\theta dr = -\rho A \tag{12.1.10}$$

where A is given in (12.1.1), and $1 - \Phi_1(x) = 1$, the expression for $\Phi_2(x)$ agrees with that in (12.1.3).

Next, we offer some discussions on the key parameters that determine $\Phi_k(x)$. Let $\alpha \triangleq x/r_0$ and $\beta \triangleq \pi r_0^2 \rho$. Parameter α is the normalized distance and parameter β is the average vertex degree. In the next paragraph we show by *mathematical induction* that *under the independence assumption, $\Phi_k(x)$ is only determined by the normalized distance and the average vertex degree.*

For $k = 1$, it is immediate that $\Phi_1(x)$ is a function of α only. Suppose that $\Phi_n(x)$ can be expressed as a function of α and β for $n \leq k$, i.e., $\Phi_n(x) = \Upsilon_n(\alpha, \beta)$. Then, when $n = k + 1$, the first term on the right side of (12.1.9) is

$$1 - \sum_{i=1}^{k} \Phi_i(x) = 1 - \sum_{i=1}^{k} \Upsilon_i(\alpha, \beta) \tag{12.1.11}$$

Let $\mu = r/r_0$, the integral in the second term on the right side of 12.1.9 becomes

$$\begin{aligned} &\int_{x-r_0}^{x+r_0} 2\Upsilon_k(\frac{r}{r_0}, \beta)\rho r\theta dr \\ &= \int_{x/r_0-1}^{x/r_0+1} 2\Upsilon_k(\mu, \beta)\rho\mu r_0\theta r_0 d\mu \\ &= \frac{\beta}{\pi} \int_{\alpha-1}^{\alpha+1} 2\Upsilon_k(\mu, \beta) \arccos \frac{\alpha^2 + \mu^2 - 1}{2\alpha\mu} \mu d\mu \end{aligned} \tag{12.1.12}$$

From (12.1.9), (12.1.11), and (12.1.12), we have $\Phi_{k+1}(x) = \Upsilon_{k+1}(\alpha, \beta)$. Hence, the hypothesis is also valid when $n = k + 1$.

12.1.1 Simulation Studies

In this subsection, we use simulations to establish the accuracy of the theoretical analysis in the presence of the boundary effect and the shortcomings of the independence assumption. In the simulation, nodes are distributed on a square of size $a \times a$, where $a = 20$, according to a homogenous Poisson point process with node density ρ. We vary the average vertex degree, i.e., $\pi r_0^2 \rho$, while keeping the node density ρ fixed and each value of the average vertex degree represents a different scenario. Each scenario is repeated 100 times and the average result is shown. The analytical results are obtained through numerical integration using the adaptive quadrature algorithm [149, pp. 27–41], which calculates more points only in regions of rapid functional variation and less points where the integrand is varying slowly, and hence obtains accurate numerical results with moderate computational complexity.

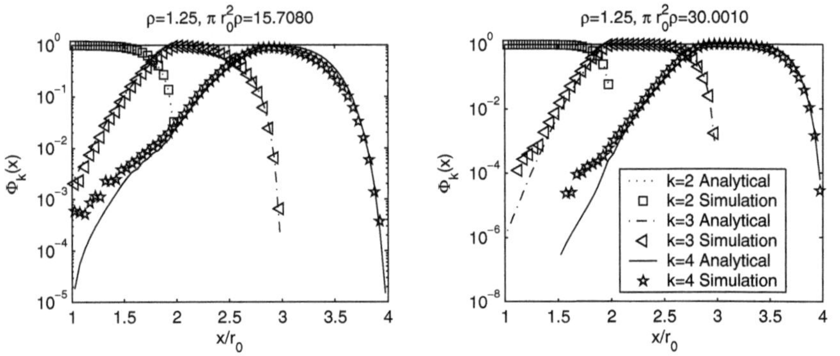

Fig. 12.3 Probability that two random nodes separated by a known distance x are k-hop neighbors for $k = 2, 3$, and 4

Figure 12.3 displays $\Phi_k(x)$ for $k = 2, 3$ and 4.

For $k = 2$, we can see that the simulation results and the analytical results agree very well, which indicates that (12.1.3) is an accurate expression of $\Phi_2(x)$. However, for $k = 3$ and 4, there are slight discrepancies between the analytical results and the corresponding simulation results, as shown in Fig. 12.3. The discrepancies are attributable to the *boundary effect* and the *independence assumption* used in the analysis. Note that the discrepancy for the small probabilities, e.g., $\Phi_k(x) \sim 10^{-5}$, is due to the accuracy of the numerical integration. Because the analytical result is given in a recursive form, it is also possible that the accuracy of the numerical integration decreases for large k. However, this was not found to be the major cause of the discrepancy in the simulation. The impact of the boundary effect is illustrated in Fig. 12.4. In Fig. 12.4, we keep $\pi r_0^2 \rho$ constant and vary the ratio r_0/a to calculate the mean absolute difference (MAD) between the analytical results and the simulation results. The mean absolute difference is the average value of the absolute differences, i.e., $MAD = \frac{1}{N} \sum_{i=1}^{N} |Ana_i - Sim_i|$, where Ana_i and Sim_i are the i-th analytical result and its corresponding simulation result respectively, and N is the number of results selected to calculate MAD. The larger MAD is, the greater the discrepancy is.

12.2 Probabilities of k-Hop Connection Under the Random Connection Model

In this section, we consider a wireless network with nodes identically and independently distributed following a homogeneous Poisson distribution with known density ρ in a given two-dimensional area, denoted by A. A node at $\boldsymbol{x}_2 \in A$ is directly connected to a node at $\boldsymbol{x}_1 \in A$ with probability $g(\boldsymbol{x}_2 - \boldsymbol{x}_1)$ independently of other pairs of nodes, i.e., following the random connection model. Under the above settings, we investigate the conditional probability $\Pr(k|\boldsymbol{x})$ that a node located at $\boldsymbol{x}$ apart from another node is connected to the other node in k hops for some positive integer k.

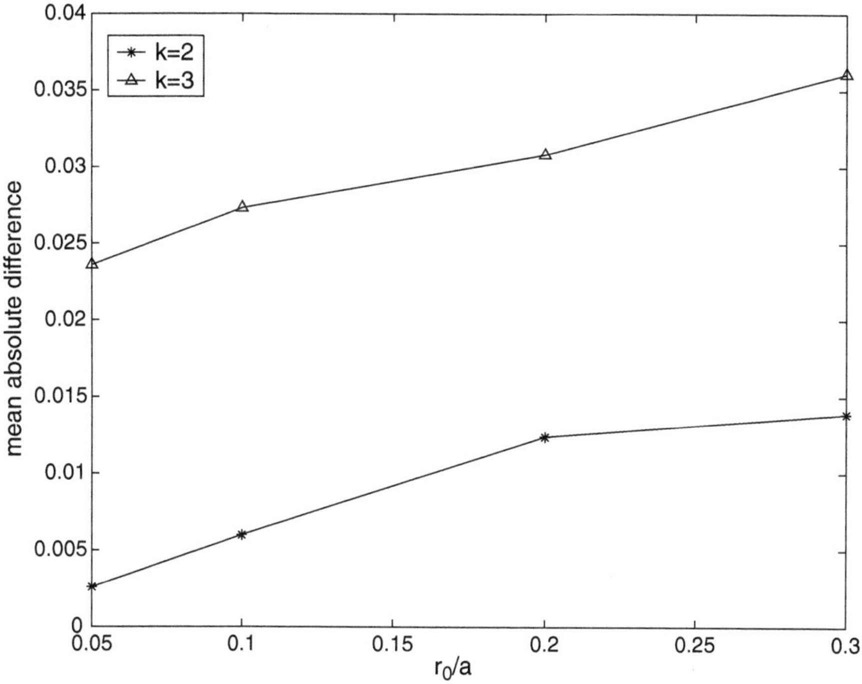

Fig. 12.4 The mean absolute difference (MAD) between the analytical results and the corresponding simulation results for different values of r_0/a. The average vertex degree $\pi r_0^2 \rho$ is kept constant at 15.7080. When r_0/a increases, a larger percentage of nodes are located near the boundary and the impact of the boundary effect is more pronounced. Thus the discrepancy between the analytical and the simulation results increases

Without loss of generality, consider that there is a node located at the origin and the origin is in A. The probability that a node located at $\boldsymbol{x}$ is directly connected to the node at the origin is given by $g(\boldsymbol{x})$. Due to the independence of connections between nodes, the set of nodes directly connected to the node at the origin, denoted by K_1, can be shown to have an inhomogeneous Poisson distribution with density $\rho g(\boldsymbol{x})$. Obviously $\Pr(k = 1|\boldsymbol{x}) = g(\boldsymbol{x})$. It follows that the set of nodes *not* directly connected to the node at the origin, denoted by $\overline{K_1}$, has an inhomogeneous Poisson distribution with density $\rho(1 - g(\boldsymbol{x}))$.

Due to the inhomogeneous Poisson distribution of K_1, the probability that there is a node in K_1 within a differential area $dA_{\boldsymbol{x}}$ centered at $\boldsymbol{x}$ is $\rho g(\boldsymbol{x})\, dA_{\boldsymbol{x}}$. Without loss of generality, we assume that the node in $dA_{\boldsymbol{x}}$ is located at $\boldsymbol{x}$. The probability that there are more than one nodes in the differential area $dA_{\boldsymbol{x}}$ can be ignored due to the Poisson distribution of nodes. Therefore, the probability that a node at $\boldsymbol{y} \notin dA_{\boldsymbol{x}}$ is *not* directly connected to *any* of the nodes within K_1 is given by

$$\lim_{dA_x \to 0} \prod_{dA_x \subset A, dA_x \cap \{y\} = \emptyset} \left[(1 - g(\boldsymbol{y} - \boldsymbol{x})) \rho g(\boldsymbol{x}) dA_x + (1 - \rho g(\boldsymbol{x}) dA_x) \right] \quad (12.2.1)$$

$$= \lim_{dA_x \to 0} e^{\sum_{dA_x \subset A, dA_x \cap \{y\} = \emptyset} \log(1 - g(y-x)\rho g(x) dA_x)} \quad (12.2.2)$$

$$= \lim_{dA_x \to 0} e^{\sum_{dA_x \subset A, dA_x \cap \{y\} = \emptyset} (-g(y-x)\rho g(x) dA_x)} \quad (12.2.3)$$

$$= e^{-\int_A \rho g(y-x) g(x) dx} \quad (12.2.4)$$

In (12.2.1), the term $(1 - g(\boldsymbol{y} - \boldsymbol{x})) \rho g(\boldsymbol{x}) dA_x$ represents the probability that there is a node in dA_x from K_1, (i.e., the $\rho g(\boldsymbol{x}) dA_x$ term) *and* the node at $\boldsymbol{y}$ is not directly connected to that node in dA_x at location $\boldsymbol{x}$ (i.e., the $1 - g(\boldsymbol{y} - \boldsymbol{x})$ term). The term $1 - \rho g(\boldsymbol{x}) dA_x$ represents the probability that there is *no* node in dA_x from K_1.

It follows from (12.2.4) that the probability that the node at $\boldsymbol{y}$ is directly connected to *at least one* of the nodes in K_1, denoted by $g_2(\boldsymbol{y})$, is given by

$$g_2(\boldsymbol{y}) = 1 - e^{-\int_A \rho g(y-x) g(x) dx} \quad (12.2.5)$$

A node at $\boldsymbol{y}$ is connected to the node at the origin in exactly two hops if and only if it is not directly connected to the node at the origin *and* it is directly connected to at least one node in K_1. Therefore,

$$\Pr(k = 2 | \boldsymbol{x}) = (1 - g(\boldsymbol{x})) g_2(\boldsymbol{x})$$

For consistency in notation we also use $g_1(\boldsymbol{x})$ for $g(\boldsymbol{x})$. Due to the spatial dependence problem mentioned in previous subsection, which will be explained in greater detail in the next section, the event that a node is directly connected to another node in k hops and the event that a third node is directly connected to the same node in k hops are *dependent* for $k \geq 2$. In this subsection, we ignore such dependence and consider the above two events to be *approximately* independent. Using the independence approximation, it then follows that the set of nodes that are directly connected to the node at the origin in exactly two hops, denoted by K_2, has an *approximate* inhomogeneous Poisson distribution with density $\rho(1 - g(\boldsymbol{x})) g_2(\boldsymbol{x})$. The set of nodes that are *not* connected to the node at the origin within two hops, denoted by $\overline{K_1 + K_2}$, has an approximate inhomogeneous Poisson distribution with density $\rho(1 - g_1(\boldsymbol{x}))(1 - g_2(\boldsymbol{x}))$.

Using the same steps leading to (12.2.5) and the fact that K_2 has an *approximate* inhomogeneous Poisson distribution with density $\rho(1 - g(\boldsymbol{x})) g_2(\boldsymbol{x})$, it can be shown that the probability that a node at $\boldsymbol{y}$ is directly connected to *at least one* of the nodes in K_2, denoted by $g_3(\boldsymbol{y})$, is given by

$$g_3(\boldsymbol{y}) = 1 - e^{-\int_A \rho g_1(y-x)(1-g_1(x)) g_2(x) dx} \quad (12.2.6)$$

It then follows that

$$\Pr(k = 3|\boldsymbol{x}) = (1 - g_1(\boldsymbol{x}))(1 - g_2(\boldsymbol{x}))\, g_3(\boldsymbol{x})$$

Using the independence approximation, it can be shown that the set of nodes within $\overline{K_1 + K_2}$ that are directly connected to at least one node in K_2, denoted by K_3, has an approximate inhomogeneous Poisson distribution with density $\rho(1 - g_1(\boldsymbol{x}))(1 - g_2(\boldsymbol{x}))\, g_3(\boldsymbol{x})$.

By recursion, for a positive integer $l > 1$, it can be shown that (adopting the independence assumption)

$$\Pr(k = l|\boldsymbol{x}) = g_l(\boldsymbol{x}) \prod_{i=1}^{l-1} (1 - g_i(\boldsymbol{x})) \tag{12.2.7}$$

where

$$g_l(\boldsymbol{y}) = 1 - e^{-\int_A \rho g_1(\boldsymbol{y}-\boldsymbol{x}) g_{l-1}(\boldsymbol{x}) \prod_{i=1}^{l-2}(1-g_i(\boldsymbol{x}))d\boldsymbol{x}}$$

Given knowledge of $\Pr(k|\boldsymbol{x})$ and $\Pr(\boldsymbol{x})$, i.e., the distribution of the displacement between two randomly selected nodes in A [144], $\Pr(\boldsymbol{x}|k)$ and $\Pr(k)$ can be readily obtained using Bayes' formula.

Note that the independence approximation is only needed for the computation of $\Pr(k = l|\boldsymbol{x})$ with $l > 2$, and is not required for the computation of $\Pr(k = 1|\boldsymbol{x})$ and $\Pr(k = 2|\boldsymbol{x})$. Therefore, the results on $\Pr(k = l|\boldsymbol{x})$ in (12.2.7) are accurate for $l = 1, 2$ and are an approximation only for $l > 2$.

12.2.1 *Simulation Studies*

In this subsection we use simulations to establish the accuracy of the analytical result on $\Pr(k|\boldsymbol{x})$. In the simulation, nodes are deployed on a 2000×2000 square area according to a homogeneous Poisson point process with density ρ. Two widely used connection models, i.e., the unit disk connection model and the log-normal connection model, are used as specific examples of the general connection g. Under the unit disk connection model, the transmission range r is 100, i.e., $g(\boldsymbol{x}) = 1$ for $\|\boldsymbol{x}\| \leq 100$ and $g(\boldsymbol{x}) = 0$ for $\|\boldsymbol{x}\| > 100$. Simulations are conducted for a number of node densities but only the results for a node density which gives an average node degrees 40 are shown. Results using other node densities showed similar trend. Recall that under the log-normal connection model, a node B is directly connected to another node C if the power received from C at B, whose propagation follows the log-normal model, is greater than a given threshold, P_T. It then follows that under the log-normal model

$$g(\boldsymbol{x}) = \Pr\left(P_r(\|\boldsymbol{x}\|) \geq P_T\right) = \int_{10\alpha \log_{10} \frac{\|x\|}{r_0}} \frac{1}{\sqrt{2\pi}\sigma} e^{-\frac{z^2}{2\sigma^2}} dz$$

where $r_0 = d_0 10^{\frac{P_t - PL_0(d_0) - P_T}{10\alpha}}$ and P_r is the received power in dB milliwatts, P_t is the transmitted power in dB milliwatts, $\|\boldsymbol{x}\|$ is the Euclidean distance between the two nodes, $PL_0(d_0)$ is the reference path loss in dB at a reference distance d_0, α is the path loss exponent, and σ is the standard deviation of the log-normal fading. The path loss exponent depends on the environment and can vary between 2 in free space to 6 in urban areas and the value of σ can be as large as 12. Several values of α and σ have been used in the simulation, but only the result for $\alpha = 4$ and $\sigma = 4$ is shown because other results have similar accuracy. Under the log-normal connection model, r_0 is chosen to be 100, while ρ is chosen to give the same average degree as that in the unit disk connection model. Every point shown in the plots is the average value from 2000 simulations. As the number of instances of random networks used in the simulation is large, the confidence interval is too small to be distinguishable and hence is ignored in the following plots. The analytical result in this subsection is compared with the analytical result in the previous subsection obtained under the unit disk connection model in Fig. 12.5, and the analytical result from [153] obtained under the log-normal connection model in Fig. 12.6, respectively. Note that under an omnidirectional model, such as the unit disk connection model and the log-normal connection model considered in the simulation, $\Pr(k|\boldsymbol{x}) = \Pr(k|\ \|\boldsymbol{x}\|)$.

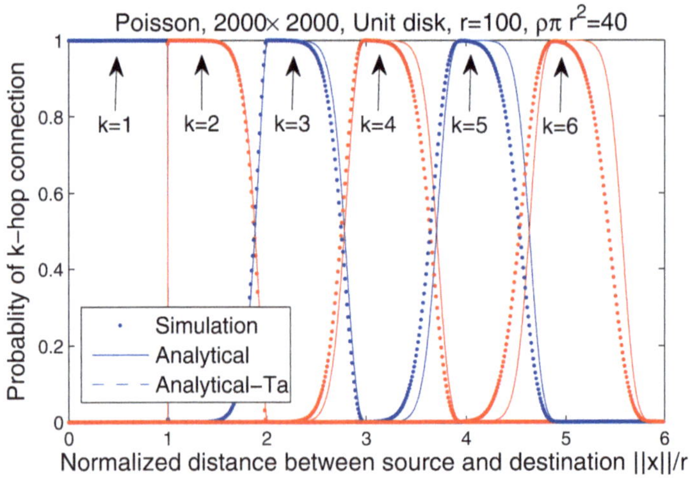

Fig. 12.5 Conditional probability $\Pr(k|\ \|\boldsymbol{x}\|)$ under the unit disk connection model for $k = 1$ to 6. Analytical-Ta is the result from Sect. 12.1. The result in this subsection is marginally more accurate than the result in Sect. 12.1 for $k > 2$ however the two analytical results are mostly indistinguishable

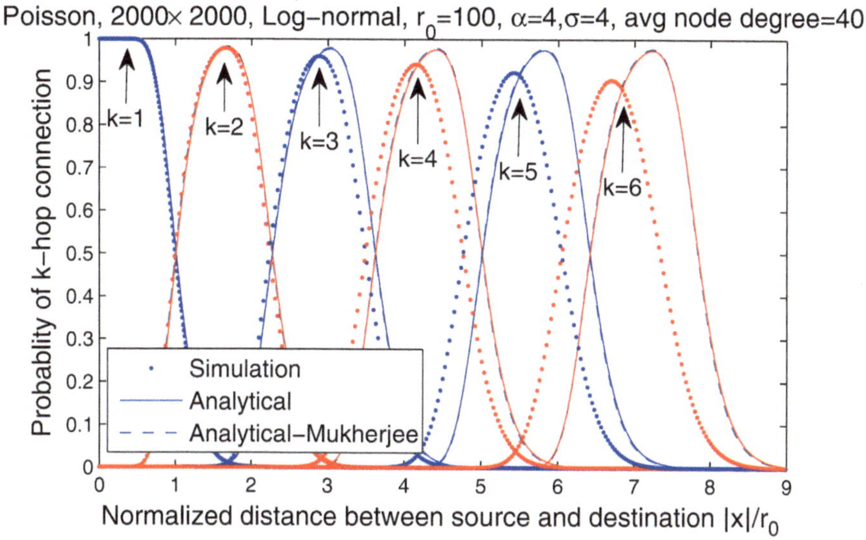

Fig. 12.6 Conditional probability $\Pr(k|\ \|\boldsymbol{x}\|)$ under the log-normal connection model for $k = 1$ to 6. Analytical-Mukherjee is the result from [153]. Our result is marginally more accurate than the result in [153] for $k > 2$ however the two analytical results are mostly indistinguishable

As shown in Figs. 12.5 and 12.6, the result in this subsection is marginally more accurate than the result in Sect. 12.1 obtained under the unit disk connection model and the result in [153] obtained under the log-normal connection model due to the inclusion of the boundary effect in the analysis. However, unsurprisingly, in most cases the analytical result in this subsection is almost indistinguishable from the previous results, reflecting the negligible impact of the boundary effect on the analysis of $\Pr(k|\boldsymbol{x})$. However analysis in this subsection is applicable for a wider class of wireless channel models whereas the results obtained in Sect. 12.1 assuming the unit disk connection model and in [153] assuming the log-normal connection model are only valid for the particular channel model being considered.

The discrepancy between the analytical result and the simulation result starts to appear for $k > 2$. This is attributable to the spatial dependence problem and the associated independence approximation mentioned earlier. The discrepancy between the analytical and the simulation results is larger for a larger value of k. This is caused both by the independence approximation and by the accumulation of errors in the recursive procedure for computing $\Pr(k|\boldsymbol{x})$. Such errors appear to be acceptable for some applications [62, 153, 203–205] but not necessarily for all.

12.3 Probabilities of k-Hop Connection in Wireless Networks Subject to Fading

In this section, we continue to investigate the probabilities of k-hop connections or hop count statistics in wireless networks assuming more realistic communication channels. A particular focus of this section is on quantifying the impact of the spatial dependence problem on the hop count statistics. As mentioned in the previous two sections, a major technical obstacle in the analysis of hop counts statistics is the so-called *spatial dependence problem*. The spatial dependence problem arises because in a wireless network the event that a randomly chosen node is k hops apart from a particular node is not independent of the event that another randomly chosen node is i hops apart from the same node for $i \leq k$. It follows that an accurate analysis on the conditional probability $\Pr(k|x)$ needs to consider all previous hops, which makes the analysis complicated. In this section, a significant improvement on the accuracy of computing $\Pr(k|x)$ is achieved by considering the positions of previous two hop nodes, compared to earlier results considering the positions of previous one hop nodes only. Furthermore, intuitively as more previous hops are taken into account, the resulting improvement on the accuracy of computing $\Pr(k|x)$ when an extra previous hop is considered will reduce. In this section, we show that considering the positions of previous two hops nodes is enough to provide an accurate estimate of $\Pr(k|x)$.

More specifically, in this section, we consider a wireless network where nodes are identically and independently distributed on a square according to a homogeneous Poisson point process with a known density ρ.

We consider that every node has the same transmission power. The simplest radio propagation model is the unit disk communication model. Under the unit disk model, the power attenuates with the Euclidean distance x from a transmitter like $x^{-\eta}$, where η is the path loss exponent. The path loss exponent can vary from 2 in free space to 6 in urban areas. The received signal strength (RSS) at a receiver separated by Euclidean distance x from the transmitter is $P_u(x) = CP_t x^{-\eta}$, where C is a constant, P_t is the transmission power. A transmission is successful if and only if the RSS exceeds a given threshold $P_{\min}$. Therefore, the required transmission power P_t allowing a transmission range r_0 is $P_t = C_1 r_0^{\eta}$, where $C_1 = P_{\min}/C$.

The unit disk connection model is simple but unrealistic. In reality the RSS may have significant variations around the mean value, because of both large scale variation (i.e., shadowing) and small-scale fading. Considering a typical type of shadowing, i.e., the log-normal shadowing, the RSS attenuation (in dB) follows a normal distribution with respect to the distance x between the transmitter and the receiver:

$$10 \log_{10} \left(\frac{P_l(x)}{CP_t x^{-\eta}} \right) \sim Z$$

where $P_l(x)$ is the RSS in the log-normal shadowing model and Z is a zero-mean Gaussian distributed random variable with standard deviation σ. When $\sigma = 0$ the model reduces to the unit disk connection model. In practice the value of σ is often computed from measured data and can be as large as 12 [172]. Denote by $q(z)$ the probability density function of the shadowing fades; then:

$$q(z) = \frac{1}{\sigma\sqrt{2\pi}} \exp\left(-\frac{z^2}{2\sigma^2}\right)$$

Furthermore, we assume that the shadowing fades Z between all pairs of transmitting node and receiving node are independently and identically distributed and the link is symmetric. In some environments, the assumption of independence of connections may not be accurate while in other environments, e.g., open space, it is a reasonable assumption. For example, it is generally accepted that if a pair of transmitters are separated by more than $\lambda/4$, where λ is the wavelength, their signals at a common receiver can be regarded as statistically independent. Furthermore, it was shown [171] that if a pair of receivers are separated by more than λ, their received signals from a common transmitter are only weakly correlated, with a correlation coefficient less than 0.15. At a typical frequency of 5GHz, $\lambda = 0.06$m. Thus the requirement on the separation of nodes can be easily fulfilled. We also note that although field measurements in real applications seem to indicate that connections between different pairs of geographically/frequency proximate wireless nodes are correlated [4, 186], the independence assumption is generally considered appropriate for far-field transmission and has been widely used in the literature under many channel models including the log-normal shadowing model [22, 153, 172].

Shadowing makes the RSS vary around its mean value over space, while the small-scale fading makes the RSS vary around its mean value over time. In this section, we consider a general model of small-scale fading, i.e., the *Nakagami-m fading* [180]. By choosing different values for the parameter m in the Nakagami-m fading model, the results include several widely used fading distributions as special cases, e.g., Rayleigh distribution (by setting $m = 1$) and one-sided Gaussian distribution (by setting $m = 1/2$). Subject to Nakagami-m fading, the RSS per symbol, ω, is distributed according to a Gamma distribution given by the following probability density function [180]:

$$\zeta(\omega) = \frac{m^m \omega^{m-1}}{\bar{\omega}\Gamma(m)} \exp\left(-\frac{m\omega}{\bar{\omega}}\right), \quad \omega \geq 0 \tag{12.3.1}$$

where $\Gamma(\cdot)$ is the standard Gamma function and $\bar{\omega} = P_l(x)$ is the mean RSS (over time), which is determined by path loss and shadowing.

We firstly conduct the analysis assuming the unit disk connection model, then we introduce the analysis in the more realistic *Log-normal-Nakagami* connection model, which takes into account statistical variations of RSS around the mean value due to both log-normal shadowing and Nakagami-m fading.

Furthermore, in a wireless network, how to optimally set the transmission power/range of nodes to minimize energy consumption in packet transmissions is an interesting problem often encountered in the design and management of wireless networks. If the transmission power is high, there is a high chance that a packet from a source can reach its destination but a significant amount of energy may be wasted. Alternatively, if the transmission power is low, there is a high chance that a packet cannot reach its destination due to the network being disconnected. The energy consumed in intermediate relay nodes to deliver the packet may also be wasted because the packet will be dropped eventually. In Chap. 4, it was established that significant energy savings can be achieved by requiring most nodes but not all nodes to be connected. Based on the above considerations, in this section, considering a sparse network in which there is not necessarily a path between all pairs of nodes, we study the energy efficiency of the network by deriving the effective energy consumption per successfully transmitted packet in end-to-end packet transmissions and on that basis, analyzing the optimum transmission power which minimizes the effective energy consumption. The established results also apply to a connected network.

In the following paragraphs, we further introduce the energy consumption model and the routing algorithm being considered in order to evaluate the energy efficiency.

12.3.1 Per Hop Energy Consumption

Assume that the time spent on transmitting a packet of unit size over single hop is a constant T_t, and all nodes transmit at the same power P_t, which results in a transmission range (without shadowing) of r_0 in the unit disk connection model. It follows that the energy consumed in transmitting a packet over single hop is:

$$Eng(r_0) = \frac{T_t P_t + Eng_c}{1 - \alpha(r_0)} = \frac{C_2 r_0^\eta + Eng_c}{1 - \alpha(r_0)} \tag{12.3.2}$$

where $C_2 = C_1 T_t$ is a constant, Eng_c is another constant which includes the processing power consumption and receiving power consumption in each node and $\alpha(r_0) = \frac{2WN_b}{(W+1)^2 + 2WN_b}$ is the packet error rate [36], W_{min} is the minimum contention window size, and N_b is the average node degree: $N_b = \rho\pi r_0^2$. Packet collision can increase the energy consumption, due to the consequent retransmission of a packet, especially when the transmission range is large. To evaluate this effect, we implement simulations in Sect. 12.3.6 using parameters shown in [36], i.e., $W_{min} = 64$. The values of C_2 and Eng_c are dependent on hardware specifications.

12.3.2 Routing Algorithm

In addition to its impact on the wireless channel between two nodes, fading also affects the performance of higher layer protocols, e.g., routing protocols. Here, we consider the issue by analyzing the performance of a wireless network using the greedy forwarding routing algorithm, as a typical example of distributed routing algorithms. The greedy forwarding routing belongs to the category of geographic routing algorithms and is a widely used routing algorithm for wireless networks. Using greedy forwarding, each node makes routing decisions independently of other nodes by using its own location information, the location information of its neighboring nodes and the location of the source and the destination. Greedy forwarding has shown great potential in wireless networks because of its distributed nature, low control overhead and capability of adapting to dynamic network topologies.

We consider a basic greedy forwarding algorithm that operates following two rules [204, 205]: 1) every node tries to forward the packet to the node within its transmission range which is closest to the destination; 2) a packet will be dropped if a node cannot find a next-hop neighbor that is closer to the destination than itself, and hence the transmission becomes unsuccessful. In the case of ties, viz., more than one nodes have the same Euclidean distance to the destination, an arbitrary one of those nodes can be chosen as the next hop node without affecting our analysis. This is because the way to settle ties does not affect the probability distribution of the remaining distance to the destination at each hop, which is the quantity used to derive our results as shown in Sect. 12.3.3.2. Note that a number of complicated recovery algorithms have been proposed to route a packet around the routing void [93]. For analytical tractability and generality of the results, we consider the basic greedy forwarding algorithm without any recovery algorithm, as in [54, 63, 204, 205].

Denote by $E_s[k_s|x_0]$ the expected number of hops for a packet to reach its destination, conditioned on the Euclidean distance between the source and the destination being x_0 *and* the transmission being successful, i.e., the packet can reach its destination. For convenience, throughout this section we use *conditioned on* x_0 for *conditioned on the Euclidean distance between the source and the destination being* x_0. Denote by $E_u[k_u|x_0]$ the expected number of hops traversed by a packet before it is dropped due to the nonexistence of a next hop node closer to the destination, conditioned on x_0 *and* the transmission being unsuccessful. In this case x_0 is the distance between the source and the *intended* destination.

It is worth noting that with the assumption that the network is connected, as used in, say, [205], the number of hops between two nodes increases as the transmission range (hence the average node degree) decreases. In the analysis of energy consumption, the assumption of a connected network results in a misleading conclusion that a smaller transmission range is always better. This conclusion is misleading because the probability of having a multi-hop path between two nodes reduces as the transmission range decreases and this important fact is not considered. Consequently, this conclusion is in sharp contrast with the result obtained in this section considering the possibility of disconnected networks where there exists an optimum transmission range that minimizes the energy consumption, as shown in Fig. 12.13. Our analysis does not rely on the assumption that the network is connected.

Consider a network with a total of N distinct source and destination pairs, where each source is separated from the associated destination by Euclidean distance x_0. Each source transmits a packet of unit size to its associated destination. Therefore, there are a total of N packets transmitted. Assume M ($M \leq N$) packets can reach their respective destinations successfully.

Define $Eng_{eff}(r_0|x_0)$ to be the effective energy consumption per successfully transmitted packet for any pair of nodes separated by Euclidean distance x_0, viz., $Eng_{eff}(r_0|x_0)$ is the total energy spent on transmitting all packets divided by the number of successfully received packets:

$$\begin{aligned} Eng_{eff}(r_0|x_0) & \\ &= \frac{MEng(r_0)E_s[k_s|x_0] + (N-M)Eng(r_0)E_u[k_u|x_0]}{M} \\ &= Eng(r_0)\frac{\phi_s(x_0)E_s[k_s|x_0] + (1-\phi_s(x_0))E_u[k_u|x_0]}{\phi_s(x_0)} \end{aligned} \tag{12.3.3}$$

where $\phi_s(x_0) := M/N$ is the probability of successful transmission between any pair of nodes separated by x_0.

In a network where the transmission range without shadowing and fading is r_0, given the distribution of the Euclidean distance between any pair of nodes $f(x_0)$, examples of which are given in [144], the average effective energy consumption is:

$$Eng_{eff}(r_0) = \int Eng_{eff}(r_0|x_0)f(x_0)dx_0 \tag{12.3.4}$$

The effective energy consumption is a measure of the energy spent on each successfully transmitted packet. A lower Eng_{eff} means a higher energy efficiency. We use Eng_{eff} as the metric to investigate the energy efficiency in end-to-end packet transmissions.

12.3.3 Analysis Assuming the Unit Disk Model

In this subsection, we analyze the hop count statistics, in particular the probability $\Pr(k|x)$, and the effective energy consumption assuming the unit disk connection model. We start with the calculations of the probability that two arbitrary nodes are k hops apart for $k \geq 3$ using greedy forwarding. The analysis for $k = 1, 2$ is straightforward.

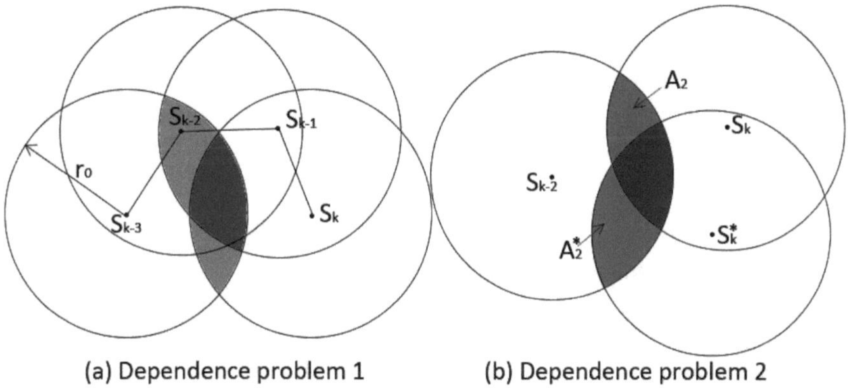

Fig. 12.7 An illustration of the spatial dependence problems in the hop count statistics using a unit disk connection model. S_k is the kth hop node, where r_0 is the transmission range

12.3.3.1 Spatial Dependence Problem

Before going into the analysis, we introduce in detail the spatial dependence problem in the analysis of hop count statistics using the unit disk connection model as an example. The same problem also exists in other models. Generally, there are two types of spatial dependence problems.

First, it can be shown that the event that a randomly chosen node is a k-th hop node (S_k) from a randomly chosen source node (S) is not independent of the event that another randomly chosen node is an i-th hop node for $1 \leq i < k$. Denote by $C(S_k, r)$ the disk centered at S_k with a radius r. As shown in the example in Fig. 12.7a, the fact that S_{k-1} is a $k-1$-th hop node from a source node S (not shown in the figure) implies that there is at least one node in the area $C(S_{k-3}, r_0) \cap C(S_{k-1}, r_0)$. On the other hand, S_k is a k-th hop node from S implies that there is no node in the area $C(S_{k-3}, r_0) \cap C(S_k, r_0)$; otherwise, S_k will become a $k-2$-th hop node. An overlap of the two areas implies that the event that S_k is a kth hop node and the event that S_{k-1} is a $k-1^{th}$ hop node are not independent.

Secondly, it can be shown that the event that a randomly chosen node is a k-th hop node from S is not independent of the event that another randomly chosen node is a k-th hop node from S. As shown in the example in Fig. 12.7b, S_k is a k-th hop node from S implies that there is at least one node (the $k-1$-th hop node) in the area $A_2 = C(S_{k-2}, r_0) \cap C(S_k, r_0)$. Another node S_k^* is a k-th hop node from the same S implies that there is at least one node (the $k-1$-th hop node) in the area $A_2^* = C(S_{k-2}, r_0) \cap C(S_k^*, r_0)$. An overlap of the two areas implies that the event that S_k is a k-th hop node and the event that S_k^* is a k-th hop node are not independent. In this subsection, a significant improvement on the accuracy of $\Pr(k|x)$ is obtained by reducing the inaccuracy associated with the first type of spatial dependence problem. The second type of spatial dependence problem can be handled by a similar technique. Specifically, when considering the area covered by the transmission range of a k-th hop node, we need to consider the overlap of the area covered by the transmission range of the k-th hop node and the area covered by the transmission range of other k-th hop nodes. However, it can be seen later that the result is fairly accurate after a proper handling of the first type of spatial dependence problem, so that specific handling of this second spatial dependence problem is effectively not warranted.

12.3.3.2 Distribution of the Remaining Distance

Denote by $A(x, r_1, r_2)$ the intersectional area of two disks with distance x between centers and radii r_1 and r_2, respectively. The size of the area is:

$$A(x,r_1,r_2) = \begin{cases} \min(\pi r_1^2, \pi r_2^2) & \text{for} \quad x \le |r_1 - r_2| \\ r_1^2 \arccos\left(\frac{x^2+r_1^2-r_2^2}{2xr_1}\right) + r_2^2 \arccos\left(\frac{x^2+r_2^2-r_1^2}{2xr_2}\right) & \\ \quad -\frac{1}{2}\sqrt{\left[(r_1+r_2)^2 - x^2\right]\left[x^2 - (r_1-r_2)^2\right]} & \text{for} \quad |r_1 - r_2| < x < r_1 + r_2 \\ 0 & \text{otherwise} \end{cases} \tag{12.3.5}$$

Define x_k to be the remaining Euclidean distance between the k-th hop node (S_k) and the destination (D). Define $A_1 = A(x_{k-1}, r_0, x_k)$ to be the intersectional area of the disks $C(S_{k-1}, r_0)$ and $C(D, x_k)$. Similarly, we have $A_2 = A(x_{k-2}, r_0, x_k)$. Next we record the form of $\frac{\partial A(x,r_1,r_2)}{\partial r_2}$, which will be used later. For $|r_1 - r_2| < x < r_1 + r_2$:

$$\frac{\partial A(x,r_1,r_2)}{\partial r_2} = \frac{-r_1^2}{\sqrt{1-S^2}}\left(\frac{\partial S}{\partial r_2}\right) + 2r_2 \arccos(T) + r_2^2\left(\frac{-1}{\sqrt{1-T^2}}\right)\left(\frac{\partial T}{\partial r_2}\right) - \frac{1}{4\sqrt{W}}\left(\frac{\partial W}{\partial r_2}\right) \tag{12.3.6}$$

where

$$S = \frac{x^2 + r_1^2 - r_2^2}{2xr_1}$$

$$T = \frac{x^2 + r_2^2 - r_1^2}{2xr_2}$$

$$W = ((r_1 + r_2)^2 - x^2)(x^2 - (r_1 - r_2)^2)$$

Define $f(x_k, k|x_0)$ to be the joint probability density function of the remaining Euclidean distance to the destination from S_k being x_k *and* the packet having been successfully forwarded k hops, conditioned on x_0. Due to the spatial dependence problem, $f(x_k, k|x_0)$ depends on the remaining distances of all previous hop nodes, i.e., $x_{k-1}, x_{k-2}, \ldots, x_0$. In this subsection, we consider no more than two previous hops and the justification is given in Sect. 12.3.4.

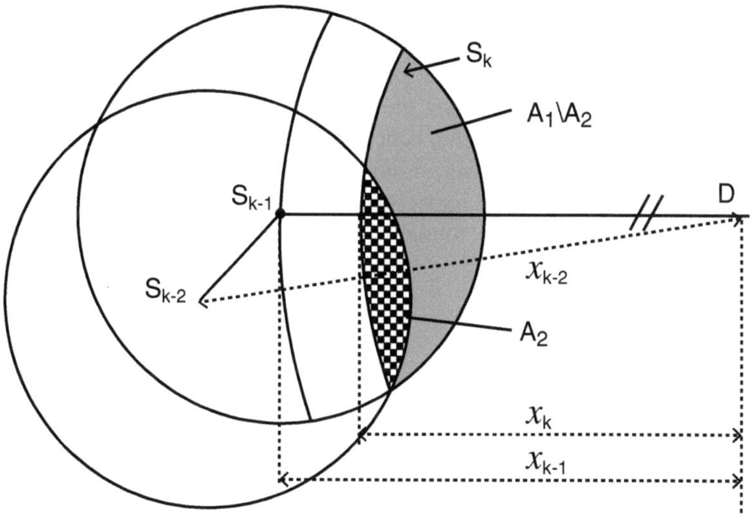

Fig. 12.8 Possible positions for the node at the k-th hop, denoted by S_k, are located on the arc, considering the positions of S_{k-1} and S_{k-2}, which are the nodes at the $k-1$-th hop and $k-2$-th hop from the source, respectively. Parameters A_1, A_2, x_k, x_{k-1} and x_{k-2} are described in the following text

Define $g(x_k, k|x_{k-1}, x_{k-2}, k-1)$ to be the joint probability density function of the remaining Euclidean distance to the destination at S_k being x_k *and* the packet having been successfully forwarded k hops, conditioned on $\mathcal{B}$, where $\mathcal{B}$ is the event that the remaining distances at S_{k-1} and S_{k-2} are x_{k-1} and x_{k-2}, respectively *and* the packet has been successfully forwarded $k-1$ hops. Note that a packet has been successfully forwarded $k-1$ hops necessarily means that it has been successfully forwarded i hops for $i \leq k-1$. Accordingly, define the cumulative distribution function of the remaining distance at the kth hop node to be $\Pr(X_k \leq x_k, k|x_{k-1}, x_{k-2}, k-1)$. Ignoring the boundary effect, whose impact will be discussed in detail later, the cumulative distribution function is equal to the probability that there is at least one node in area $A_1 \setminus A_2$ as indicated by the uniform-shaded area in Fig. 12.8. The area A_2 needs to be excluded because if there is a node in this area, that node will be closer to the destination than S_{k-1}, which violates the condition that S_{k-1} is the $k-1$-th hop node using greedy forwarding. We approximate the size of $A_1 \setminus A_2$ by $A_1 - A_2$. This approximation will greatly simplify the calculation while giving a sufficiently accurate result, as validated in Sect. 12.3.6. Using the approximation, it can be shown that

$$\begin{aligned}
&\Pr(X_k \leq x_k, k|x_{k-1}, x_{k-2}, k-1) \\
&\quad = 1 - \exp\left(-\rho(A(x_{k-1}, r_0, x_k) - A(x_{k-2}, r_0, x_k))\right) \qquad (12.3.7)
\end{aligned}$$

For any two nodes close to the border, the intersectional area of the transmission ranges of the two nodes may be partially located outside the network area, which causes an error in computing the size of the area $A_1 \backslash A_2$ in (12.3.7). This effect is due to the boundary effect. Ignoring the boundary effect may generally cause an overestimation on the size of $A_1 \backslash A_2$, hence an overestimation on the probability of finding the next hop node. However, simulation results in Sect. 12.3.6 show that the boundary effect has very limited impact on the accuracy of the analytical results.

Taking the derivative of the cumulative distribution function with respect to x_k, we have

$$\begin{aligned} g(x_k, k|x_{k-1}, x_{k-2}, k-1) &= \frac{\partial \Pr(X_k \leq x_k, k|x_{k-1}, x_{k-2}, k-1)}{\partial x_k} \\ &= \rho \left(\frac{\partial A(x_{k-1}, r_0, x_k)}{\partial x_k} - \frac{\partial A(x_{k-2}, r_0, x_k)}{\partial x_k} \right) \times \exp\left(-\rho(A(x_{k-1}, r_0, x_k) - A(x_{k-2}, r_0, x_k))\right) \end{aligned} \tag{12.3.8}$$

where the partial differentiations are given in (12.3.6).

Define $h(x_k, x_{k-1}, k|x_0)$ to be the joint probability density function of the remaining Euclidean distances at the k-th hop node and $k-1$-th hop node being x_k and x_{k-1}, respectively, *and* the packet having been successfully forwarded k hops, conditioned on x_0.

For $k = 1$, it is straightforward that:

$$f(x_1, k = 1|x_0) = \rho \frac{\partial A(x_0, r_0, x_1)}{\partial x_1} e^{-\rho A(x_0, r_0, x_1)}$$

For convenience, $f(x_1, k = i|x_0)$ is denoted by $f(x_1, i|x_0)$ hereafter. Based on the above result, for $k = 2$ we have:

$$h(x_2, x_1, 2|x_0) = g(x_2, 2|x_1, x_0, 1) f(x_1, 1|x_0)$$

For $k > 2$, $h(x_k, x_{k-1}, k|x_0)$ can be calculated recursively:

$$h(x_k, x_{k-1}, k|x_0) = \int_{r_0}^{x_0} g(x_k, k|x_{k-1}, x_{k-2}, k-1) h(x_{k-1}, x_{k-2}, k-1|x_0) dx_{k-2} \tag{12.3.9}$$

Finally for $k > 1$ we have:

$$f(x_k, k|x_0) = \int_{r_0}^{x_0} h(x_k, x_{k-1}, k|x_0) dx_{k-1} \tag{12.3.10}$$

12.3.3.3 Hop Count Statistics

Define $\Pr(k|x_0)$ to be the probability that the destination can be reached at the k-th hop conditioned on x_0. The destination can be reached at the k-th hop if the $k-1$-th hop node is within the transmission range of the destination. Therefore,

$$\Pr(k|x_0) = \int_0^{r_0} f(x_{k-1}, k-1|x_0)dx_{k-1} \tag{12.3.11}$$

12.3.3.4 Results for Successful Transmissions

Denote by $\Pr_s(k_s|x_0)$ the conditional probability that a packet can reach its destination at the k_s-th hop, conditioned on x_0, *and* the transmission is successful. It follows that $\Pr(k_s|x_0) = \Pr_s(k_s|x_0)\phi_s(x_0)$, and

$$\begin{aligned} \sum_{k_s=1}^{\infty} k_s \Pr(k_s|x_0) &= \phi_s(x_0) \sum_{k_s=1}^{\infty} k_s \Pr_s(k_s|x_0) \\ &= \phi_s(x_0) E_s[k_s|x_0] \end{aligned} \tag{12.3.12}$$

where $\phi_s(x_0)$ and $E_s[k_s|x_0]$ have been given earlier in the subsection. In reality, an upper bound on k_s can be found beyond which $\Pr(k_s|x_0)$ is 0. So (12.3.12) and other similar equations only need to be computed for a finite range of k_s.

An end-to-end packet transmission is successful if a packet can reach the destination at any number of hops. Therefore:

$$\phi_s(x_0) = \sum_{k_s-1}^{\infty} \Pr(k_s|x_0) \tag{12.3.13}$$

12.3.3.5 Results for Unsuccessful Transmissions

Define $\phi_u(k|x_0)$ to be the probability of a packet having been successfully forwarded k hops from the source toward the destination x_0 apart, but *not* reaching the destination in k hops, which distinguishes $\phi_u(k|x_0)$ from $\Pr(k|x_0)$. Therefore:

$$\phi_u(k|x_0) = \int_0^{x_0} f(x_k, k|x_0)dx_k \tag{12.3.14}$$

Based on the example introduced earlier in the subsection, we further assume that only M_k out of N packets reach the kth hop nodes. Then, $\phi_u(k|x_0) = M_k/N$. At the next hop, there are three possibilities for each of these M_k packets: 1) a packet reaches the destination at the next hop; 2) a packet makes another hop without reaching the destination; 3) the packet is dropped because a next-hop node cannot be found. Let W_{k+1} and M_{k+1} be the number of packets for which the first and second possibilities apply.

Define $\psi(k_u|x_0)$ to be the probability of the packets being dropped at the k_u-th hop. Then:

$$\begin{aligned}\psi(k_u|x_0) &= \frac{M_{k_u} - M_{k_u+1} - W_{k_u+1}}{N} \\ &= \phi_u(k_u|x_0) - \phi_u(k_u+1|x_0) - \Pr(k_u+1|x_0)\end{aligned} \tag{12.3.15}$$

The average number of hops for unsuccessful transmissions between a source and a destination separated x_0 apart is the expected value of k_u whose probability density function is given by $\psi(k_u|x_0)$. Similar to the way to derive (12.3.12), we have:

$$\sum_{k_u=1}^{\infty} k_u \psi(k_u|x_0) = (1 - \phi_s(x_0))E_u[k_u|x_0]$$

Given the above analysis, the effective energy consumption can be computed using (12.3.4), which is shown in Sect. 12.3.6. Furthermore, the above results can also be useful in the analysis of delay, throughput or reliability of end-to-end packet transmissions [54, 153], as well as localization [54, 62].

12.3.4 *Impact of Spatial Dependence Problem*

In this subsection, we considered that the remaining distance at the kth hop node (S_k) depends on the remaining distance at previous two hops nodes (S_{k-1} and S_{k-2}). Due to the spatial dependence problem, it can be shown that an accurate analysis of the hop count statistics requires all previous hops to be considered, but the calculation is more complicated than if an independence assumption is made. Previous research, e.g., [205], usually considered the dependence on only previous one hop. In this subsection we study the impact of the spatial dependence problem on the accuracy of the $\Pr(k|x)$.

Define A_m to be the intersectional area of the disk centered at S_{k-m} with radius r_0 and the disk centered at D with radius x_k. Therefore, the precise area that should be considered in the calculation of (12.3.7) is $A = A_1 \setminus (A_2 \cup A_3 \cup \ldots \cup A_k)$ instead of $A_1 \setminus A_2$.

Consider only previous one hop, then $A \approx A_1$. Consider only previous two hops, then $A \approx A_1 \setminus A_2 = A_1 \underline{-A_1 \cap A_2}$. Consider only previous three hops, then

$$\begin{aligned}A &\approx A_1 \setminus (A_2 \cup A_3) = A_1 - A_1 \cap (A_2 \cup A_3) \\ &= A_1 - A_1 \cap A_2 \underline{- A_1 \cap A_3 + A_1 \cap A_2 \cap A_3}\end{aligned} \tag{12.3.16}$$

The underlined terms are the additional terms introduced when considering one more previous hop. In considering the previous m hops instead of previous $m-1$ hops, the improvement is bounded by a term determined by $A_1 \cap A_m$. Furthermore, it is evident that $x_{k-1} < x_{k-2} < \ldots < x_0$. Therefore, $A_1 > A_2 > \ldots > A_k$ and the size of $A_1 \cap A_m$ is dominated by the size of A_m.

Define $h(x_k, x_{k-m}, k|x_0)$ to be the joint probability density function of the remaining Euclidean distances at the kth hop node and $k-m^{th}$ hop node being x_k and x_{k-m}, respectively, *and* the packet having been successfully forwarded k hops, conditioned on x_0. Then, the expected size of A_m at the k-th hop can be calculated by:

$$E[A_m, k|x_0] = \int_0^{x_0} \int_{x_k}^{x_k+r_0} A(x_{k-m}, r_0, x_k) h(x_k, x_{k-m}, k|x_0) dx_{k-m} dx_k \quad (12.3.17)$$

For $m = 1$, $h(x_k, x_{k-1}, k|x_0)$ can be calculated using (12.3.9). For $m = 2$, we have:

$$h(x_k, x_{k-2}, k|x_0) = \int_{r_0}^{x_0} g(x_k, k|x_{k-1}, x_{k-2}, k-1) h(x_{k-1}, x_{k-2}, k-1|x_0) dx_{k-1} \quad (12.3.18)$$

For $m \geq 3$, the calculation becomes more intricate but approximately $h(x_k, x_{k-m}, k|x_0) \approx f(x_k, k|x_0) f(x_{k-m}, k-m|x_0)$, where $f(x_k, k|x_0)$ is given by (12.3.10). This approximation is valid because the distance between S_k and S_{k-m} generally increases as m increases, hence the size of the overlapping area decreases. Therefore, the correlation between $f(x_k, k|x_0)$ and $f(x_{k-m}, k-m|x_0)$ reduces as m increases.

Based on the approach introduced above, Fig. 12.9 shows the results for the average size of A_m when the source node and the destination node are separated by distance $x_0 = 10r_0$. The simulation parameters are introduced in Sect. 12.3.6. It is evident that the size of A_m, $m \geq 3$, is negligibly small (less than 1% of the size of the area covered by the transmission range) compared to the size of A_1 and A_2. It validates the claim that the improvement made by taking previous m hops into consideration will become marginal for $m \geq 3$, which explains our choice of considering two previous hops only.

Our results suggest that the accuracy of the analysis on $\Pr(k|x)$ can be significantly improved by considering previous two hops, compared to considering previous one hop only. However, moving beyond two hops results in marginal improvement in accuracy of the analysis. Therefore, a conclusion can be drawn that the location of nodes three or more hops away provides little information for a node to determine its geometric relationship with other nodes. This conclusion provides analytical support for observations, to this point unsupported by analysis, in routing, localization, and network security that taking into account the (location or link status) information of two-hop neighbors can significantly improve the routing [133] (respectively, localization [137], network security [116]) performance compared with using one-hop neighborhood information only. However beyond two hops, taking into account more neighborhood information only has marginal impact. Therefore, many distributed routing, localization, and network security protocols use two-hop neighborhood information.

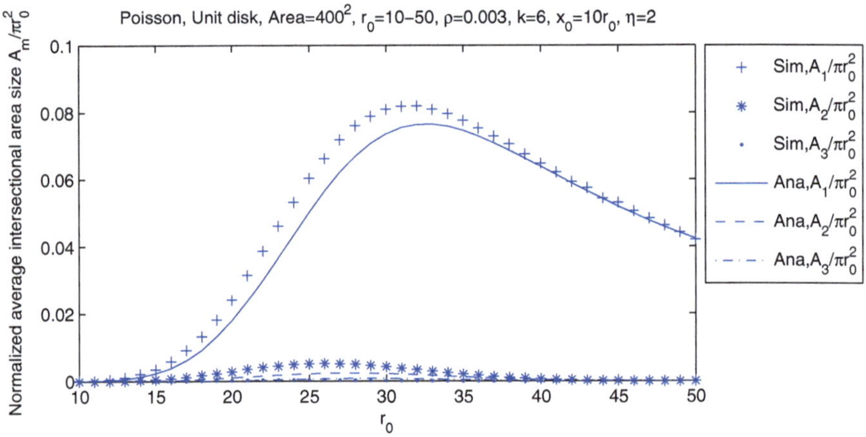

Fig. 12.9 Simulation (Sim) and analytical (Ana) results for the normalized average intersectional area size under the unit disk connection model. AreaA_m is the intersectional area of the disk centered at S_{k-m} with radius r_0 and the disk centered at D with radius x_k

12.3.5 Analysis Assuming the Log-Normal-Nakagami Model

In this subsection, we further investigate the hop count statistics assuming the log-normal-Nakagami model for communication channels. The technique to incorporate the impacts of both shadowing and small-scale fading is through the use of the random split property of a Poisson point process.

12.3.5.1 Random Split of a Poisson Point Process

First, we introduce a random variable named the *Nakagami fades* Ω_0, which follows the Gamma distribution with mean 1. The probability density function of Ω_0 is

$$\zeta_0(\omega_0) = \frac{m^m \omega_0^{m-1}}{\Gamma(m)} \exp(-m\omega_0), \quad \omega_0 \geq 0 \tag{12.3.19}$$

where m is introduced in (12.3.1).

It can be shown that the random variable $P_l(x)\Omega_0$ follows the Gamma distribution with mean $P_l(x)$, where $P_l(x) = CP_t x^{-\eta} 10^{Z/10}$ is the RSS given by the log-normal shadowing model. Then, in the Log-normal-Nakagami model, the RSS at a receiver at distance x from the transmitter is $P_N(x) = P_l(x)\Omega_0 = CP_t x^{-\eta} 10^{Z/10}\Omega_0$, where Z is a zero-mean Gaussian distributed random variable and Ω_0 is a Gamma distributed random variable with mean 1.

According to the random split property of a Poisson point process ([155], see also Sect. 1.3), the subset of nodes whose RSS from a particular transmitting node with shadowing fades $Z \in [z, z + dz]$ *and* Nakagami fades $\Omega_0 \in [\omega_0, \omega_0 + d\omega_0]$

are identically and independently distributed following a Poisson point process with density $\rho q(z)dz\zeta_0(\omega_0)d\omega_0$. Using the splitting of the Poisson point process, we can study the sub-process by the same technique used in the unit disk connection model.

Remark 175 The aforementioned technique can be readily extended to other communication models, e.g., the class of random connection models.

Define $\Pr_l(k|x_0)$ to be the probability that two arbitrary nodes separated by Euclidean distance x_0 are k hops apart using greedy forwarding in the Log-normal-Nakagami model. We start with $k = 1$.

12.3.5.2 Probability of Direct Connection

Under the Log-normal-Nakagami model, two nodes separated by distance x are directly connected if and only if the RSS exceeds a given threshold $P_{\min}$. Without shadowing and small-scale fading, the model reduces to the unit disk connection model where $P_{\min} = CP_t r_0^{-\eta}$. With shadowing and fading, we have:

$$\begin{aligned}\Pr(P_N(x) \geq P_{\min}) &= \Pr(CP_t x^{-\eta} 10^{Z/10}\Omega_0 \geq CP_t r_0^{-\eta}) \\ &= \Pr\left(Z \geq 10\eta \log_{10}\left(\frac{x}{r_0\Omega_0^{1/\eta}}\right)\right) \qquad (12.3.20) \\ &= \Pr\left(x \leq r_0\Omega_0^{\frac{1}{\eta}} \exp\left(\frac{Z\log 10}{10\eta}\right)\right) \qquad (12.3.21)\end{aligned}$$

Thus two nodes are directly connected if either of the following two conditions is satisfied: 1) given the distance x and Nakagami fades value ω_0, two nodes are directly connected if and only if the (random) shadowing fades $Z \geq 10\eta\log_{10}(\frac{x}{r_0\omega_0^{1/\eta}})$; 2) given that the shadowing fades z and Nakagami fades value ω_0, two nodes are directly connected if and only if their distance $x \leq r_0\omega_0^{1/\eta}\exp(\frac{z\log 10}{10\eta})$.

Based on the first condition, the probability of having a direct connection between two arbitrary nodes separated by x_0 is:

$$\begin{aligned}\Pr_l(k = 1|x_0) &= \int_0^\infty \int_{10\eta\log_{10}\left(\frac{x}{r_0\omega_0^{1/\eta}}\right)}^\infty q(z)dz\zeta_0(\omega_0)d\omega_0 \\ &= \int_0^\infty \frac{1}{2}\left(1 - erf\left(\frac{10\eta\log_{10}\left(\frac{x}{r_0\omega_0^{1/\eta}}\right)}{\sqrt{2\sigma^2}}\right)\right)\zeta_0(\omega_0)d\omega_0 \qquad (12.3.22)\end{aligned}$$

where $erf(.)$ is the error function.

Remark 176 Without small-scale fading, viz., considering the log-normal shadowing model only, the probability of having a direct connection between two arbitrary nodes separated by x_0 is $\left(\frac{1}{2}\left(1 - erf(\frac{10\eta \log_{10}(\frac{x}{r_0})}{\sqrt{2\sigma^2}})\right)\right)$. Similarly, the following analysis can be reduced to the analysis without small-scale fading by simply removing the integral with respect to ω_0.

In order to derive $\Pr_l(k|x_0)$ for $k > 1$, we use the second condition to study the probability of a direct connection. Define $r_N(z_S, \omega_S)$ to be the transmission range of a transmitter (S) conditioned on the shadowing fades and Nakagami fades being z_S and ω_S, respectively. Then:

$$r_N(z_S, \omega_S) = r_0 \omega_S^{1/\eta} \exp\left(\frac{z_S \log 10}{10\eta}\right) \tag{12.3.23}$$

Therefore, any node, whose RSS from the transmitter (S) has shadowing fades $Z_S \in [z_S, z_S + dz_S]$ and Nakagami fades $\Omega_S \in [\omega_S, \omega_S + d\omega_S]$, is directly connected to S if and only if its Euclidean distance to the transmitter is smaller than or equal to $r_N(z_S, \omega_S)$. This allows us to apply the analysis used in the unit disk connection model.

12.3.5.3 Distribution of the Remaining Distance

Define area size $A_1 = A(x_{k-1}, r_N(z_1, \omega_1), x_k)$ and $A_2 = A(x_{k-2}, r_N(z_2, \omega_2), x_k)$, where $A(x, r_1, r_2)$ and x_k are defined in Sect. 12.3.3. With a bit abuse of symbols, we use the same symbol to denote the size of an area and the area itself. Define $f_l(x_k, k|x_0)$, $g_l(x_k, k|x_{k-1}, x_{k-2}, k-1)$, the event $\mathcal{B}_l$, $\Pr_l(X_k \le x_k, k|x_{k-1}, x_{k-2}, k-1)$ and $h_l(x_k, x_{k-1}, k|x_0)$ analogously as in Sect. 12.3.3 and use the subscript l to mark the corresponding probabilities in the Log-normal-Nakagami model. We shall derive $\Pr_l(X_k \le x_k, k|x_{k-1}, x_{k-2}, k-1)$ by studying the following two events. Denote by $\mathcal{C}$ the event that there is at least one node whose Euclidean distance to the destination is smaller than x_k *and* has a direct connection to S_{k-1} *and* has no direct connection to S_{k-m} for $m \in [2, k]$ where S_0 is the source node. Denote by $\mathcal{D}$ the event that the node S_{k-1} is not directly connected to the destination. Events $\mathcal{C}$ and $\mathcal{D}$ are independent because of the independence of the shadowing and Nakagami fades. It is evident that:

$$\Pr_l(X_k \le x_k, k|x_{k-1}, x_{k-2}, k-1) = \Pr(\mathcal{C}|\, \mathcal{B}_l) \times \Pr(\mathcal{D}|\, \mathcal{B}_l) \tag{12.3.24}$$

We start with the analysis of event $\mathcal{C}$. In this paragraph we only consider the subset of nodes whose RSS from S_{k-1} has fades $Z_1 \in [z_1, z_1 + dz_1]$ and $\Omega_1 \in [\omega_1, \omega_1 + d\omega_1]$, and whose RSS from S_{k-2} has fades $Z_2 \in [z_2, z_2 + dz_2]$ and $\Omega_2 \in [\omega_2, \omega_2 + d\omega_2]$. Due to the independence of the fades and the property of Poisson point process, these nodes are distributed following a homogeneous Poisson point process with density $\rho q(z_1) q(z_2) dz_1 dz_2 \zeta_0(\omega_1) \zeta_0(\omega_2) d\omega_1 d\omega_2$. Denote by $\mathcal{E}$ the

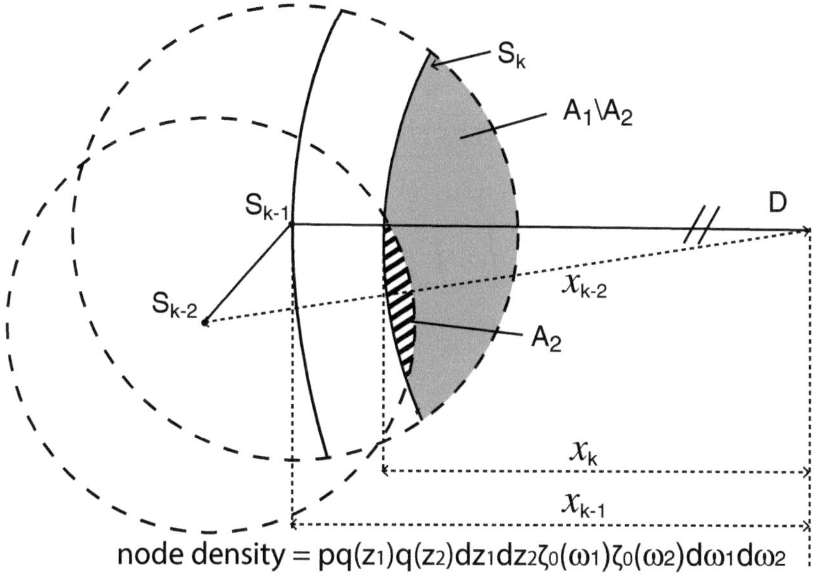

Fig. 12.10 Possible positions for the k-th hop node (S_k) are located on the arc. Consider the nodes whose RSS from S_{k-1} has fades $Z_1 \in [z_1, z_1 + dz_1]$ and $\Omega_1 \in [\omega_1, \omega_1 + d\omega_1]$, and whose RSS from S_{k-2} has fades $Z_2 \in [z_2, z_2 + dz_2]$ and $\Omega_2 \in [\omega_2, \omega_2 + d\omega_2]$. The *dashed-line circles* represent the transmission range of S_{k-1} (respectively, S_{k-2}) conditioned on the above values of shadowing and Nakagami fades. Parameters A_1 and A_2 are described in the following

event that $Z_1 \in [z_1, z_1 + dz_1]$ *and* $Z_2 \in [z_2, z_2 + dz_2]$ *and* $\Omega_1 \in [\omega_1, \omega_1 + d\omega_1]$ *and* $\Omega_2 \in [\omega_2, \omega_2 + d\omega_2]$. $\Pr(\mathcal{C}, \mathcal{E}|\mathcal{B}_l)$ is equal to the probability that there is at least one node in area $A_1 \setminus A_2$, as shown in Fig. 12.10. We approximate the size of area $A_1 \setminus A_2$ by $(A_1 - A_2)^+$, where $(A_1 - A_2)^+ = \max\{0, A_1 - A_2\}$ Note that through this approximation we ignored some rare events that cause $A_1 - A_2 < 0$, which can possibly occur when $r_N(z_2, \omega_2)$ is much larger than $r_N(z_1, \omega_1)$. In contrast, under the unit disk connection model, it is always the case that $A_1 - A_2 \geq 0$. Considering this subset of nodes only, $1 - \Pr(\mathcal{C}, \mathcal{E}|\,\mathcal{B}_l)$ is equal to $1 - \exp(-(A_1 - A_2)^+ \rho q(z_1)q(z_2)dz_1dz_2\zeta_0(\omega_1)\zeta_0(\omega_2)d\omega_1 d\omega_2)$, which is the probability that there is no node in area $A_1 \setminus A_2$. Note that A_1 depends on z_1 and ω_1; while A_2 depends on z_2 and ω_2.

Then, considering all subset of nodes, we have

$$\Pr(\mathcal{C}|\mathcal{B}_l) = 1 - \prod_{z_1, z_2 \in (-\infty, +\infty), \omega_1, \omega_2 \in (0, +\infty)} \exp(-(A_1 - A_2)^+ \rho \, q(z_1)q(z_2)dz_1dz_2\zeta_0(\omega_1)\zeta_0(\omega_2)d\omega_1 d\omega_2 \tag{12.3.25}$$

$$= 1 - \exp(-\int_0^{\infty}\int_0^{\infty}\int_{-\infty}^{\infty}\int_{-\infty}^{\infty}(A_1 - A_2)^+ \rho \, q(z_1)q(z_2)dz_1dz_2\zeta_0(\omega_1)\zeta_0(\omega_2)d\omega_1 d\omega_2) \tag{12.3.26}$$

Since the event $\mathcal{D}$ only depends on x_{k-1}, we have

$$\Pr(\mathcal{D}|\,\mathcal{B}_l) = 1 - \Pr_l(1|x_{k-1}) \tag{12.3.27}$$

Then, substitute (12.3.26) and (12.3.27) into (12.3.24),

$$\begin{aligned} &\Pr_l(X_k \le x_k, k|x_{k-1}, x_{k-2}, k-1) \\ &= (1 - \exp(-\int_0^\infty \int_0^\infty \int_{-\infty}^\infty \int_{-\infty}^\infty (A_1 - A_2)^+ \rho \\ &q(z_1)q(z_2)dz_1dz_2\zeta_0(\omega_1)\zeta_0(\omega_2)d\omega_1 d\omega_2)) \times (1 - \Pr_l(1|x_{k-1})) \end{aligned} \tag{12.3.28}$$

By Leibniz integral rule,

$$\begin{aligned} &g_l(x_k, k|x_{k-1}, x_{k-2}, k-1) \\ &= \frac{\partial \Pr_l(X_k \le x_k, k|x_{k-1}, x_{k-2}, k-1)}{\partial x_k} \\ &= \int_0^\infty \int_0^\infty \int_{-\infty}^\infty \int_{-\infty}^\infty \frac{\partial (A_1 - A_2)^+}{\partial x_k} \rho q(z_1)q(z_2)dz_1dz_2\zeta_0(\omega_1)\zeta_0(\omega_2)d\omega_1 d\omega_2 \\ &\times \exp(-\int_0^\infty \int_0^\infty \int_{-\infty}^\infty \int_{-\infty}^\infty (A_1 - A_2)^+ \rho q(z_1)q(z_2) \\ &\zeta_0(\omega_1)\zeta_0(\omega_2)dz_1dz_2d\omega_1 d\omega_2)(1 - \Pr_l(1|x_{k-1})) \end{aligned} \tag{12.3.29}$$

where $\partial A_1/\partial x_k$ and $\partial A_2/\partial x_k$ can be calculated by (12.3.6).

It is straightforward that for $k = 1$, we have

$$\begin{aligned} &f_l(x_1, 1|x_0) \\ &= \int_0^\infty \int_{-\infty}^\infty \frac{\partial A(x_0, r_N(z_1), x_1)}{\partial x_1} \rho q(z_1)dz_1\zeta_0(\omega_1)d\omega_1 \\ &\times \exp(-\int_0^\infty \int_{-\infty}^\infty A(x_0, r_N(z_1), x_1)\rho q(z_1)dz_1 \\ &\zeta_0(\omega_1)d\omega_1)(1 - \Pr_l(1|x_0)) \end{aligned} \tag{12.3.30}$$

For $k = 2$, the probability density function of the remaining distance of the previous hop is given by (12.3.30). Therefore:

$$h_l(x_2, x_1, 2|x_0) = g_l(x_2, 2|x_1, x_0, 1)f_l(x_1, 1|x_0) \tag{12.3.31}$$

For $k > 2$, the joint probability density function of x_k and x_{k-1} is calculated recursively:

$$h_l(x_k, x_{k-1}, k|x_0) = \int_0^{x_0} g_l(x_k, k|x_{k-1}, x_{k-2}, k-1)h_l(x_{k-1}, x_{k-2}, k-1|x_0)dx_{k-2} \tag{12.3.32}$$

Finally for $k \geq 2$, we have

$$f_l(x_k, k|x_0) = \int_0^{x_0} h_l(x_k, x_{k-1}, k|x_0)dx_{k-1} \tag{12.3.33}$$

12.3.5.4 Hop Count Statistics

Because of shadowing and small-scale fading, the destination can be possibly reached in single hop no matter how far the remaining distance from that hop is. Therefore, for $k \geq 2$,

$$\Pr_l(k|x_0) = \int_0^{x_0} \Pr_l(1|x_{k-1})f_l(x_{k-1}, k-1|x_0)dx_{k-1} \tag{12.3.34}$$

Remark 177 Based on the above results, we can calculate the average number of hops for successful and unsuccessful transmissions, the probability of successful transmissions and the effective energy consumption by the same technique used in the unit disk connection model.

12.3.6 Simulation Studies

In this subsection, we report on simulations to establish the accuracy of the analytical results. In the simulations, nodes are deployed on a 400×400 square following a homogeneous Poisson point process with density $\rho = 0.003$. The boundary effect is included in the simulation but it is shown to have a limited impact on the results. The route between two nodes is determined by the basic greedy forwarding algorithm. The transmission range r_0 is varied from 10 to 50, which results in the average node degree varying from around 1 to 24. Note that r_0 is the transmission range without shadowing and small-scale fading. The value of r_0 can be specified by the network designer via adjusting the transmission power and the receiver gain. The existence of a direct wireless link between an arbitrary pair of nodes will be further affected by shadowing and small-scale fading. Several values of the standard deviation in log-normal shadowing model have been used in our simulations, but only the results for $\sigma = 4$ are shown in this chapter because other results show a similar trend. Furthermore, we only include the results for $C_2 = 0.01$ and $Eng_c = 0.02$ (in (12.3.2)) as an example and the value of Eng_c is found to have very limited impact on the results. In order to distinguish the impact on the network performance of different parameters, the packet error rate is not included (i.e., setting $\alpha = 0$) except in Fig. 12.14 and the small-scale fading is not included except in Figs. 12.12 and 12.13b. Every point shown in the simulation result is the average value from 3000 simulations. As the number of instances of random networks used in the simulation is large, the confidence interval is too small to be distinguishable and hence is ignored in the following plots.

12.3.6.1 Hop Count Statistics

Figure 12.11 shows the probability that two arbitrary nodes separated by distance x_0 are k hops apart using greedy forwarding under the unit disk connection model and the log-normal connection model, respectively. As shown in Fig. 12.11, the analytical results have a good match with the simulation results, which verifies the accuracy of our analysis in both the unit disk connection model and the log-normal shadowing model.

In addition, we can see that the accuracy is significantly improved by considering previous two hops compared with that considering only previous one hop. Furthermore, it can be seen in Fig. 12.11 that the improvement of accuracy will be marginal if more than two previous hops are considered, which also confirms the analysis in Sect. 12.3.4. We expect this observation to be applicable to many other areas (e.g., routing, localization, network security) and the approach used for shedding the independence assumption can be seen in a broader context. Specifically, our approach for shedding the independence assumption is to show that one can improve the accuracy of $\Pr(k|x)$ by taking into account the location of previous m-hops nodes ($1 \leq m \leq k-1$). However the improvement becomes marginal as $m > 2$. This suggests that the location of nodes three or more hops away provides little information in determining the geometric relationship of a node with other nodes in the network. This observation further confirms our assertion in Sect. 12.3.4.

It is also interesting to see that packets can be transmitted to a longer distance under the log-normal shadowing model than that under the unit disk model, at the same number of hops. This is because log-normal shadowing introduces a Gaussian variation of the transmission range around the mean value, and with a higher chance a node can find a next-hop neighbor closer to the destination. This phenomenon has also been observed in other studies of connectivity.

Figure 12.12 shows the probability that two arbitrary nodes separated by distance x_0 are k hops apart in the Log-normal-Nakagami model when the Nakagami parameter $m = 1$. Therefore, the corresponding network is subject to log-normal shadowing and Rayleigh fading. The result shown in Fig. 12.12 verifies the accuracy of our analysis. Furthermore, it can be seen that Rayleigh fading reduces the probability that two nodes are connected by a path with k hops. This can be explained by the exponentially distributed RSS over the mean value caused by the Rayleigh fading which reduces the probability of direct connection. Therefore, Rayleigh fading has a negative impact on the network connectivity. A similar result is also observed in the next section.

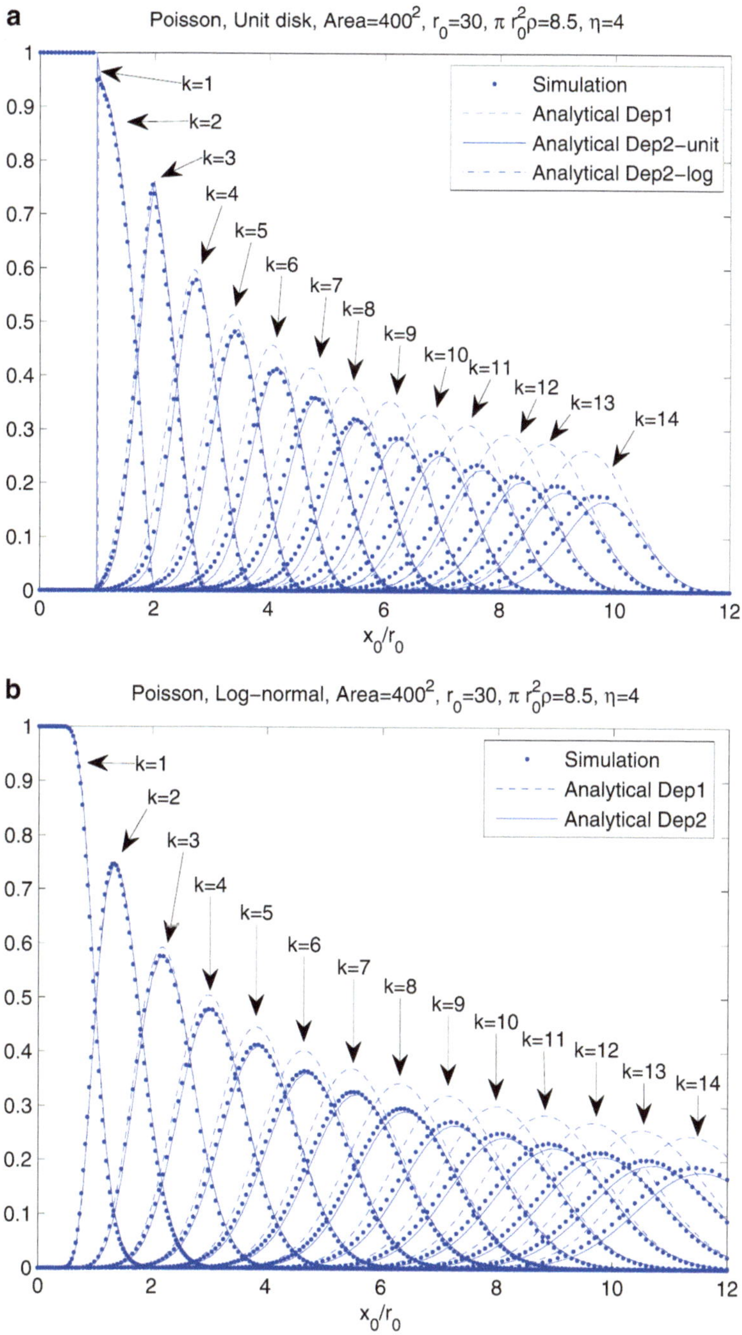

Fig. 12.11 The probability that two arbitrary nodes separated by Euclidean distance x_0 are k hops apart. *Dep*1 stands for the result calculated by considering the dependency on previous one hop. *Dep*2 is the result from this chapter . (**a**) In the unit disk model, Dep2-unit is the result under the unit disk model, while Dep2-log is the result under log-normal connection model by setting $\sigma = 0$. (**b**) Dep2-log is indistinguishable in the plot because the curve fully agrees with Dep2-unit. In the log-normal shadowing model

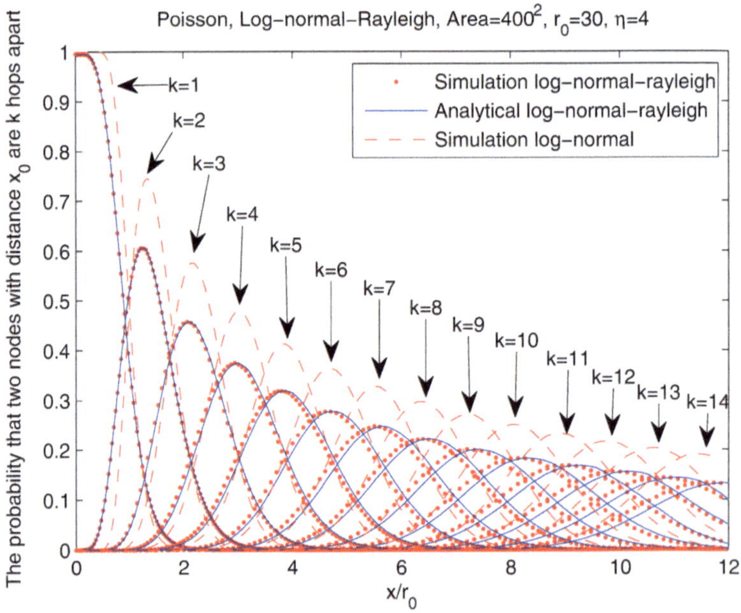

Fig. 12.12 The probability that two arbitrary nodes separated by distance x_0 are k hops apart in a network subject to log-normal shadowing and Rayleigh fading

12.3.6.2 Effective Energy Consumption

Figure 12.13 shows the probability of successful transmissions and Eng_{eff}. It can be seen that, unsurprisingly, the probability of successful transmissions increases from nearly 0 to nearly 1 as r_0 increased from 10 to 50. In contrast, the effective energy consumption could hardly have been predicted by heuristic reasoning, and needs more explanations.

Take the results under the unit disk connection model as example. When r_0 is small, the network is made up of a large number of small components. An increase in r_0 will cause an increase in the size (number of nodes) of the components and also a reduction in the number of components. Therefore, the average number of hops for unsuccessful transmission increases, and the energy wasted on unsuccessful transmissions also increases. Consequently, there is an initial increase in Eng_{eff} with the increase in r_0. As r_0 further increases, although the average number of hops for successful/unsuccessful transmission still increases, the energy wasted on unsuccessful transmissions starts to decrease as more source–destination pairs become connected. The balance of the two effects causes Eng_{eff} to peak at $r_0 \approx 19$. Above this transmission range, the decrease in wasted energy starts to dominate, which causes a subsequent decrease in Eng_{eff}. As r_0 increases further, the average number of hops approaches its maximum and the energy wasted on unsuccessful transmissions also reduces to a small amount. These cause Eng_{eff} to reach its

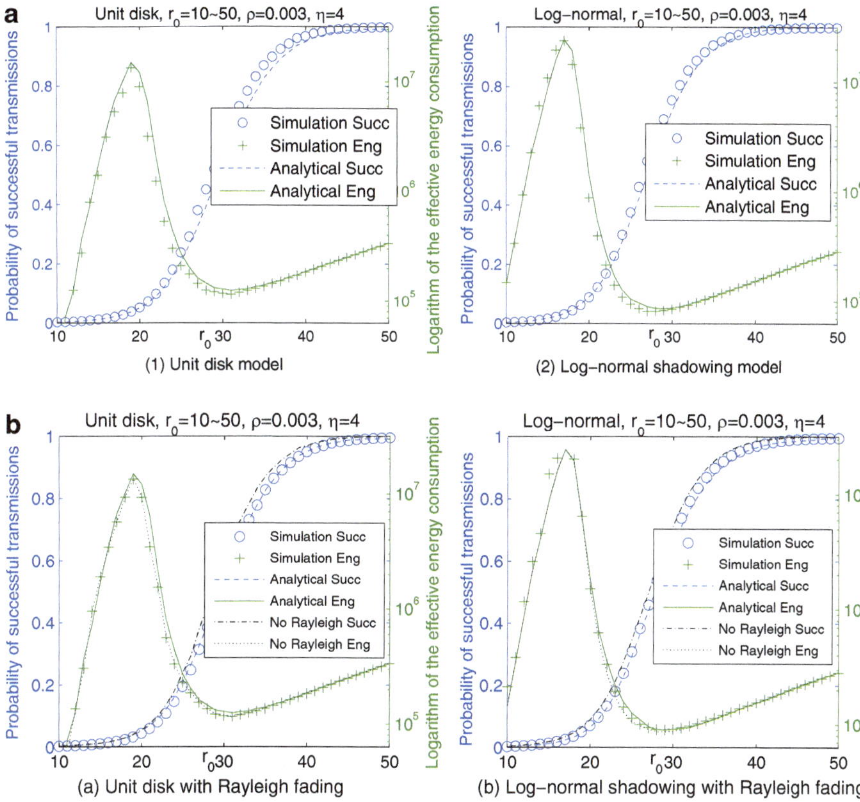

Fig. 12.13 Probability of successful transmissions (Succ) and effective energy consumption (Eng) in the network. Subfigure (**a**) shows the results without small-scale fading and (**b**) shows the results with Rayleigh fading. Furthermore, "No Rayleigh" is the result without small-scale fading shown in (**b**) for comparison

minimum at $r_0 \approx 31$. Above this transmission range, most source–destination pairs are connected as shown in Fig. 12.13(a.1). Another effect starts to dominate. That is, the increase in r_0 causes the increase in the per-hop energy consumption (like r_0^2) and the decrease in the number of hops (approximately like $1/r_0$). The net effect is an increase in Eng_{eff} with the increased r_0. Most previous studies have only considered this last stage of the relation between the energy consumption and the transmission range and therefore cannot give a complete understanding of the energy efficiency in end-to-end packet transmissions.

It is interesting to note that the energy optimizing transmission range is around 31, which corresponds to a network with most (around 70%) source–destination pairs connected but not all of them. Note that for $r_0 < 20$, Eng_{eff} may be smaller than the minimum Eng_{eff}. However at such value of r_0 most source–destination pairs are disconnected using greedy forwarding and no meaningful service can be provided by the wireless network. In order for more than 99% source–destination pairs to be connected, r_0 has to be larger than 47 and Eng_{eff} will increase to more than 225% of its minimum value in the unit disk connection model. A similar result can also be found in the log-normal shadowing model and the models with Rayleigh fading. Therefore, significant energy savings can be obtained by requiring most nodes, instead of all nodes, in the network to be connected. This observation also agrees with the results in Chap. 4. In addition, our result gives the amount of energy that can be saved. Purely from an energy-saving perspective and without consideration of other implications, this interesting result shows that the most energy-efficient topology control algorithms should be designed to let 70% (under this network setting) of the source–destination pairs be connected at the same time. The result sheds insight on the design of large wireless multi-hop networks where energy efficiency is a important issue.

Furthermore, Fig. 12.13b shows that the probability of successful transmissions is slightly lower in a network with Rayleigh fading compared to a network without Rayleigh fading. This confirms our assertion in the previous subsection.

Figure 12.14 shows the effective energy consumption with a nonzero packet error rate as shown in (12.3.2). The packet error rate increases from 0.004 when $r_0 = 10$ to 0.40 when $r_0 = 50$. It can be seen that as the transmission range increases, the tail of the effective energy consumption increases faster than its error-free counterpart. This is because an increase in the transmission range causes an increase in the number of neighbors and also an increase in the distance between the transmitter and the receiver. This in turn increases the packet error rate and the energy consumption.

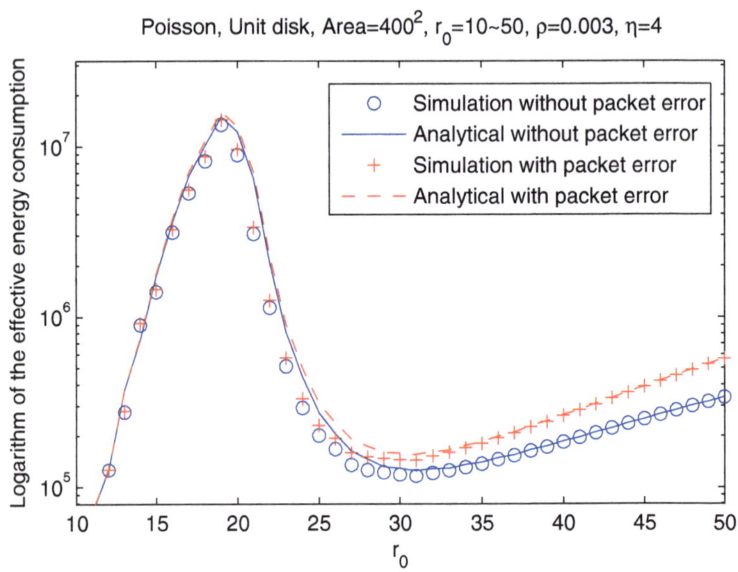

Fig. 12.14 The effective energy consumption subject to packet error

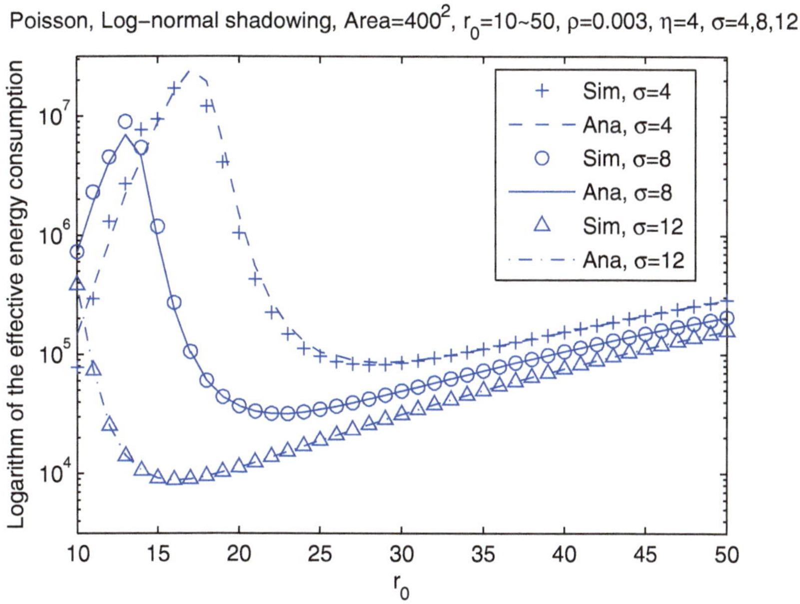

Fig. 12.15 The effective energy consumption under the log-normal shadowing model with various values of standard deviations

Therefore, when the packet error rate is nonzero, the energy optimizing transmission range becomes smaller as can be seen in Fig. 12.14.

Figure 12.15 shows the effective energy consumption under the log-normal shadowing model with various values of standard deviations. It can be seen that a larger variance in the log-normal shadowing model leads to a lower energy consumption and a smaller optimum transmission range. This is because a larger variance provides a larger probability for a node to forward the packet to a further node that is closer to the intended destination, which is similar to the observation obtained in Sect. 12.3.6.1.

12.3.6.3 Impact of Node Density and Path Loss Exponent on the Optimum Transmission Range

Figure 12.16 illustrates the impact of node density and path loss exponent on the optimum transmission range under the unit disk connection model. It can be seen that an increase in the node density will cause a decrease in the optimum transmission range. This is because an increase in the node density without reduction in the transmission range causes an increase in the average number of neighbors as well as the probability of successful transmissions between two nodes. Figure 12.16 also shows that a higher path loss exponent will result in a smaller optimum transmission range. This is because an increase in the path loss exponent will cause an increase in the per-hop energy consumption, as given by (12.3.2).

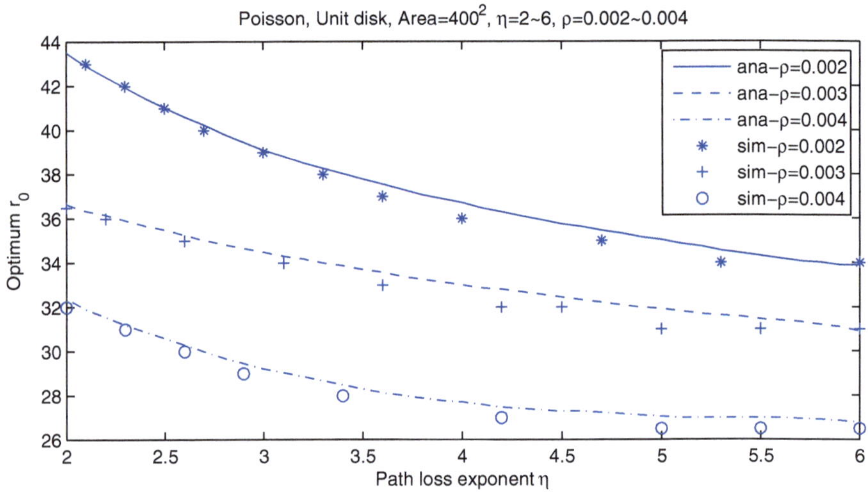

Fig. 12.16 Impact of node density and path loss exponent on the optimum transmission range

Therefore, under a higher value of the path loss exponent it is more energy efficient to have smaller components, hence a smaller optimum transmission range.

It has been shown that the probability $\Pr(k|x)$ and the energy consumption are affected by the node density and path loss exponent. Our analysis fully captures these effects and sheds insight on the design of a wireless multi-hop network.

12.4 Notes and Further Readings

In this chapter, we studied connectivity of two-dimensional small to medium sized wireless networks by investigating the associated hop count statistics. We started with networks assuming the unit disk connection model and then moved on to investigating networks assuming the general random connection model. Finally, we investigated wireless networks assuming more realistic Log-normal-Nakagami communication model with a particular focus on quantifying the impact of the spatial dependence problem on the hop count statistics. The spatial dependence problem arises because in a wireless network the event that a randomly chosen node is k hops apart from a particular node is not independent of the event that another randomly chosen node is i hops apart from the same node for $i \leq k$. It follows that an accurate analysis on the conditional probability $\Pr(k|x)$ needs to consider all previous hops. The spatial dependence problem is a major technical obstacle in the analysis of hop counts statistics. We showed that considering the positions of previous two hops nodes is sufficient to provide an accurate estimate of $\Pr(k|x)$.

The hop count statistics were first investigated by Chandler [38] in 1989. He analyzed the probability that two randomly chosen nodes separated by a known distance can communicate in k or less hops where nodes are uniformly distributed over a plane. However the aforementioned spatial dependence problem was incorrectly ignored in his analysis. Zorzi *et al.* [204, 205] proposed a greedy forwarding algorithm for a network where nodes are Poissonly distributed in the coverage area of a transmitting node. They studied an upper and a lower bound on the average number of hops between two nodes separated by a known Euclidean distance. In [45], Contla and Stojmenovic considered position based routing schemes for a wireless multi-hop network where nodes are uniformly distributed on a square. They studied the average number of hops between an arbitrary pair of source–destination nodes. Dulman *et al.* [63] investigated the probability $\Pr(k|x)$ by estimating the expected progress per hop using greedy forwarding. They considered the impact of the Euclidean distance between neighboring nodes in the previous hop on the progress in the current hop. Both [205] and [63] were established on the assumption that a packet can always reach the destination using greedy forwarding, which may not be true in some networks.

The aforementioned results are all based on the unit disk connection model. Considering the log-normal connection model, Hekmat and Mieghem [100] showed, through simulations, that the probability of a network being connected increases with increasing value of the shadowing parameter, which is the ratio between the standard deviation of shadowing and the path loss exponent. Mukherjee and Avidor [153] considered the impact of the log-normal shadowing on the probability $\Pr(k|x)$ in a wireless network where nodes are Poissonly distributed in a disk.

Chapter 13
A New Measure of Wireless Network Connectivity

Abstract Despite intensive research in the area of network connectivity, there is an important category of problems that remain unsolved: how to characterize and measure the quality of connectivity of a wireless network which has a realistic number of nodes, not necessarily large enough to warrant the use of asymptotic analysis, and which has unreliable connections, reflecting the inherent unreliability of wireless communications? The quality of connectivity measures how easily and reliably a packet sent by a node can reach another node. It complements the use of capacity to measure the quality of a network in saturated traffic scenarios and provides an intuitive measure of the quality of (end-to-end) network connections. In this chapter, we introduce a probabilistic connectivity matrix as a tool to measure the quality of network connectivity. Some interesting properties of the probabilistic connectivity matrix and their connections to the quality of connectivity are demonstrated. We demonstrate that the largest magnitude eigenvalue of the probabilistic connectivity matrix, which is positive, can serve as a good measure of the quality of network connectivity. Furthermore, we provide a flooding algorithm whereby the nodes repeatedly flood the network with packets, and by measuring just the number of packets a given node receives, the node is able to asymptotically estimate this largest eigenvalue.

In the previous chapters, we have studied connectivity of large static networks, connectivity of large and highly dynamic networks, and connectivity of small to medium sized networks. Despite intensive research on network connectivity, there is an important category of problems that remain unsolved: how to characterize and measure the *quality of connectivity* of a wireless network, which has a realistic number of nodes, *not necessarily* large enough to warrant the use of asymptotic analysis, and which has unreliable connections, reflecting the inherent unreliability of wireless communications? The quality of connectivity measures how easily and reliably a packet sent by a node can reach another node. It complements the use of *capacity* to measure the quality of a network in saturated traffic scenarios and provides an intuitive measure of the quality of (end-to-end) network connections.

In this chapter, we introduce a *probabilistic connectivity matrix* as a tool to measure the quality of network connectivity. Some interesting properties of the probabilistic connectivity matrix, and their connections to the quality of connectivity

G. Mao, *Connectivity of Communication Networks*,
DOI 10.1007/978-3-319-52989-9_13

will be demonstrated. We demonstrate that the largest magnitude eigenvalue of the probabilistic connectivity matrix, which is positive, can serve as a good measure of the quality of network connectivity.

The precise computation of the elements of this connectivity matrix, given the individual link quality measures and the network topology, involves significant computation. As an alternative, we provide a *flooding algorithm* that computes the largest magnitude eigenvalue in a decentralized fashion using experimental data, with multiple experiments to allow some averaging. Using the algorithm, the topology and link probabilities do not need to be known in order to determine the largest magnitude eigenvalue. This new measure constitutes the first that can be determined for moderate to small size networks.

13.1 Motivation for a New Connectivity Measure

As manifested in the previous chapters of the book, there are two general approaches to studying the connectivity problem. The first, spearheaded by the seminal work of Penrose [165] and Gupta and Kumar [92], is based on an asymptotic analysis of large random networks, which considers a network of n nodes that are identically and independently distributed on an area with an underlying uniform distribution. A pair of nodes are directly connected if and only if their Euclidean distance is smaller than or equal to a given threshold $r(n)$, independent of other connections. Some interesting results are obtained on the value of $r(n)$ required for the above network to be *asymptotically almost surely* connected as $n \to \infty$. In [185], these results are extended to provide the radius for k-connectivity. In Chaps. 2 and 3, we extended the above results by Penrose and Gupta and Kumar from the unit disk connection model to a random connection model, in which any pair of nodes separated by a displacement $\boldsymbol{x}$ are directly connected with probability $g(\boldsymbol{x})$, independent of other connections. The analytical techniques used in this approach have some intrinsic connections to continuum percolation theory [143], which is usually based on a network setting with nodes Poissonly distributed in an infinite area and studies the conditions required for the network to have a connected component containing an infinite number of nodes (in other words, the network *percolates*).

The second approach is based on a deterministic setting and studies connectivity and other topological properties of a network using algebraic graph theory. Specifically, consider a network with a set of n nodes. Its properties can be studied using its underlying graph $G(V,E)$, where $V \triangleq \{v_1,\ldots,v_n\}$ denotes the vertex set and E denotes the edge set. The underlying graph is obtained by representing each node in the network uniquely using a vertex and the converse. An undirected edge exists between two vertices if and only if there is a direct connection (or link) between the associated nodes. Note that in this chapter, we limit our discussions to a *simple graph* (network) where there is at most one edge (link) between a pair of vertices (nodes) and an undirected graph. Define an *adjacency matrix* A_G of the graph $G(V,E)$ to be a symmetric $n \times n$ matrix whose $(i,j)^{th}$, $i \neq j$, entry is equal to one if there is an edge between v_i and v_j and is equal to zero otherwise. Furthermore, the diagonal entries of A_G are all equal to zero. The *eigenvalues of the graph* $G(V,E)$ are defined to be the eigenvalues of A_G. The network connectivity information, e.g., connectivity and k-connectivity, is entirely contained in its adjacency matrix. Many interesting connectivity and topological properties of the network can be obtained by investigating the eigenvalues of its underlying graph. For example, let $\mu_1 \geq \ldots \geq \mu_n$ be the eigenvalues of a graph G. If $\mu_1 = \mu_2$, then G is disconnected. If $\mu_1 = -\mu_n$ and G is not empty, then at least one connected component of G is nonempty and bipartite [101, p. 28–6]. If the number of distinct eigenvalues of G is r, then G has a *diameter* of at most $r-1$ [25]. Some researchers have also studied properties of the underlying graph using its Laplacian matrix [148], where the Laplacian matrix of a graph G is defined as $L_G \triangleq D - A_G$ and D is a diagonal matrix with degrees of vertices in G on the diagonal. Particularly, the *algebraic connectivity* of a graph G is the second-smallest eigenvalue of L_G and it is greater than 0 if and only if G is a connected graph. Furthermore, the algebraic connectivity is also known to be a good indicator of the convergence rate of consensus algorithms [159]. Chapter 1 provides a brief introduction of the algebraic graph theory tools. We refer readers to [25] and [80] for a comprehensive treatment of the topic. Reference [101] provides a concise summary of major results in the area. The adjacency matrix, the Laplacian matrix and their associated parameters mainly focus on describing the connectivity between vertices with direct connections. As demonstrated later in this section, it is not trivial to use these tools to quantify the quality of end-to-end connections (especially when the existence of a direct connection between two nodes becomes probabilistic), which is of paramount concern in many communication applications. In this chapter, we develop the probabilistic connectivity matrix, a concept defined later in the chapter, to fill this theoretical gap.

Despite the aforementioned research, an important category of problems remain unsolved: how to measure the *quality of connectivity* of a wireless network which has a realistic number of nodes, *not necessarily* large enough to warrant the use of asymptotic analysis, and has unreliable connections, reflecting the inherent unreliable characteristics of wireless communications? The quality of connectivity measures how easily and reliably packets sent by a node can reach another. It complements the use of *capacity* to measure the quality of a network in saturated traffic scenarios and provides an intuitive measure of the quality of (end-to-end) network connections. The following paragraphs elaborate on the above question using two examples of networks with a fixed number of nodes and known transmission power.

Example 178 Assume that the wireless propagation model of a network is known and its characteristics have been quantified through *a priori* measurements or empirical estimation. Furthermore, a link exists between two nodes if and only if the received signal strength from one node at the other node, whose propagation follows the wireless propagation model and the signal strength is random, e.g., due to fading and shadowing, is greater than or equal to a predetermined threshold *and* the same is also true in the opposite direction. One can then find the probability that a link exists between two nodes at two fixed location: it is determined by the probability that the received signal strength is greater than or equal to the pre-determined threshold. Two related questions can be asked: a) if these nodes are deployed at a set of known location, what is the quality of connectivity of the network, measured by the probability that there is a path between any two nodes, as compared to node deployment at another set of location? b) how can one optimize the node deployment to maximize the quality of connectivity?

Example 179 The transmission between a pair of nodes with a direct connection, say v_i and v_j, may fail with a known probability, say $1 - a_{ij}$, quantifying the inherent unreliable characteristics of wireless communications. There are no direct connections between some pairs of nodes because the probability of successful transmission between them is too low to be acceptable. How should one measure the quality of connectivity of such a network, in the sense that a packet transmitted from one node can easily and reliably reach another node via a multi-hop path. Will single "good" path between a pair of nodes be preferable to multiple "bad" paths? These questions are illustrated in Figs. 13.1 and 13.2.

In this chapter, we introduce and explore the use of a *probabilistic connectivity matrix*, a concept to be defined later in Sect. 13.2, as a tool to measure the quality of network connectivity. Some key properties of the probabilistic connectivity matrix and their connections to the quality of connectivity are demonstrated. Armed with certain inequalities derived later and assuming a symmetric network, we further derive several properties of the eigenvalues of the probabilistic connectivity matrix. First we show that in a connected network, i.e., where there is a path of nonzero probability between every pair of nodes, the largest magnitude eigenvalue, which is positive, does indeed quantify the quality of network connectivity. Should the network be disconnected, then we show that it naturally partitions into connected components. Specifically, there is a path of nonzero probability between any two nodes in a connected component, but all inter-component paths have zero probability. In this case the probabilistic connectivity matrix is block diagonal, each diagonal block in turn being the connectivity matrix of a particular component. In this case the largest magnitude eigenvalue provides the connectivity measure of this component. We show also that the matrix is positive semidefinite, and is in fact positive definite, unless there is a path in the network that has probability one.

We further demonstrate that increasing a link probability increases the largest eigenvalue of the component to which the link belongs. Exploring the positive semi-definiteness of this matrix we provide an algorithm that computes the largest eigenvalue in a decentralized fashion using experimental measurements on the network, including averaging over a number of experiments. Specifically this *flooding* algorithm requires the nodes to repeatedly flood the network with packets, and by measuring just the number of packets a given node receives, the node is able to asymptotically estimate this largest eigenvalue without knowing any element of the probabilistic connectivity matrix or the number of packets received by the other nodes.

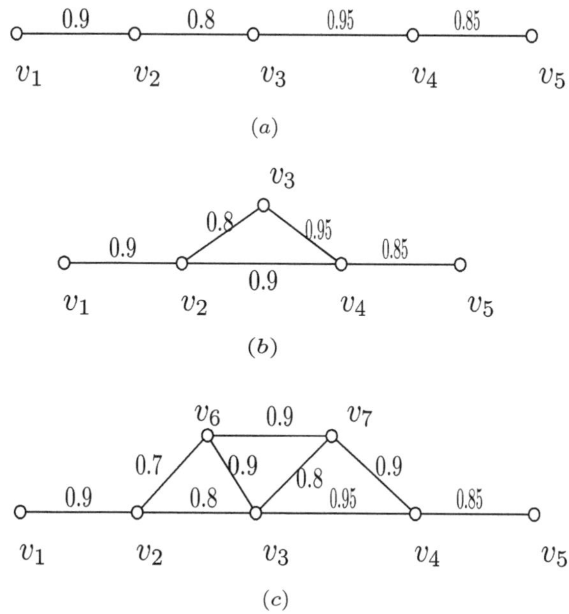

Fig. 13.1 An illustration of networks with different quality of connectivity. A *solid line* represents a direct connection between two nodes and the number beside the line represents the corresponding transmission successful probability. The networks shown in (**a**), (**b**), and (**c**) are all connected networks but not 2-connected networks, i.e., their connectivity cannot be differentiated using the k-connectivity concept. However intuitively the quality of the network in (**b**) is better than that of the network in (**a**) because of the availability of the additional high-quality link between v_2 and v_4 in (**b**). The quality of the network in (**c**) is even better because of the availability of the additional nodes and the associated high-quality links, hence additional routes, if these additional nodes act as relay nodes only. If these additional nodes also generate their own traffic, it is uncertain whether the quality of the network in (**c**) is better or not. Therefore, it is important to develop a measure to quantitatively compare the quality of connectivity (for the networks in (**a**) and (**b**)) and to evaluate the benefit of additional nodes on connectivity (for the network in (**c**))

13.2 Probabilistic Connectivity Matrix and Its Properties

In this section we define the network to be studied, its probabilistic adjacency matrix and probabilistic connectivity matrix, give an approach to computing the probabilistic connectivity matrix, and discuss its properties.

Consider a network of n nodes. For some pair of nodes, an edge (or link) may exist with a non-negligible probability. The edges are considered to be undirected. That is, if a node v_i is connected to a node v_j, then the node v_j is also connected to the node v_i. Furthermore, as is commonly done in the area, it is assumed that the event that there is an edge between a pair of nodes and the event that there is an edge between another distinct pair of nodes (which may include one node in common with the first pair) are independent. In addition to such spatial independence, we

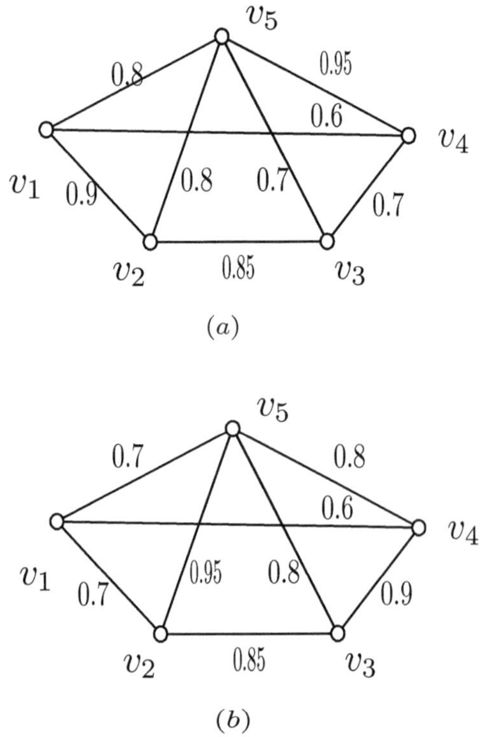

Fig. 13.2 The networks shown in (**a**) and (**b**) have the same topology but different link quality. It is difficult to compare the quality of the two networks

also assume *temporal independence*; specifically that each edge event is identically and independently distributed over time, e.g., due to fading and shadowing. This temporal independence is needed for some later results, and is formalized in that part.

Denote the underlying graph of the above network by $G(V,E)$, where $V = \{v_1, \ldots, v_n\}$ is the vertex set and $E = \{e_1, \ldots, e_m\}$ is the edge set, which contains the set of *all possible* edges, i.e., all vertex pairs for which the probability of being directly connected is nonzero. Here the vertices and the edges are indexed from 1 to n and from 1 to m, respectively. For convenience, in some parts of this chapter we also use the symbol e_{ij} to denote an edge between vertices v_i and v_j when there is no confusion. We associate with each edge e_i, $i \in \{1, \ldots m\}$, an indicator random variable I_i such that $I_i = 1$ if the edge e_i exists; $I_i = 0$ if the edge e_i does not exist. The indicator random variables I_{ij}, $i \neq j$ and $i, j \in \{1, \ldots n\}$, are defined analogously. Furthermore, we use $(I_i, i \in \{1, 2, \ldots, m\})$ to denote a particular instance of the indicator random variables associated with an instance of the random edge set.

In the following, we give a definition of the probabilistic adjacency matrix, differing mildly from that of Brooks *et al.* [32] as described further below:

Definition 180 The probabilistic adjacency matrix of $G(V,E)$, denoted by A_G, is a $n \times n$ matrix whose $(i,j)^{th}$, $i \neq j$, entry $a_{ij} \triangleq \Pr\left(I_{ij} = 1\right)$ and its diagonal entries are all equal to 1.

Due to the undirected property of an edge mentioned above, A_G is a symmetric matrix, i.e., $a_{ij} = a_{ji}$. Note that the diagonal entries of A_G are defined to be 1, which is different from the usual convention in the literature, e.g., [32]. In Chap. 8 we have discussed the implication of this definition in the context of mobile ad hoc networks. This treatment of the diagonal entries reflects the fact that if a node in the network finds the wireless channel busy, it can store a packet (or equivalently transmit the packet to itself) until the channel is free. A pair of nodes v_i and v_j are said to be *directly connected* if the associated a_{ij} is greater than 0.

The probabilistic connectivity matrix is defined in the following way:

Definition 181 The probabilistic connectivity matrix of $G(V,E)$, denoted by Q_G, is an $n \times n$ matrix whose $(i,j)^{th}$, $i \neq j$, entry is the probability that there exists a path between vertices v_i and v_j, and its diagonal entries are all equal to 1.

As a ready consequence of the symmetry of A_G, Q_G is also a symmetric matrix. Furthermore, the following property of Q_G can be easily obtained from the above definition. The lemma refers to the *direct sum* between matrices, defined as $A \bigoplus B = \textbf{\textit{diag}}\{A,B\}$.

Lemma 182 *Suppose A_G defined in Definition 180 is symmetric. Then, the probabilistic connectivity matrix Q_G is a symmetric nonnegative matrix. If it has a zero element, then there is an ordering of vertices under which Q_G is a direct sum of positive matrices.*

Proof Symmetry of Q_G follows from the symmetry of A_G. Nonnegativity of Q_G follows from the fact that its diagonal elements are one and the rest are probabilities. Now suppose for some i,j, $q_{ij} = q_{ji} = 0$ but that for some k, $q_{ik} = q_{ki} \neq 0$. This indicates that all paths between v_i and v_j have zero probability (henceforth, v_i and v_j are not connected) but at least one between v_k and v_i has a nonzero probability (v_k and v_i are connected). Thus $q_{kj} = q_{jk} = 0$ as otherwise there is a path between v_k and v_j and consequently between v_i and v_k that has nonzero probability, violating the assumption that $q_{ij} = q_{ji} = 0$. Thus, one can partition the vertex set V into sets V_l, such that all nodes in V_l are connected to each other but are not connected to any node in V_m, $m \neq l$. Order the vertices so that for each l those of V_l are consecutive. The resulting Q_G is clearly a direct sum of positive matrices. □

Remark 183 We call the network connected if Q_G is positive, as there is then a nonzero probability that a path exists between any two nodes. Lemma 182 and its proof also formalize the fact that a network that is not connected partitions into disjoint components, each of which is connected, but all paths between nodes from different components have probability zero (we are not distinguishing conceptually between the notion that a link or path may not exist, and the notion that a link or path always has zero probability).

Given the probabilistic adjacency matrix A_G, the probabilistic connectivity matrix Q_G is fully determined. However the computation of Q_G is not trivial because for a pair of vertices v_i and v_j, there may be multiple paths between them and some of the paths may share common edges, i.e., paths are not *independent* or are *spatially correlated.* In the rest of this subsection, we give a method to compute Q_G.

13.2.1 Computation of the Probabilistic Connectivity Matrix

We now indicate in rather formal language the conceptual basis of computing the probabilistic connectivity matrix Q_G.

Let $Q_G|\,(I_i, i \in \{1,2,\ldots,m\})$ be the connectivity matrix of G *conditioned on* a particular instance of the indicator random variables $I_1,\ldots,I_m$ associated with an instance of the random edge set. The $(i,j)^{th}$ entry of $Q_G|\,(I_i, i \in \{1,2,\ldots,m\})$ is either 0, when there is no path between v_i and v_j, or 1 when there exists a path between v_i and v_j. The diagonal entries of $Q_G|\,(I_i, i \in \{1,2,\ldots,m\})$ are always 1. Conditioned on $(I_i, i \in \{1,2,\ldots,m\})$, $G(V,E)$ is just a deterministic graph. Therefore, the entries of $Q_G|\,(I_i, i \in \{1,2,\ldots,m\})$ can be efficiently computed using a search algorithm, such as breadth-first search. Given $Q_G|\,(I_i, i \in \{1,2,\ldots,m\})$, Q_G can be computed using the following:

$$Q_G = E_{(I_i, i \in \{1,2,\ldots,m\})}\left(Q_G|\,(I_i, i \in \{1,2,\ldots,m\})\right) \tag{13.2.1}$$

where the expectation is taken over all possible instances of $(I_i, i \in \{1,2,\ldots,m\})$.

Using the technique introduced in the previous paragraph, the probabilistic connectivity matrix of the three networks in Fig. 13.1 and two networks in Fig. 13.2, denoted by Q_{1a}, Q_{1b}, Q_{1c}, Q_{2a} and Q_{2b} respectively, can be computed. For example,

$$Q_{2a} = \begin{bmatrix} 1.0000 & 0.9876 & 0.9744 & 0.9823 & 0.9880 \\ 0.9876 & 1.0000 & 0.9812 & 0.9856 & 0.9916 \\ 0.9744 & 0.9812 & 1.0000 & 0.9780 & 0.9827 \\ 0.9823 & 0.9856 & 0.9780 & 1.0000 & 0.9926 \\ 0.9880 & 0.9916 & 0.9827 & 0.9926 & 1.0000 \end{bmatrix} \tag{13.2.2}$$

$$Q_{2b} = \begin{bmatrix} 1.0000 & 0.9603 & 0.9571 & 0.9540 & 0.9614 \\ 0.9603 & 1.0000 & 0.9918 & 0.9854 & 0.9961 \\ 0.9571 & 0.9918 & 1.0000 & 0.9879 & 0.9936 \\ 0.9540 & 0.9854 & 0.9879 & 1.0000 & 0.9878 \\ 0.9614 & 0.9961 & 0.9936 & 0.9878 & 1.0000 \end{bmatrix} \tag{13.2.3}$$

A comparison of the entries of Q_{2a} and Q_{2b} leads to intuitive and quantitative conclusion on the quality of end-to-end paths between any pair of nodes in the two networks in Fig. 13.2a and b. In the rest of this chapter, we shall further establish properties of the probabilistic connectivity matrix that facilitates the analysis of network quality and connectivity.

The approach suggested in the last paragraph is essentially a brute-force approach to computing Q_G. More efficient algorithms can be possibly designed to compute Q_G. Indeed in Sect. 13.3 we suggest an approach to simplifying the computation of Q_G via a recursive procedure exploring the property of Q_G. Since the main focus here is on exploring the properties of Q_G that facilitate the connectivity analysis, an extensive discussion of designing computationally efficient algorithms to compute Q_G is left out.

That said, the complications in computing Q_G are mitigated by the fact that a measure of connectivity developed in this chapter can also be estimated using experimental data without explicitly obtaining the elements of Q_G. This measure is the largest eigenvalue of Q_G. As shown in the chapter it can be asymptotically estimated in a completely decentralized fashion without knowing the entries of Q_G or the link probabilities and network topology.

Remark 184 For simplicity, the terms used in our discussion are based on the problems in Example 178. The discussion, however, can be easily adapted to the analysis of the problems in Example 13.2. For example, if a_{ij} is defined to be the probability that a direct transmission between nodes v_i and v_j is successful, the $(i,j)^{th}$ entry of the probabilistic connectivity matrix Q_G computed using (13.2.1) then gives the probability that a transmission from v_i to v_j via a multi-hop path is successful under the best routing algorithm, which can always find a shortest and error-free path between from v_i to v_j if it exists, or alternatively, the probability that a packet flooded from v_i can reach v_j where each node receiving the packet only broadcasts the packet to its directly connected neighbors once. Therefore, the $(i,j)^{th}$ entry of Q_G can be used as a quality measure of the end-to-end paths between v_i and v_j, which takes into account the fact that availability of an extra path between a pair of nodes can be explored to improve the probability of successful transmissions.

13.2.2 Some Key Inequalities for Connection Probabilities

The entries of the probabilistic connectivity matrix give an intuitive idea about the overall quality of end-to-end paths in a network. In this subsection, we provide some important inequalities that may facilitate the analysis of the quality of connectivity. Some of these inequalities are explored in the next subsection to establish some key properties of the probabilistic connection matrix itself.

We first introduce some concepts and results that are required for the further analysis of the probabilistic connectivity matrix Q_G.

For a random graph with a given set of vertices, a particular event is *increasing* if the event is preserved when more edges are added into the graph. An event is *decreasing* if its complement is increasing.

The following theorems summarizing a relevant form of the FKG inequality and BK inequality, respectively, will be used:

Theorem 185 ([143, Theorem 1.4] FKG Inequality) *If events A and B are both increasing events or decreasing events depending on the state of finitely many edges, then*

$$\Pr(A \cap B) \geq \Pr(A)\Pr(B)$$

Theorem 186 ([17, 143, Theorem 1.5] BK Inequality) *If events A and B are both increasing events depending on the state of finitely many edges, then*

$$\Pr(A \Box B) \leq \Pr(A)\Pr(B)$$

where for two events A and B, $A \Box B$ denotes the event that there exist two disjoint *sets of edges such that the first set of edges guarantees the occurrence of A and the second set of edges guarantees the occurrence of B.*

Denote by ξ_{ij} the event that there is a path between vertices v_i and v_j, $i \neq j$. Denote by ξ_{ikj} the event that there is a path between vertices v_i and v_j *and* that path passes through the third vertex v_k, where $k \in \Gamma_n \backslash \{i, j\}$ and Γ_n is the set of indices of all vertices. Denote by η_{ij} the event that there is an edge between vertices v_i and v_j. Denote by π_{ikj} the event that there is a path between vertices v_i and v_k *and* there is a path between vertices v_k and v_j, where $k \in \Gamma_n \backslash \{i, j\}$. Obviously

$$\pi_{ikj} \Rightarrow \xi_{ij} \tag{13.2.4}$$

It is clear from the above definitions that

$$\xi_{ij} = \eta_{ij} \cup \left(\cup_{k \neq i,j} \xi_{ikj}\right) \tag{13.2.5}$$

Let q_{ij}, $i \neq j$, be the $(i, j)^{th}$ entry of Q_G, i.e., $q_{ij} = \Pr\left(\xi_{ij}\right)$. The following theorem is obtained from the FKG inequality and the above definitions.

Theorem 187 *For two distinct indices $i, j \in \Gamma_n$ and $\forall k \in \Gamma_n \backslash \{i, j\}$*

$$q_{ij} \geq \max_{k \in \Gamma_n \backslash \{i,j\}} q_{ik} q_{kj} \tag{13.2.6}$$

$$q_{ij} \leq 1 - \left(1 - a_{ij}\right) \prod_{k \in \Gamma_n \backslash \{i,j\}} \left(1 - q_{ik} q_{kj}\right) \tag{13.2.7}$$

where $a_{ij} = \Pr\left(\eta_{ij}\right)$.

Proof We first prove inequality (13.2.6). It follows readily from the above definitions that the event ξ_{ij} is an increasing event. Due to (13.2.4) and the FKG inequality:

$$\Pr\left(\xi_{ij}\right) \geq \Pr\left(\pi_{ikj}\right) \geq \Pr\left(\xi_{ik}\right)\Pr\left(\xi_{kj}\right) \tag{13.2.8}$$

The conclusion follows.

Now we prove the second inequality (13.2.7). We shall first show that $\xi_{ikj} \Leftrightarrow \xi_{ik}\square\xi_{kj}$. That is, the occurrence of the event ξ_{ikj} is a sufficient and necessary condition for the occurrence of the event $\xi_{ik}\square\xi_{kj}$.

Using the definition of ξ_{ikj}, occurrence of ξ_{ikj} means that there is a path between vertices v_i and v_j *and* that path passes through vertex v_k. It follows that there exist a path between vertex v_i and vertex v_k *and* a path between vertex v_k and vertex v_j *and* the two paths do not have edge(s) in common. Otherwise, it will contradict the definition of ξ_{ikj}, noting that the definition of a path requires its edges to be distinct. Therefore, $\xi_{ikj} \Rightarrow \xi_{ik}\square\xi_{kj}$. Likewise, $\xi_{ikj} \Leftarrow \xi_{ik}\square\xi_{kj}$ also follows directly from the definitions of ξ_{ikj}, ξ_{ik}, ξ_{kj} and $\xi_{ik}\square\xi_{kj}$. Consequently

$$\Pr\left(\xi_{ikj}\right) = \Pr\left(\xi_{ik}\square\xi_{kj}\right) \leq \Pr\left(\xi_{ik}\right)\Pr\left(\xi_{kj}\right) \tag{13.2.9}$$

where the inequality is a direct result of the BK inequality.

Note that the event $\cup_{k\in\Gamma_n\backslash\{i,j\}}\xi_{ikj}$ and the event η_{ij} are independent because the existence of a direct connection between v_i and v_j has no impact on the event $\cup_{k\in\Gamma_n\backslash\{i,j\}}\xi_{ikj}$. Therefore, using (13.2.5) and independence of edges (used in the second step)

$$\begin{aligned} q_{ij} &= \Pr\left(\eta_{ij} \cup \left(\cup_{k\in\Gamma_n\backslash\{i,j\}}\xi_{ikj}\right)\right) \\ &= 1-\left(1-a_{ij}\right)\Pr\left(\cap_{k\in\Gamma_n\backslash\{i,j\}}\overline{\xi_{ikj}}\right) \\ &\leq 1-\left(1-a_{ij}\right)\prod_{k\in\Gamma_n\backslash\{i,j\}}\Pr\left(\overline{\xi_{ikj}}\right) && (13.2.10) \\ &\leq 1-\left(1-a_{ij}\right)\prod_{k\in\Gamma_n\backslash\{i,j\}}\left(1-q_{ik}q_{kj}\right) && (13.2.11) \end{aligned}$$

where in (13.2.10), FKG inequality and the obvious fact that $\overline{\xi_{ikj}}$ is a decreasing event are used and the last step is from (13.2.9). □

Remark 188 Inequality (13.2.6) also provides another proof of a key relationship used in Lemma 182. Specifically, if $q_{ij} = 0$, then this inequality implies that at least one among q_{ik} and q_{kj} must be zero. Likewise if neither q_{ik} nor q_{kj} is zero, then $q_{ij} > 0$.

When there is no edge between vertices v_i and v_j, the upper and lower bounds in Theorem 186 reduce to

$$\max_{k\in\Gamma_n\backslash\{i,j\}} q_{ik}q_{kj} \leq q_{ij} \leq 1-\prod_{k\in\Gamma_n\backslash\{i,j\}}\left(1-q_{ik}q_{kj}\right) \tag{13.2.12}$$

The above inequality sheds insight on how the quality of paths between a pair of vertices is related to the quality of paths between other pairs of vertices. It can be possibly used to determine the most effective way of improving the quality of a particular set of paths by improving the quality of a particular (set of) edge(s), or equivalently what can be reasonably expected from an improvement of a particular edge on the quality of end-to-end paths. Furthermore, an immediate consequence of this inequality is that: if $q_{ij} = 0$, then at least one of q_{ik} and q_{kj} must be 0 for all $k \neq i, j$.

The following lemma further shows that the occurrence of a certain relation among entries of the probabilistic connectivity matrix Q_G can be used to derive some topological information of the graph.

Lemma 189 *If $q_{ij} = q_{ik}q_{kj}$ for distinct vertices v_i, v_j and v_k, the vertex set V of the graph $G(V, E)$ can be divided into three nonempty and nonintersecting sub-sets V_1, V_2, and V_3 such that $v_i \in V_1$, $v_j \in V_3$ and $V_2 = \{v_k\}$* and *any possible path between a vertex in V_1 and a vertex in V_2 must pass through v_k, and the converse. Furthermore, for any pair of vertices v_l and v_m, where $v_l \in V_1$ and $v_m \in V_3$, $q_{lm} = q_{lk}q_{km}$.*

Proof Using (13.2.8) in the second step, it follows that

$$\begin{aligned} q_{ij} &= \Pr\left(\xi_{ij} \backslash \pi_{ikj}\right) + \Pr\left(\pi_{ikj}\right) \\ &\geq \Pr\left(\xi_{ij} \backslash \xi_{ikj}\right) + q_{ik}q_{kj} \end{aligned}$$

Therefore, $q_{ij} = q_{ik}q_{kj}$ implies that $\Pr\left(\xi_{ij} \backslash \pi_{ikj}\right) = 0$ or equivalently $\xi_{ij} \Leftrightarrow \pi_{ikj}$

Furthermore, $\Pr\left(\xi_{ij} \backslash \pi_{ikj}\right) = 0$ implies that a *possible* path (i.e., a path with a nonzero probability) connecting v_i and v_k and a *possible* path connecting v_k and v_j cannot have any edge in common. Otherwise a path from v_i to v_j, bypassing v_k, exists with a nonzero probability which implies $\Pr\left(\xi_{ij} \backslash \xi_{ikj}\right) > 0$. The conclusion follows readily that if $q_{ij} = q_{ik}q_{kj}$ for three distinct vertices v_i, v_j, and v_k, the vertex set V of the underlying graph $G(V, E)$ can be divided into three nonempty and nonoverlapping sub-sets V_1, V_2 and V_3 such that $v_i \in V_1$, $v_j \in V_3$ and $V_2 = \{v_k\}$ *and* a path between a vertex in V_1 and a vertex in V_2, if exists, must pass through v_k.

Furthermore, for any pair of vertices v_l and v_m, where $v_l \in V_1$ and $v_m \in V_3$, it is easily shown that $\Pr(\xi_{lm} \backslash \pi_{lkm}) = 0$. Due to independence of edges and further using the fact that $\Pr(\xi_{lm} \backslash \pi_{lkm}) = 0$, it can be shown that

$$\Pr(\xi_{lm}) = \Pr(\pi_{lkm}) = \Pr(\xi_{lk}) \Pr(\xi_{km}) \tag{13.2.13}$$

where (13.2.13) results due to the fact that under the condition of $\Pr(\xi_{lm} \backslash \pi_{lkm}) = 0$, a path between vertices v_l and v_k and a path between vertices v_k and v_m cannot possibly have any edge in common. □

An implication of Lemma 189 is that for any three distinct vertices, v_i, v_j, and v_k, if a relationship $q_{ij} = q_{ik}q_{kj}$ holds, vertex v_k must be a *critical* vertex whose removal will render the graph disconnected.

13.3 The Largest Eigenvalue of the Probabilistic Connectivity Matrix

We now establish a measure of the quality of network connectivity. Just as the eigenvalues of the adjacency matrix provide a deterministic measure of connectivity, we now provide a series of arguments supporting the contention that a similar property can be ascribed to certain eigenvalues of the probabilistic connectivity matrix Q_G.

From Lemma 182, Q_G is a nonzero nonnegative matrix. Thus from the Perron-Frobenius Theorem [75], its largest magnitude eigenvalue, known as the *Perron-Frobenius eigenvalue,* is real and positive. Furthermore, as Q_G is symmetric, all its eigenvalues are real, and its largest magnitude eigenvalue $\lambda_{\max}(Q_G)$ is also its largest singular value. Also from the Perron-Frobenius Theorem, should the network be connected, i.e., Q_G is positive as opposed to just nonnegative, this eigenvalue is simple.

We now argue that $\lambda_{\max}(Q_G)$ quantifies the quality of network connectivity. Indeed, suppose that the i-th node v_i transmits x_i number of *distinct* packets in a time interval. This means that v_i floods the packet across the entire network and each node receiving the packet only broadcasts the packet once to its directly connected neighbors. If the same packet is received more than once by the same node, it is counted as one packet. Let $x = [x_1, \cdots, x_n]^\top$ and let y_i denote the expected number of packets received by the i-th node, $y = [y_1, \cdots, y_n]^\top$. Then by definition: $y = Q_G x$. As the basic purpose of any network is to transport packets from some nodes in the network to some others, a measure of connectivity that naturally arises is *the largest size of* y *relative to* x. One measure of the size of y is its 2-norm, denoted by $\|y\|_2$. Then, as Q_G is symmetric and nonnegative,

$$\begin{aligned}\max_{\|x\|_2 \neq 0} \frac{\|y\|_2}{\|x\|_2} &= \max_{\|x\|_2 \neq 0} \frac{\sqrt{y^\top y}}{\sqrt{x^\top x}} = \max_{\|x\|_2 \neq 0} \frac{\sqrt{x^\top Q_G^\top Q_G x}}{\sqrt{x^\top x}} \\ &= \max_{\|x\|_2 \neq 0} \sqrt{\frac{x^\top Q_G^2 x}{x^\top x}} = \lambda_{\max}(Q_G)\end{aligned}$$

It is well known that for a symmetric Q_G, the maximum ratio is attained when x is the eigenvector associated with the eigenvalue $\lambda_{\max}(Q_G)$. Observe also from Perron-Frobenius theory [75] that as Q_G is nonnegative, the eigenvector associated with $\lambda_{\max}(Q_G)$ has all entries of the same sign, without loss of generality nonnegative. Thus the largest value of $\max_{\|x\|_2 \neq 0} \frac{\|y\|_2}{\|x\|_2}$ is itself attained by a x with nonnegative elements. Thus indeed one can strengthen the equality above to state that:

$$\max_{\|x\|_2 \neq 0, x_i \geq 0} \frac{\|y\|_2}{\|x\|_2} = \lambda_{\max}(Q_G)$$

Consequently, $\lambda_{\max}(Q_G)$ is a natural measure of network connectivity.

There are two other approaches to characterizing $\lambda_{\max}(Q_G)$: min-max and max-min flow gain:

$$\max_{x>0}\min_i \frac{y_i}{x_i} \text{ and } \min_{x>0}\max_i \frac{y_i}{x_i}.$$

Regardless of whether Q_G is symmetric, its largest magnitude eigenvalue, obeys min-max and max-min type relations through the Collatz-Wielandt equalities (see Corollary 8.1.31 in [102]). In particular,

$$\max_{x>0}\min_i \frac{y_i}{x_i} = \lambda_{\max}(Q_G) = \min_{x>0}\max_i \frac{y_i}{x_i}.$$

The case of using $\lambda_{\max}(Q_G)$ as a measure of connectivity is further supported by the following observation. When Q_G is positive as opposed to just nonnegative, $\lambda_{\max}(Q_G)$ strictly increases with increasing values of its off-diagonal elements [75]. If, on the other hand, it has zero elements, then on the face of it, it is merely nondecreasing. However, recall from Lemma 182 and Remark 183 that if there are zero entries in Q_G, the network partitions into disjoint connected components represented by graphs $G_i(V_i, E_i)$, and Q_G itself can be expressed as $Q_G = \bigoplus_{i=1}^{l} Q_{G_i}$, with Q_{G_i} all positive. Should an element of a particular Q_{G_i} increase, then so must its largest eigenvalue. On the other hand, for $v_i \in V_i$ and $v_j \in V_j$, $q_{ij} = 0$. Should now this become positive, then we argue that with $G'_{ij} = \left(V_i \bigcup V_j, E_i \bigcup E_j\right)$, $\lambda_{\max}(Q_{G'_{ij}})$ does indeed strictly increase. Indeed suppose the new $q_{ij} = q > 0$. Then, from Lemma 182, for every $0 < q_{ij} < q$, the resulting $Q_{G'_{ij}}$ is positive and the result follows.

We next establish the remarkable fact that in fact Q_G is a *positive semidefinite* matrix. The implications of the positive semi-definiteness of Q_G will be explored later. At the core of the development leading to this result is the following fact.

Lemma 190 *Each off-diagonal entry of the probabilistic connectivity matrix Q_G is a multiaffine function of a_{ij}. A multiaffine function is affine in each variable when the other variables are fixed.*

Proof Consider an arbitrary off-diagonal entry, q_{kl} of Q_G. This is the probability that there is a path between vertices v_k and v_l. This event is ξ_{kl}. Enumerate the distinct events constituting a path between v_k and v_l, listing first those not containing edge e_{ij} as $\bar{\xi}_{1,kl}, \ldots \bar{\xi}_{s,kl}$ and then those containing edge e_{ij} as $\bar{\xi}_{s+1,kl} \cap \eta_{ij}, \cdots, \bar{\xi}_{t,kl} \cap \eta_{ij}$. Of course, the event that a path exists is the intersection of the events η_{pq} for the edges e_{pq} along the path. Evidently,

$$\xi_{kl} = \bar{\xi}_{1,kl} \cup \cdots \cup \bar{\xi}_{s,kl} \cup (\bar{\xi}_{1,kl} \cap \eta_{ij}) \cup \cdots \cup (\bar{\xi}_{t,kl} \cap \eta_{ij}) \tag{13.3.1}$$

Because every event η_{ij} is independent of all the other edge connection events, it is easy to verify that q_{kl} is equal to

$$\Pr\left(\bar{\xi}_{1,kl} \cup \cdots \cup \bar{\xi}_{t,kl}\right) a_{ij} + \Pr\left(\bar{\xi}_{1,kl} \cup \cdots \cup \bar{\xi}_{s,kl}\right)(1 - a_{ij}) \tag{13.3.2}$$

Since the probabilities multiplying a_{ij} and $1 - a_{ij}$ in (13.3.2) are probabilities of events independent of the event η_{ij}, they do not depend on a_{ij}. Thus if we hold a_{pq} with $\{i,j\} \neq \{p,q\}$ constant, q_{kl} is an affine function of a_{ij}. The same applies to every off-diagonal element of A_G. The result follows. □

Note that $\Pr(\bar{\xi}_{1,kl} \cup \cdots \cup \bar{\xi}_{t,kl})$ is the probability of a path between vertices v_k and v_l with the original network modified by eliminating any link between vertices $\{v_i, v_j\}$, while $\Pr(\bar{\xi}_{s+1,kl} \cup \cdots \cup \bar{\xi}_{t,kl})$ is the probability of a path between the same vertices with the original network modified by imposing a perfect connection ($a_{ij} = 1$) between vertices v_i and v_j (equivalently the two vertices are merged); the latter is obviously greater than or equal to the former. The associated matrices are themselves probabilistic connectivity matrices.

Due to this multiaffine property, for $k, l, i, j \in \{1, \ldots n\}$, where $k \neq l$ and $i \neq j$, the following holds:

$$q_{lk} = c_1 a_{ij} + c_2 \tag{13.3.3}$$

where c_1 and c_2 are in $[0, 1]$, are determined by the state of the set of edges in $E \setminus \{e_{ij}\}$ only, and are not affected by the state of e_{ij}; $c_2 = 0$ implies that v_l and v_k will be disconnected without the edge e_{ij}. Thus e_{ij} is a *critical* edge for the end-to-end paths between the vertices v_l and v_k. $c_1 = 0$ implies that the state of the edge e_{ij} is irrelevant for the end-to-end paths between v_l and v_k. In fact, c_1 measures the *criticality* of the edge e_{ij} to the end-to-end paths between v_l and v_k.

Using the multiaffine property, a more efficient algorithm for computing Q_G than the one suggested earlier using (13.2.1) can be constructed. Particularly, the probabilistic connectivity matrix of a network forming a tree can be easily computed. Therefore, the algorithm may start by first identifying a spanning tree in $G(V, E)$ and computing the associated probabilistic connectivity matrix. Then, the edges in E but outside the spanning tree can be added recursively and the corresponding probabilistic connectivity matrix updated using (13.3.3). Since the computational complexity of Q_G depends on $2^{|E|}$, let l be the number of edges in the spanning tree, the computational complexity improves approximately by a factor of 2^l compared with the algorithm using (13.2.1) directly. The key purpose of this chapter is to postulate and justify as valid, a measure of network connectivity and to formulate a procedure for estimating this measure, without having to explicitly obtain Q_G. Therefore, we leave out discussions for constructing computationally efficient algorithms for computing Q_G from the a_{ij} and network topology. That said, the following remark is also instructive.

Remark 191 Several papers have explored multiaffine variations. These include the design of adaptive estimation algorithms [49–51] and stability analysis [5, 52, 200]. All explore the fact that variations are individually affine in each variable as long as the other variables are fixed. The fact that there is an increasing relationship between the elements of Q_G and $\lambda_{\max}(Q_G)$ and the latter depends multiaffinely on the probabilities a_{ij} suggests the following obvious optimization. Modify one or more a_{ij} under suitable constraints to maximize $\lambda_{\max}(Q_G)$. The multiaffine dependence of q_{ij} on the a_{ij}, *together with the fact that Q_G is positive semi-definite*, promises to provide several avenues for such optimization.

The basis for these calculations is likely to be the following observation. If $Q_G = a_{ij}Q_{1G} + Q_{2G}$ with Q_{1G}, Q_{2G} independent of a_{ij}, and if x is a positive eigenvector of Q_G associated with the maximum eigenvalue $\lambda_{\max}(Q_G)$, then it is easily seen that $\frac{\partial \lambda_{max}}{\partial a_{ij}} = \frac{x^\top Q_{1G} x}{x^\top x}$.

We now establish that Q_G is positive semidefinite.

Theorem 192 *The matrix $Q_G = Q_G^\top \in \Re^{n\times n}$ is a positive semi-definite matrix. It is not positive definite if and only if there exist $i \neq j$, such that $q_{ij} = 1$.*

We prove this theorem at the end of this section. For the moment we discuss its implications. One in particular is its use in the analysis of the flooding algorithm of the next subsection. There are also implications to the level of connectivity. Let $\lambda_{\max}(Q_G) \geq \lambda_2(Q_G) \geq \ldots \geq \lambda_{\min}(Q_G) \geq 0$ be the eigenvalues of Q_G. As all diagonal elements of Q_G are one, the trace of Q_G and hence $\lambda_{\max}(Q_G) + \lambda_2(Q_G) \ldots + \lambda_{\min}(Q_G)$ is equal to n. Thus as an easy consequence of Theorem 192, $n \geq \lambda_{\max}(Q_G) \geq 1$ and $1 \geq \lambda_{\min}(Q_G) \geq 0$. In the best case, Q_G is a matrix with all entries equal to 1. Then, $\lambda_{\max}(Q_G) = n$ and $\lambda_2(Q_G) = \cdots = \lambda_{\min}(Q_G) = 0$. In the worst case, when no node is connected to any other, Q_G is an identity matrix. Then, $\lambda_{\max}(Q_G) = \lambda_2(Q_G) = \ldots = \lambda_{\min}(Q_G) = 1$. Consider also the following consequence of Lemma 182.

Lemma 193 *Suppose for all i, j, $a_{ij} \in \{0, 1\}$. Then, there is a relabeling of vertices under which Q_G is a direct sum of matrices whose elements are all ones.*

Proof From Lemma 182 under a reordering of vertices $Q_G = \bigoplus_i Q_{Gi}$ with Q_{Gi} all positive. As all $a_{ij} \in \{0, 1\}$, there is an edge between v_i and v_j surely when $a_{ij} = 1$; or there is no edge between v_i and v_j surely when $a_{ij} = 0$. Thus either there is a path between v_i and v_j surely or there is no path between v_i and v_j surely, i.e., for all i, j, $q_{ij} \in \{0, 1\}$. Thus every element of every Q_{Gi} is 1. □

This lemma thus characterizes Q_G when $a_{ij} \in \{0, 1\}$ for all i, j, i.e., the network is effectively deterministic. In this case, there is an ordering of vertices for which Q_G is a direct sum of square matrices of all ones. If there are m such summands, then $n - m$ eigenvalues of Q_G are 0. Of course, as noted above, in the extreme case where all $a_{ij} = 1$, there are $n - 1$ zero eigenvalues. *This also suggests that the proximity of $\lambda_{\min}(Q_G)$ to zero in a connected network is a measure of connectivity, as is the number of eigenvalues that are close to zero when the network is not connected.*

13.3.1 Proof of Theorem 192

We are now ready to prove Theorem 192. To prove Theorem 192 we prove in turn that (A) each Q_G is positive semidefinite; (B) that should any $q_{ij} = 1$ for $i \neq j$ then Q_G cannot be positive definite; and that (C) if for all $i \neq j$, $0 \leq q_{ij} < 1$, then Q_G is positive definite. First we recount Corollary 2.1 of [52] which explores the facts that all convex combinations of positive semidefinite matrices are positive semidefinite, and that multiaffine functions are affine in each variable, if the others are fixed.

Lemma 194 *Suppose for integers n and N, $P(\alpha) \in \Re^n$ is a multiaffine function of the elements of $\alpha = [\alpha_1, \cdots, \alpha_N]^\top$. Then, $P(\alpha)$ is positive semidefinite for all $\alpha_i \in [\alpha_i^-, \alpha_i^+]$ and $i \in \{1, \cdots, N\}$ if and only if it is positive semidefinite for all $\alpha_i \in \{\alpha_i^-, \alpha_i^+\}$ and $i \in \{1, \cdots, N\}$.*

Proof of (A) As matrices of all ones are positive semidefinite, Lemma 193 proves that Q_G is positive semidefinite whenever for all i, j, $a_{ij} \in \{0, 1\}$. The result follows from Lemmas 190 and 194.

Proof of (B) This follows from the following lemma and the fact that a matrix with two identical rows cannot be positive definite.

Lemma 195 *Suppose for some $i \neq j$, $q_{ij} = 1$. Then, row i and row j of Q_G are identical, as are columns i and j.*

Proof Note that Q_G is a symmetric matrix. Thus it suffices to show that the row property holds. One has

$$q_{ij} = q_{ji} = q_{ii} = q_{jj} = 1 \tag{13.3.4}$$

Now consider any $k \notin \{i, j\}$. Using Theorem 187 and (13.3.4), $q_{ik} \geq q_{ij}q_{jk} = q_{jk}$ and $q_{jk} \geq q_{ij}q_{ik} = q_{ik}$. Thus $q_{jk} = q_{ik}$. □

Proof of (C) Denote $N = \frac{n(n-1)}{2}$; $\mathcal{A} \in \Re^N$ a vector whose elements are $0 \leq a_{ij} < 1$, $i > j$; $\mathcal{A}_l \in \Re^N$ the vector whose first l elements equal the corresponding elements of $\mathcal{A}$ and the rest are zeros; $\mathcal{A}_l^+ \in \Re^N$ the vector whose $(l+1)$-th element is one and the rest identical to $\mathcal{A}_l$; and $Q_G(\mathcal{A})$ the Q_G formed when the a_{ij} are the elements of $\mathcal{A}$. As $\mathcal{A}_N = \mathcal{A}$ it suffices to show that $Q_G(\mathcal{A}_l)$ is positive definite for all $l \in \{0, \cdots, N\}$.

Use induction on l. Note that for every l, there is a $\alpha_l \in (0, 1]$ such that $\mathcal{A}_{l+1} = \alpha_l \mathcal{A}_l + (1 - \alpha_l)\mathcal{A}_l^+$. Because of Lemma 190 and the fact that only the $(l + 1)$-th element of the three vectors $\mathcal{A}_{l+1}$, $\mathcal{A}_l$ and $\mathcal{A}_l^+$ differ from each other, there holds:

$$Q_G(\mathcal{A}_{l+1}) = \alpha_l Q_G(\mathcal{A}_l) + (1 - \alpha_l)Q_G(\mathcal{A}_l^+), \ \alpha_l \in (0, 1]. \tag{13.3.5}$$

As $\mathcal{A}_0 = 0$, $Q_G(\mathcal{A}_0) = I$ and is positive definite. Suppose for some $l \in \{0, \cdots, N-1\}$, $Q_G(\mathcal{A}_l)$ is positive definite. From (A), $Q_G(\mathcal{A}_l^+)$ is positive semidefinite. Thus (13.3.5) implies that $Q_G(\mathcal{A}_{l+1})$ is positive definite.

13.4 A Decentralized Algorithm for Finding the Largest Eigenvalue

We now describe an algorithm for computing $\lambda_{\max}(G)$ in a decentralized fashion without having to know Q_G or even the individual link probabilities. We do require the ability to experiment by introducing packets repeatedly at nodes, and measuring how many arrive at their intended destinations. For this reason, we call the algorithm the *flooding algorithm*.

Section 13.4.1 provides a recursion and a theorem that provide the conceptual basis for the algorithm. Section 13.4.2 explains the theorem by exposing certain properties of positive matrices. Section 13.4.3 explains how the near convergence of this conceptual algorithm can be locally detected at each node. The recursion in principle requires that Q_G be known. Section 13.4.4 provides the flooding algorithm that under the temporal independence of the links, implements this algorithm in a completely decentralized fashion, without having to know Q_G. Section 13.4.5 discusses some practical issues and convergence rates. Section 13.4.6 has simulations. Section 13.4.7 proves a theorem in Sect. 13.4.3.

13.4.1 A Basic Recursion

We begin with a theorem on the conceptual recursion.

Theorem 196 *Suppose* $Q_G = Q_G^\top \in \Re^{n\times n}$ *is positive. Consider* $z[k] = [z_1[k], \cdots, z_n[k]]^\top$ *and the recursion,*

$$z[k+1] = Q_G z[k] \tag{13.4.1}$$

with $z[0]$ *strictly positive. Then, for all* $i \in \{1, \cdots, n\}$,

$$\lim_{k\to\infty} \frac{z_i[k+1]}{z_i[k]} = \lambda_{\max}(Q_G) \tag{13.4.2}$$

Thus $z[k]$ converges to a positive eigenvector of Q_G associated with its maximum eigenvalue. Furthermore, (13.4.1) induces $\frac{z_i[k+1]}{z_i[k]}$, *locally seen at each node*, to converge to $\lambda_{\max}(Q_G)$.

Many variations of this theorem appear in the literature [83, 99, 184]. In most cases it is proved under an additional normalization, namely replacing (13.4.1) by:

$$z[k+1] = Q_G \frac{z[k]}{\|z[k]\|} \tag{13.4.3}$$

Such a normalization militates against our eventual goal of decentralization as its implementation requires each node to know the state of all other nodes. We still omit the proof of Theorem 196. Instead we recount properties of positive matrices that explain this result and help derive an important refinement.

13.4.2 Properties of the Recursion

Consider the *projective metric* [99], $p(x, y)$, between two positive vectors x and y with elements x_i and y_i,

$$p(x, y) = \log\left[\frac{\max_i \frac{x_i}{y_i}}{\min_i \frac{x_i}{y_i}}\right] \tag{13.4.4}$$

Evidently $p(x, y) \geq 0$ with equality if and only if for a scalar α, $x = \alpha y$. This metric is *scale invariant*, i.e., for all positive scalar α, β

$$p(\alpha x, \beta y) = p(x, y) \tag{13.4.5}$$

For a strictly positive matrix such as Q_G there is a $0 \leq \tau < 1$ such that for all positive x, y, $p\left(Q_G x, Q_G y\right) \leq \tau p(x, y)$ [99]. *In fact τ is independent of x and y and depends only on Q_G.*

Call $\lambda_{\max}(Q_G)$ the Perron-Frobenius eigenvalue of Q_G and associated eigenvectors Perron-Frobenius eigenvectors. Then, for a positive Q_G, as Perron-Frobenius eigenvectors are positive to within a scaling, with $\eta = [\eta_1, \cdots, \eta_n]^\top$ a positive Perron-Frobenius eigenvector, using (13.4.5) in (13.4.1) one has:

$$\begin{aligned} p\left(z[k+1], \eta\right) &= p\left(z[k+1], \lambda_{\max}\left(Q_G\right)\eta\right) \\ &= p\left(Q_G z[k], Q_G \eta\right) \\ &\leq \tau p\left(z[k], \eta\right) \end{aligned} \tag{13.4.6}$$

Thus as $0 \leq \tau < 1$,

$$\lim_{k \to \infty} p\left(z[k], \eta\right) = 0 \tag{13.4.7}$$

Therefore, for every $\epsilon_n > 0$, there exists k_1 such that for all $k \geq k_1$,

$$0 \leq \log\left[\frac{\max_i \frac{z_i[k]}{\eta_i}}{\min_i \frac{z_i[k]}{\eta_i}}\right] \leq \log(1 + \epsilon_n) \tag{13.4.8}$$

Then, the following lemma connects (13.4.2) to (13.4.8).

Lemma 197 *Suppose the probabilistic connectivity matrix $Q_G \in \Re^{n\times n}$ is symmetric and positive, and $\eta = [\eta_1, \cdots, \eta_n]^\top$ a Perron-Frobenius eigenvector of Q_G with all elements strictly positive. Consider (13.4.1) with positive $z[0]$. Suppose that for some $\beta \geq 0$ there exists a k_0 such that for all $k \geq k_0$,*

$$1 \leq \frac{\max_{i\in\{1,\cdots,n\}} \frac{z_i[k]}{\eta_i}}{\min_{i\in\{1,\cdots,n\}} \frac{z_i[k]}{\eta_i}} \leq 1 + \beta \tag{13.4.9}$$

Then, for all $i \in \{1, \cdots, n\}$ and $k \geq k_0$

$$\left|\frac{z_i[k+1]}{z_i[k]} - \lambda_{\max}(Q_G)\right| \leq \beta\lambda_{\max}(Q_G)$$

Proof As Q_G and $z[0]$ are positive so is $z[k]$. Consider any k for which (13.4.9) holds. At such a k define $\alpha = \min_i \frac{z_i[k]}{\eta_i}$. Then, for all $i \in \{1, \cdots, n\}$, there holds

$$\alpha\eta_i \leq z_i[k] \leq (1+\beta)\alpha\eta_i \tag{13.4.10}$$

Define $\xi[k] = z[k] - \alpha\eta$ and $(Q_G\xi[k])_i$ as the i-th element of $Q_G\xi[k]$. Because of (13.4.10), $\xi[k]$ is nonnegative. Thus, as Q_G is positive, $Q_G\xi[k]$ is nonnegative and for each $i \in \{1, \cdots, n\}$,

$$\begin{aligned} 0 &\leq (Q_G\xi[k])_i \\ &= (Q_G(z[k] - \alpha\eta[k]))_i \\ &\leq (Q_G((1+\beta)\alpha - \alpha))\eta[k])_i \\ &= \beta\alpha\lambda_{\max}(Q_G)\eta_i \end{aligned} \tag{13.4.11}$$

As $z[k+1] = Q_G\xi[k] + \alpha\lambda_{\max}(Q_G)\eta$, and $\xi[k]$ is nonnegative, from (13.4.11) for all $i \in \{1, \cdots, n\}$, there thus holds:

$$\alpha\lambda_{\max}(Q_G)\eta_i \leq z_i[k+1] \leq \alpha(1+\beta)\lambda_{\max}(Q_G)\eta_i \tag{13.4.12}$$

Hence (13.4.10) and (13.4.12) provide:

$$-\frac{\beta\lambda_{\max}(Q_G)}{1+\beta} \leq \frac{z_i[k+1]}{z_i[k]} - \lambda_{\max}(Q_G) \leq \beta\lambda_{\max}(Q_G)$$

□

Identify ϵ_n in (13.4.8) with β in Lemma 197. Then, there exists a k_1 such that for all $k \geq k_1$ and all $i \in \{1, \cdots, n\}$:

$$\left|\frac{z_i[k+1]}{z_i[k]} - \lambda_{\max}(Q_G)\right| \leq \epsilon_n\lambda_{\max}(Q_G) \tag{13.4.13}$$

This constitutes an explanation if not a proof of Theorem 196. The convergence is monotonic in the sense of (13.4.6).

Remark 198 Should Q_G be nonnegative as opposed to positive, the ratio for the i-th element of $z[k]$ will converge to the largest eigenvalue of the probabilistic connectivity matrix of the component to which the corresponding nodes belong.

13.4.3 Local Detection of Convergence

Since the convergence in (13.4.2) is asymptotic, we now explore whether each node can detect near convergence locally. Indeed the next theorem states that should n successive ratios $\frac{z_i[k+1]}{z_i[k]}$ be close enough for *any given* i, then this ratio must be close to $\lambda_{\max}(Q_G)$ and will remain close in subsequent iterations.

Theorem 199 *Under the conditions of Theorem 196, consider, for some* $c > 0$, $i \in \{1, \cdots, n\}$, $\delta > 0$ *and* k_0 *the* n *inequalities:*

$$\left| \frac{z_i[k+1]}{z_i[k]} - c \right| \leq \delta, \quad \forall k \in \{k_0, k_0+1, \cdots, k_0+n-1\}. \tag{13.4.14}$$

Then, for every $\epsilon > 0$, *there exists a* δ^* *such that for all* $0 < \delta \leq \delta^*$, *(13.4.14) implies for all* $k \geq k_0$

$$\left| \frac{z_i[k+1]}{z_i[k]} - \lambda_{\max}(Q_G) \right| \leq \epsilon \tag{13.4.15}$$

We shall prove this Theorem in Sect. 13.4.7. While this theorem does permit the i-th node to conclude if its ratios are close to the postulated connectivity measure, the question remains on whether this node can also conclude that all other nodes are also close to convergence. We now argue that though this is not true in general, it is true for general values of the probabilities a_{ij}, and hence also for general networks.

To see this, suppose for sufficiently small ϵ, (13.4.15) holds for $i = 1$. Were one to be able to conclude that this implied that $p(z[k_0], \eta)$ were small, η being a Perron-Frobenius eigenvector of Q_G, then one can conclude that (13.4.15) would hold for all i, but possibly different, albeit small ϵ. So the issue boils down to whether (13.4.15) implies a correspondingly small $p(z[k_0], \eta)$?

Though a small $p(z[k_0], \eta)$, implies a small ϵ in (13.4.15), the reverse, is generally but not always true. For all $k \geq k_0$,

$$z_1[k] = e_1^\top Q_G^{k-k_0} z[k_0] \tag{13.4.16}$$

where $e_1 = [1, 0, \cdots, 0]^\top$. Should the pair $[Q_G, e_1^\top]$ be *completely observable* [39], i.e.,

$$W = \left[e_1^\top, e_1^\top Q_G, \cdots, e_1^\top Q_G^{n-1}\right]^\top \tag{13.4.17}$$

be nonsingular then the $z[k_0]$ leading to the n-successive samples in (13.4.16) is unique. In such a case a small ϵ in (13.4.15), with $i = 1$, forces a small $p(z[k_0], \eta)$. *Consequently, each node can detect near convergence of the ratios at all other nodes, from the near convergence of its own ratios.*

For every $n > 2$, we now provide example networks, that (a) for a particular choice of the probabilities a_{ij} yield a Q_G for which $[Q_G, e_1^\top]$ is not completely observable; and (b) for a particular choice of the probabilities a_{ij} yield a Q_G for which $[Q_G, e_1^\top]$ is completely observable.

In particular (a) shows that there are networks for which a node cannot conclude that the near convergence of its ratios implies that other nodes are near convergence. What is more important from a practical point of view is (b), that shows that almost all choices of a_{ij} yield networks for which near convergence at one node implies near convergence at all. This is so as Q_G and hence W in (13.4.17) is polynomial in the a_{ij}. Thus, either W is singular for *all values of* a_{ij} or it is nonsingular for *general values*. The network is as follows.

Example 200 For $n > 2$, choose the $a_{ij} = a_{ji}$ as follows. For some $1 \geq r_i > 0$ and $i \in \{1, \cdots, n-1\}$ there holds: $a_{1,i+1} = r_i$. For all $i \in \{2, \cdots, n-1\}$ and $j \in \{i+1, \cdots, n\}$, $a_{ij} = 0$. Under the independence assumption,

$$q_{ij} = \begin{cases} r_{j-1} & i = 1 \text{ and } j \in \{2, \cdots, n\} \\ r_i r_j & i \in \{2, \cdots, n-1\} \text{ and } j \in \{i+1, \cdots, n\} \end{cases} \tag{13.4.18}$$

Thus, e.g., for $n = 4$ one has

$$Q = \begin{bmatrix} 1 & r_1 & r_2 & r_3 \\ r_1 & 1 & r_1 r_2 & r_1 r_3 \\ r_2 & r_1 r_2 & 1 & r_2 r_3 \\ r_3 & r_1 r_3 & r_2 r_3 & 1 \end{bmatrix}$$

The next Lemma proves both (a) and (b) above.

Lemma 201 *For $n > 2$, consider under $0 < r_i < 1$, the symmetric probabilistic connectivity matrix with diagonal elements $q_{ii} = 1$ and the remaining elements as in (13.4.18). Then, with $e_1 = [1, 0, \cdots, 0]^\top$, the pair $[Q_G, e_1^\top]$ is completely observable if and only if the r_is are all distinct.*

Proof By the Popov-Belevitch-Hautus test [39], $[Q_G, e_1^\top]$ is a completely observable pair if and only if for all scalar complex λ:

$$\mathbf{rank}\left(\begin{bmatrix} e_1 & \lambda I - Q_G \end{bmatrix}\right) = n \tag{13.4.19}$$

With $r = [r_1, \cdots, r_{n-1}]^\top$ and $R = \mathbf{diag}\{r_i^2\}_{i=1}^{n-1}$, (13.4.18) is

$$QG = \begin{bmatrix} 1 & r^\top \\ r & I - R - rr^\top \end{bmatrix} \tag{13.4.20}$$

Suppose the r_is are distinct, but to establish a contradiction, $[Q_G, e_1^\top]$ is not completely observable, i.e., (13.4.19) is violated. Then, there exists a scalar complex λ and nonzero $f \in \Re^{(n-1)}$ such that

$$r^\top f = 0 \tag{13.4.21}$$

and $\left((\lambda - 1)I + R - rr^\top\right)f = 0$; i.e.,

$$((\lambda - 1)I + R)f = 0 \tag{13.4.22}$$

As $r_i > 0$, $\forall i$, from (13.4.21) at least two elements of f, without loss of generality f_1 and f_2, are nonzero. Thus (13.4.22) yields

$$\lambda = 1 - r_i^2 \quad \forall i \in \{1, 2\} \tag{13.4.23}$$

which is impossible as $r_1^2 \neq r_2^2$, establishing a contradiction.

Now suppose at least two elements of r, without loss of generality, r_1 and r_2, are equal. Choose $f = [0, 1, -1, 0, \cdots, 0]^\top$ and the scalar λ as in (13.4.23). Then, clearly $e_1^\top f = 0$. Furthermore,

$$(\lambda I - Q_G)f = \left[1 \; r_1 \; r_2 \; 0_{n-3}^\top\right]^\top (r_1 - r_2) = 0,$$

where 0_{n-3} is the zero vector in $\Re^{n-3}$ (empty if $n = 3$). Thus (13.4.19) is violated and $[Q_G, e_1^\top]$ is not completely observable. □

Note that for $n = 2$, $[Q_G, e_1^\top]$ is completely observable if and only if $q_{12} \neq 0$. We have effectively shown that for almost all networks, local detection of near convergence implies near convergence of all nodes.

13.4.4 The Flooding Algorithm

Observe, (13.4.1) requires that the i-th node knows all the q_{ij} as well as all elements of $z[k]$. We now provide an algorithm that sidesteps this need and can be used in our probabilistic network setting provided the transmissions at different time slots are identically and independently distributed. Formally, we make the following assumption:

The indicator random variables I_i defined before Definition 180 are identically and independently distributed across transmission slots.

This assumption permits us to postulate a flooding algorithm that asymptotically approximates (13.4.1) in a totally decentralized fashion. Suppose, for some k, $z[k]$ has been obtained, and in a series of simultaneous experiments the i-th node *floods* the network with $x_i = z_i[k]$ number of distinct packets. Then, the expected number of packets received by this node is the i-th entry of $Q_G z[k]$. Now suppose for some K, each node repeats this flooding operation K times. Denote by $z[k, m], i \in \{1, \ldots, n\}, m \in \{1, \ldots, K\}$, the number of packets received by node v_i in the m-th repetition. Then, because of assumption of the independence of I_is, by the law of large numbers, for sufficiently large K:

$$z[k+1] \approx \frac{1}{K} \sum_{m=1}^{K} z[k, m] \tag{13.4.24}$$

There are clearly two approximations inherent in (13.4.24). Firstly, implicitly for non-integer $z[k]$, we quantize to the nearest vector of integers. Secondly (13.4.24) represents a better approximation as K grows. We comment on the size of K in Sect. 13.4.5.

Accordingly, the flooding algorithm we postulate is as follows: for some K, $l = 0$ and positive vector $y[l, K]$, let the i-th node flood the network with $y_i[l, K]$ number of packets. Every node repeats this experiment K times in the l-th iteration. The number of packets transmitted by the i-th node in the $(l+1)$-th iteration is the number of packets averaged over K transmissions, received by it in the l-th iteration. Then,

$$\lim_{l \to \infty} \left\{ \lim_{K \to \infty} \frac{y_i[l+1, K]}{y_i[l, K]} \right\} = \lambda_{\max}(Q_G) \tag{13.4.25}$$

In principle, the number of packets from a node increases by a factor approximately equal to $\lambda_{\max}(Q_G)$ in each iteration of (13.4.1). In a large network, this leaves open the risk that after a modest number of iterations, the number of packets becomes very large. As explained in Sect. 13.4.5, this may require larger values of K for the approximation in (13.4.24) to be sufficiently good. The implementation of (13.4.3), rather than just (13.4.1) would avoid this difficulty. However, the normalization by $\|z[k]\|$ in (13.4.3) does not permit a decentralized implementation. Instead we propose an *optional* renormalization to combat this challenge. Specifically, should the $y_i[l, K]$ exceed a pre-specified threshold at a particular node i, then this node must divide the number of packets it transmits by a pre-specified factor. It can then piggyback this scaling information in every packet it transmits, so that all the other nodes are alerted of this scaling, and scale the number of packets they transmit by the *the same factor*. If the pre-designated threshold is chosen to be sufficiently large, the chance of missing this scaling information is negligible. As only the convergence of ratios are at issue, there is no resulting impact on convergence speed to speak of. As argued later, this option is rarely needed.

Despite quantization and approximate averaging, simulations in Sect. 13.4.6 show that relatively small l and K suffice for the ratios $\frac{y_i[l+1,K]}{y_i[l,K]}, i \in \{1, \ldots, n\}$, to converge to a value that is very close to $\lambda_{\max}(Q_G)$.

13.4.5 Practical Issues and Convergence Rates

To avoid the effect of network delays, packets must be accumulated over large intervals. The convergence speed of (13.4.1) is measured by $\lambda_2(Q_G)/\lambda_{\max}(Q_G)$, where $\lambda_2(Q_G)$ is the second largest eigenvalue of Q_G. *Inter alia*, this suggests faster convergence in highly connected networks. To see why, observe that as Q_G is positive semidefinite, and its trace is always n, $\lambda_2(Q_G)$ is upper bounded by $n - \lambda_{\max}(Q_G)$. Thus $\lambda_{\max}(Q_G)$ lower bounds the convergence rate.

The slowest part of the convergence is determined by the law of large numbers. In fact K is proportional to the variance of the independently and identically distributed variables being averaged. As Q_G is positive semidefinite and has trace n, $\lambda_{\max}(Q_G) \geq 1$. Thus, in (13.4.1) $z_i[k]$ is potentially unbounded *though ratios of successive values are not.* Nonetheless the flooding algorithm does not estimate these ratios directly, but rather estimates the $z_i[k]$.

Just as the $z_i[k]$, $y_i[l, K]$ grow in size with l. Larger they are, the larger their initial variance. This in turn correspondingly increases the required K, thus slowing convergence. This underscores the importance of the renormalization proposed in Sect. 13.4.4 and used in the simulations. There are other mechanisms of renormalization one may invoke. For example, for some predetermined integer m all nodes scale down $y_i[l, K]$ by a factor C whenever l is a multiple of m.

Actually, in practice renormalization is rarely needed. As shown in the simulations in Sect. 13.4.6, in networks with even moderate connectivity, convergence is so rapid that it can be detected well before packet growth becomes unmanageable. In networks with low connectivity, $\lambda_{\max}(Q_G)$ is relatively small, and larger number of iterations can be sustained before packet growth becomes so large as to require normalization.

13.4.6 Simulation Studies

The simulation shown in Figs. 13.3 and 13.4 involves six nodes, and $K = 10$. Within just seven iterations, the ratio (13.4.25) converges to within half a percent of the true $\lambda_{\max}(Q_G)$.

Figure 13.5 considers a network with 50 nodes where a_{ij}s, $1 \leq i < j \leq 50$, are drawn uniformly from $[0, P]$. Varying P, which controls network connectivity, illustrates the effect of connectivity to convergence speed. Note that when the number of nodes is equal to 50, the number of edges is equal to 1225. It becomes computationally prohibitive to compute Q_G and $\lambda_{\max}(Q_G)$ whose computational complexity increases approximately with the number of edges according to $2^{|E|}$ with $|E|$ being the number of edges. Therefore, in the figure we use $\frac{y_i[10]}{y_i[9]}$ averaged over 50 nodes as an approximation of $\lambda_{\max}(Q_G)$. Furthermore, as explained in Sect. 13.4.4, to make the algorithm more efficient, whenever the number of packets flooded by a

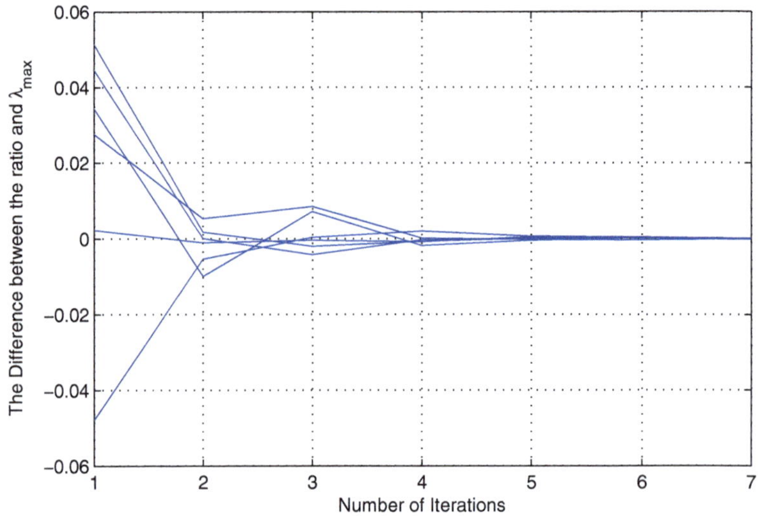

Fig. 13.3 An illustration of the convergence of the ratio $\frac{y_i[l+1]}{y_i[l]}$ to $\lambda_{\max}(Q_G)$. The simulation result is obtained from a random network with six nodes. a_{ij}s, $1 \leq i < j \leq 6$, are drawn uniformly, randomly, and independently from [0, 1]. K is chosen to be 10. The horizontal axis is the number of iterations and the vertical axis is the difference between $\frac{y_i[l+1]}{y_i[l]}$ and $\lambda_{\max}(Q_G)$. Since there are six nodes, six curves are shown in the figure corresponding to the value of $\frac{y_i[l+1]}{y_i[l]}$ for each of the six nodes

node in an iteration exceeds 5000, the number of packets flooded by all nodes in the next iteration is divided by a common factor equal to the number of nodes.

A feature of note is that foreshadowed at the end of Sect. 13.4.5. Observe in Fig. 13.5 that even with $P = 0.5$, representing a network of moderate connectivity, convergence is virtually immediate. When $P > 0.5$, this convergence occurs by $l = 1$, obviating the need for renormalization.

13.4.7 Proof of Theorem 199

We conclude this subsection by proving Theorem 199 which requires the following lemma.

Lemma 202 *Suppose $F = F^\top \in \Re^{n\times n}$ is positive and $h \in \Re^n$ is nonnegative. Suppose also that there exists a $\psi \in \Re^n$ such that:*

$$\left[h^\top, h^\top F, \cdots, h^\top F^{n-1}\right]^\top \psi = 0 \tag{13.4.26}$$

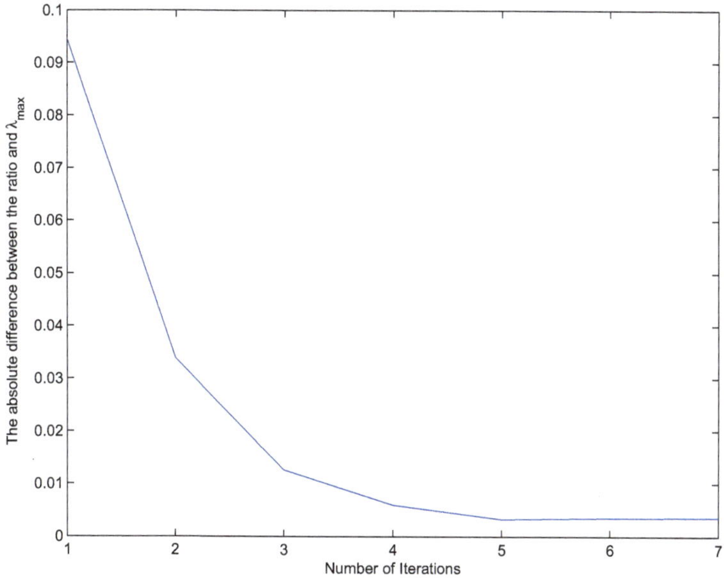

Fig. 13.4 A further illustration of the convergence of the ratio $\frac{y_i[l+1]}{y_i[l]}$ to $\lambda_{\max}(Q_G)$. The simulation result is obtained from a random network with six nodes. a_{ij}s, $1 \le i < j \le 6$, are drawn uniformly, randomly, and independently from [0, 1]. K is chosen to be 10. The horizontal axis is the number of iterations and the vertical axis is the average absolute difference between $\frac{y_i[l+1]}{y_i[l]}$ and $\lambda_{\max}(Q_G)$, i.e., $|\frac{y_i[l+1]}{y_i[l]} - \lambda_{\max}(Q_G)|$ averaged over six nodes. Furthermore the simulation is repeated 50 times and each point in the curve corresponds to the average value over 50 simulations

Consider any eigenvector ω_i of F, other than the Perron-Frobenius eigenvector, and a nonzero $\gamma \in \Re^n$ that is given by $\alpha\psi + \beta\omega_i$ for some constants α, β. Then, γ must have at least one element negative and another positive.

Proof As $F = F^\top$, its eigenvalues are real and *the eigenvectors can be chosen to form an orthonormal basis*. Suppose $\lambda_1 > \lambda_2 \ge \cdots \ge \lambda_n$, where the strictness of the first inequality is a consequence of F being positive. Suppose ω_i is a unit norm eigenvector corresponding to λ_i, with at least one element positive. From the Perron-Frobenius theorem ω_1 is positive.

Suppose γ is a linear combination of ψ with some ω_i, $i \in \{2, \cdots, n\}$. To establish a contradiction, suppose all elements of $\gamma \neq 0$ are nonnegative. Define the orthogonal matrix $U = \begin{bmatrix}\omega_1 & \Omega\end{bmatrix}$ with $\Omega = \begin{bmatrix}\omega_2, & \cdots, & \omega_n\end{bmatrix}$. Observe: $\psi = UU^\top\psi = \sum_{i=1}^{n} \omega_i \left(U^\top\psi\right)_i$, where $\left(U^\top\psi\right)_i$ denotes the i-th element of $U^\top\psi$. Now consider two cases.

Case I $\left(U^\top\psi\right)_1 = 0$: Then, ψ is in the range space of Ω. As γ is a linear combination of ψ and a column of Ω, γ is in the range space of Ω as well. Now as every column of Ω is orthogonal to ω_1, so must be γ. Then as ω_1 is positive, γ cannot be nonnegative and nonzero, establishing a contradiction.

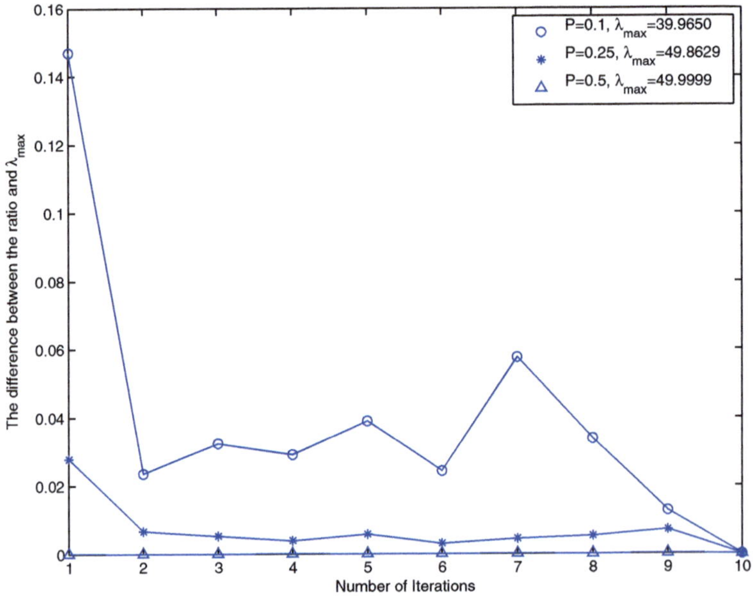

Fig. 13.5 A further illustration of the convergence of the ratio $\frac{y_i[l+1]}{y_i[l]}$ to $\lambda_{\max}(Q_G)$. The simulation result is obtained from a random network with 50 nodes. a_{ij}s, $1 \leq i < j \leq 50$, are drawn uniformly, randomly, and independently from $[0, P]$. K is chosen to be 10. The horizontal axis is the number of iterations and the vertical axis is the average absolute difference between $\frac{y_i[l+1]}{y_i[l]}$ and $\lambda_{\max}(Q_G)$, i.e., $|\frac{y_i[l+1]}{y_i[l]} - \lambda_{\max}(Q_G)|$ averaged over 50 nodes. Furthermore the simulation is repeated 10 times and each point in the curve corresponds to the average value over 10 simulations. As P increases above 0.5, the ratio converges to the true value of $\lambda_{\max}(Q_G) = 50$ immediately in the first iteration

Case II $\left(U^\top \psi\right)_1 \neq 0$: Observe that $F = U \Lambda U^\top$, with $\Lambda = \mathbf{diag}\{\lambda_1, \cdots, \lambda_n\}$. Thus using [39], (13.4.26) implies for all t:

$$0 = h^\top e^{Ft}\psi = h^\top U e^{\Lambda t} U^\top \psi = \sum_{i=1}^{N} \left(h^\top U\right)_i \left(U^\top \psi\right)_i e^{\lambda_i t}$$

As $\lambda_1 \neq \lambda_i$, for all $i \in \{2, \cdots, n\}$, this in particular implies that $\left(h^\top U\right)_1 \left(U^\top \psi\right)_1 = 0$, i.e., $0 = \left(h^\top U\right)_1 = h^\top \omega_1$. As $h \neq 0$ is nonnegative and ω_1 is positive, this cannot be true.

□

We now prove Theorem 199 by showing in turn the following: for small enough δ, (A) c in (13.4.14) is close to an eigenvalue of Q_G; (B) that this is $\lambda_{\max}(Q_G)$; and (C) that subsequent ratios $\frac{z_i[k+1]}{z_i[k]}$ remain close to $\lambda_{\max}(Q_G)$.

Proof of (A) With e_i a vector with i-th element 1 and rest 0,

$$z_i[k] = e_i^\top Q_G^k z[0] \ \ \forall k \geq 0 \tag{13.4.27}$$

Because of (13.4.14), there exist $|\delta_i| < \delta$, such that for all $k \in \{k_0, \cdots, k_0+n\}$, there holds:

$$z_i[k] = \left\{ \prod_{j=k_0}^{k} (c + \delta_j) \right\} z_i[k_0] \tag{13.4.28}$$

Suppose the characteristic polynomial of Q_G is given by: $\det(\lambda I - Q_G) = \lambda^n - \sum_{i=0}^{n-1} \alpha_i \lambda^i$. Then

$$Q_G^n = \sum_{i=0}^{n-1} \alpha_i Q_G^i \tag{13.4.29}$$

□

From (13.4.28), (13.4.27) and (13.4.29), there obtains

$$\begin{aligned}
&\left\{ \prod_{j=k_0}^{n} (c + \delta_j) \right\} z_i[k_0] \\
&= z_i[k_0 + n] \\
&= e_i^\top Q_G^{k_0+n} z[0] \\
&= e_i^\top \left(\sum_{l=0}^{n-1} \alpha_l Q_G^l \right) Q_G^{k_0} z[0] \\
&= \sum_{l=0}^{n-1} \alpha_l e_i^\top Q_G^{k_0+l} z[0] \\
&= \sum_{l=0}^{n-1} \alpha_l z_i[l + k_0] \\
&= \left(\sum_{l=0}^{n-1} \alpha_l \left\{ \prod_{j=k_0}^{k_0+l} (c + \delta_j) \right\} \right) z_i[k_0]
\end{aligned}$$

A positive $z[0]$ implies $z[k]$ is positive for all $k > 0$. Thus:

$$\left\{ \prod_{i=k_0}^{k_0+n} (c + \delta_i) \right\} = \left(\sum_{l=0}^{n-1} \alpha_l \left\{ \prod_{i=k_0}^{k_0+l} (c + \delta_i) \right\} \right) \tag{13.4.30}$$

As the roots of a monic polynomial vary continuously with its coefficients, with $\lambda_{\max}(Q_G) = \lambda_1 > \lambda_2 \geq \cdots \geq \lambda_n$ the eigenvalues of Q_G, for every $\epsilon > 0$ there exists a δ^* such that for all $0 < \delta \leq \delta^*$ under (13.4.14)

$$c \in \bigcup_{i=1}^{n} [\lambda_i - \epsilon, \lambda_i + \epsilon] \tag{13.4.31}$$

Proof of (B) We shall now show that in fact for every $\epsilon > 0$ there exists a δ^* such that for all $0 < \delta \leq \delta^*$ under (13.4.14) $c \in [\lambda_1 - \epsilon, \lambda_1 + \epsilon]$.

Suppose instead that for some $l \in \{2, \cdots, n\}$, $c \in [\lambda_l - \epsilon, \lambda_l + \epsilon]$. As (13.4.27) holds for all $k \in \{k_0, \cdots, k_0 + n - 1\}$, under (13.4.17) we have $\left[z_i[k_0] \cdots z_i[k_0 + n - 1]\right] = z^\top[k_0]W^\top$. Suppose χ is an eigenvector of Q_G corresponding to λ_l, and χ_i is its i-th element. Then: $\left[\chi_i \cdots \lambda_l^{k_0+n-1}\chi_i\right] = \chi^\top W^\top$. Then, a standard continuity argument shows that for every ϵ there exists a δ^* such that for all $0 < \delta \leq \delta^*$ under (13.4.14)

$$z[k_0] = \psi + \chi + e, \ \ \|e\| \leq \epsilon, \text{ and } W\psi = 0 \tag{13.4.32}$$

As $z[0]$ is positive, so is $z[k_0]$. Yet, because of Lemma 202, $\psi + \chi$ has at least one negative element. Thus, because of (13.4.32) for sufficiently small ϵ, $z[k_0]$ has at least one negative element.

Proof of (C) Thus with η a Perron-Frobenius eigenvector of Q_G, and ψ obeying (13.4.32), for every ϵ there exists a δ^* such that for all $0 < \delta \leq \delta^*$ under (13.4.14), (13.4.32) holds. Thus

$$e_i^\top Q_G^m \psi = 0; \ \ \forall m \tag{13.4.33}$$

Now consider the alternative recursion: $s[k+1] = Q_G s[k]; \ \ s[k_0] = \eta + e$. Because of (13.4.33) for all $k \geq k_0$,

$$z_i[k] = s_i[k] \tag{13.4.34}$$

Furthermore, as η is a Perron-Frobenius eigenvector, for every ϵ there exists a δ^* such that for all $0 < \delta \leq \delta^*$ under (13.4.14), $p(s[k_0], \eta) \leq \log(1+\epsilon)$. Consequently from (13.4.6) and Lemma 197 for every ϵ there exists a δ^* such that for all $0 < \delta \leq \delta^*$ under (13.4.14) the following holds for all $j \in \{1, \cdots, n\}$ and $k \geq k_0$: $\left|\frac{s_j[k+1]}{s_j[k]} - \lambda_{\max}(Q_G)\right| \leq \epsilon\lambda_{\max}(Q_G)$. The result follows as this also holds for $j = i$, $\lambda_{\max}(Q_G)$ is finite and (13.4.34).

13.5 Notes and Further Readings

In this chapter, we have considered the probabilistic connectivity matrix Q_G as a tool to measure the quality of network connectivity. Key properties of this matrix and their relation to the quality of network connectivity have been demonstrated. In particular, the off-diagonal entries of the probabilistic connectivity matrix provide a measure of the quality of end-to-end connections. We have provided theoretical analysis supporting the use of the largest eigenvalue of Q_G as a measure of the quality of overall network connectivity. Our analysis compares networks with the same number of nodes. For networks with different number of nodes, the largest eigenvalue of Q_G normalized by the number of nodes may be used as the quality metric. A flooding algorithm is presented for experimentally estimating the largest eigenvalue in a decentralized fashion, without knowledge of the individual link probabilities or the network topology.

Inequalities between the entries of the probabilistic connectivity matrix have been established. These may provide insights into the correlations between quality of end-to-end connections. We have also shown that Q_G is positive semidefinite and its off-diagonal entries are multiaffine functions of link probabilities. These two properties should facilitate optimization and robust network design, e.g., determining the link that maximally impacts network quality, and determining quantitatively the relative criticality of a link to either a particular end-to-end connection or to the entire network.

We assume that the links are symmetric *and* independent. We expect that our analysis can be extended with nontrivial work to the case where the assumption on symmetric links is removed. We conjecture that the largest singular value, as opposed to the largest eigenvalue of Q_G is a more appropriate measure of connectivity. Relaxing the independence assumption requires more work. Yet, we are encouraged by the fact that the elements of Q_G, being probabilities of union of edge events, are multiaffine functions of the a_{ij} and the conditional link probabilities, as $P(A \bigcup B) = P(A) + P(B) - P(B|A)P(A)$. Thus we still expect all the results in Sect. 13.3 to hold, though the proof may be nontrivial. In real applications link correlations may arise due to both physical layer correlations and correlations caused by traffic congestion.

Another implicit assumption in the chapter is that traffic is uniformly distributed and traffic between every source–destination pair is equally important. If this is not the case, a weighted version of the probabilistic connectivity matrix can be contemplated. Whether our results can be extended to a weighted probabilistic connectivity matrix is an open issue.

The research most related to the work in this chapter is possibly the work of Brooks *et al.* [32]. In [32] Brooks *et al.* considered a probabilistic version of the adjacency matrix and defined a *probabilistic adjacency matrix* as a $n \times n$ square matrix M whose (i,j)-th entry m_{ij} represents the probability of having a direct connection between distinct nodes i and j, and $m_{ii} = 0$. They observed that the probability that there exists at least one walk of length z between nodes i and j is m_{ij}^z, where m_{ij}^z is the (i,j)-th entry of $M \otimes M \otimes \cdots \otimes M$ (z times). Here $C \triangleq A \otimes B$ is defined by $C_{ij} = 1 - \prod_{l \neq i,j} \left(1 - A_{il}B_{lj}\right)$ where A_{ij}, B_{ij} and C_{ij} are the (i,j)-th entries of the $n \times n$ square matrix A, B, and C, respectively and the operator $\otimes$ is associative, so that powers are well-defined. Obviously, the existence of a walk implies the existence of a path and conversely. Furthermore, the existence of a walk of length z implies the existence of a path of length smaller than or equal to z. Considering that in a walk, an edge may appear more than once whereas in a path, all edges are distinct, it is not trivial to use their result to derive the probability of existence of a path or the probability of existence of a path of a particular length.

Part IV
Applications of Connectivity Studies

Chapter 14
Applications of Connectivity Studies

Abstract Connectivity studies play a vital role in the design, deployment, and management of a network. Furthermore, connectivity of a network conveys topological information of the network, which can be subsequently used to infer topology-related information of the network, e.g., location of nodes and boundary of the network. In this chapter, we give three examples on the applications of the connectivity studies. The first example is on the analysis of two key performance measures, viz., the access probability and the connectivity probability, of vehicular networks. The analysis reveals the tradeoffs between key system parameters, such as distance between vehicular infrastructure, e.g., base stations (BS) and road side units, vehicle density, transmission ranges of a BS and a vehicle, and their collective impact on the access probability and the connectivity probability under different communication channel models. The analysis enables network designers and operators to effectively improve network planning, deployment, and radio resource management. In the second example, we demonstrate the use of connectivity information to estimate the distance between a pair of neighboring nodes. Such distance estimates can be subsequently used to derive the relative location of nodes. In the third example, we introduce a category of connectivity-based wireless localization schemes that estimate the location of nodes directly using the connectivity information, without the need to first estimating the distances between particular pairs of neighboring nodes.

Any communication network is essentially used to transport packets from some nodes to other nodes in the network. Connectivity is a prerequisite to achieving this function. Therefore, connectivity studies play a vital role in the design, deployment, and management of a network. Furthermore, in wireless networks, a direct connection can only occur between two nearby nodes. Therefore, a node being directly connected to another node reveals information on the relative positions of the two nodes. More generally, connectivity of a network conveys topological information of the network, which can be subsequently used to infer topology-related information of the network, e.g., location of nodes and boundary of the network.

G. Mao, *Connectivity of Communication Networks*,
DOI 10.1007/978-3-319-52989-9_14

In this chapter, we give three examples on the applications of the connectivity studies. The first example is on the analysis of two key performance measures, viz., the access probability and the connectivity probability, of vehicular networks. The analysis reveals the tradeoffs between key system parameters, such as distance between vehicular infrastructure, e.g., base stations (BS) and road side units, vehicle density, transmission ranges of a BS and a vehicle, and their collective impact on the access probability and the connectivity probability under different communication channel models. The analysis enables network designers and operators to effectively improve network planning, deployment, and radio resource management. In the second example, we demonstrate the use of connectivity information to estimate the distance between a pair of neighboring nodes. Such distance estimates can be subsequently used to derive the relative location of nodes. In the third example, we introduce a category of connectivity-based wireless localization schemes that estimate the location of nodes directly using the connectivity information, without the need to first estimating the distances between particular pairs of neighboring nodes.

14.1 Access and Connectivity Properties of Vehicular Networks

Vehicular networks are a type of application-oriented networks deployed along roads for safety and emergency information dissemination, entertainment content distribution, and road traffic data collection and communication. A vehicular network is a hybrid wireless network that supports both infrastructure-based and ad hoc communications. Specifically, vehicles on the road can communicate with each other through a multi-hop ad hoc path. They can also access the Internet and other broadband services via the roadside infrastructure, i.e., base stations (BSs), road side units (RSUs), or access points (APs) along the road. When a vehicle moves outside the radio coverage area of a BS, it will identify and use its neighboring vehicles, if exist, as relays to access the roadside infrastructure. These types of Vehicle-to-Vehicle (V2V) and Vehicle-to-Infrastructure (V2I) communications have received significant interests from both academia and industry.

In this section, we shall develop an analytical model to study the access probability (for user satisfaction analysis) and the connectivity probability (for service coverage analysis) for infrastructure-based vehicular networks, wherein both one-hop (direct access) and two-hop (via a relay) communications between a vehicle and the infrastructure (i.e., a BS or a RSU) are supported. A general random connection model is first used to investigate the impact of different system parameters, i.e., inter-BS distance (or BS density), vehicle density, radio coverage ranges of BSs and vehicles, on key performance metrics, i.e., user access probability and service connectivity probability. The analysis is then applied to two widely used communication channel models, i.e., the unit disk connection model and

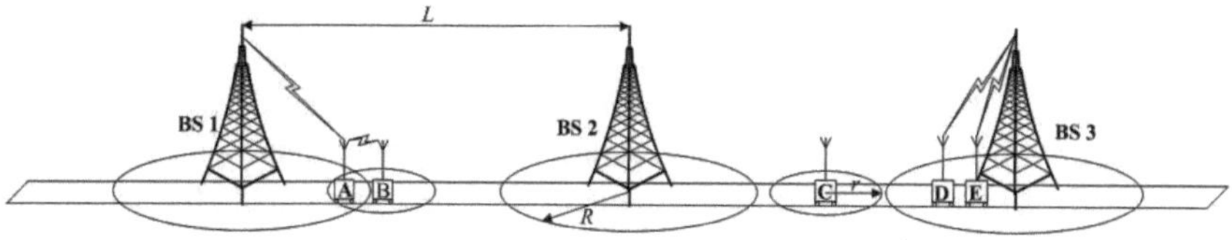

Fig. 14.1 An infrastructure-based vehicular network

the log-normal connection model, as specific examples of the general random connection model. This research enables us to improve the access probability and the connectivity probability in vehicular networks, and thereby support reliable V2I and V2V data transmissions in a number of applications and services, such as emergency messaging service, mobile Internet access, and on-road entertainments.

14.1.1 System Model

We consider an infrastructure-based vehicular network, as shown in Fig. 14.1, where a number of BSs are uniformly deployed along a long road, while other vehicles are distributed on the road randomly according to a homogeneous Poisson point process. We analyze the access probability, i.e., the probability that an arbitrary vehicle can access its nearby BSs within two hops, and the connectivity probability, i.e., the probability that all vehicles can access at least one BS within two hops, of the network by investigating a subnetwork bounded by two adjacent base stations. Here the limitation to two hops comes from the relevant technical standards, e.g., IEEE 802.11p standard, and the fact that in a wireless multi-hop network, end-to-end communication quality deteriorates significantly with even a modest increase in the number of hops. Let L be the Euclidean distance (in meters) between two adjacent BSs and ρ be the vehicle density measured in vehicles per meter. Since the vehicles are assumed to be Poissonly distributed with density ρ, discussion on the distribution of the number of vehicles on the road is only meaningful if we restrict to the (random) number of vehicles in a specific section of the road, and we call any section of the road a *road segment*. For a road segment with length x, the number of vehicles on that road segment is then a Poisson random variable with mean ρx. It follows that the probability that there are k vehicles on a road segment of x meters is given as

$$f(k,x) = \frac{(\rho x)^k e^{-\rho x}}{k!}, \; k \geq 0 \tag{14.1.1}$$

Since we investigate a subnetwork bounded by two adjacent BSs, the probability that there are k vehicles on the road segment bounded by two adjacent BSs is then $f(k,L)$.

Assuming a general channel model $\mathcal{C}$, let $g_v^{\mathcal{C}}(x)$ be the probability that two vehicles separated by a Euclidean distance x are directly connected. Similarly, denote by $g_b^{\mathcal{C}}(x)$ the probability that a vehicle and a BS separated by a Euclidean distance x are directly connected. We assume that the event that a pair of vehicles (or a vehicle and a BS) are directly connected is independent of the event that another distinct pair of vehicles (or a vehicle and a BS) are directly connected. That is, the event that two vehicles (or in the similar case, between a vehicle and a BS) are directly connected is only determined by the location of the two vehicles and is not affected by the presence or absence of connections between other pairs of vehicles. Although field measurements in real applications seem to indicate that connections between different pairs of geographically/frequency proximate wireless nodes are correlated [4, 41, 186], the independence assumption is generally considered appropriate for far-field transmission and has been widely used in the literature under many channel models including log-normal connection model.

We also assume that $g_b^{\mathcal{C}}(x) \geq g_v^{\mathcal{C}}(x)$ for any $x \geq 0$. This assumption is justified because it is often the case that a BS can not only transmit at a larger transmission power than a vehicle, it can also be equipped with more sophisticated antennas, which make it more sensitive to the transmitted signal from a vehicle.

14.1.2 Analysis of Access and Connectivity Probabilities

Assume that the subnetwork being considered is placed at $[0, L]$. The two BSs at both ends of the subnetwork are labeled as BS1 and BS2 and are at 0 and L respectively. Denote by $G(L, \rho, \mathcal{C})$ the subnetwork with length L, vehicle density ρ, and channel model $\mathcal{C}$. We investigate the access probability p_a that an arbitrary vehicle in $G(L, \rho, \mathcal{C})$ can access either BS (either BS1 or BS2). We also investigate the probability p_c that all vehicles in $G(L, \rho, \mathcal{C})$ are connected to at least one of the BSs at both ends of the subnetwork.

A vehicle is said to be located at x if its Euclidean distance to BS1 is x. The probabilities that a vehicle located at x is not directly connected to BS1 and BS2 are $1 - g_b^{\mathcal{C}}(x)$ and $1 - g_b^{\mathcal{C}}(L - x)$ respectively. Because the event that a vehicle is not directly connected to BS1 and the event that the same vehicle is not directly connected to BS2 are independent, the probability that the vehicle is directly connected to either BS1 or BS2 is then

$$p_1(x) = 1 - (1 - g_b^{\mathcal{C}}(x))(1 - g_b^{\mathcal{C}}(L - x)) \tag{14.1.2}$$

In order to derive p_a and p_c we need the following lemmas.

Lemma 203 *Let K_1 be the set of vehicles in the subnetwork $G(L, \rho, \mathcal{C})$ which are directly connected to either BS1 or BS2, then K_1 has an inhomogeneous Poisson distribution with density $\rho p_1(x)$ where $p_1(x)$ is given by (14.1.2).*

Lemma 203 is an easy consequence of Theorem 3.

Lemma 204 *Let $p_2(x)$ be the probability that a vehicle located at x in $G(L, \rho, \mathcal{C})$ is directly connected to at least one vehicle in K_1, then*

$$p_2(x) = 1 - e^{-\int_0^L g_v^{\mathcal{C}}(\|x-y\|)\rho p_1(y)dy} \tag{14.1.3}$$

where $p_1(y)$ is given by (14.1.2).

Proof Imagine we partition $[0, L]$ into L/dy nonoverlapping intervals of differential length dy. Since dy is a very small value, the probability that there exist more than one vehicles within each interval of length dy can be ignored and the probability that there exists exactly one vehicle within dy is ρdy. The probability that there exists a vehicle in $[y, y + dy]$ which is also in K_1 is then given by $\rho p_1(y)dy$. Note that the vehicles at x and y are directly connected to each other with probability $g_v^{\mathcal{C}}(\|x-y\|)$. Therefore, the probability that a vehicle at x is directly connected to a vehicle in K_1 and is located in $[y, y + dy]$ is $g_v^{\mathcal{C}}(\|x - y\|)\rho p_1(y)dy$.

Let $h(x, y)$ denotes the probability that the vehicle at x is *not* directly connected to any of the vehicles in K_1 located within $[0, y]$. Because the events that distinct pairs of vehicles are directly connected are independent, the event that the vehicle at x is *not* directly connected to any of the vehicles in K_1 located within $[0, y]$ is independent of the event that the same vehicle is not directly connected to the vehicle in K_1 located within $[y, y + dy]$ (if there is any). We have

$$h(x, y + dy) = h(x, y)(1 - g_v^{\mathcal{C}}(\|x - y\|)\rho p_1(y)dy) \tag{14.1.4}$$

where the second term on the right-hand side of (14.1.4) is the complement of the probability that a vehicle at x is directly connected to a vehicle in K_1 and located in $[y, y + dy]$. Equation (14.1.4) leads to

$$dh(x, y) = -h(x, y)g_v^{\mathcal{C}}(\|x - y\|)\rho p_1(y)dy \tag{14.1.5}$$

Therefore, the probability that a vehicle at x is not directly connected to any vehicle in K_1 is

$$h(x) = e^{-\int_0^L g_v^{\mathcal{C}}(\|x-y\|)\rho p_1(y)dy} \tag{14.1.6}$$

The result follows. □

The following two theorems give the access probability p_a and the connectivity probability p_c respectively.

Theorem 205 *Denote by $p_a(x)$ the access probability of a vehicle at x, i.e., the probability that the vehicle at x is connected to either BS1 or BS2 in at most two hops. Then*

$$p_a(x) = 1 - (1 - p_1(x))(1 - p_2(x)) \tag{14.1.7}$$

where $p_1(x)$ is given by (14.1.2) and $p_2(x)$ is given by (14.1.3).

Proof The result follows readily from the observation that the event that a vehicle at x is directly connected to either BS1 or BS2 is independent of the event that the same vehicle is directly connected to at least one vehicle in K_1. □

Theorem 206 *Denote by p_c the connectivity probability of $G(L, \rho, \mathcal{C})$, i.e., the probability that all vehicles in the subnetwork $G(L, \rho, \mathcal{C})$ are connected to either BS1 or BS2 in at most two hops. Assume that the event that a vehicle is connected to either BS1 or BS2 in at most two hops is independent of the event that another vehicle is connected to either BS1 or BS2 in at most two hops. Then*

$$p_c = e^{-\int_0^L \rho(1-p_a(x))dx} \tag{14.1.8}$$

where $p_a(x)$ is given by (14.1.7).

Proof Let K_2 be the set of vehicles in $G(L, \rho, \mathcal{C})$ which are connected to either BS1 or BS2 in exactly two hops. Together with the definition of K_1 in Lemma 203, let $\overline{K_1 + K_2} = K \backslash (K_1 + K_2)$ be the set of vehicles in $G(L, \rho, \mathcal{C})$ which are *not* connected to either BS1 or BS2 in at most two hops. Apply the thinning procedure for K, i.e., consider a realization of K and remove each vehicle located at x independently from this realization with probability $p_a(x)$. The resulting set of vehicles can be viewed as a realization of $\overline{K_1 + K_2}$ under our assumption that the event that one vehicle is connected to either BS in two hops is independent of the event that another vehicle is connected to either BS in two hops, and the probability that vehicle at x is connected to either BS in two hops is $p_a(x)$. Using the same technique as that used in the proof of Lemma 203, it can be readily shown that $\overline{K_1 + K_2}$ has an inhomogeneous Poisson distribution with density $\rho(1 - p_a(x))$. Then, all vehicles $G(L, \rho, \mathcal{C})$ are connected to either BS1 or BS2 in at most two hops if and only if $\mathcal{N}(\overline{K_1 + K_2}) = 0$ where $\mathcal{N}(\overline{K_1 + K_2})$ denotes the number of vehicles in the set $\overline{K_1 + K_2}$. The result follows. □

Note that Theorem 206 only gives an approximate result for the connectivity probability because of the independence assumption. The following lemma proves, in a way, that the event that a vehicle is connected to either BS1 or BS2 in at most two hops is *not* independent of the event that another vehicle is connected to either BS1 or BS2 in at most two hops.

Lemma 207 *Let $h(x) = 1 - p_2(x)$ be the probability that a vehicle at x is not directly connected to any vehicle in K_1. Let $h(x_1, x_2)$ be the probability that two vehicles, at x_1 and x_2 respectively, are not directly connected to any vehicle in K_1. Then, $h(x_1, x_2) \geq h(x_1)h(x_2)$.*

Proof Let $h(x_1, x_2; y)$ denotes the probability that two vehicles, at x_1 and x_2 respectively, are *not* directly connected to any vehicle in K_1 located in $[0, y]$. Using the similar argument in (14.1.4), we have

$$h(x_1, x_2; y + dy) = h(x_1, x_2; y)k(x_1, x_2; y) \tag{14.1.9}$$

where

$$k(x_1, x_2; y) = (1 - g_v^{\mathcal{C}}(\|x_1 - y\|))(1 - g_v^{\mathcal{C}}(\|x_2 - y\|))\rho p_1(y)dy + (1 - \rho p_1(y)dy)$$

The first term on the right-hand side of $k(x_1, x_2; y)$ is the probability there is a vehicle in K_1 located in $[y, y + dy]$ and both vehicles in x_1 and x_2 are not directly connected to it. The second term on the right-hand side of $k(x_1, x_2; y)$ is the probability that there is no vehicle in K_1 located in $[y, y + dy]$. Expanding the right-hand side of $k(x_1, x_2; y)$ we have

$$\begin{aligned} k(x_1, x_2; y) &= 1 - g_v^{\mathcal{C}}(\|x_1 - y\|)\rho p_1(y)dy \\ &\quad - g_v^{\mathcal{C}}(\|x_2 - y\|)\rho p_1(y)dy \\ &\quad + g_v^{\mathcal{C}}(\|x_1 - y\|)g_v^{\mathcal{C}}(\|x_2 - y\|)\rho p_1(y)dy \end{aligned}$$

Using the same approach in Lemma 204, we obtain

$$\begin{aligned} h(x_1, x_2) &= e^{-\int_0^L \left[g_v^{\mathcal{C}}(\|x_1-y\|)+g_v^{\mathcal{C}}(\|x_2-y\|)\right]\rho p_1(y)dy} \\ &\quad \times e^{\int_0^L g_v^{\mathcal{C}}(\|x_1-y\|)g_v^{\mathcal{C}}(\|x_2-y\|)\rho p_1(y)dy} \\ &\geq e^{-\int_0^L \left[g_v^{\mathcal{C}}(\|x_1-y\|)+g_v^{\mathcal{C}}(\|x_2-y\|)\right]\rho p_1(y)dy} \\ &= h(x_1)h(x_2) \end{aligned}$$

where the last step results from (14.1.6). □

Before obtaining the exact result of the connectivity probability, we introduce some properties in the following lemma.

Lemma 208 *Let $p_c(\mathbf{y})$ be the connectivity probability of $G(L, \rho, \mathcal{C})$ conditioned on that the number of vehicles directly connected to either BS is n and they are located at $\mathbf{y} = \{y_1, y_2, \cdots, y_n : 0 \leq y_i \leq L, 1 \leq i \leq n\}$. Let $p_Y(\mathbf{y})$ be the probability density function of $\mathbf{y}$ conditioned on that there are n vehicles directly connected to either BS. The following properties hold.*

$$(i) p_Y(\mathbf{y}) = \prod_{i=1}^{n} \frac{p_1(y_i)}{\int_0^L p_1(x)dx} \tag{14.1.10}$$

$$(ii) p_c(\mathbf{y}) = e^{-\int_0^L \rho(1-p_1(x))\prod_{i=1}^{n}(1-g_v^{\mathcal{C}}(\|x-y_i\|))dx} \tag{14.1.11}$$

Proof For $n = 1, p_Y(y_1) = \frac{p_1(y_1)}{\int_0^L p_1(x)dx}$ is the probability density function that a vehicle in K_1 is located at y_1. Since the events corresponding to $p_1(y_i)$ and $p_1(y_j)$ respectively are mutually independent for $i \neq j$, the result follows for (14.1.10).

For (14.1.11), note that a vehicle at x is not connected to any BS in at most two hops if it is not directly connected to any BS (the probability is $1 - p_1(x)$) *and* it is not directly connected to vehicles which are located at $\mathbf{y}$ given that these vehicles are in K_1 (the probability is $1 - g_v^{\mathcal{C}}(\|x - y_i\|)$ for $1 \leq i \leq n$). That is, vehicle at x cannot access any BS in at most two hops with probability

$$(1 - p_1(x)) \prod_{i=1}^{n} (1 - g_v^{\mathcal{C}}(\|x - y_i\|)) \tag{14.1.12}$$

Equation (14.1.12) is valid when $x \notin \mathbf{y}$. When $x = y_j$ for arbitrary j, we assume that $g_v^{\mathcal{C}}(0) = 1$. This implies that $p_a(x|\mathbf{y}) = 0$. This treatment allows (14.1.12) to be still valid when $x \in \mathbf{y}$.

Applying the thinning procedure and the technique used in Lemma 203, we have the number of vehicles which are neither directly connected to any BS nor directly connected to any of the vehicles at $\mathbf{y}$ is an inhomogeneous Poisson random variable with density $\rho(1 - p_1(x)) \prod_{i=1}^{n}(1 - g_v^{\mathcal{C}}(\|x - y_i\|))$. The result readily follows. □

Theorem 209 *Denote by p_c the connectivity probability of $G(L, \rho, \mathcal{C})$, i.e., the probability that all vehicles in the subnetwork $G(L, \rho, \mathcal{C})$ are connected to either BS1 or BS2 in at most two hops. Then,*

$$p_c = \sum_{n=0}^{\infty} \Pr(\mathcal{N}(K_1) = n) \left[\int_{[0,L]^n} p_c(\mathbf{y}) p_Y(\mathbf{y}) d\mathbf{y} \right] \tag{14.1.13}$$

where $p_c(\mathbf{y})$ and $p_Y(\mathbf{y})$ are given by Lemma 208, $\Pr(\mathcal{N}(K_1) = n)$ *is given by Lemma 203. When $n = 0$, we declare*

$$\begin{aligned} &\int_{[0,L]^n} p_c(\mathbf{y}) p_Y(\mathbf{y}) d\mathbf{y} \bigg|_{n=0} \\ &= p_c(\mathbf{y}) p_Y(\mathbf{y})\big|_{n=0} \\ &= e^{-\int_0^L \rho(1-p_1(x))dx} \end{aligned}$$

Proof Equation (14.1.13) directly follows from the total probability theorem, so the details are omitted here. □

Equation (14.1.13) gives an exact formula for the connectivity probability which does not rely on the assumption that the event that a vehicle is connected to either BS in two hops and the event that another vehicle is connected to either BS in two hops are independent. However (14.1.13) is much more complicated than the approximate result in (14.1.8). In many situations, (14.1.8) provides a reasonably accurate result for the connectivity probability. Therefore, we include both results.

14.1.3 Performance Evaluation Under Specific Connection Models

Based on the analysis in Sect. 14.1.2, we further derive and compare in this subsection the access probability and connectivity probability performance under two specific connection models, i.e., the unit disk connection model and the log-normal connection model.

14.1.3.1 Unit Disk Connection Model

In the unit disk connection model $\mathcal{U}$, assume that two vehicles are directly connected if and only if their Euclidean distance is less than or equal to r; assume that a vehicle and a BS are directly connected if and only if their Euclidean distance is not more than R. In other words,

$$g_v^{\mathcal{U}}(x) = \begin{cases} 1 & \text{if } x \le r \\ 0 & \text{otherwise,} \end{cases} \qquad g_b^{\mathcal{U}}(x) = \begin{cases} 1 & \text{if } x \le R \\ 0 & \text{otherwise.} \end{cases}$$

where r and R are predetermined values, known as the transmission ranges. Typically we have $R > r$. Applying the above equations into (14.1.2), (14.1.3), and (14.1.7) we obtain the access probability under the unit disk model $\mathcal{U}$:

(1) For $0 < L \le 2R$,
we have $p_1(x) = 1$ implies that $p_a(x) = 1$ for $x \in [0, L]$. Hence,

$$p_a = 1$$

(2) For $2R < L \le 2R + r$,
we have $p_1(x)$ is 0 when $x \in (R, L - R)$, and 1 otherwise. When $x \in (R, L - R)$, (14.1.3) becomes

$$\begin{aligned} p_2(x) &= 1 - e^{-\int_0^R g_v^{\mathcal{C}}(\|x-y\|)\rho dy - \int_{L-R}^L g_v^{\mathcal{C}}(\|x-y\|)\rho dy} \\ &= 1 - e^{-\int_{x-r}^R \rho dy - \int_{L-R}^{x+r} \rho dy} \\ &= 1 - e^{-\rho(2R+2r-L)} \end{aligned}$$

So substituting $p_2(x)$ into (14.1.7), it is obtained that

$$\begin{aligned} p_a &= \frac{2R}{L} + \frac{L-2R}{L}(1 - e^{-\rho(2R+2r-L)}) \\ &= 1 - \frac{L-2R}{L} e^{-\rho(2R+2r-L)} \end{aligned}$$

(3) For $2R + r < L \le 2R + 2r$,
we have for $x \in (R, L - R)$, (14.1.3) becomes

$$\begin{aligned} p_2(x) &= 1 - e^{-\int_0^R g_v^{\mathcal{C}}(\|x-y\|)\rho dy - \int_{L-R}^L g_v^{\mathcal{C}}(\|x-y\|)\rho dy} \\ &= 1 - e^{-\int_{x-r}^R \rho dy - \int_{L-R}^{x+r} \rho dy} \end{aligned}$$

So substituting $p_2(x)$ into (14.1.7), it follows that

$$\begin{aligned} p_a &= \frac{2R}{L} + \frac{1}{L}\int_R^{L-R-r} (1 - e^{-\rho(R+r-x)})dx \\ &+ \frac{1}{L}\int_{L-R-r}^{R+r} (1 - e^{-\rho(2R+2r-L)})dx \\ &+ \frac{1}{L}\int_{R+r}^{L-R} (1 - e^{-\rho(R+r+x-L)})dx \\ &= 1 + \frac{2}{\rho L}(e^{-\rho r} - e^{-\rho(2R+2r-L)}) \\ &- \frac{2R+2r-L}{L}e^{-\rho(2R+2r-L)} \end{aligned}$$

(4) For $L > 2R + 2r$,
From (14.1.3)

$$p_2(x) = \begin{cases} 1 - e^{-\int_{x-r}^{R} \rho dy} = 1 - e^{-\rho(R+r-x)} & \text{when } x \in (R, R+r] \\ 1 - e^{-\int_{L-R}^{x+r} \rho dy} = 1 - e^{-\rho(R+r+x-L)} & \text{when } x \in [L-R-r, L-R) \\ 0 & \text{when } x \in (R+r, L-R-r) \end{cases}$$

Substituting $p_2(x)$ into (14.1.7):

$$\begin{aligned} p_a &= \frac{2R}{L} + \frac{1}{L}\int_R^{R+r} (1 - e^{-\rho(R+r-x)})dx \\ &+ \frac{1}{L}\int_{L-R-r}^{L-R} (1 - e^{-\rho(R+r+x-L)})dx \\ &= \frac{2R+2r}{L} + \frac{2(e^{-\rho r} - 1)}{\rho L} \end{aligned}$$

To derive the equations for the connectivity probability (exact result), we first look at Lemma 208. Under the unit disk connection model, $p_1(x)$ is 1 when $x \in [0, R] \cup [L-R, L]$ and zero otherwise. Hence, (14.1.10) becomes

$$p_Y(\mathbf{y}) = \frac{1}{(\min(2R, L))^n} \tag{14.1.14}$$

when $y_i \in [0, R] \cup [L-R, L]$, $\forall y_i$, and zero otherwise. Equation (14.1.11) becomes

$$p_c(\mathbf{y}) = e^{-\int_R^{L-R} \rho \prod_{i=1}^n (1 - g_v^C(\|x - y_i\|))dx} \tag{14.1.15}$$

Note that $\prod_{i=1}^{n}(1 - g_v^{\mathcal{C}}(\|x - y_i\|))$ is 1 when $\|x - y_i\| > r$ for all y_i. For $L \leq 2R$, we can easily obtain p_c from (14.1.13) by substituting (14.1.14) and (14.1.15) into it (will be shown later). To obtain the result for $L > 2R$, the following transformation will simplify the arithmetic work.

Let S_a (and S_b) be the set of vehicles in $[0, R]$ (and $[L-R, L]$) which, by definition, are also in K_1. Let $\mathcal{N}(S_a)$ (and $\mathcal{N}(S_b)$) be the number of vehicles in S_a (and S_b). Note that $S_a \cup S_b = K_1$ and $\mathcal{N}(S_a) + \mathcal{N}(S_b) = \mathcal{N}(K_1)$. Let y_a (and y_b) be the location of the vehicle in S_a (and S_b) which is furthest from BS1 (and BS2). That is,

$$y_a = \begin{cases} 0 & \text{if } S_a = \emptyset \\ \max\{y : y \in S_a\} & \text{otherwise,} \end{cases} \tag{14.1.16}$$

$$y_b = \begin{cases} L & \text{if } S_b = \emptyset \\ \min\{y : y \in S_b\} & \text{otherwise.} \end{cases} \tag{14.1.17}$$

Therefore, the cumulative distribution function of p_a is

$$\begin{aligned} \Pr(y_a \leq y_{max}) &= \Pr(y_i \leq y_{max}, \forall y_i \in S_a) \\ &= (\frac{y_{max}}{R})^{n_a} \quad \text{for } n_a = \mathcal{N}(S_a) \geq 1 \end{aligned}$$

With (14.1.16), defining $y_a = 0$ when $\mathcal{N}(S_a) = 0$, we have the probability density function of y_a as

$$f_a(y_a; n_a) = \begin{cases} \frac{n_a}{R}(\frac{y_a}{R})^{n_a - 1} & \text{if } n_a \geq 1 \\ \delta(y_a) & \text{if } n_a = 0 \end{cases}$$

Similarly, we have the probability density function of y_b as

$$f_b(y_b; n_b) = \begin{cases} \frac{n_b}{R}(\frac{L - y_b}{R})^{n_b - 1} & \text{if } n_b \geq 1 \\ \delta(L - y_b) & \text{if } n_b = 0 \end{cases}$$

With y_a and y_b, we can rewrite (14.1.15) into

$$p_c(y_a, y_b) = e^{-\int_{\max\{R, y_a + r\}}^{\min\{L-R, y_b - r\}} \rho dx} \tag{14.1.18}$$

and (14.1.13) can be transformed into

$$p_c = \sum_{n_a=0}^{\infty} \sum_{n_b=0}^{\infty} \Pr(\mathcal{N}(S_a) = n_a) \Pr(\mathcal{N}(S_b) = n_b)$$

$$\left[\int_0^R \int_{L-R}^{L} p_c(y_a, y_b) f_a(y_a; n_a) f_b(y_b; n_b) dy_b dy_a\right] \tag{14.1.19}$$

for $L > 2R$. Equation (14.1.19) can be further simplified under different cases. For $n_a > 0$ and $n_b > 0$, (14.1.19) becomes

$$p_c^{(n_a>0,n_b>0)}$$

$$= \int_0^R \int_{L-R}^L p_c(y_a, y_b) \left[\sum_{n_a=1}^{\infty} \sum_{n_b=1}^{\infty} \Pr(\mathcal{N}(S_a) = n_a) \right.$$

$$\left. \Pr(\mathcal{N}(S_b) = n_b) f_a(y_a; n_a) f_b(y_b; n_b) \right] dy_b dy_a$$

$$= \int_0^R \int_{L-R}^L p_c(y_a, y_b) \left[\sum_{n_a=1}^{\infty} \sum_{n_b=1}^{\infty} \frac{(\rho R)^{n_a}}{n_a!} e^{-\rho R} \right.$$

$$\left. \frac{(\rho R)^{n_b}}{n_b!} e^{-\rho R} \frac{n_a}{R} (\frac{y_a}{R})^{n_a-1} \frac{n_b}{R} (\frac{L-y_b}{R})^{n_b-1} \right] dy_b dy_a$$

$$= \int_0^R \int_{L-R}^L p_c(y_a, y_b) \rho^2 e^{-2\rho R}$$

$$\left[\sum_{n_a=1}^{\infty} \frac{(\rho y_a)^{n_a-1}}{(n_a-1)!} \right] \left[\sum_{n_b=1}^{\infty} \frac{(\rho(L-y_b))^{n_b-1}}{(n_b-1)!} \right] dy_b dy_a$$

$$= \int_0^R \int_{L-R}^L p_c(y_a, y_b) \rho^2 e^{-2\rho R} e^{\rho y_a} e^{\rho(L-y_b)} dy_b dy_a \qquad (14.1.20)$$

For $n_a = 0$ and $n_b > 0$, (14.1.19) becomes

$$p_c^{(n_a=0,n_b>0)}$$

$$= \sum_{n_b=1}^{\infty} e^{-\rho R} \frac{(\rho R)^{n_b}}{n_b!} e^{-\rho R} \left[\int_{L-R}^L p_c(0, y_b) \frac{n_b}{R} (\frac{L-y_b}{R})^{n_b-1} dy_b \right]$$

$$= \int_{L-R}^L p_c(0, y_b) \rho e^{-2\rho R} \sum_{n_b=1}^{\infty} \frac{(\rho(L-y_b))^{n_b-1}}{(n_b-1)!} dy_b$$

$$= \int_{L-R}^L p_c(0, y_b) \rho e^{-2\rho R} e^{\rho(L-y_b)} dy_b \qquad (14.1.21)$$

With similar steps (omitted here) we can obtain for $n_a > 0$ and $n_b = 0$, (14.1.19) becomes

$$p_c^{(n_a>0,n_b=0)} = \int_0^R p_c(y_a, L) \rho e^{-2\rho R} e^{\rho y_a} dy_a \qquad (14.1.22)$$

Note that it can be shown that (14.1.22) is equivalent to (14.1.21) by letting $y_b = L - y_a$, then

$$p_c^{(n_a>0,n_b=0)} = -\int_{L}^{L-R} p_c(L-y_b, L)\rho e^{-2\rho R} e^{\rho(L-y_b)} dy_b$$
$$= \int_{L-R}^{L} p_c(0, y_b)\rho e^{-2\rho R} e^{\rho(L-y_b)} dy_b$$

where $p_c(L-y_b, L) = p_c(0, y_b)$. Finally, for $n_a = 0$ and $n_b = 0$,

$$\begin{aligned} p_c^{(n_a=0,n_b=0)} &= e^{-\rho R} e^{-\rho R} p_c(0, L) \\ &= e^{-2\rho R} e^{-\int_R^{L-R} \rho dx} \\ &= e^{-2\rho R} e^{-\rho(L-2R)} \\ &= e^{-\rho L} \end{aligned} \tag{14.1.23}$$

Using (14.1.20), (14.1.21), (14.1.22), and (14.1.23), we can obtain the connectivity probability as follows. Due to the lengthy (but straightforward) steps involved to derive the results, we omit the intermediate steps and only include the results of (14.1.20) and (14.1.21).

(1) For $0 < L \leq 2R$,
$p_Y(\mathbf{y}) = \frac{1}{L^n}$ from (14.1.10) and $p_c(\mathbf{y}) = 1$ from (14.1.11) implies that

$$p_c = \sum_{n=0}^{\infty} \Pr(\mathcal{N}(K_1) = n) = 1$$

(2) For $2R < L \leq 2R + r$,

$$\begin{aligned} p_c^{(n_a>0,n_b>0)} &= 1 + e^{-\rho L} - 2e^{-\rho R} + e^{-\rho(3R+r-L)} \\ &\quad + \left(-\frac{1}{4} - \frac{1}{2}\rho(L-2R)\right) e^{-\rho(2R+2r-L)} \\ &\quad - e^{-\rho(L+r-R)} + \frac{1}{4} e^{-\rho(L+2r-2R)} \end{aligned}$$

$$\begin{aligned} p_c^{(n_a=0,n_b>0)} &= -e^{-\rho L} + e^{-\rho R} - \frac{1}{2} e^{-\rho(3R+r-L)} \\ &\quad + \frac{1}{2} e^{\rho(L+r-R)} \end{aligned}$$

$$\begin{aligned} p_c &= 1 + \frac{1}{4} e^{-\rho(L+2r-2R)} \\ &\quad + (-\frac{1}{4} - \frac{1}{2}\rho(L-2R)) e^{-\rho(2R+2r-L)} \end{aligned}$$

(3) For $2R + r < L \leq 2R + 2r$,

$$p_c^{(n_a>0,n_b>0)}$$
$$= 1 + e^{-\rho L} + \frac{1}{2}e^{-\rho(L-2R)} - e^{-\rho(L+r-R)}$$
$$+ \left(-\frac{3}{4} - \frac{1}{2}\rho(2R + 2r - L)\right) e^{-\rho(2R+2r-L)}$$
$$+ \frac{1}{4}e^{-\rho(L+2r-2R)} - e^{-\rho(L-R-r)}$$

$$p_c^{(n_a=0,n_b>0)}$$
$$= -e^{-\rho L} + \frac{1}{2}e^{-\rho(L+r-R)} + \frac{1}{2}e^{-\rho(L-R-r)}$$

$$p_c = 1 + \frac{1}{2}e^{-\rho(L-2R)} + \frac{1}{4}e^{-\rho(L+2r-2R)}$$
$$+ \left(-\frac{3}{4} - \frac{1}{2}\rho(2R + 2r - L)\right) e^{-\rho(2R+2r-L)}$$

(4) For $L > 2R + 2r$,

$$p_c^{(n_a>0,n_b>0)}$$
$$= e^{-\rho L} + \frac{1}{4}e^{-\rho(L+2r-2R)} - e^{-\rho(L+r-R)}$$
$$+ \frac{1}{4}e^{-\rho(L-2R-2r)} - e^{-\rho(L-R-r)} + \frac{1}{2}e^{-\rho(L-2R)}$$

$$p_c^{(n_a=0,n_b>0)}$$
$$= -e^{-\rho L} + \frac{1}{2}e^{-\rho(L+r-R)} + \frac{1}{2}e^{-\rho(L-R-r)}$$

$$p_c = \frac{1}{4}e^{-\rho(L+2r-2R)} + \frac{1}{4}e^{-\rho(L-2R-2r)} + \frac{1}{2}e^{-\rho(L-2R)}$$

14.1.3.2 Log-Normal Connection Model

In the log-normal connection model $\mathcal{L}$, the received power (in dBmW) at a destination vehicle is given by

$$p_{rx} = p_0 - 10\alpha \log_{10} \frac{l}{d_0} + N_\sigma \tag{14.1.24}$$

where p_{rx} is the received power (in dBmW) at the destination vehicle; p_0 is the power (in dBmW) at a reference distance d_0; α is the path loss exponent; N_σ is a Gaussian random variable with zero mean and variance σ^2; l is the Euclidean distance between the two vehicles (or a vehicle and a BS depending on the context). A source vehicle can establish a direct connection to a destination vehicle if the received power at the destination vehicle p_{rx} is greater than or equal to a certain threshold power p_{th}^v. Similarly, a source vehicle can establish a direct connection to a destination BS if the received power at the destination BS p_{rx} is greater than or equal to a certain threshold power p_{th}^b. We further assume that wireless connections between vehicles, and between vehicles and BSs, are symmetric. Note that when $\sigma = 0$, the log-normal connection model reduces to the unit disk connection model. Due to this fact, we assign $p_{th}^v = p_0 - 10\alpha \log_{10} \frac{r}{d_0}$, $p_{th}^b = p_0 - 10\alpha \log_{10} \frac{R}{d_0}$ so that the results under log-normal connection model can be compared with the results under the unit disk connection model later. It can be shown that under the log-normal connection model

$$g_v^{\mathcal{L}}(x) = \Pr(p_{rx} \geq p_{th}^v) = Q\left(\frac{10\alpha}{\sigma}\log_{10}\frac{x}{r}\right) \tag{14.1.25}$$

where function $Q(y) = \frac{1}{\sqrt{2\pi}}\int_y^\infty e^{-\frac{x^2}{2}}dx$ is the tail probability of the standard normal distribution. Similarly, $g_b^{\mathcal{L}}(x) = Q(\frac{10\alpha}{\sigma}\log_{10}\frac{x}{R})$. When $\sigma = 0$, $g_v^{\mathcal{L}}(x) = \Pr(x \leq r)$, $g_b^{\mathcal{L}}(x) = \Pr(x \leq R)$, and the log-normal connection model becomes the unit disk connection model as expected.

The access probability can then be obtained for different values of α and σ by computing (14.1.7) using numerical integration. The approximate and exact results for the connectivity probability can be obtained by computing (14.1.8) and (14.1.13) using numerical integration.

14.1.4 Analytical and Simulation Results

14.1.4.1 Unit Disk Connection Model

Figure 14.2 shows the access probability assuming different values of L and ρ. The simulation results are obtained from 40,000 randomly generated network topologies. As the number of instances of random networks used in the simulation is very large, the confidence interval is too small to be distinguishable and hence ignored in this plot as well as other plots. As shown in the figure, the access probability decreases with L when L exceeds some limits. For small ρ, the access probability decreases as soon as $L > 2R$. That is because when the vehicle density ρ is low, a vehicle is either directly connected to a BS or disconnected, i.e., cannot reach any BS in at most two hops. It is difficult for the vehicle to find a one-hop relay in its range via which it can access a BS if it is not within the transmission range of any BS. However, for large ρ, it is easier for the vehicle, which is not within the transmission range of any BS, to find a one-hop relay to access the BS. In general

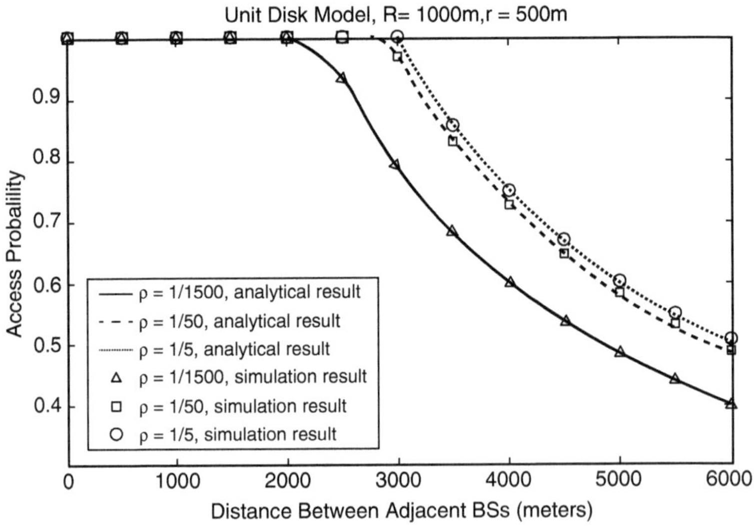

Fig. 14.2 Access probability with varying value of L under the unit disk connection model with $R = 1000\text{m}$, $r = 500\text{m}$, $\rho = 1/5, 1/50, 1/500$ vehicles/m respectively

the access probability increases with an increase in ρ, and the reason is that when the vehicle density increases, the probability increases for vehicles in the gap of the transmission ranges of BSs to find a neighbor within the transmission range of a BS to act as a relay.

Similarly, Fig. 14.3 shows the connectivity probability under different values of L and ρ. The approximate analytical result is shown to be reasonably close to the exact analytical result and the simulation result. The figure shows that when $L \leq 2R + r = 2500$ m, it is easy for all vehicles to be connected to either BS in at most two hops, hence the connectivity probability is high. As L gets larger, it becomes more difficult for all vehicles to be connected to the BSs due to the larger possible distances between the vehicles and the BSs. This causes a drop in the connectivity probability, and the connectivity probability tends to zero as L goes to infinity. The transition of the connectivity probability from 1 to 0 gets sharper as the vehicle density increases. As ρ goes to infinity, the transition happens at the critical distance $L = 2R + 2r = 3000$ m, below which the network is disconnected with a high probability and above which the network is connected with a high probability. Furthermore, networks with a larger ρ have a higher connectivity probability than networks with a smaller ρ when L is small. This is because when the vehicle density is large, it is easier for vehicles not directly connected to a BS to find a vehicle within its communication range and is directly connected to a BS to act as a relay. When L is large, the networks with a larger ρ have a lower connectivity probability than the networks with a smaller ρ. This is because at large values of L when the vehicle density is large it is easier to have at least one vehicle which is located too far from the BSs to be connected to a BS in at most two hops.

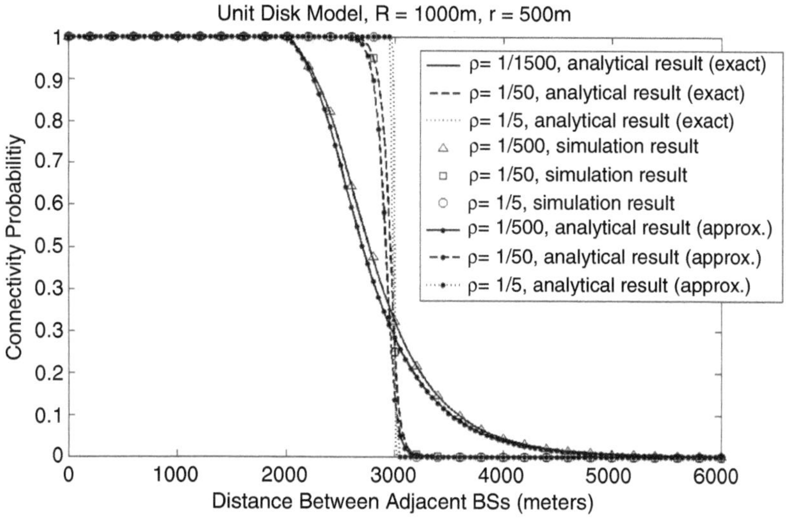

Fig. 14.3 Connectivity probability with varying value of L under the unit disk connection model with $R = 1000$m, $r = 500$m, $\rho = 1/5, 1/50, 1/500$ vehicles/m respectively

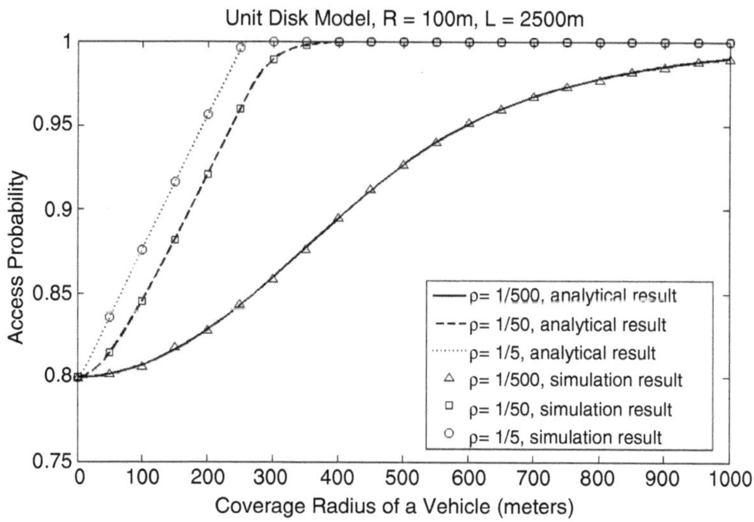

Fig. 14.4 Access probability with varying value of r under the unit disk connection model with $R = 1000$m, $L = 2500$m, $\rho = 1/5, 1/50, 1/500$ vehicles/m respectively

Figure 14.4 shows how the transmission range of the vehicles r affect the access probability. It shows that the access probability increases with r, and when ρ is large enough, the access probability could be quite close to 1. Figure 14.4 shows again that the access probability increases with an increase in ρ.

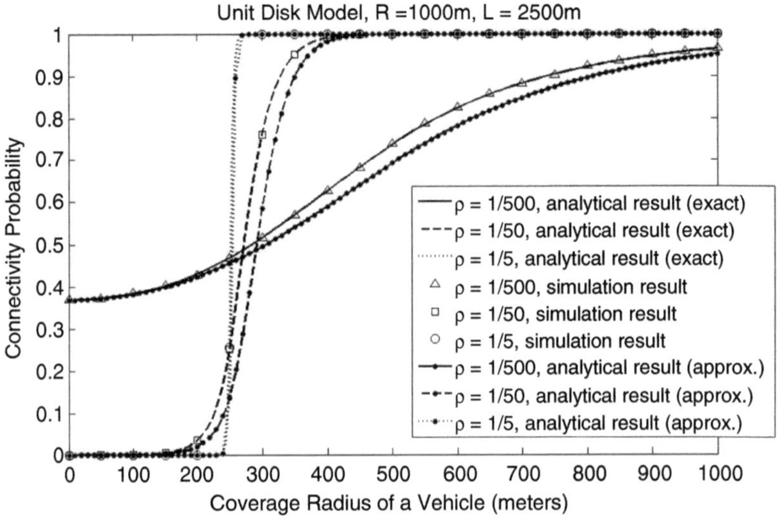

Fig. 14.5 Connectivity probability with varying value of r under the unit disk connection model with $R = 1000$m, $L = 2500$m, $\rho = 1/5, 1/50, 1/500$ vehicles/m respectively

With a similar setup, Fig. 14.5 shows the sensitivity of the connectivity probability to r. For a large ρ, around a certain value of r, a small increase in r will result in a dramatic increase in the connectivity probability from near 0 to near 1, i.e., the well-known phase transition phenomenon which has been discussed in Chap. 6. From the figure, it can be seen that such phenomenon does not exist for small ρ. Figure 14.5 also shows a scenario where there may be a significant gap between the approximate and exact results for connectivity probability.

Figure 14.6 supported our conclusion that an increase in ρ will improve the access probability as it shows that the access probability monotonically increases with ρ. While ρ is relatively small and the width of the gap region not directly covered by any of the BSs is relatively large, the access probability will be low. Therefore, in this circumstance, network operator should consider deploying more BSs for better connectivity and greater access probability.

14.1.4.2 Log-Normal Connection Model

Figure 14.7 shows the access probability under the log-normal shadowing model. In general, it is easier for the vehicles in the subnetwork to get access to any BS under the log-normal model. As σ increases, the access probability improves. The improvement in access probability is more pronounced for high vehicle density.

Figure 14.8 shows the connectivity probability under the log-normal model when the vehicle density is low ($\rho = \frac{1}{500}$ vehicles/m). As the vehicle density increases, the computational complexity involved in numerically computing the exact result

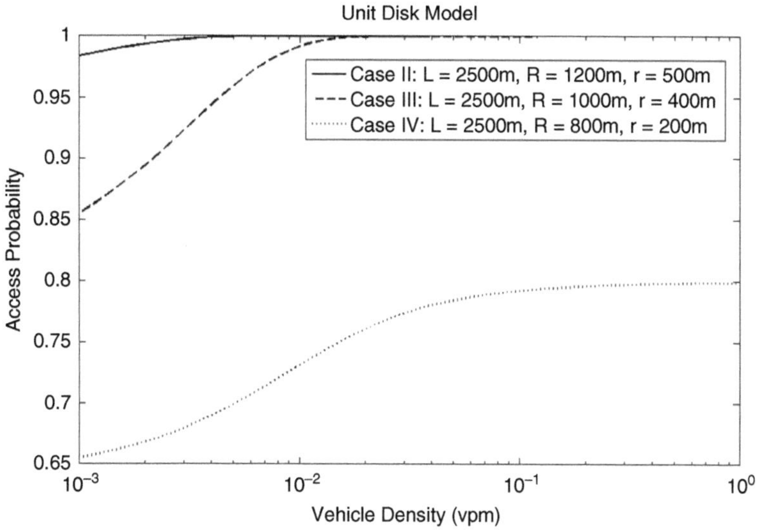

Fig. 14.6 Access probability with varying value of ρ changing the unit disk connection model where L, R, r are constants

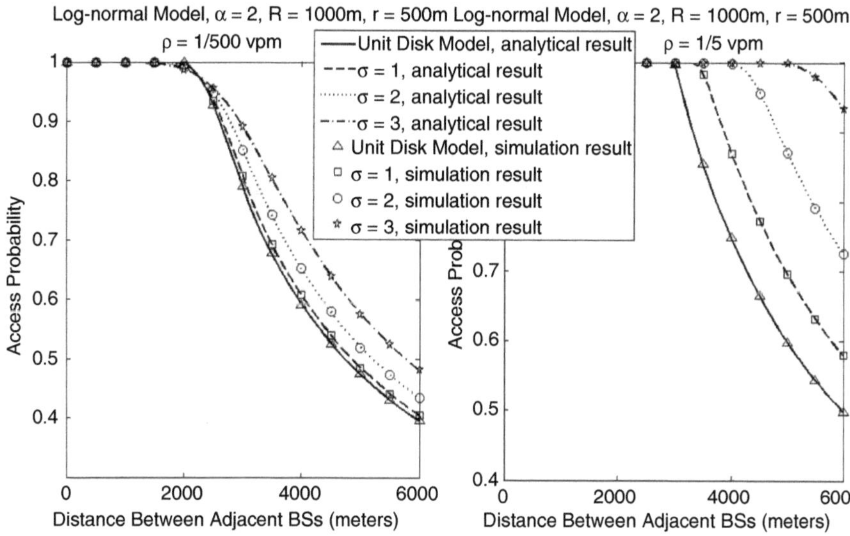

Fig. 14.7 Access probability with varying value of L under the log-normal connection model where R = 1000m, r = 500m, $\rho = 1/5, 1/500$ vehicles/m respectively under different values of σ. R (r) is the transmission range of a BS (vehicle) ignoring shadowing effect, i.e., $\sigma = 0$

increases very quickly. As such, we only provide the exact analytical results for low vehicle density. Furthermore, Fig. 14.8 shows that the approximate analytical results are reasonably close to the true values when the vehicle density is low. However, as

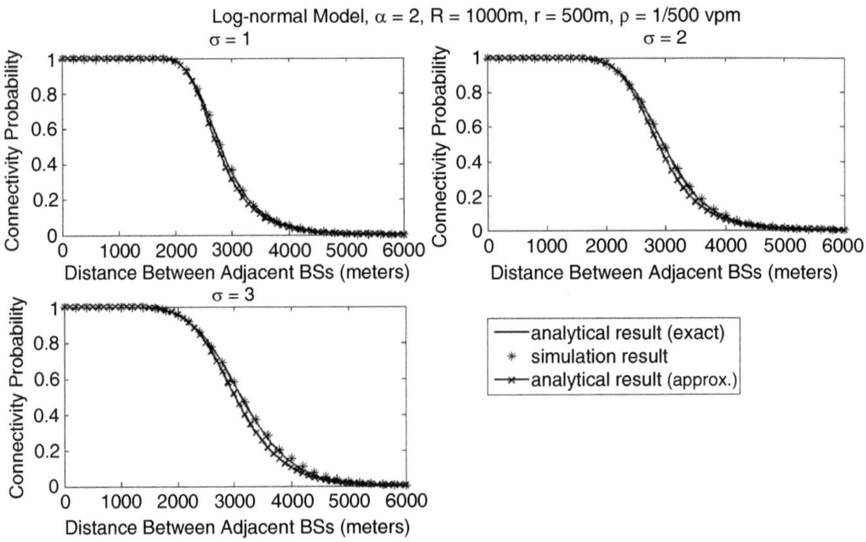

Fig. 14.8 Connectivity probability with varying value of L under the log-normal connection model where $R = 1000$m, $r = 500$m, $\rho = 1/500$ vehicles/m under different values of σ. R (r) is the transmission range of a BS (vehicle) ignoring shadowing effect, i.e., $\sigma = 0$

shown in Fig. 14.9, the discrepancy between the approximate results and the true values can be significant when the vehicle density is high ($\rho = \frac{1}{50}, \frac{1}{5}$ vehicles/m). In general, the approximate analytical result always underestimate the simulation result. Same situation can be observed for the results obtained under the unit disk connection model. This can be explained by Lemma 207 that a vehicle is more likely to be able to access to any BS where there is another vehicle nearby that can access to the BSs. Because of the independence assumption used in obtaining the approximate analytical result, the approximate result will underestimate the true value.

14.2 Distance Estimation via Connectivity

In this section, we present a technique that employs a maximum-likelihood estimator (MLE) to estimate distances between a pair of neighboring nodes in static wireless networks using their local connectivity information, namely the numbers of their common and non-common one-hop neighbors. We develop the distance estimation technique assuming the general random connection model. Then, we take the log-normal connection model as an example to demonstrate the implementation of the proposed technique. The impact of uncertainties in the log-normal connection model is examined and the bias and standard deviation of distance estimates produced by the technique are numerically evaluated. The proposed technique,

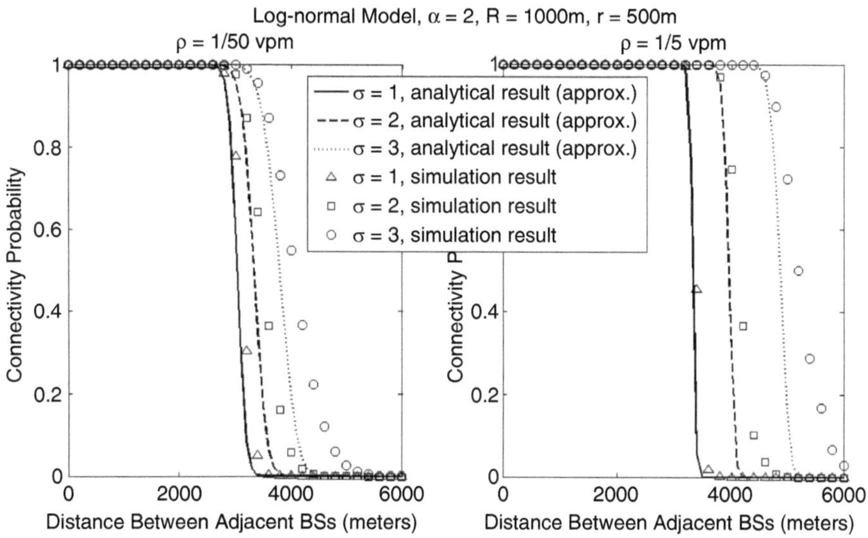

Fig. 14.9 Connectivity probability with varying value of L under the log-normal connection model where $R = 1000$m, $r = 500$m, $\rho = 1/5, 1/50$ vehicles/m respectively under different values of σ. R (r) is the transmission range of a BS (vehicle) ignoring shadowing effect, i.e., $\sigma = 0$

though not comparable to the fine-grained distance estimation techniques like time of arrival and time difference of arrival, outperforms the well-known received signal strength based technique for estimation of long distances. Moreover, using the Cramér-Rao lower bound (CRLB), we analyze the influences of various factors on distance estimation from local connectivity information, and derive useful guidelines on implementing the proposed technique in reality. Finally, the proposed technique is validated using real measurement data.

14.2.1 The Connectivity-Based Distance Estimation Method

In this subsection, we present the technique of estimating distances via connectivity. Specifically, we consider a static wireless network on an infinite plane where nodes are distributed according to a homogeneous Poisson point process of density λ. Assuming a random connection model, a pair of nodes separated a Euclidean distance x are directly connected with probability $g(x)$, independently of other pairs of nodes. As two widely used special cases of the random connection model, the unit disk connection model and the log-normal connection model are also considered. We further assume that the connections are symmetric, i.e., if a node v is directly connected to a node u, the node u is also directly connected to the node v.

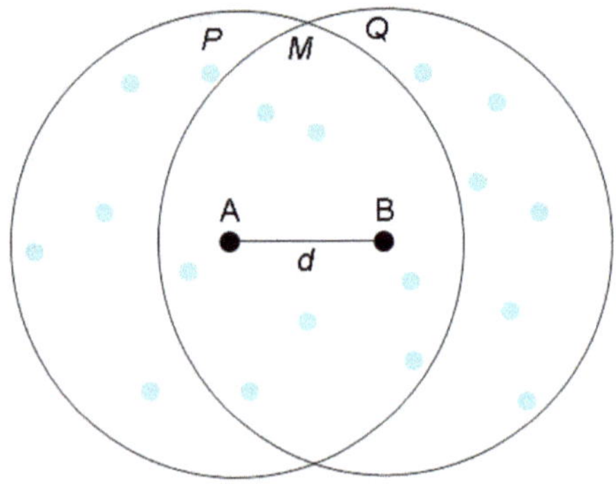

Fig. 14.10 An illustration of the coverage of two nodes under the unit disk connection model

14.2.1.1 Estimating Distances Under the Unit Disk Connection Model

Let us first consider a unit disk connection model with a transmission range r and then extend to the more general random connection model. Consider two neighboring nodes A and B with coordinates (x_A, y_A) and (x_B, y_B) respectively. Denote by d $(d \leq r)$ the Euclidean distance between the two nodes. Let $D(A, r)$ and $D(B, r)$ be two disks with the same radius r and centered at node A and node B respectively. The two disks represent respectively the coverage of node A and node B, as shown in Fig. 14.10. The two disks intersect and create three disjoint regions: $D(A, r) \setminus D(B, r)$, $D(A, r) \cap D(B, r)$, and $D(B, r) \setminus D(A, r)$. Let $S = \pi r^2$ and $f(d)$ be the size of the overlapping region $D(A, r) \cap D(B, r)$. It can be readily established that

$$f(d) = \frac{2S}{\pi} \arccos(\frac{d}{2r}) - d\sqrt{r^2 - \frac{d^2}{4}} \tag{14.2.1}$$

It is obvious that the nodes residing in $D(A, r) \cap D(B, r)$ are common one-hop neighbors of A and B, and the nodes residing in $D(A, r) \setminus D(B, r)$ $(D(B, r) \setminus D(A, r))$ are non-common one-hop neighbors of A (B). Define three random variables M, P, and Q to be the numbers of the above three categories of neighbors respectively. Due to the Poisson distribution of nodes, the three random variables are mutually independent and Poissonly distributed with means $\lambda f(d)$, $\lambda(S - f(d))$, and $\lambda(S - f(d))$ respectively. The actual values of M, P, and Q can be easily obtained after A and B exchange their neighborhood information. Based on the observations of M, P, and Q, an MLE for estimating d is summarized as follows:

Theorem 210 *M, P, and Q are mutually independent Poisson random variables with means $\lambda f(d)$, $\lambda(S - f(d))$ and $\lambda(S - f(d))$, respectively. If $f(d)$ is invertible and S is a nonzero constant, then the MLE for d, termed $\hat{d}$, is*

$$\hat{d} = \begin{cases} f^{-1}(S) & \text{if } M = P = Q = 0 \\ f^{-1}(\hat{\rho} S) & \text{otherwise} \end{cases}$$

where $\hat{\rho} = \frac{2M}{2M+P+Q}$.

Proof Note that observations of M, P, and Q provide a measured triplet $\phi = [m\ p\ q]$ where m, p, q are nonnegative integers. The unknown parameters are $\theta = [d\ \lambda]$. By formulating the likelihood function, we can derive that the MLE for d is the solution for d in the following equation set

$$\begin{cases} \frac{m}{f(d)} - \frac{p+q}{S-f(d)} + \lambda = 0 \\ \frac{m+p+q}{\lambda} - (2S - f(d)) = 0 \end{cases}$$

By eliminating λ, we can obtain

$$2mS = (2m + p + q)f(d) \tag{14.2.2}$$

If $2m + p + q > 0$, $\hat{d} = f^{-1}(\frac{2m}{2m+p+q}S)$; otherwise, the solution for d is not well defined. Note however $f(d) = S$ maximizes the likelihood, thus we have $\hat{d} = f^{-1}(S)$. This completes the proof of the theorem. □

Note that the actual value of λ is not needed in obtaining $\hat{d}$. Furthermore, the derivation of the MLE suggests that as long as nodes in a *local* region which covers the communication coverage of two neighboring nodes admits a uniform density, the proposed method is applicable. In addition, if the least squares method, instead of the MLE, is applied here, the resulting expression of the distance estimator is actually the same as that in Theorem **210**.

14.2.1.2 Extension Under the General Random Connection Model

Under the general random connection model defined by $g(d)$, M, P, and Q are still used to denote the numbers of common and non-common one-hop neighbors associated with two nodes. First, we can compute their expectations as follows:

$$\begin{aligned} E(M+P) &= E(M+Q) \\ &= \lambda \int_{-\infty}^{\infty}\int_{-\infty}^{\infty} g(\sqrt{(x-x_B)^2+(y-y_B)^2})dxdy \end{aligned} \tag{14.2.3}$$

$$E(M) = \lambda \int_{-\infty}^{\infty}\int_{-\infty}^{\infty} g(\sqrt{(x-x_A)^2+(y-y_A)^2})g(\sqrt{(x-x_B)^2+(y-y_B)^2})dxdy \tag{14.2.4}$$

Then, from the integral boundedness condition on $g(d)$, i.e., (1.1.9), it follows that $E(M) < E(M + P) < \infty$. Unlike the unit disk connection model where the independence among M, P, and Q is straightforward due to having three disjoint regions, the random connection model does not necessarily lead to such three disjoint regions. The following theorem establishes the mutual independence of the three random variables.

Theorem 211 *Consider a wireless network on an infinite plane with nodes distributed according to a homogeneous Poisson point process of density λ and a pair of nodes are directly connected following the random connection model.*

Considering two randomly chosen nodes in the network, let M be the number of their common one-hop neighbors, and P and Q be the numbers of their non-common one-hop neighbors. Then, M, P, and Q are mutually independent Poisson random variables.

The results in Theorem 211 can be readily proved by noting that the nodes are distributed following homogeneous Poisson point process and that given the location of the two nodes, say A and B, each of the three events: (1) a node is connected to A but not B; (2) a node is connected to B but not A; (3) a node is connected to both A and B can be regarded as an outcome of the independent thinning of the original homogeneous Poisson point process.

Under the random connection model, parameters S and $f(d)$ can be generalized to be linked to the expectations of M, P, and Q, instead of the areas defined under the unit disk connection model, and can be formulated as follows:

$$S = \int_{-\infty}^{\infty}\int_{-\infty}^{\infty} g(\sqrt{(x-x_B)^2 + (y-y_B)^2})dxdy \tag{14.2.5}$$

$$f(d) = \int_{-\infty}^{\infty}\int_{-\infty}^{\infty} g(\sqrt{(x-x_A)^2 + (y-y_A)^2})g(\sqrt{(x-x_B)^2 + (y-y_B)^2})dxdy \tag{14.2.6}$$

In this way, if S and $f(d)$ satisfy the conditions in Theorem 210, the MLE obtained assuming the unit disk connection model can be directly applied here under the random connection model.

Note that, in reality, wireless networks are deployed in regions of finite areas, and thus the expectations of M, P, and Q associated with two nodes, especially those near the network boundaries, cannot be derived by simply integrating over an infinite plane to compute S and $f(d)$. This is due to the boundary effect. In this study, we concentrate on the theoretical foundations of the proposed method and ignore the impact of the boundary effect.

Prior to implementing the proposed method, it is a prerequisite to know the wireless channel, i.e., $g(d)$, such that the quantity S, the function $f(d)$, and its inverse can be determined offline and then programmed into each node. After that, due to the simple mechanism of the proposed method, a distributed protocol can be easily designed for collecting and exchanging local connectivity information by each node through local broadcast. Once a node obtains neighborhood information of all its immediate neighbors, the node is able to estimate the distances from its immediate neighbors using the inverse of $f(d)$, S, and the MLE in Theorem 210.

14.2.1.3 Distance Estimation Under the Log-Normal Connection Model

In this subsection, we provide the method to estimate the distance between nodes under the log-normal connection model assuming that the parameters characterizing the log-normal model, i.e., σ_{dB} the standard deviation of the log-normal fading measured in dB unit, r the equivalent transmission range when $\sigma_{dB} = 0$ and α the path loss exponent, are known.

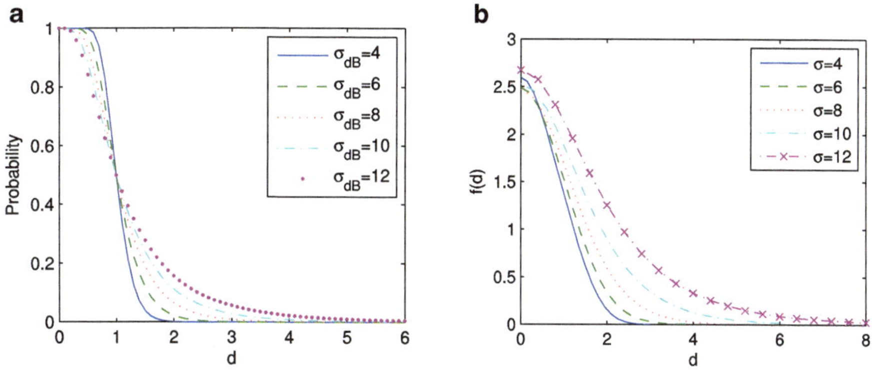

Fig. 14.11 The functions $g(d)$ and $f(d)$ under the log-normal connection model with $\alpha = 4$ and $r = 1$. (**a**) The connection function $g(d)$ under the log-normal connection model. (**b**) The function $f(d)$ under the log-normal connection model

We shall demonstrate how to determine the three parameters $g(d)$, S, and $f(d)$ required in the MLE to estimate the distance.

Under the log-normal connection model, the probability that the two nodes separated by a Euclidean distance d are directly connected is given by

$$g(d) = \int_{k \log \frac{d}{r}}^{\infty} \frac{e^{-\frac{z^2}{2\sigma_{dB}^2}}}{\sqrt{2\pi}\sigma_{dB}} dz \tag{14.2.7}$$

where $k = \frac{10\alpha}{\log 10}$.

We plot $g(d)$ with respect to d for different values of σ_{dB} given $\alpha = 4$ and $r = 1$ in Fig. 14.11a. It can be seen that the smaller is d, the higher is the probability that a direct connection exists. Furthermore, a larger σ_{dB} tends to inhibit a direct connection for a smaller d but promotes a direction connection for a larger d.

In view of the restrictions on $g(d)$, it is straightforward to obtain $\lim_{d\to\infty} g(d) = 0$. As such, for an arbitrarily small and positive ε, there exists d_{th} such that $g(d) < \varepsilon$ if $d > d_{th}$. That is, nodes whose distances to a particular node are larger than d_{th} can hardly connect with this node directly. As such, d_{th} is a surrogate of the transmission range. This phenomenon can be observed in Fig. 14.11a.

14.2.1.4 Formulating *S* and *f*(*d*)

In Chap. 12, the expectation $E(M + P)$ (or $E(M + Q)$) has been studied. Using the result, the equivalent S under the log-normal connection model is given by:

$$S = \pi r^2 e^{\frac{2\sigma_{dB}^2}{k^2}} \tag{14.2.8}$$

Further using (14.2.4), (14.2.6), and (14.2.7), we can derive the formula for $f(d)$ under the log-normal connection model. By letting $\alpha = 4$ and $r = 1$, we plot $f(d)$ with respect to different values of d and σ_{dB} in Fig. 14.11b. As can be seen in the figure, $f(d)$ is a monotonically decreasing function of d and invertible. Hence, Theorem 210 is applicable under the log-normal connection model. But the closed-form formula for $f(d)$ and its inverse are difficult to obtain. Alternatively, we can establish a piecewise linear function to approximate its inverse such that for each affine segment, a linear regression model can be used to predict d.

Considering the fact that two nodes with distance larger than d_{th} can hardly connect with each other directly, we restrict the distance estimates to be between 0 and d_{th}. But in real applications, $\hat{\rho}S$ and S may exceed $[f(d_{th}), f(0)]$ and consequently $\hat{d}$ may exceed $[0, d_{th}]$. Therefore, we adapt the distance estimator as follows

$$\hat{d} = \begin{cases} 0 & \text{if } M = P = Q = 0 \text{ or } \hat{\rho}S > f(0) \\ f^{-1}(\hat{\rho}S) & \text{if } f(d_{th}) \leq \hat{\rho}S \leq f(0) \\ d_{th} & \text{if } \hat{\rho}S < f(d_{th}) \end{cases} \tag{14.2.9}$$

14.2.2 Performance Analysis of the Distance Estimator

In this subsection, we evaluate the performance of the proposed method under the log-normal connection model from several different perspectives.

14.2.2.1 Impact of Imprecise Knowledge of Parameters α and σ_{dB}

In the proposed method, the parameters α and σ_{dB} are supposed to be known precisely. It is interesting to know the robustness of the estimation in the presence of imperfect knowledge of the two values. To answer this question, we define

$$\rho_{\alpha,\sigma_{dB}}(d) = \frac{f(d)}{S} \tag{14.2.10}$$

where $f(d)$ and S are computed using (14.2.6), (14.2.8), and (14.2.7) given α and σ_{dB}. The distance estimator in (14.2.9) then becomes

$$\hat{d} = \begin{cases} 0 & \text{if } M = P = Q = 0 \text{ or } \hat{\rho} > \rho_{\alpha,\sigma_{dB}}(0) \\ \rho^{-1}_{\alpha,\sigma_{dB}}(\hat{\rho}) & \rho_{\alpha,\sigma_{dB}}(d_{th}) \leq \hat{\rho} \leq \rho_{\alpha,\sigma_{dB}}(0) \\ d_{th} & \hat{\rho} < \rho_{\alpha,\sigma_{dB}}(d_{th}) \end{cases} \tag{14.2.11}$$

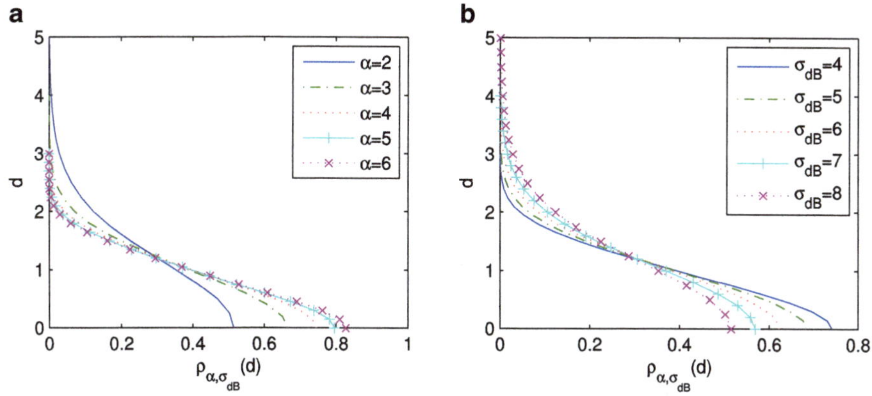

Fig. 14.12 The inverse function of $\rho_{\alpha,\sigma_{dB}}(d)$ under the log-normal connection model. (**a**) $\sigma_{dB} = 4$. (**b**) $\alpha = 4$

As shown in the above equation, imprecise knowledge of α and/or σ_{dB} results in an incorrect value of the function $\rho^{-1}_{\alpha,\sigma_{dB}}(\hat{\rho})$, which in turn affects the accuracy of the distance estimate $\hat{d}$. We plot the function $\rho^{-1}_{\alpha,\sigma_{dB}}(\hat{\rho})$ with respect to different values of α and σ_{dB} in Fig. 14.12.

Supposing σ_{dB} is known to be exactly 4, we further investigate the impact of the uncertainty in the knowledge of α. As shown in Fig. 14.12a, $\rho^{-1}_{\alpha,\sigma_{dB}}(\hat{\rho})$ is much more sensitive to a small α than to a large α; in other words, for a small α, an imprecise value of α tends to degrade the accuracy of the distance estimate $\hat{d}$ more seriously than for a large α. Moreover, if α is overestimated, an underestimated $\hat{d}$ will be produced for a small d but an overestimated $\hat{d}$ for a large d, and the converse. However, $\rho^{-1}_{\alpha,\sigma_{dB}}(\hat{\rho})$ does not demonstrate the same sensitivity to σ_{dB} as that observed for α, as illustrated in Fig. 14.12b. We can therefore conclude that, if σ_{dB} is overestimated, then an overestimated $\hat{d}$ will be produced for a small d but an underestimated $\hat{d}$ for a large d, and the converse.

14.2.2.2 Bias and Standard Deviation

According to Theorem 210, all possible values of $\hat{\rho}$ are rational numbers within [0, 1], so that $\hat{d}$ is a discrete random variable and its j-th moment is given as follows

$$E\left(\hat{d}^j\right) = \sum_a \left[\hat{d}^j \Pr(\hat{d} = a)\right] \tag{14.2.12}$$

We divide the range of $\hat{d}$, i.e., $[0, d_{th}]$, into w equal-length intervals: $\mathcal{I}_1 = [z_0, z_1), \cdots, \mathcal{I}_w = [z_{w-1}, z_w]$ with $z_i = \frac{i \times d_{th}}{w}$. Given a sufficiently large w, $\hat{d}$ is approximately constant over each interval, denoted by $\tilde{d}_i$. Then, we can approximately reformulate (14.2.12) as

$$E(\hat{d}^j) \approx \sum_{i=1}^{w} \left[(\tilde{d}_i)^j \Pr(\hat{d} \in \mathcal{I}_i) \right] \tag{14.2.13}$$

Towards the probability associated with the i-th interval $\mathcal{I}_i$, we have

$$\Pr\left(\hat{d} \in \mathcal{I}_i\right) = \begin{cases} \Pr\left(f\left(z_1\right) < \hat{\rho}S \leq S\right) & \text{if } i = 1 \\ \Pr\left(f\left(z_i\right) < \hat{\rho}S \leq f\left(z_{i-1}\right)\right) & \text{if } 1 < i < w \\ \Pr\left(0 \leq \hat{\rho}S \leq f\left(z_{w-1}\right)\right) & \text{if } i = w \end{cases}$$

By letting $Y = P + Q$, we have

$$\Pr(b < \hat{\rho}S < c) = \sum_{y=0}^{\infty} \left[\Pr(b < \hat{\rho}S < c|Y = y) \Pr(Y = y)\right] \tag{14.2.14}$$

which makes it possible for us to numerically evaluate the moments of $\hat{d}$ and thus the bias and the standard deviation.

Let μ be the expected number of one-hop neighbors of a node, namely $\mu = E(M + P) = E(M + Q)$, and the values of λ with respect to different σ_{dB} and μ are listed in Table 14.1. For better presentation, the connectivity index (or equivalently the average node degree) μ will be used in the following discussions instead of the node density λ.

Given $\alpha = 4$, $r = 1$, $w = 1000$, and μ varying from 5 to 40, Fig. 14.13 depicts the numerical bias and the standard deviation associated with the proposed method and the corresponding simulation results. The two groups of results are highly consistent, and the comparatively non-smooth aspect of some of the curves, e.g., Fig 14.13a, is attributable to the fact that all observations of M, P, and Q are necessarily integers, and such observations are used in determining the curves.

It can be seen from Fig. 14.13 that the proposed method is obviously *biased*. The absolute bias is much less than the standard deviation in most cases; except for $\sigma_{dB} = 0$, the absolute bias and the standard deviation are comparable to true distances, especially for short distances and sparse networks. Particularly, when $\mu = 5$, their values are extraordinarily large and nearly twice the corresponding

Table 14.1 The values of λ with respect to different values of σ_{dB} and μ when $\alpha = 4$

	μ				
σ_{dB}	5	10	20	30	40
0	1.59	3.18	6.37	9.55	12.73
4	1.43	2.86	5.73	8.59	11.45
8	1.04	2.08	4.17	6.25	8.33
12	0.61	1.23	2.45	3.68	4.90

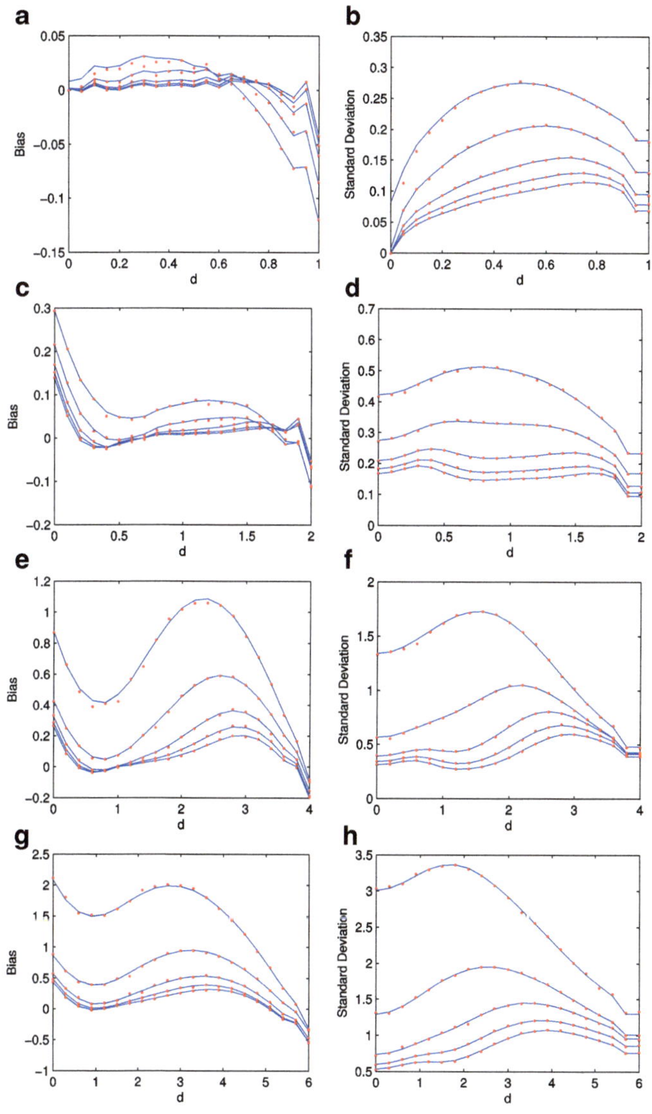

Fig. 14.13 The bias and the standard deviation of distance estimation from numerical evaluations (*solid lines*) and simulations (*dashed lines*) with $\mu = 5, 10, 20, 30, 40$ and $\alpha = 4, r = 1$. For the standard deviation, a larger μ corresponds to a line to the bottom; for the bias, a larger μ corresponds to a line with the bias closer to 0. (**a**) Bias ($\sigma_{dB} = 0$, $d_{th} = 1$). (**b**) Standard deviation ($\sigma_{dB} = 0$, $d_{th} = 1$). (**c**) Bias ($\sigma_{dB} = 4$, $d_{th} = 2$). (**d**) Standard deviation ($\sigma_{dB} = 4$, $d_{th} = 2$). (**e**) Bias ($\sigma_{dB} = 8$, $d_{th} = 4$). (**f**) Standard deviation ($\sigma_{dB} = 8$, $d_{th} = 2$). (**g**) Bias ($\sigma_{dB} = 12$, $d_{th} = 6$). (**h**) Standard deviation ($\sigma_{dB} = 12$, $d_{th} = 6$)

values when $\mu = 10$. Moreover, with μ increasing, the standard deviation always reduces, while the absolute bias reduces in most cases. An intuitive explanation is that with μ increasing, the variances of the ratios $2M/E(2M)$ and $(2M + P + Q)/E(2M + P + Q)$ both decrease, the variance of $\hat{\rho}$ is reduced, and so is the variance of $\hat{d}$. As mentioned in the previous subsection, when the average node degree is fixed, a large σ_{dB} results in more direct connections between distant nodes but reduces direct connections between nearby nodes. As a result, the geometric relationship implied by direct connections becomes less accurate. Hence, the larger is σ_{dB}, the worse are both the bias and the standard deviation.

14.2.2.3 Root Mean Square Error

As a performance measure, the root mean square error (RMSE) is defined to be the square root of the sum of the square bias and variance of estimation errors. We plot the RMSE of $\hat{d}$ produced by the proposed method in Fig. 14.14. As shown in the figure, the RMSE decreases with μ increasing and σ_{dB} decreasing, which is consistent with the trend exhibited by the variations of the bias and the standard deviation of $\hat{d}$ with μ and σ_{dB}. When d is near 0, the RMSE is extraordinarily large compared to the true value of d, implying that the proposed method fails to provide reasonable estimates for short distances. This underperformance with short distances limits the use of the proposed method in practice, and is due to a mixed impact of the following facts:

- It is evident that the variance of $2M + P + Q$ is constant irrespective of the value of d, but the variance of $2M$ increases with d decreasing. As a result, $\hat{\rho}$, i.e., $\frac{2M}{2M+P+Q}$, is more likely to have larger variances when d is smaller.
- As depicted in Fig. 14.12, $\rho^{-1}_{\alpha,\sigma_{dB}}(\hat{\rho})$ is quite sensitive to $\hat{\rho}$ when d is small. Namely, a small perturbation in $\hat{\rho}$ leads to a large change in $\hat{d}$ and thus a large distance estimation error.
- In light of (14.2.11), $\hat{d}$ is roughly set to be 0 when $\hat{\rho}$ is greater than $\rho_{\alpha,\sigma_{dB}}(0)$, but a small d often causes $\hat{\rho}$ to be within $[\rho_{\alpha,\sigma_{dB}}(0), 1]$ and hence resulting in the underperformance.

To conclude, for short distances, the non-smooth aspect and the sensitivity to $\hat{\rho}$ of the function defined in (14.2.11) are responsible for the underperformance.

Under the log-normal connection model with $\sigma_{dB} > 0$, distance estimation can also be realized by using the received signal strength (RSS) measurements. The bias and variance of the resulting distance estimate (denoted by $\hat{d}_{RSS}$) are provided in [42], so that we can compute the RMSE of $\hat{d}_{RSS}$ and compare it with that of $\hat{d}$ in Fig. 14.14. It can be seen that: (1) the RMSE of $\hat{d}_{RSS}$ increases in direct proportion to d, but that of $\hat{d}$ appears to have comparatively small variations with d increasing; (2) the proposed method outperforms the RSS method for long distances by a large margin.

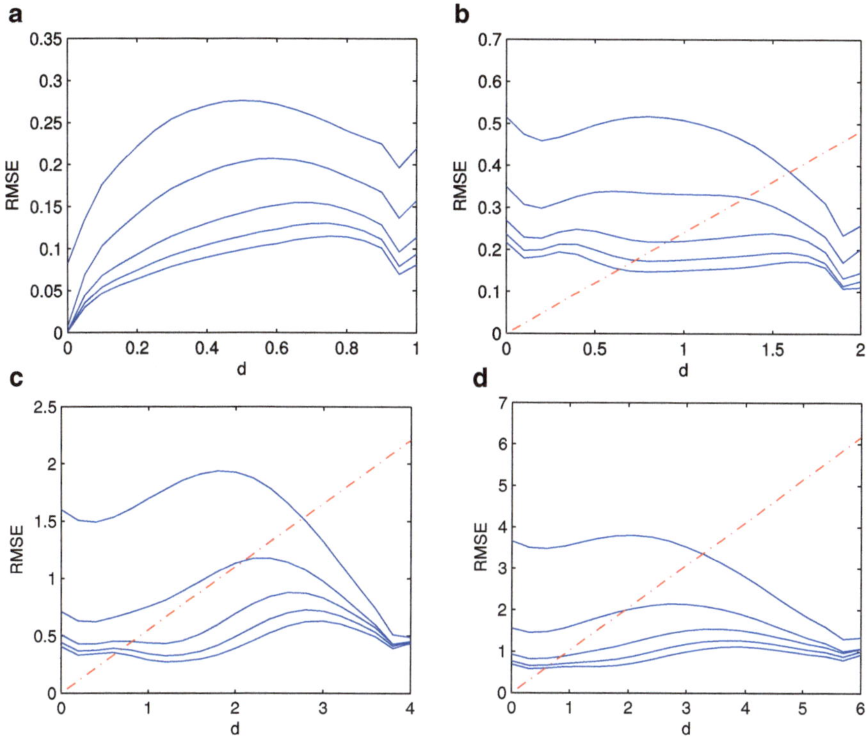

Fig. 14.14 The RMSE of $\hat{d}_{RSS}$ (*dashed lines*) and $\hat{d}$ (*solid lines*; $\mu = 10, 20, 30, 40$ with $\alpha = 4, r = 1$; a larger μ corresponds to a curve to the bottom.) under the log-normal connection model .(**a**) $\sigma_{dB} = 0$, $d_{th} = 1$. (**b**) $\sigma_{dB} = 4$, $d_{th} = 2$. (**c**) $\sigma_{dB} = 8$, $d_{th} = 4$. (**d**) $\sigma_{dB} = 12$, $d_{th} = 6$

14.2.3 Analysis Based on the CRLB

In this subsection, we formulate the Cramér-Rao lower bound (CRLB) regarding the distance estimation problem via connectivity, i.e., estimating d from M, P, and Q, under the log-normal connection model. In estimation theory and statistics, the CRLB, named in honor of Harald Cramér and Calyampudi Radhakrishna Rao, expresses a lower bound on the variance of estimators of a deterministic parameter [114]. In its simplest form, the bound states that the variance of any unbiased estimator is at least as high as the inverse of the Fisher information.

For this estimation problem, the unknown parameters are d and λ. The Fisher Information Matrix (FIM) for this estimation problem, denoted by $FIM(d, \lambda)$, is

$$FIM(d, \lambda) = \begin{pmatrix} \lambda \left(f'(d)\right)^2 \left(\frac{1}{f(d)} + \frac{2}{S - f(d)}\right) & -f'(d) \\ -f'(d) & \frac{2S - f(d)}{\lambda} \end{pmatrix} \tag{14.2.15}$$

where $f(d)$ is differentiable to the first order in (14.2.6). The evaluation of the FIM relies on the computation of the first derivative of $f(d)$. In the following paragraphs, we establish the existence of $f'(d)$

First, consider the following expression

$$\lim_{\varepsilon\to 0}\left(-\frac{1}{\varepsilon}\left[g(\sqrt{x^2+d^2-2xd\cos\theta})-g(\sqrt{x^2+(d+\varepsilon)^2-2x(d+\varepsilon)\cos\theta})\right]\right)$$

$$=\frac{k(x\cos\theta-d)e^{\frac{-(k(\log\sqrt{x^2+d^2-2xd\cos\theta}-\log r))^2}{2\sigma_{dB}^2}}}{\sqrt{2\pi}\sigma_{dB}(x^2+d^2-2xd\cos\theta)}$$

which is bounded for $x\in[0,+\infty)$. Then, note that the derivative of $f(d)$ can be formulated as

$$f'(d)=\int_0^{\infty}\int_0^{2\pi}g(x)x\lim_{\varepsilon\to 0}\left(-\frac{1}{\varepsilon}\left[g(\sqrt{x^2+d^2-2xd\cos\theta})\right.\right.$$
$$\left.\left.-g(\sqrt{x^2+(d+\varepsilon)^2-2x(d+\varepsilon)\cos\theta})\right]\right)d\theta dx \qquad (14.2.16)$$

Because $\int_0^{\infty}\int_0^{2\pi}g(x)xd\theta dx$ is equal to $E(M+P)$ and is convergent, $f'(d)$ is also convergent.

It follows from the expression for the FIM in (14.2.15) and the existence of $f'(d)$ that the CRLB for d, denoted by $CRLB(d)$, is

$$CRLB(d)=\frac{(S-f(d))(2S-f(d))f(d)}{2\lambda S^2(f'(d))^2} \qquad (14.2.17)$$

Though the CRLB is only valid for unbiased distance estimates and the proposed method is known to be biased, the bound is still helpful to understand the essential features of the distance estimation problem. In what follows, we shall discuss the influences of various parameters on the CRLB.

14.2.3.1 Influence of λ

It is clear that the CRLB is inversely proportional to λ. In other words, a better estimation accuracy can be attained in dense wireless networks, which is intuitive and is also consistent with that shown in Fig. 14.14. Hence, it is desirable to apply the proposed method in dense networks.

14.2.3.2 Influence of d

Using (14.2.17), it is difficult to directly observe the influence of d on the CRLB because we do not have the closed-form formulas for $f(d)$ and $f'(d)$ except for the case of $\sigma_{dB} = 0$. However, note that the numerator of (14.2.17) is bounded in a narrow range. Therefore, if the denominator can be very small, the CRLB will be seriously affected by the denominator. Based on Fig. 14.11b, we can obtain some intuitive knowledge about the key component in the denominator, i.e., $f'(d)$. As shown in Fig. 14.11b, when d increases from 0, $|f'(d)|$ first experiences an increase and then decreases after d grows above a certain value which differs from σ_{dB}; when d increases further, $|f'(d)|$ continuously decreases and approaches 0. Hence, it is postulated that the CRLB will experience a rise when d increases.

As shown in Fig. 14.13, in the cases of $\sigma_{dB} = 8, 12$, the standard deviation of the distance estimation exhibits an apparent rise when d is larger than a certain value, and then drops when d approaches d_{th}. The reason behind such a drop is that we restrict the maximum distance estimate to be d_{th}, so that estimates for d near d_{th} are improved. In the case of $\sigma_{dB} = 4$, $|f'(d)|$ is not so close to 0 when d is near d_{th}, but is comparatively small when d is near 0; as a result, the expected rise does not happen.

14.2.3.3 Influences of r

As manifested in (14.1.25), parameter r in the log-normal connection model includes the impact of the transmission power of wireless nodes. The following theorem establishes the impact of the parameter r on the CRLB:

Theorem 212 *Consider the CRLB in (14.2.6), (14.2.8), and (14.2.17) and suppose that only d and r are variables. Then, the CRLB for a given distance d with $r = r_0$ is equal to the CRLB for a distance $\frac{d}{r_0}$ with $r = 1$.*

Proof Regarding r as a variable, we substitute the notations as follows: $S \to S(r)$, $f(d) \to f(r, d)$, $f'(d) \to \frac{\partial f(r,d)}{\partial d}$, $CRLB(d) \to CRLB(r, d)$, $g(d) \to g(r, d)$. According to (14.2.7), we have

$$g(r, dr) = \int_{k \log d}^{\infty} \frac{e^{-\frac{z^2}{2\sigma_{dB}^2}}}{\sqrt{2\pi}\sigma_{dB}} dz = g(1, d)$$

Using (14.2.6), we have

$$\begin{aligned} f(r, dr) &= \int_0^{\infty} \int_0^{2\pi} g(r, x) g(r, \sqrt{x^2 + (dr)^2 - 2xdr\cos\theta}) x d\theta dx \\ &= \int_0^{\infty} \int_0^{2\pi} g(r, xr) g(r, r\sqrt{x^2 + d^2 - 2xd\cos\theta}) r^2 x d\theta dx \end{aligned}$$

$$= r^2 \int_0^{\infty} \int_0^{2\pi} g(1,x) g(1, \sqrt{x^2 + d^2 - 2xd\cos\theta}) x d\theta dx$$
$$= r^2 f(1,d)$$

Moreover, we can obtain

$$\left.\frac{\partial f(x,y)}{\partial y}\right|_{x=1,y=d} = \frac{1}{r} \times \left.\frac{\partial f(x,y)}{\partial y}\right|_{x=r,y=dr} \tag{14.2.18}$$

Using the relationship that $S(r) = r^2 S(1)$ (based on (14.2.8)), (14.2.17), and the above equations, it can be obtained that

$$CRLB(r, dr) = CRLB(1, d) \tag{14.2.19}$$

or equivalently,

$$CRLB(r, d) = CRLB(1, \frac{d}{r}).$$

□

Theorem 212 reveals that: (1) the CRLB is mainly determined by the ratio $\frac{d}{r}$; (2) the value field of the CRLB is invariant no matter how large the parameter r is. Therefore, if the parameter r is increased (by increasing the transmission power), distant nodes will become more easily connected, so that estimates for long distances will be available. However the CRLB will not exceed the value field of the CRLB associated with the original small value of r. Consequently, estimates for long distances will generally have less relative errors (i.e., the ratio of the estimation error to the true value d) than those for short distances.

Further note that, because the CRLB is not a monotonic function of d, tuning r does not necessarily increase or decrease the corresponding CRLB associated with a given value of d. Increasing r however results in more one-hop neighbors for each node and consequently more distance estimates, though any distance estimate is not necessarily improved. This increase in the number of distance estimates will benefit other applications, e.g., wireless network localization. This feature can be explored in the implementation of the proposed method. Considering that in static wireless networks the procedure of estimating distances is usually executed only once and probably in the beginning of the network lifetime, the ratio r can be initially set a high value to achieve a high connectivity or average node degree, and then is tuned to be a normal value when the distance estimation is completed. As a result, more estimates of long distances with comparatively good accuracies will be available.

14.2.4 *Practical Implementation of the Distance Estimation Technique*

In this subsection, we improve the proposed method when dealing with short distances and then test it in a practical environment.

Given α and σ_{dB}, define $\epsilon_{\alpha,\sigma_{dB}}$ to be the RMSE when $d = 0$. As illustrated in Fig. 14.14, the RMSE of distance estimates produced by the proposed distance estimation technique experiences small variations as d increases from 0 up to d_{th}, so that if $d \geq \epsilon_{\alpha,\sigma_{dB}}$ the RMSE tends to be under d, implying that relatively good performance is attained. Moreover, based on the analysis in Sect. 14.2.2.3, we focus on the function defined in (14.2.11) with $\hat{\rho} \in [\rho_{\alpha,\sigma_{dB}}(\epsilon_{\alpha,\sigma_{dB}}), 1]$, and reformulate it by a linear function

$$\frac{(1-\hat{\rho})\epsilon_{\alpha,\sigma_{dB}}}{1-\rho_{\alpha,\sigma_{dB}}(\epsilon_{\alpha,\sigma_{dB}})}$$

which smoothly transforms any $\hat{\rho}$ between $\rho_{\alpha,\sigma_{dB}}(\epsilon_{\alpha,\sigma_{dB}})$ and 1 to a distance estimate between 0 and $\epsilon_{\alpha,\sigma_{dB}}$. Consequently, (14.2.11) is updated to be

$$\hat{d} = \begin{cases} 0 & \text{if} \quad M = P = Q = 0 \\ \frac{(1-\hat{\rho})\epsilon_{\alpha,\sigma_{dB}}}{1-\rho_{\alpha,\sigma_{dB}}(\epsilon_{\alpha,\sigma_{dB}})} & \text{if} \quad \hat{\rho} > \rho_{\alpha,\sigma_{dB}}(\epsilon_{\alpha,\sigma_{dB}}) \\ \rho^{-1}_{\alpha,\sigma_{dB}}(\hat{\rho}) & \text{if} \quad \rho_{\alpha,\sigma_{dB}}(d_{th}) \leq \hat{\rho} \leq \rho_{\alpha,\sigma_{dB}}(\epsilon_{\alpha,\sigma_{dB}}) \\ d_{th} & \text{if} \quad \hat{\rho} < \rho_{\alpha,\sigma_{dB}}(d_{th}). \end{cases} \tag{14.2.20}$$

Now we test the distance estimation technique using real measurement data. In [162], a wireless sensor network consisting of 44 nodes was deployed in a real environment and RSS measurements between any two nodes were recorded. Based on the measurement data, we can simulate a realistic environment to implement the distance estimation technique. According to [162], $\alpha = 2.3$, $\sigma_{dB} = 3.92$, and the reference power $p_0 = -37.47$dBm (see (14.1.24)). To proceed with the experiment, we also need to specify the threshold power P_{th} (see (14.1.25)), which essentially "defines" whether two nodes are connected, and $\epsilon_{\alpha,\sigma_{dB}}$ in (14.2.20). After that, we can compute the connection function $g(d)$, and on that basis, obtain the distance estimators based on (14.2.11) and (14.2.20) respectively. To remove the impact of the boundary effect, we consider only the four nodes near the center of the deployment region, i.e., nodes 15, 23, 24, 25 shown in Fig. 14.15, and estimate the inter-node distances between the four nodes using the originally proposed method and the method with the adjustment for short distances.

In this experiment, by letting $\epsilon_{\alpha,\sigma_{dB}}$ be $0.5r$ and raising P_{th} from -61dBm to -52dBm, the average distance estimation errors incurred by the original and the improved methods are listed in Table 14.2. According to the distance estimates produced by the RSS method, which were provided in [162], we compute the corresponding average distance estimation error to be 1.07 m. As shown in the

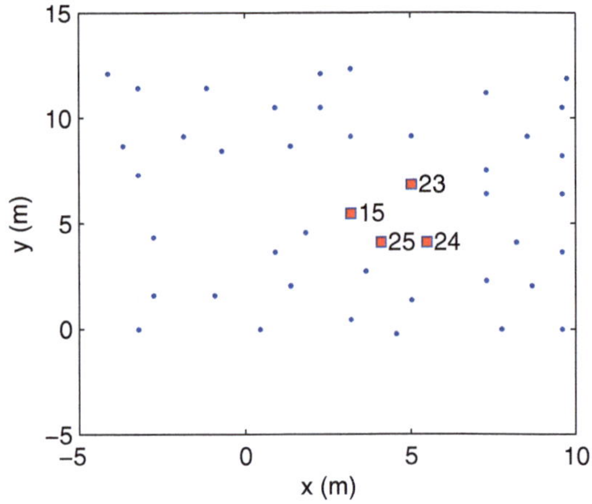

Fig. 14.15 An illustration of the true location of nodes in the measurements

Table 14.2 Distance estimation using real measurements with different settings of P_{th}

P_{th}(dBm)	$r(m)$	ND	DEE (m)	DEE Improved (m)
−52	4.28	8.50	1.00	0.44
−53	4.73	11.25	1.49	0.56
−54	5.23	14.00	1.95	0.95
−55	5.78	17.25	1.78	0.74
−56	6.38	20.75	1.81	0.70
−57	7.05	27.25	2.00	1.08
−58	7.80	31.75	2.08	1.41
−59	8.62	34.50	2.11	1.57
−60	9.52	37.00	2.15	1.72
−61	10.53	38.75	2.14	1.67

Note: average node degree (ND), average distance estimation error (DEE)

table, (1) the improved method always outperforms the original method; (2) the original method outperforms the RSS-based method only in the case of P_{th}(dBm)= −52dBm, while the improved method outperforms the RSS-based method when P_{th}(dBm) is between −56dBm and −52dBm; (3) though the average node degree increases when P_{th} decreases, the average estimation error obtained with the improved method increases in general, a phenomenon which is attributable to the boundary effect.

14.3 Connectivity Based Localization

In the last section, we have demonstrated the use of connectivity information to estimate distances between neighboring nodes. On that basis, a wireless localization scheme can be readily designed that uses the estimated inter-node distances to derive the location of nodes. In this section, we further introduce a category of connectivity-based wireless localization schemes that estimate the location of nodes directly using the connectivity information, without the need to first estimate the distances between particular pairs of neighboring nodes.

Wireless network localization algorithms estimate the location of wireless nodes with initially unknown location information by using knowledge of the absolute location of a few nodes and inter-node measurements such as distance, bearing, and connectivity. Nodes with known location information are called *anchors* and their location can be obtained by using a global positioning system (GPS), or by installing anchors at points with known coordinates. In applications requiring a global coordinate system, these anchors will determine the location of the wireless network in the global coordinate system. In applications where a local coordinate system suffices, these anchors define the local coordinate system to which all other nodes are referred. Because of constraints on the cost and size of wireless devices, energy consumption, implementation environment (e.g., GPS is not accessible in some environments), and the deployment of nodes (e.g., wireless nodes may be randomly scattered in the region), most nodes do not know their location. These nodes with unknown location information are called non-anchor nodes and their coordinates will be estimated by the wireless network localization algorithm.

There is a distinct category of localization algorithms, called connectivity-based or "range free" localization algorithms, which use the connectivity information, i.e., "who is within the communications range of whom" [179] to estimate the location of the non-anchor nodes. The principle of these algorithms is: a wireless node being in the transmission range of another wireless node defines a proximity constraint between both nodes, which can be explored for localization. Bulusu *et al.* [33] and Niculescu *et al.* [158] developed distributed connectivity-based localization algorithms. Shang *et al.* [179] and Doherty *et al.* [56] developed centralized connectivity-based localization algorithms.

In [33], Bulusu *et al.* defined a connectivity metric, which is the ratio of the number of transmitter signals successfully received to the total number of signals from that transmitter, to measure the quality of communication for a specific transmitter-receiver pair. A receiver at an unknown location uses the centroid of its reference points as its location estimate, where a reference point is a transmitter with a known location and whose connectivity metric exceeds a certain threshold (90% in [33]). An experiment was conducted in a $10m \times 10m$ outdoor parking lot using four reference points placed at the four corners of the $10m \times 10m$ square. The $10m \times 10m$ square was subdivided into 100 smaller $1m \times 1m$ grids and the receivers were placed at the grid points. Experimental results showed that for over 90% of the data points, the localization error falls within 30% of the separation distance between two adjacent reference points.

The "DV(distance vector)-hop" approach developed by Niculescu *et al.* [158] starts with all anchors flooding their location to other nodes in the network. The messages are propagated hop-by-hop and there is a hop-count in the message. Each node maintains an anchor information table and counts the least number of hops that it is away from an anchor. When an anchor receives a message from another anchor, it estimates the average distance of one hop using the location of both anchors and the hop-count, and sends it back to the network as a correction factor. When receiving the correction factor, a non-anchor node is able to estimate its distance to anchors and performs trilateration to estimate its location. The algorithm was tested using simulation with a total of 100 nodes uniformly distributed in a circular region of diameter 10. The average node degree, i.e., average number of neighbors per node, is 7.6. Simulation results showed that the algorithm has a mean error of 45% transmission range with 10% anchors; and has a reduced mean error of about 30% transmission range when the percentage of anchors increases above 20%.

Shang *et al.* [179] developed a centralized algorithm by using multi-dimensional scaling (MDS). MDS was originally used in psychometrics and psychophysics and it is a set of data analysis techniques that displays the structure of distance-like data as a geometric picture. In their algorithm, the shortest paths, measured in the number of hops, between all pairs of nodes are first computed, which are used to construct a distance matrix for MDS. Then, MDS is applied to the distance matrix and an approximate value of the relative coordinates of each node is obtained. Finally, the relative coordinates are transformed to the absolute coordinates by aligning the estimated relative coordinates of anchors with their absolute coordinates. The location estimates obtained using earlier steps can be refined using a least-squares minimization. Simulation was conducted using 100 nodes uniformly distributed in a square of size 10×10 and four anchors randomly placed in the region. The average node degree is 10. Simulation results showed a localization error of 0.35. Shang *et al.* further improved their algorithm in [178] by dividing the entire sensor network into overlapping local regions. Localization is performed in individual regions using the earlier described procedures. Then, these local maps are patched together to form a global map by using common nodes shared between adjacent regions. The improved algorithm can achieve better performance on irregularly shaped networks by avoiding the use of distance information between far away nodes. The improved algorithm can also be implemented in a distributed fashion.

In the centralized algorithm of Doherty *et al.* [56], the connectivity-based localization problem is formulated as a convex optimization problem and solved using existing algorithms for solving linear programs and semidefinite programming (SDP) algorithms. Semidefinite programs are a generalization of the linear programs and have the form:

$$\text{Minimize} \quad \mathbf{c}^T\mathbf{x} \tag{14.3.1}$$

$$\text{Subject to:} \quad \mathbf{F}(x) = \mathbf{F}_0 + \mathbf{x}_1\mathbf{F}_1 + \cdots + \mathbf{x}_n\mathbf{F}_n \tag{14.3.2}$$

$$\mathbf{A}\mathbf{x} < \mathbf{b} \tag{14.3.3}$$

$$\mathbf{F}_i = \mathbf{F}_i^T \tag{14.3.4}$$

where $\mathbf{x} = [\mathbf{x}_1, \mathbf{x}_2, \ldots, \mathbf{x}_n]^T$ and $\mathbf{x}_i$ represents the coordinate vector of node i, i.e., $\mathbf{x}_i = [x_i, y_i]$. The quantities $\mathbf{A}$, $\mathbf{b}$, $\mathbf{c}$, and $\mathbf{F}_i$ are all known. The inequality (14.3.3) is known as a linear matrix inequality (LMI). A connection between nodes i and j

can be represented by a "radial constraint" on the node location: $||\mathbf{x}_i - \mathbf{x}_j|| \leq R$, where R is the transmission range. This constraint is a convex constraint and can be transformed into an LMI using Schur complements [56]. A solution to the coordinates of the non-anchor nodes satisfying the radial constraints can be obtained by leaving the objective function $\mathbf{c}^T\mathbf{x}$ blank and solving the problem. Because there may be many possible coordinates of the non-anchor nodes satisfying the constraints, the solution may not be unique. If we set the element of $\mathbf{c}$ corresponding to x_i (or y_i) to be 1 (or -1) and all other elements of $\mathbf{c}$ to be zero, the problem becomes a constrained maximization (or minimization) problem. A lower bound or an upper bound on x_i (or y_i) satisfying the radial constraints can be computed, from which a rectangular box bounding the location estimates of the non-anchor nodes can be obtained. The algorithm was tested using simulation with a total of 200 nodes randomly placed on a square of size $10R \times 10R$ and the average node degree is 5.7 [56]. Simulation results showed that the mean location error is a monotonically decreasing function of the number of anchors. When the number of anchors is small, the estimated location is as poor as a random guess of the node's coordinates. The mean location error reduces to R when the number of anchors increases to 18; it reduces to $0.5R$ when the number of anchors increases to 50.

In comparison with other localization algorithms, the most attractive feature of the connectivity-based localization algorithms is their simplicity. However they can only provide a coarse grained estimate of each node's location, which means that they are only suitable for applications requiring an approximate location estimate only. Also the localization error is highly dependent on the node density of the network, the number of anchors, and the network topology. The location error will be larger in a network with a smaller node density, fewer anchors, or irregular network topology. There are many later extension of the aforementioned basic connectivity-based localization algorithms to deal with these issues to improve the localization accuracy [125, 132, 135, 187].

14.4 Notes and Further Readings

In this chapter, we gave three examples to illustrate the applications of connectivity studies. Obviously, the applications of the connectivity studies go much further beyond the three examples. Throughout this book and the literature, we can find extensive applications of connectivity studies in determining the capacity of a network, in determining the energy consumption of a network, in determining the lifetime of a network, in determining the information propagation speed of dynamic networks, and in radio resource management. These in turn lead to the optimum design, deployment, and management of a network.

Particularly, it was demonstrated in [138, 142, 195] that the capacity scaling law of large wireless networks observed in [91] can be easily explained by the increase in the average number of hops (hence the increase in the portion of bandwidth spent on relaying traffic) as the network becomes larger. In [204, 205], the analysis on the hop count statics was used in estimating the energy consumption, the network lifetime, and the reliability of end-to-end packet transmission. The connectivity analysis can also be used to estimate the effective energy consumption and help the network designers to choose the optimum transmission range/power to minimize the energy consumption, or to set the transmission range/power to provide a guaranteed performance on end-to-end packet transmissions.

Appendix

Landau's Order Notation

For two real functions f and h, the following order notations are used throughout the book:

- $f(z) = O_z(h(z))$ if and only if there exists a positive constant C and a real number z_0 such that for $z \geq z_0$, $|f(z)| \leq C|h(z)|$;
- $f(z) = o_z(h(z))$ if and only if $\lim_{z\to\infty} \frac{f(z)}{h(z)} = 0$;
- $f(z) = \omega_z(h(z))$ if and only if $h(z) = o_z(f(z))$;
- $f(z) = \Theta_z(h(z))$ if and only if there exist a sufficiently large z_0 and two positive constants c_1 and c_2 such that for any $z > z_0$, $c_1 h(z) \geq f(z) \geq c_2 h(z)$;
- $f(z) \sim_z h(z)$ if and only if $\lim_{z\to\infty} \frac{f(z)}{h(z)} = 1$;

The above definition applies whether the argument z is continuous or discrete, e.g., assuming integer values.

When we do not need to emphasize the dependence on the argument z, we omit the argument in the order notation.

Frequently Used Statistical Inequalities and Functions

Chebyshev's Inequality

Let X be a real-valued random variable with finite expected value μ and finite nonzero variance σ^2. Then, for any real number $k > 0$,

$$\Pr(|X - \mu| \geq k\sigma) \leq \frac{1}{k^2}$$

G. Mao, *Connectivity of Communication Networks*,
DOI 10.1007/978-3-319-52989-9

Markov Inequality

Let X be a nonnegative random variable and $a > 0$, then

$$\Pr(X \geq a) \leq \frac{E(X)}{a}$$

Jensen Inequality

Let X be a real-valued random variable and f be a convex function, then

$$f(E(X)) \leq E(f(x))$$

Fortuin–Kasteleyn–Ginibre (FKG) Inequality

The FKG inequality is a correlation inequality and states that for two increasing functions g and h and a real-valued random variable X

$$\int_{\Re} g(x)\, h(x) f_X(x)\, dx \geq \int_{\Re} g(x) f_X(x)\, dx \int_{\Re} h(x) f_X(x)\, dx$$

where $f_X(x)$ is the probability density function of X.

Q-Function

The Q-function is the tail probability of the standard normal distribution and is defined as

$$Q(x) = \int_x^{\infty} \frac{1}{\sqrt{2\pi}} e^{-\frac{y^2}{2}}\, dy$$

Error Function

For a normally distributed random variable X with mean 0 and variance $\frac{1}{2}$, the error function $erf(x)$ describes the probability of X falling in the range $[-x, x]$. The error function is defined as

$$erf(x) = \frac{1}{\sqrt{\pi}} \int_{-x}^{x} e^{-y^2}\, dy$$

Lambert W Function

The Lambert-W function is defined to be the multivalued inverse of the function $f(z) = ze^z$. In other words,

$$z = f^{-1}(ze^z) = W(ze^z)$$

By substituting $x = ze^z$ into the above equation, we get the defining equation for the Lambert W function:

$$x = W(x)\, e^{W(x)}$$

Since the function f is not injective, W is multivalued except at 0. If we restrict W to be real-valued, W is defined only for $x \geq -1/e$ and is double-valued on $(-1/e, 0)$. The additional constraint $W \geq -1$ defines a single-valued function $W_0(x)$. We have $W_0(0) = 0$ and $W_0(-1/e) = -1$. Meanwhile, the lower branch has $W \leq -1$ and is denoted $W_{-1}(x)$. It decreases from $W_{-1}(-1/e) = -1$ to $W_{-1}(0^-) = -\infty$.

The Lambert W relation cannot be expressed in terms of elementary functions.

Bibliography

1. Agarwal, A., Little, T.: Impact of asymmetric traffic densities on delay tolerant vehicular ad hoc networks. In: IEEE Vehicular Networking Conference, pp. 1–8 (2009)
2. Agarwal, A., Starobinski, D., Little, T.: Analytical model for message propagation in delay tolerant vehicular ad hoc networks. In: IEEE VTC Spring, pp. 3067–3071 (2008)
3. Agarwal, A., Starobinski, D., Little, T.D.C.: Phase transition of message propagation speed in delay-tolerant vehicular networks. IEEE Trans. Intell. Transp. Syst. **13**(1), 249–264 (2012)
4. Agrawal, P., Patwari, N.: Correlated link shadow fading in multi-hop wireless networks. IEEE Trans. Wirel. Commun. **8**(8), 4024–4036 (2009). URL 10.1109/TWC.2009.071293
5. Anderson, B.D.O., Kraus, F., Mansour, M., Dasgupta, S.: Easily testable sufficient conditions for the robust stability of systems with multilinear parameter dependence. Automatica **31**(1), 25–40 (1995)
6. Apostol, T.M., Mnatsakanian, M.A.: Sums of squares of distances in m-space. Am. Math. Month. **110**(6), 516–526 (2003)
7. Arratia, R., Goldstein, L., Gordon, L.: Poisson approximation and the chen-stein method. Stat. Sci. **5**(4), 403–434 (1990)
8. Aspnes, J., Eren, T., Goldenberg, D.K., Morse, A.S., Whiteley, W., Yang, Y.R., Anderson, B.D.O., Belhumeur, P.N.: A theory of network localization,. IEEE Trans. Mob. Comput. **5**(12), 1663–1678 (2006)
9. Aste, T., Weaire, D.: The Pursuit of Perfect Packing. Institute of Physics Publishing, Bristol (2000)
10. Avin, C., Lotker, Z., Pasquale, F., Pignolet, Y.A.: A note on uniform power connectivity in the sinr model. Algorithmic Asp. Wirel. Sens. Netw. **5804**, 116–127 (2009)
11. Baccelli, E., Jacquet, P., Mans, B., G. Rodolakis pp. 1743–1756, .: Highway vehicular delay tolerant networks: Information propagation speed properties. IEEE Trans. Inf. Theory **58**(3), 1743–1756 (2012)
12. Balister, P., Bollobas, B., Sarkar, A., Walters, M.: Connectivity of random k-nearest-neighbour graphs. Adv. Appl. Probab. **37**(1), 1–24 (2005)
13. Balister, P., Bollobas, B., Sarkar, A., Walters, M.: A critical constant for the k nearest neighbour model. Adv. Appl. Probab. **41**(1), 1–12 (2009)
14. Banerjee, N., Corner, M.D., Towsley, D., Levine, B.N.: Relays, base stations, and meshes: Enhancing mobile networks with infrastructure. In: ACM International Conference on Mobile Computing and Networking, pp. 81–91 (2008)
15. Barbour, A.D., Holst, L., Jason, S.: Poisson Approximation. Oxford University Press, Oxford (2003)
16. Benyamina, D., Hafid, A., Gendreau, M.: Wireless mesh networks design - a survey. IEEE Commun. Surv. Tutorials **14**(2), 299–310 (2012)

G. Mao, *Connectivity of Communication Networks*,
DOI 10.1007/978-3-319-52989-9

17. Berg, J.v.d., Kesten, H.: Inequalities with applications to percolation and reliability. J. Appl. Probab. **22**(3), 556–569 (1985)
18. Bettstetter, C.: Mobility modeling in wireless networks: Categorization, smooth movement, and border effects. SIGMOBILE Mobile Comput. Commun. Rev. **5**(3), 55–66 (2001)
19. Bettstetter, C.: On the minimum node degree and connectivity of a wireless multihop network. In: The 3rd ACM International Symposium on Mobile Ad Hoc Networking and Computing, pp. 80–91 (2002)
20. Bettstetter, C.: Failure-resilient ad hoc and sensor networks in a shadow fading environment. In: IEEE/IFIP International Conference on Dependable Systems and Networks, pp. 1–7 (2004)
21. Bettstetter, C.: On the connectivity of ad hoc networks. Comput. J. **47**(4), 432–447 (2004)
22. Bettstetter, C., Hartmann, C.: Connectivity of wireless multihop networks in a shadow fading environment. Wirel. Netw. **11**(5), 571–579 (2005)
23. Bettstetter, C., Zangl, J.: How to achieve a connected ad hoc network with homogeneous range assignment: an analytical study with consideration of border effects. In: The 4th International Workshop on Mobile and Wireless Communications Network, pp. 125–129 (2002)
24. Bhadra, S., Ferreira, A.: Complexity of connected components in evolving graphs and the computation of multicast trees in dynamic networks. In: Lecture Notes in Computer Science, vol. 2865, pp. 259–270
25. Biggs, N.L.: Algebraic Graph Theory. Cambridge University Press, Cambridge (1974)
26. Blough, D.M., Leoncini, M., Resta, G., Santi, P.: The k-neighbors approach to interference bounded and symmetric topology control in ad hoc networks. IEEE Trans. Mob. Comput. **5**(9), 1267–1282 (2006)
27. Bollobas, B.: Random Graphs, 2nd edn. Cambridge University Press, Cambridge (2001)
28. Bollobas, B.: Modern Graph Theory, 2nd edn. Springer, New York (2002)
29. Bollobás, B., Riordan, O.: Percolation. Cambridge University Press, Cambridge (2006)
30. Boyd, S., Vandenberghe, L.: Convex Optimization. Cambridge University Press, Cambridge (2004)
31. Bradonjic, M., Hagberg, A., Percus, A.G.: Giant component and connectivity in geographical threshold graphs. In: Procedings of Algorithms and Models for the Web-Graph, pp. 209–216 (2007)
32. Brooks, R.R., Pillai, B., Racunas, S., Rai, S.A.R.S.: Mobile network analysis using probabilistic connectivity matrices. IEEE Trans. Syst. Man Cybern. C Appl. Rev. **37**(4), 694–702 (2007)
33. Bulusu, N., Heidemann, J., Estrin, D.: Gps-less low-cost outdoor localization for very small devices. IEEE Pers. Commun. **7**(5), 28–34 (2000)
34. Busson, A., Chelius, G.: Point processes for interference modeling in csma/ca ad-hoc networks. In: Proceedings of the 6th ACM Symposium on Performance Evaluation of Wireless Ad hoc, Sensor, and Ubiquitous Networks, pp. 33–40 (2009)
35. Camp, T., Boleng, J., Davies, V.: A survey of mobility models for ad hoc network research. Wirel. Commun. Mob. Comput. **2**(5), 483–502 (2002)
36. Carvalho, M.M., Garcia-Luna-Aceves, J.J.: Delay analysis of ieee 802.11 in single-hop networks. In: IEEE International Conference on Network Protocols, pp. 146–155 (2003)
37. Cerf, R., Cirillo, E.N.M.: Finite size scaling in three-dimensional bootstrap percolation. Ann. Probab. **27**(4), 1837–1850 (1999)
38. Chandler, S.A.G.: Calculation of number of relay hops required in randomly located radio network. Electron. Lett. **25**(24), 1669–1671 (1989)
39. Chen, C.: Linear System Theory and Design, 4th edn. The Oxford Series in Electrical and Computer Engineering. Oxford University Press, Oxford (2012)
40. Chen, P., Chen, K.: Information epidemics in complex networks with opportunistic links and dynamic topology. In: IEEE GLOBECOM, pp. 1–6 (2010)
41. Cheng, X., Wang, C.X., Laurenson, D.I., Salous, S., Vasilakos, A.V.: An adaptive geometry-based stochastic model for non-isotropic mimo mobile-to-mobile channels. IEEE Trans. Wirel. Commun. **8**(9), 4824–4835 (2009)

42. Chitte, S., Dasgupta, S., Ding, Z.: Distance estimation from received signal strength under log-normal shadowing: Bias and variance. IEEE Signal Process. Lett. **16**(3), 216–218 (2009)
43. Clementi, A., Pasquale, F., Silvestri, R.: Opportunistic manets: mobility can make up for low transmission power. IEEE/ACM Trans. Netw. **21**(2), 610–620 (2013)
44. Conti, M., Giordano, S.: Multihop ad hoc networking: The reality. IEEE Commun. Mag. **45**(4), 88–95 (2007)
45. Contla, P.A., Stojmenovic, M.: Estimating hop counts in position based routing schemes for ad hoc networks. Telecommun. Syst. **22**(1), 109–118 (2003)
46. Conway, J., Sloane, N.J.A.: Sphere Packings, Lattices and Groups (Grundlehren der mathematischen Wissenschaften). Springer, New York (2010)
47. Corless, R.M., Gonnet, G.H., Hare, D.E.G., Jeffrey, D.J., Knuth, D.E.: On the lambert w function. Adv. Comput. Math. **5**(1), 329–359 (1996)
48. Cvetković, D.M., Doob, M., Sachs, H.: Spectra of Graphs, 3rd edn. Springer, New York (1995)
49. Dasgupta, S., Anderson, B.: Physically based parameterizations for designing adaptive algorithms. Automatica **23**(4), 469–477 (1987)
50. Dasgupta, S., Anderson, B., Kaye, R.J.: Output error identification methods for partially known systems. Int. J. Control **43**(1), 177–191 (1985)
51. Dasgupta, S., Anderson, B., Kaye, R.J.: Identification of physical parameters in structured systems. Automatica **24**(2), 217–225 (1988)
52. Dasgupta, S., Chockalingam, C., Fu, M., Anderson, B.: Lyapunov functions for uncertain systems with applications to the stability of time varying systems. IEEE Trans. Circuits Syst. I Fundam. Theory Appl. **41**(2), 93–105 (1994)
53. David, H.A.: Order Statistics. Wiley, New York (1970)
54. De, S., Caruso, A., Chaira, T., Chessa, S.: Bounds on hop distance in greedy routing approach in wireless ad hoc networks. Int. J. Wirel. Mob. Comput. **1**(2), 131–140 (2006)
55. Desai, M., Manjunath, D.: On the connectivity in finite ad hoc networks. IEEE Commun. Lett. **6**(10), 437–439 (2002)
56. Doherty, L., pister, K., El Ghaoui, L.: Convex position estimation in wireless sensor networks. In: IEEE INFOCOM, pp. 1655–1663 (2001)
57. Dousse, O., Baccelli, F., Thiran, P.: Impact of interferences on connectivity in ad hoc networks. IEEE/ACM Trans. Netw. **13**(2), 425–436 (2005)
58. Dousse, O., Franceschetti, M., Macris, N., Meester, R., Thiran, P.: Percolation in the signal to interference ratio graph. J. Appl. Probab. **43**(2), 552–562 (2006)
59. Dousse, O., Mannersalo, P., Thiran, O.: Latency of wireless sensor networks with uncoordinated power savings mechanisms. In: Proceedings of the 5th ACM International Symposium on Mobile Ad hoc Networking and Computing, pp. 109–120 (2004)
60. Dousse, O., Thiran, P.: Connectivity vs capacity in dense ad hoc networks. In: IEEE INFOCOM, pp. 476–486 (2004)
61. Dousse, O., Thiran, P., Hasler, M.: Connectivity in ad-hoc and hybrid networks. In: IEEE INFOCOM, pp. 1079–1088
62. Dulman, S., Havinga, P.: Statistically enhanced localization schemes for randomly deployed wireless sensor networks. In: Proceedings of the Intelligent Sensors, Sensor Networks and Information Processing Conference, pp. 403–410 (2004)
63. Dulman, S., Rossi, M., Havinga, P., Zorzi, M.: On the hop count statistics for randomly deployed wireless sensor networks. Int. J. Sens. Netw. **1**(1/2), 89–102 (2006)
64. Feeney, L.M., Nilsson, M.: Investigating the energy consumption of a wireless network interface in an ad hoc networking environment. In: IEEE INFOCOM, vol. 3, pp. 1548–1557 (2001)
65. Fekete, S.P., Schmidt, C., Wegener, A., Hellbruck, H., Fischer, S.: Empowered by wireless communication: Distributed methods for selforganizing traffic collectives. ACM Trans. Autonomous Adaptive Syst. **5**(3), 1–30 (2010)
66. Ferreira, A.: Building a reference combinatorial model for manets. IEEE Netw. **18**(5), 24–29 (2004)

67. Foh, C.H., Lee, B.S.: A closed form network connectivity formula for one-dimensional MANETs. In: IEEE International Conference on Communications, vol. 6, pp. 3739–3742 (2004)
68. Fracchia, R., Meo, M.: Analysis and design of warning delivery service in intervehicular networks. IEEE Trans. Mob. Comput. **7**(7), 832–845 (2008)
69. Franceschetti, M., Dousse, O., Tse, D.N.C., Thiran, P.: Closing the gap in the capacity of wireless networks via percolation theory. IEEE Trans. Inf. Theory **53**(3), 1009–1018 (2007)
70. Franceschetti, M., Meester, R.: Critical node lifetimes in random networks via the chen-stein method. IEEE Trans. Inf. Theory **52**(6), 2831–2837 (2006)
71. Franceschetti, M., Meester, R.: Random Networks for Communication. Cambridge University Press, Cambridge (2007)
72. Friedgut, E., Kalai, G.: Every monotone graph property has a sharp threshold. Am. Math. Soc. **124**(10), 2993–3002 (1996)
73. Friedman, R., Viana, A.C.: Gossiping on manets: The beauty and the beast. ACM Operating Syst. Rev. **41**(5), 67–74 (2007)
74. Gallager, R.: A perspective on multiaccess channels. IEEE Trans. Inf. Theory **31**(2), 124–142 (1985)
75. Gantmacher, F.: The Theory of Matrices, vol. 2. Chelsea Publishing, New York (2000)
76. Gerla, M.: From battlefields to urban grids: New research challenges in ad hoc wireless networks. Pervasive Mob. Comput. **1**(1), 77–93 (2005)
77. Ghasemi, A., Nader-Esfahani, S.: Exact probability of connectivity in one-dimensional ad hoc wireless networks. IEEE Commun. Lett. **10**(4), 251–253 (2006)
78. Ghasemi, A., Sousa, E.S.: Asymptotic performance of collaborative spectrum sensing under correlated log-normal shadowing. IEEE Commun. Lett. **11**(1), 34–36 (2007)
79. Godehardt, E., Jaworski, J.: On the connectivity of a random interval graph. Random Struct. Algorithm **9**(1 and 2), 137–161 (1996)
80. Godsil, C., Royle, G.F.: Algebraic Graph Theory. Springer, New York (2001)
81. Goeckel, D., Benyuan, L., Towsley, D., Liaoruo, W., Westphal, C.: Asymptotic connectivity properties of cooperative wireless ad hoc networks. IEEE J. Sel. Areas Commun. **27**(7), 1226–1237 (2009). URL 10.1109/JSAC.2009.090918
82. Goel, A., Rai, S., Krishnamachari, B.: Sharp thresholds for monotone properties in random geometric graphs. In: Proceedings of the Thirty-sixth Annual ACM Symposium on Theory of Computing, pp. 580–586 (2004)
83. Golub, G.H., Loan, C.F.V.: Matrix Computations, 3rd edn. Johns Hopkins Studies in Mathematical Sciences. Johns Hopkins University Press (1996)
84. Grimmett, G., Stirzaker, D.: Probability and Random Processes, 3rd edn. Oxford University Press, Oxford (2001)
85. Grinstead, C.M., Snell, J.L.: Introduction to Probability. American Mathematical Society, Providence. Chapman and Hall/CRC (1997)
86. Gross, J.L., Yellen, J.: Handbook of Graph Theory, 1st edn. CRC Press, Chapman and Hall/CRC (2004)
87. Grossglauser, M., Tse, D.N.C.: Mobility increases the capacity of ad hoc wireless networks. IEEE/ACM Trans. Netw. **10**(4), 477–486 (2002)
88. Gu, B., Hong, X.: Critical phase of connectivity in wireless network expansion. In: IEEE GLOBECOM, pp. 1–5 (2010)
89. Gudmundson, M.: Correlation model for shadow fading in mobile radio systems. Electron. Lett. **27**(23), 2145–2146 (1991)
90. Guerin, R.A.: Channel occupancy time distribution in a cellular radio system. IEEE Trans. Veh. Technol. **36**(3), 89–99 (1987)
91. Gupta, P., Kumar, P.: The capacity of wireless networks. IEEE Trans. Inf. Theory **46**(2), 388–404 (2000)
92. Gupta, P., Kumar, P.R.: Critical Power for Asymptotic Connectivity in Wireless Networks, pp. 547–566. Systems and Control: Foundations and Applications. Birkhauser, Boston (1998)

93. G.Xing, Lu, C., Pless, R., Huang, Q.: Impact of sensing coverage on greedy geographic routing algorithms. IEEE Trans. Parallel Distrib. Syst. **17**(4), 348–360 (2006)
94. Haenggi, M., Andrews, J.G., Baccelli, F., Dousse, O., Franceschetti, M.: Stochastic geometry and random graphs for the analysis and design of wireless networks. IEEE J. Sel. Areas Commun. **27**(7), 1029–1046 (2009)
95. Haenggi, M., Ganti, R.K.: Interference in large wireless networks. Foundations Trends Netw. **3**(2), 127–248 (2009)
96. Hall, P.: Introduction to the Theory of Coverage Processes. Wiley Series in Probability and Statistics. Wiley, New York (1988)
97. Han, G., Makowski, A.: Poisson convergence can yield very sharp transitions in geometric random graphs. In: Proceedings of Inaugural Workshop on Informaition Theory and Appllications, pp. 1–5 (2006)
98. Harris, T.E.: The Theory of Branching Processes. Springer-Verlag (1963)
99. Hartfiel, D.J.: Nonhomogeneous Matrix Products. World Scientific Publishing Company, Singapore (2002)
100. Hekmat, R., Mieghem, P.V.: Connectivity in wireless ad-hoc networks with a log-normal radio model. Mob. Netw. Appl. **11**(3), 351–360 (2006)
101. Hogben, L.: Handbook of Linear Algebra, 1st edn. Discrete Mathematics and Its Applications. Chapman and Hall/CRC, Boca Raton (2007)
102. Horn, R.A., Johnson, C.R.: Matrix Analysis. Cambridge University Press, Cambridge (1985)
103. Hsu, E.P., Varadhan, S.R.S.: Probability Theory and Applications, vol. 6. American Mathematical Society, Providence (1999)
104. Hu, W., Bulusu, N., Chou, C., Jha, S., Taylor, A., Tran, V.N.: The design and evaluation of a hybrid sensor network for cane-toad monitoring. ACM Trans. Sensor Netw. **5**(4), 4:1–4:28 (2009)
105. Ingraham, D., Beresford, R., Kaluri, K., Ndoh, M., Srinivasan, K.: Wireless sensors: Oyster habitat monitoring in the bras d'or lakes. In: IEEE 1st International Conference on Distributed Computing in Sensor Systems, pp. 399–400 (2005)
106. Isham, V.: Stochastic models for epidemics. Tech. rep., University College London (2004)
107. Jachymski, J.: König chains for submultiplicative functions and infinite products of operators. Trans. Am. Math. Soc. **361**(11), 5967–5981 (2009)
108. Jackson, B., Jordan, T.: Connected rigidity matroids and unique realizations of graphs. J. Comb. Theory B **94**(1), 1–29 (2005)
109. Jacquet, P., Mans, B., Rodolakis, G.: Information propagation speed in mobile and delay tolerant networks. In: IEEE INFOCOM, pp. 244–252 (2009)
110. Jagers, P.: Branching Processes with Biological Applications. Wiley, New York (1975)
111. Janson, S., Luczak, T., Ruciski, A.: Random Graphs. Wiley, New York (2000)
112. Juang, P., Oki, H., Wang, Y., Martonosi, M., Peh, L., Rubenstein, D.: Energy-efficient computing for wildlife tracking: Design tradeoffs and early experiences with zebranet. ACM SIGARCH Comput. Archit. News **30**(5), 96–107 (2002)
113. Kashyap, A., Khuller, S., Shayman, M.: Relay placement for higher order connectivity in wireless sensor networks. In: IEEE INFOCOM, pp. 1–12 (2006)
114. Kay, S.: Fundamentals of Statistical Signal Processing: Estimation Theory. Prentice Hall, Englewood Cliffs (1993)
115. Khabbazian, M., Bhargava, V.K.: Efficient broadcasting in mobile ad hoc networks. IEEE Trans. Mob. Comput. **8**(2), 231–245 (2009)
116. Khalil, I., Bagchi, S., Shroff, N.: Liteworp: A lightweight countermeasure for the wormhole attack in multihop wireless networks. In: Proceedings of the 2005 International Conference on Dependable Systems and Networks, pp. 612–621 (2005)
117. Khelil, A.: Mobility-aware buffering for delay-tolerant ad hoc broadcasting. Simul. Ser. **38**(3), 145–154 (2006)
118. Kong, Z., Yeh, E.M.: Analytical lower bounds on the critical density in continuum percolation. In: International Symposium on Modeling and Optimization in Mobile, Ad Hoc and Wireless Networks and Workshops (WiOpt), pp. 1–6 (2007)

119. Kong, Z., Yeh, E.M.: Connectivity and latency in large-scale wireless networks with unreliable links. In: IEEE INFOCOM, pp. 394–402 (2008)
120. Korniss, G., White, C.J., Rikvold, P.A., Novotny, M.A.: Dynamic phase transition, universality, and finite-size scaling in the twodimensional kinetic ising model in an oscillating field. Phys. Rev. E **63**(1), 016,120–1 – 016,120–15 (2000)
121. Krishnamachari, B., S. Wicker, R.B., Fernandez, C.: On the complexity of distributed self-configuration in wireless networks. Telecommun. Syst. **22**(1–4), 33–59 (2003)
122. Krishnamachari, B., Wicker, S.B., Bejar, R.: Phase transition phenomena in wireless ad hoc networks. In: IEEE Globecom, vol. 5, pp. 2921–2925 (2001)
123. Kumar, S., Raghavan, V.S., Deng, J.: Medium access control protocols for ad hoc wireless networks: A survey. Ad Hoc Netw. **4**(3), 326–358 (2006)
124. Lebhar, E., Lotker, Z.: Unit disk graph and physical interference model: Putting pieces together. In: IEEE International Symposium on Parallel and Distributed Processing, pp. 1–8 (2009)
125. Lederer, S., Wang, Y., Gao, J.: Connectivity-based localization of large scale sensor networks with complex shape. In: IEEE INFOCOM, pp. 789–797 (2008)
126. Leutzbach, W.: Introduction to the Theory of Traffic Flow. Springer, New York (1988)
127. Li, J., Chigan, C.: Delay-aware transmission range control for vanets. In: IEEE GLOBECOM, pp. 1–6 (2010)
128. Li, J.S., Kao, H.C., Ke, J.D.: Voronoi-based relay placement scheme for wireless sensor networks. IET Commun. **3**(4), 530–538 (2009)
129. Li, P., Zhang, C., Fang, Y.: Asymptotic connectivity in wireless ad hoc networks using directional antennas. IEEE/ACM Trans. Netw. **17**(4), 1106–1117 (2009)
130. Li, S.: Concise formulas for the area and volume of a hyperspherical cap. Asian J. Math. Stat. **4**(1), 66–70 (2011)
131. Li, Y., Yang, Y.: Asymptotic connectivity of large-scale wireless networks with a log-normal shadowing model. In: IEEE VTC, pp. 1–5 (2010)
132. Lim, H., Hou, J.: Distributed localization for anisotropic sensor networks. ACM Trans. Sensor Netw. **5**(2), 1–26 (2009)
133. Lo, N., Kuo, H.Y.: Two-hops neighbor-aware routing protocol in mobile ad hoc networks. In: Lecture Notes in Computer Science, vol. 4611/2007, pp. 340–349 (2008)
134. Lorenz, C.D., May, R., Ziff, R.M.: Similarity of percolation thresholds on the HCP and FCC lattices. J. Stat. Phys. **98**(3), 961–970 (2000)
135. Ma, D., Meng, J., Wang, B., Lim, H.: A novel approach towards source-to-destination distance estimation in wireless sensor networks. In: Proceedings of the Intelligent Sensors, Sensor Networks and Information Processing, pp. 463–467 (2009)
136. Mainwaring, A., Culler, D., Polastre, J., Szewczyk, R., Anderson, J.: Wireless sensor networks for habitat monitoring. In: ACM international Workshop on Wireless Sensor Networks and Applications, pp. 88–97 (2002)
137. Mallery, C.J., Medidi, S., Medidi, M.: Relative localization with 2-hop neighborhood. In: International Symposium on a World of Wireless, Mobile and Multimedia Networks, pp. 1–4 (2008)
138. Mao, G., Anderson, B.: Capacity of large wireless networks with generally distributed nodes. IEEE Trans. Wirel. Commun. **13**(3), 1678–1691 (2014)
139. Mao, G., Anderson, B.D.: On the asymptotic connectivity of random networks under the random connection model. In: IEEE INFOCOM, pp. 631–639 (2011)
140. Mao, G., Anderson, B.D.O., Fidan, B.: Path loss exponent estimation for wireless sensor network localization. Comput. Netw. **51**(10), 2467–2483 (2007)
141. Mao, G., Fidan, B., Anderson, B.D.O.: Wireless sensor network localization techniques. Comput. Netw. **51**(10), 2529–2553 (2007)
142. Mao, G., Lin, Z., Ge, X., Yang, Y.: Towards a simple relationship to estimate the capacity of static and mobile wireless networks. IEEE Trans. Wirel. Commun. **12**(8), 3883 – 3895 (2013)
143. Meester, R., Roy, R.: Continuum Percolation. Cambridge Tracts in Mathematics. Cambridge University Press, Cambridge (1996)

144. Miller, L.E.: Distribution of link distances in a wireless network. J. Res. Natl. Inst. Stand. Technol. **106**(2), 401–412 (2001)
145. Miorandi, D.: The impact of channel randomness on coverage and connectivity of ad hoc and sensor networks. IEEE Trans. Wirel. Commun. **7**(3), 1062–1072 (2008)
146. Miorandi, D., Altman, E.: Coverage and connectivity of ad hoc networks presence of channel randomness. In: IEEE INFOCOM, pp. 491–502 (2005)
147. Miorandi, D., Altman, E.: Connectivity in one-dimensional ad hoc networks: a queueing theoretical approach. Wirel. Netw. **12**(5), 573–587 (2006)
148. Mohar, B.: Laplace eigenvalues of graphs - a survey. J. Discrete Math. **109**, 171–183 (1992)
149. Moin, P.: Fundamentals of Engineering Numerical Analysis. Cambridge University Press, Cambridge (2001)
150. Molloy, M., Reed, B.: The size of the giant component of a random graph with a given degree sequence. Comb. Probab. Comput. **7**(3), 295–305 (1998)
151. Monteiro, J., Goldman, A., Ferreira, A.: Performance evaluation of dynamic networks using an evolving graph combinatorial model. In: IEEE International Conference on Wireless and Mobile Computing, Networking and Communications, pp. 173–180 (2006)
152. Montroll, E., West, B.: On an enriched collection of stochastic processes. Fluctuation Phenom. Studies Stat. Mech. **7**, 61–175 (1979)
153. Mukherjee, S., Avidor, D.: Connectivity and transmit-energy considerations between any pair of nodes in a wireless ad hoc network subject to fading. IEEE Trans. Veh. Technol. **57**(2), 1226–1242 (2008)
154. Nain, P., Towsley, D., Liu, B., Liu, Z.: Properties of random direction models. In: IEEE INFOCOM, pp. 1897–1907 (2005)
155. Nelson, R.: Probability, Stochastic Processes, and Qqueueing Theory: The Mathematics of Computer Performance Modelling. Springer, New York (1995)
156. Németh, G., Vattay, G.: Giant clusters in random ad hoc networks. Phys. Rev. E **67**(3) (2003)
157. Nguyen, H.Q., Baccelli, F., Kofman, D.: A stochastic geometry analysis of dense ieee 802.11 networks. In: INFOCOM 2007. 26th IEEE International Conference on Computer Communications, pp. 1199–1207 (2007)
158. Niculescu, D., Nath, B.: Ad hoc positioning system (aps). In: IEEE GLOBECOM, pp. 2926–2931 (2001)
159. Olfati-Saber, R., Fax, J.A., Murray, R.M.: Consensus and cooperation in networked multi-agent systems. Proc. IEEE **95**(1), 215–233 (2007)
160. Orriss, J., Barton, S.K.: Probability distributions for the number of radio transceivers which can communicate with one another. IEEE Trans. Commun. **51**(4), 676–681 (2003)
161. Padhy, P., Martinez, K., Riddoch, A., Hart, J.K., Ong, H.L.R.: Glacial environment monitoring using sensor networks. In: Proceedings of Real-World Wireless Sensor Networks, pp. 10–14 (2005)
162. Patwari, N., Hero A.O., I., Perkins, M., Correal, N., O'Dea, R.: Relative location estimation in wireless sensor networks. IEEE Trans. Signal Process. **51**(8), 2137–2148 (2003)
163. Penrose, M.D.: On a continuum percolation model. Adv. Appl. Probab. **23**(3), 536–556 (1991)
164. Penrose, M.D.: On k-connectivity for a geometric random graph. Random Struct. Algorithm **15**(2), 145–164 (1999). URL http://dx.doi.org/10.1002/(SICI)1098-2418(199909)15:2<145::AID-RSA2>3.0.CO;2-G
165. Penrose, M.D.: Random Geometric Graphs. Oxford Studies in Probability. Oxford University Press, Oxford (2003)
166. Perkins, C.E., Royer, E.M.: Ad hoc on-demand distance vector routing. In: The 2nd IEEE Workshop On Mobile Computing Systems and Applications, pp. 90–100 (1999)
167. Philips, T.K., Panwar, S.S., Tantawi, A.N.: Connectivity properties of a packet radio network model. IEEE Trans. Inf. Theory **35**(5), 1044–1047 (1989)
168. Piorkowski, M., Sarafijanovic-Djukic, N., Grossglauser, M.: CRAWDAD data set epfl/mobility (v. 2009-02-24) (2014). URL http://crawdad.cs.dartmouth.edu/epfl/mobility
169. Qin, L., Kunz, T.: On-demand routing in manets: The impact of a realistic physical layer model. Ad-Hoc Mob. Wirel. Netw. **2865**, 37–48 (2003)

170. Raghavan, U.N., Thadakamalla, H.P., Kumara, S.: Phase transition and connectivity in distributed wireless sensor networks. In: Procedings of 13th International Conference on Advances in Computing and Communications, pp. 1–8 (2005)
171. Rajabi, S., Shahabadi, M., ArdebiliPoor, M.: Modeling of the correlation coefficients of a receive antenna array in a mimo multipath channel. In: The 2nd IEEE/IFIP International Conference in Central Asia on Internet, pp. 1–4 (2006). URL 10.1109/CANET.2006.279251
172. Rappaport, T.S.: Wireless Communications: Principles and Practice. Prentice Hall Communications Engineering and Emerging Technologies Series. Prentice Hall, Englewood Cliffs (2002)
173. Ravelomanana, V.: Extremal properties of three-dimensional sensor networks with applications. IEEE Trans. Mob. Comput. **3**(3), 246–257 (2004)
174. Redner, S.: A Guide to First-Passage Processes. Cambridge University Press, Cambridge (2001)
175. Ross, S.M.: Introduction to Probability Models. Academic Press, 11th edition (2014)
176. Rudack, M., Meincke, M., Lott, M.: On the dynamics of ad hoc networks for inter vehicle communications (ivc). In: Proceedings of the International Conference on Wireless Networks (2003)
177. Seneta, E.: Non-negative Matrices and Markov Chains. Springer Series in Statistics. Springer, New York (1973)
178. Shang, Y., Ruml, W.: Improved mds-based localization. In: IEEE INFOCOM, pp. 2640–2651 (2004)
179. Shang, Y., Ruml, W., Zhang, Y., Fromherz, M.: Localization from connectivity in sensor networks. IEEE Trans. Parallel Distrib. Syst. **15**(11), 961–974 (2004)
180. Simon, M.K., Alouini, M.S.: Digital Communication Over Fading Channels. Wiley, New York (2000)
181. Tang, A., Florens, C., Low, S.H.: An empirical study on the connectivity of ad hoc networks. In: IEEE Aerospace Conference, pp. 1333–1338 (2003)
182. Thedinger, T., Jabbar, A., Sterbenz, J.P.G.: Store and haul with repeated controlled flooding. In: International Congress on Ultra Modern Telecommunications and Control Systems and Workshops, pp. 728–733 (2010)
183. Vahdat, A., Becker, D.: Epidemic routing for partially-connected ad hoc networks. Duke Tech Report CS-2000-06 (2000)
184. Varga, R.A.: Matrix Iterative Analysis. Prentice Hall, Englewood Cliffs (1962)
185. Wan, P.J., Yi, C.W.: Asymptotic critical transmission radius and critical neighbor number for k-connectivity in wireless ad hoc networks. In: Proceedings of the 5th ACM international symposium on Mobile ad hoc networking and computing, pp. 1–8 (2004)
186. Wang, C.X., Patzold, M., Yao, Q.: Stochastic modeling and simulation of frequency-correlated wideband fading channels. IEEE Trans. Veh. Technol. **56**(3), 1050–1063 (2007). URL 10.1109/TVT.2007.895490
187. Wang, Y., Li, K., Wu, J.: Distance estimation by constructing the virtual ruler in anisotropic sensor networks. In: Proc. of the IEEE INFOCOM, pp. 1172–1180 (2010)
188. Wei, G., Guohong, C., La Porta, T., Jiawei, H.: On exploiting transient social contact patterns for data forwarding in delay-tolerant networks. IEEE Trans. Mob. Comput. **12**(1), 151–165 (2013)
189. Williams, B., Camp, T.: Comparison of broadcasting techniques for mobile ad hoc networks. In: The Third ACM International Symposium on Mobile Ad Hoc Networking and Computing, pp. 194–205 (2002)
190. Wu, H., Fujimoto, R.M., Riley, G.F., Hunter, M.: Spatial propagation of information in vehicular networks. IEEE Trans. Veh. Technol. **58**(1), 420–431 (2009)
191. Wu, H., Lee, J., Hunter, M., Fujimoto, R., Guensler, R.L., Ko, J.: Efficiency of simulated vehicle-to-vehicle message propagation in atlanta, georgia i-75 corridor. Transp. Res. Rec. J. Transp. Res. Board **1910**, 82–89 (2005)
192. Xuan, B.B., Ferreira, A., Jarry, A.: Computing shortest, fastest, and foremost journeys in dynamic networks. Int. J. Found. Comput. Sci. **14**(2), 267–285 (2003)

193. Xue, F., Kumar, P.: The number of neighbors needed for connectivity of wireless networks. Wirel. Netw. **10**(2), 169–181 (2004)
194. Yamamoto, K., Kusuda, A., Yoshida, S.: Impact of shadowing correlation on coverage of multihop cellular systems. In: IEEE International Conference on Communications, pp. 4538–4542 (2006)
195. Yang, T., Mao, G., Zhang, W., Tao, X.: Transport capacity of distributed wireless csma networks. IEEE Trans. Wirel. Commun. **13**(10), 5635–5647 (2014)
196. Yang, Y., Hu, H., Xu, J., Mao, G.: Relay technologies for wimax and lte-advanced mobile systems. IEEE Commun. Mag. **47**(10), 100–105 (2009)
197. Yi, C.W., Wan, P.J., Li, X.Y., Frieder, O.: Asymptotic distribution of the number of isolated nodes in wireless ad hoc networks with bernoulli nodes. IEEE Trans. Commun. **54**(3), 510–517 (2006)
198. Yousefi, S., Altman, E., El-Azouzi, R., Fathy, M.: Analytical model for connectivity in vehicular ad hoc networks. IEEE Trans. Veh. Technol. **57**(6), 3341–3356 (2008)
199. Yu, X., Chandra, S.: Delay-tolerant collaborations among campus wide wireless users. In: IEEE INFOCOM, pp. 2101–2109 (2008)
200. Zadeh, L., Desoer, C.A.: Linear Systems Theory. McGraw Hill, New York (1963)
201. Zanella, A., Pierobon, G., Merlin, S.: On the limiting performance of broadcast algorithms over unidimensional ad-hoc radio networks. In: Proceedings of Wireless Personal Multimedia Communications (2004)
202. Zhang, Z., Mao, G., Anderson, B.D.O.: On the information propagation in mobile ad-hoc networks using epidemic routing. In: IEEE GLOBECOM, pp. 1–6 (2011)
203. Zhao, L., Liang, Q.: Hop-distance estimation in wireless sensor networks with applications to resource allocation. EURASIP J. Wirel. Commun. Netw. **2007**, 1–8 (2007)
204. Zorzi, M., Rao, R.R.: Geographic random forwarding (geraf) for ad hoc and sensor networks: energy and latency performance. IEEE Trans. Mob. Comput. **2**(4), 349–365 (2003)
205. Zorzi, M., Rao, R.R.: Geographic random forwarding (geraf) for ad hoc and sensor networks: multihop performance. IEEE Trans. Mob. Comput. **2**(4), 337–348 (2003)

MIX
Papier aus verantwortungsvollen Quellen
Paper from responsible sources
FSC® C105338

If you have any concerns about our products,
you can contact us on
ProductSafety@springernature.com

In case Publisher is established outside the EU,
the EU authorized representative is:
Springer Nature Customer Service Center GmbH
Europaplatz 3, 69115 Heidelberg, Germany

Printed by Libri Plureos GmbH
in Hamburg, Germany